Advances in Experimental Medicine and Biology

VOLUME 844

Advances in Experimental Medicine and Biology presents multidisciplinary and dynamic findings in the broad fields of experimental medicine and biology. The wide variety in topics it presents offers readers multiple perspectives on a variety of disciplines including neuroscience, microbiology, immunology, biochemistry, biomedical engineering and cancer research. *Advances in Experimental Medicine and Biology* has been publishing exceptional works in the field for over 30 years and is indexed in Medline, Scopus, EMBASE, BIOSIS, Biological Abstracts, CSA, Biological Sciences and Living Resources (ASFA-1), and Biological Sciences. The series also provides scientists with up to date information on emerging topics and techniques.

2013 Impact Factor: 2.012

More information about this series at http://www.springer.com/series/5584

Seth J. Corey • Marek Kimmel
Joshua N. Leonard
Editors

A Systems Biology Approach to Blood

Springer

Editors
Seth J. Corey
Departments of Pediatrics
and Cell & Molecular Biology
Northwestern University Feinberg
School of Medicine and Lurie Children's
Hospital of Chicago
Chicago
Illinois
USA

Joshua N. Leonard
Northwestern University
Comprehensive Cancer Center
Evanston
Illinois
USA

Marek Kimmel
Department of Statistics
Rice University
Houston
Texas
USA

ISSN 0065-2598 ISSN 2214-8019 (electronic)
ISBN 978-1-4939-2094-5 ISBN 978-1-4939-2095-2 (eBook)
DOI 10.1007/978-1-4939-2095-2
Springer New York Heidelberg Dordrecht London

Library of Congress Control Number: 2014952025

Printed on acid-free paper

Springer is part of Springer Science+Business Media (www.springer.com)

Contents

Contributors

Mark S. Alber Department of Applied and Computational Mathematics and Statistics, University of Notre Dame, Notre Dame, IN, USA

Department of Medicine, Indiana University School of Medicine, Indianapolis, IN, USA

Robert K. Andrews Australian Centre for Blood Diseases, Monash University, Melbourne, VIC, Australia

Nabil Azhar University of Pittsburgh, Pittsburgh, PA, USA

Rosemary Braun Biostatistics Division, Department of Preventive Medicine and Northwestern Institute on Complex Systems, Northwestern University, Chicago, IL, USA

Yolande Chen RH Lurie Comprehensive Cancer Center, Northwestern University Feinberg School of Medicine, Chicago, IL, USA

Julia Chifman Department of Cancer Biology, Wake Forest School of Medicine, Winston-Salem, NC, USA

Yishan Chuang Northwestern University, Evanston, IL, USA

Lily A. Chylek Department of Chemistry and Chemical Biology, Cornell University, Ithaca, NY, USA

Seth Joel Corey Department of Pediatrics and Cell & Molecular Biology, Northwestern University Feinberg School of Medicine and Lurie Children's Hospital of Chicago, Chicago, IL, USA

Rachel M. Dudek Northwestern University, Evanston, IL, USA

Elizabeth E. Gardiner Australian Centre for Blood Diseases, Monash University, Melbourne, VIC, Australia

Taly Glaubach Department of Pediatrics, Lurie Children's Hospital of Chicago, Chicago, IL, USA

Daniel Hidalgo Department of Cancer Biology, University of Massachusetts Medical School, Worcester, MA, USA

William S. Hlavacek Los Alamos National Laboratory, Los Alamos, NM, USA

Anthony D. Ho Department of Hematology and Oncology, University Hospital of Heidelberg, Heidelberg, Germany

Alexis Kaushansky Malaria Program, Seattle Biomedical Research Institute, Seattle, WA, USA

Kenneth Kaushansky Office of the Sr. Vice President, Health Sciences, Stony Brook, NY, USA

Oleg V. Kim Department of Applied and Computational Mathematics and Statistics, University of Notre Dame, Notre Dame, IN, USA

Marek Kimmel Department of Statistics and Bioengineering, Rice University, Houston, TX, USA

Miroslav Koulnis Department of Cancer Biology, University of Massachusetts Medical School, Worcester, MA, USA

Axel Krinner Faculty of Medicine Carl Gustav Carus, TU Dresden, Institute for Medical Informatics and Biometry, Dresden, Germany

Reinhard Laubenbacher Center for Quantitative Medicine, University of Connecticut Health Center, Farmington, CT, USA

Jinzhi Lei Zhou Pei-Yuan Center for Applied Mathematics, Tsinghua University, Beijing, China

Joshua N. Leonard Technological Institute, Rm E136, Evanston, IL, USA

Northwestern University, Evanston, IL, USA

Huamin Li Department of Mathematics, University of California at Berkeley, Berkeley, CA, USA

Michael C. Mackey Department of Physiology and CAMBAM, Monteral, QC, Canada

Anna Marciniak-Czochra Interdisciplinary Center for Scientific Computing (IWR), University of Heidelberg, Heidelberg, Germany

BIOQUANT Center, University of Heidelberg, Heidelberg, Germany

Hrishikesh M. Mehta Feinberg School of Medicine, Northwestern University, Chicago, IL, USA

Alex C. Minella Northwestern University Feinberg School of Medicine, Chicago, IL, USA

Ermelinda Porpiglia Department of Cancer Biology, University of Massachusetts Medical School, Worcester, MA, USA

Pankaj Qasba Blood Diseases Branch, Division of Blood Diseases and Resources, National Heart, Lung, and Blood Institute, NIH, Bethesda, MD, USA

Tomas Radivoyevitch Epidemiology and Biostatistics, Case Western Reserve University, Cleveland, OH, USA

Ingo Roeder Faculty of Medicine Carl Gustav Carus, TU Dresden, Institute for Medical Informatics and Biometry, Dresden, Germany

Rainer K. Sachs Department of Mathematics, University of California at Berkeley, Berkeley, CA, USA

Ka Tat Siu Northwestern University Feinberg School of Medicine, Chicago, IL, USA

Merav Socolovsky Department of Cancer Biology, and Department of Pediatrics, University of Massachusetts Medical School, Worcester, MA, USA

Thomas Stiehl Interdisciplinary Center for Scientific Computing (IWR), University of Heidelberg, Heidelberg, Germany

Cristian Tomasetti Johns Hopkins School of Medicine, Baltimore, MD, USA

Suzy V. Torti Department of Molecular Biology and Biophysics, University of Connecticut Health Center, Farmington, CT, USA

Yoram Vodovotz Department of Surgery, University of Pittsburgh, Pittsburgh, PA, USA

Bridget S. Wilson Department of Pathology, University of New Mexico School of Medicine, Albuquerque, NM, USA

Part I
Basic Components

The blood system is multi-scale, from the organism to the organs to cells to intracellular signaling pathways to macromolecule interactions. Blood consists of circulating cells, cellular fragments (platelets and microparticles), and plasma macromolecules. Blood cells and their fragments result from a highly-ordered process, hematopoiesis. Definitive hematopoiesis occurs in the bone marrow, where pluripiotential stem cells give rise to multiple lineages of highly specialized cells. Highly-productive and continuously regenerative, hematopoiesis requires a microenvironment of mesenchymal cells and blood vessels.

In this first section, we shall cover the important components of blood: beginning with the microenvironment and then focusing on erythrocytes, megakaryocytes, phagocytic cells, and platelets. In Chap. 1, the editors of this volume provide a multidisciplinary overview of hematopoiesis and systems biology. This should serve to introduce hematology to the quantitative and modeling scientists as well as to introduce basic mathematical principles and Text modeling to the hematologists. In Chap. 2, Krinner and Roeder discuss the interactions among hematopoietic stem cells and the microenvironment. No other tissue undergoes the tremendous amount of regeneration and accurate specialization of diverse tissues as the blood system. Fortunately, defects in production (overproduction or underproduction) are infrequent occurrences. In Chap. 3, Socolovsky and associates focus on how erythropoietin drives production of red blood cells through basal and stress conditions. This highlights an important property of the blood system—the ability to function for the most part within a narrow range of physiological conditions and still retain the dynamic capacity to respond quick to stressful stimuli and other environmental changes. In Chap. 4, the Kaushanskys describe in detail thrombopietin's intracellular signaling that drives the differentiation of megakaryocytes. Interestingly, many of its proximal components are found activated in response to other cytokines. In Chap. 5, Alber and colleagues detail how platelets are formed from megakaryocytes and how they become activated. Platelet production and homeostasis highlights their clinical significance—a sufficient number of platelets must remain quiescent and then be able to respond briskly to bleeding. Too many platelets and too much activation result in life-threatening clots; whereas too few platelets and too little activation result in life-threatening bleeding. Lastly in this section, in Chap. 6, Corey and colleagues discuss granulocytes and monocytes, two critical components in innate immunity.

Seth Joel Corey, MD, MPH
Chicago, IL
Mark Kimmel, PhD
Houston, TX
Joshua N. Leonard, PhD
Evanston, IL

Chapter 1
Systems Hematology: An Introduction

Seth Joel Corey, Marek Kimmel and Joshua N. Leonard

Abstract Hematologists have traditionally studied blood and its components by simplifying it into its components and functions. A variety of new techniques have generated large and complex datasets. Coupled to an appreciation of blood as a dynamic system, a new approach in systems hematology is needed. Systems hematology embraces the multi-scale complexity with a combination of mathematical, engineering, and computational tools for constructing and validating models of biological phenomena. The validity of mathematical modeling in hematopoiesis was established early by the pioneering work of Till and McCulloch. This volume seeks to introduce to the various scientists and physicians to the multi-faceted field of hematology by highlighting recent works in systems biology. Deterministic, stochastic, statistical, and network-based models have been used to better understand a range of topics in hematopoiesis, including blood cell production, the periodicity of cyclical neutropenia, stem cell production in response to cytokine administration, and the emergence of drug resistance. Future advances require technological improvements in computing power, imaging, and proteomics as well as greater collaboration between experimentalists and modelers. Altogether, systems hematology will improve our understanding of normal and abnormal hematopoiesis, better define stem cells and their daughter cells, and potentially lead to more effective therapies.

Keywords Hematology · Models · Reductionist · Systems biology

S. J. Corey (✉)
Department of Pediatrics and Cell & Molecular Biology,
Northwestern University Feinberg School of Medicine, Lurie 5-107,
303 E. Superior St., Chicago, IL 60611, USA
Tel.: 312-503-6694
e-mail: coreylab@yahoo.com

M. Kimmel
Department of Statistics and Bioengineering, Rice University, 2102 Duncan Hall,
6100 Main St., Houston, TX 77005, USA
Tel.: (713) 348-5255
e-mail: kimmel@rice.edu

J. N. Leonard
Technological Institute, Rm E136, 2145 Sheridan Rd., Evanston, IL 60208-3120, USA
Tel.: 847-491-7455
e-mail: j-leonard@northwestern.edu

S. J. Corey et al. (eds.), *A Systems Biology Approach to Blood*,
Advances in Experimental Medicine and Biology 844, DOI 10.1007/978-1-4939-2095-2_1

Blood, pure and eloquent, wrote Max Wintrobe, one of the pioneers of modern hematology. His description was but a reference to lines written by the seventeenth-century English poet John Donne. Hematology, the study of blood and its components, has undergone dramatic changes over the millennia, since man first recognized its power. The first plague visited upon the Egyptians was blood, "I will strike the water of the Nile, and it will be changed into blood. The fish in the Nile will die, and the river will stink and thus the Egyptians will not be able to drink its water." What we consider so vital to human life was viewed as deleterious. Rabbis later writing commentary warned against circumcising a third son after two had died of bleeding. Contemporaneously, Hippocratic writings described blood as one of the four humors. More appreciative of its vital nature, the Greek physicians equated blood with spring and air. The Greek word for blood, *haima*, has been sustained in all things hematologic and hematopoietic.

Like other branches of medicine, hematology has undergone paradigm shifts. From the ancient Jews' and Greeks' attribution of blood to health and disease through the seventeenth century's rationalists who described its circulation through arteries and veins, to the modern physiologists of the past century, our understanding of blood and its components has advanced. The past 50 years have provided us with more intimate knowledge of its components at the subcellular level. This reductionist approach to science has now been superceded by the awareness of complex, large datasets, made possible by proteomic, flow cytometric, microarray, genomic sequencing, and epigenetics. The complexity of blood and its components is also recognized at multiple levels from the subcellular to the macroscopic, such as the environment and its effect on the organism. While physical and chemical laws have been applied to biology, limitations to their applicability and predictability are frequently encountered. Biology is dynamic.

The biomedical discipline that has been called physiology has evolved to a new approach—a modern synthesis of biochemistry, genetics, mathematics, engineering, and machine-based learning. Complex, large datasets of genes, lipids, metabolites, and proteins have made it impossible for one investigator to intuit the whole. This new, integrative field has been called systems biology. In this volume, we seek to introduce physicians and scientists, qualitative and quantitative, to the different facets of systems hematology. Systems hematology embraces this complexity, utilizing engineering principles and computational methods to build and validate models using experimental data. The approach rests on (i) defining all (or the known knowns) of the components, (ii) systematically perturbing and monitoring the components of the system, (iii) reconcile the experimentally observed responses with those predicted by the model, and (iv) designing and performing new experiments to distinguish between multiple or competing models. The goals are to understand how the system works, identify new systems-based properties, and predict outcomes.

The major obstacle to success in systems biology lies in the disciplines practiced by physicians and scientists. Major differences exist in the methods, jargon, and philosophies between quantitative scientists, the theoretical physicists, the mathematicians, the engineers, computer programmers, and experimentalists. Even within the experimentalists, there is diversity and increasing technologization, as evidenced by cell biology, molecular biology, and proteomics. Until there is a common vernacular, fundamental concepts in the fields of biology, mathematics, engineering, and computation can be understood and transdisciplinary studies can be successful.

Blood as a System

Biological systems operate at multiple levels (or scales): molecular, cellular, tissue, and organismal, and environmental. Stem cells generate differentiated blood cells through a continuous process of asymmetric stem cell division, yielding daughter cells with different capacities for renewal or differentiation. This process occurs in a specialized microenvironment. The blood system consists of highly specialized cells and plasma containing a range of proteins to regulate different processes. Among the blood cells are erythrocytes that shuttle oxygen or its waste product to and from tissues; white blood cells to fight infection and mediate inflammation; and platelets to stop bleeding. Within the compartment of white blood cells, there is variability: neutrophils to engulf foreign agents, lymphocytes to make antibodies and coordinate immunity, and monocytes to process and regulate host defense. Plasma contains more than 1000 proteins [1]. Homeostatic mechanisms insure that the right number of cells is produced, but they are sufficiently dynamic to meet the needs of environmental changes (e.g., hypoxia, infection, or bleeding). While hematologists diagnose and treat patients with anemias, immune deficiencies, leukemias and lymphomas, and hypercoagulability, it is astonishing that such high level of quality control of blood and its elements exists and that blood diseases are not more common.

Systems Properties in Hematopoiesis

Because of the facility in sampling blood or bone marrow repetitively and quantitatively, the blood system is well suited for modeling and validation. Hematopoiesis and the functioning of specialized blood cells involve complex processes that can be examined at the level of genes [2], signal transduction proteins [3], or the population distribution of diverse cell types [4]. Both deterministic and stochastic processes contribute. By viewing hematopoiesis (cell proliferation and differentiation) as a dynamic system and disease as perturbations of the system, one can learn more about both disease and physiological states.

Proliferation and loss are fundamental properties of hematopoietic stem cells and their progeny. Population dynamics offers a quantitative approach in studying them. Asymmetric division results in a stem cell dividing into either another stem cell or a more committed cell, while symmetric division yields either two stem cells or two differentiated daughter cells. These processes can be combined in a series of short steps [5–8]. Models built around these division (a)symmetries usually result in exponential cell growth, but such growth cannot be realistically sustained in vitro due to spatial and nutrient limitations. Models based on heterogeneous population account for cell proliferation and loss due to death or differentiation.

Differentiation is the other fundamental property of hematopoietic progenitor cells and requires critical processes of cell fate decision making. Decision making occurs as a result of biochemical signaling and gene regulatory networks within the cell [9], [10]. Ultimately, transcription factors determine cellular differentiation and specialization [11]. The relative contributions of instructive and permissive programming in hematopoiesis have long been debated [6, 12–23]. To describe hematopoietic stem cell renewal and differentiation, deterministic and stochastic models have been constructed. James Till, a biophysicist, and Ernest McCulloch, a physician, pioneered the study of hematopoiesis in the early 1960s through their development of a quantitative spleen colony assay, establishment of a hematopoietic stem cell, and data analysis that yielded a stochastic model of hematopoiesis [24], [25]. In their stochastic model [5], cells have two possible fates: (1) differentiate and leave the proliferative compartment or (2) undergo symmetric division forming two colony-forming cells. Each fate was assigned a probability. Drawing random numbers to determine the fate of each cell, Till and McCullouch calculated the diversity of stem cell populations after the course of several generations. Colony generation appears as a well-defined process even though individual cell-fate decisions are random. Regulation acts at the population, not cellular, level and the population of stem cells can be affected by influencing processes that define the effective probabilities of birth and death.

A cell uses complex intracellular signaling and gene regulatory networks in order to integrate the multiplicity of cues in its environment and to ultimately make a specific decision. In particular, gene regulatory networks have provided great insights into lineage commitment of hematopoietic progenitors.

Types of Mathematical Models

Different methods of modeling have been developed to describe and predict biological processes. Not all models are accurate, but some are more useful than others. Deterministic models describe the state of a system over time in the absence of random events. These always produce the same output for a given input [26]. In contrast, stochastic models describe the effects of randomness and noise on system output [27]. Statistical models use existing data to estimate a functional relationship between system input and output. Network models graph the direction and magnitude of interactions that exist between the various components in a system [28].

Deterministic models typically consist of one or more differential equations, with each equation describing the change in a system state variable over time, as it depends on other system variables and rates. If the state variable of interest is the number of cells in the population, a differential equation modeling the change in the population over time would consist of the difference between rates of cell production and rates of cell loss:

$$\frac{dN_X}{dt} = \text{(rate at which precursor of X differentiates into X)}$$
$$- \text{(rate at which X differentiates into next cell lineage)}$$
$$- \text{(rate at which X dies)} \quad (1.1)$$

where N_X is the number of cells of type X.

Each equation describes the rate of change in the number of cells of given type and maturity in the system by including terms for the rates of cell production, death, and differentiation. Once the equations are established, they are solved either analytically or numerically to determine the population's functional dependence on time. In models describing physiological conditions, the equations tend toward a steady-state solution representing system homeostasis; that is, after sufficient time has elapsed, positive and negative contributions to cell number balance and the population attains a constant level (e.g., $dN_X/dt = 0$ in Eq. 1.1). For disease-state cell populations, other types of behavior such as oscillations or uncontrolled growth are frequently modeled.

Stochastic models are employed to examine the effects of intrinsic and extrinsic randomness on a system. Intrinsic randomness arises from interactions of a finite ("small") number of discrete components, e.g., binding of a given gene's promoters (two copies per diploid genome) by transcription factor's molecules (also a limited number). Extrinsic randomness arises either from variability (genetic and phenotypic) among cells or from environmental fluctuations. The most common type of stochastic model is a Markov process, in which the future state of the system depends only on its current state and is independent of its past states. Monte Carlo simulations are an empirical method to investigate dynamics of a stochastic system, by generating repeated random trajectories and computing frequencies that estimate probability distributions.

Statistical models are sometimes confused with stochastic models. Whereas stochastic models reflect the structure of the biological system, statistical models are data driven. Statistical models can be employed even when no knowledge about system's structure exists and can generate predictions, which may be only statistically validated. However, some statistical models such as Bayesian networks may provide insights concerning the structure. Bayesian network models are built from graphs in which the states of and relationships between network elements are probabilistic. While graph theoretical models can be circular, Bayesian networks have a definite, distinct set of termini. These models have a wide range of uses. For example, a Bayesian network model could be used to predict the probabilities of certain cellular mutations based on abnormalities in protein expression levels (assuming, of course, that there is a relationship between the two). Their structure and necessary constants have to be estimated based on data. Though popular, Bayesian networks suffer from the possible reversal of causality [29].

Network models have recently gained popularity in the social, physical, and biological sciences from the widespread application of graph theory, an area of mathematics that investigates the relationships between the objects of a group [30]. Graph theory lends itself to visual representations making it an appealing tool for biologists investigating phenomena ranging from the interactions between populations in an ecosystem to the interactions between molecular species involved in a signaling pathway. At its simplest, a graph is a map of all known system components or system states and their possible interactions or transitions. Circles (nodes) represent components and states, and lines and arrows (branches or edges) represent relationships between nodes. Graphs help portray topological structures such as loops. Complex dynamics can arise from relatively few interacting components [31], and network maps are widely used to help visualize the interactions. Building upon existing graph theoretical notation, an international group has developed Systems Biology Graphical Notation to standardize the visual representations used to describe biological interaction networks [32].

Current Status of Systems Biology

The success of systems analysis of hematopoiesis will depend upon technological breakthroughs and collaborations between the biological and physical sciences that yield accurate predictions and emergent properties. With each discipline using a different language, this is easier said than done. Changes in undergraduate, graduate, and medical curricula must be implemented to train a new generation of biomedical researchers fluent in quantitative or engineering disciplines [33–35]. Systems biology requires a balance between models sufficiently complex to describe a system and yet simple enough to be clinically useful. Understanding large quantities of data well enough to validate a model is especially challenging. The development of Systems Biology Markup Language (SBML) has made it easier to develop biology-oriented software packages, such as COPASI, Simmune, MetaCore, and Cytoscape, which aid model building and data analysis [32, 36–39]. Since 2001, the number of such packages developed for systems biology has grown from 5 to over 170. With computational power becoming ever greater and cheaper, the number and diversity of such software packages will only increase, bringing within their scope models that may not be impossible to validate with current technology. At present, most models of hematopoiesis are built at a single scale, e.g., cellular or molecular. The future lies in building models that span multiple scales, incorporating more of the connections that exist between them and thereby being able to account for some of the complexity that arises from the connections. Among the fundamental questions in normal and leukemic hematopoiesis that systems biology will address are: integration of signaling pathways, circuits, and networks that determine cell fate, multi-scale modeling of stem cell plasticity, synthesis of genetic and epigenetic data, global analysis of phosphoproteins, dynamics of hematopoiesis in the bone marrow microenvironment presented in three-dimensional imaging, and cellular engineering to expand selective blood cell compartments for therapy. The complexity or density of experimental data will demand a systems approach. More in-depth coverage may be found in the few textbooks of systems biology and bioinformatics that have appeared, none solely devoted to hematologic topics [40–43].

References

1. Qian WJ, Monroe ME, Liu T, et al. Quantitative proteome analysis of human plasma following in vivo lipopolysaccharide administration using 16O/18O labeling and the accurate mass and time tag approach. Mol Cell Proteomics. 2005;4(5):700–9.
2. Laslo P, Pongubala JM, Lancki DW, Singh H. Gene regulatory networks directing myeloid and lymphoid cell fates within the immune system. Semin Immunol. 2008;20(4):228–35.
3. Hlavacek WS, Faeder JR, Blinov ML, Posner RG, Hucka M, Fontana W. Rules for modeling signal-transduction systems. Sci STKE. 2006;17(344):re6.
4. Roeder I, Loeffler M. A novel dynamic model of hematopoietic stem cell organization based on the concept of within-tissue plasticity. Exp Hematol. 2002;30(8):853–61.
5. Till JE, McCulloch EA, Siminovitch L. A stochastic model of stem cell proliferation, based on the growth of spleen colony-forming cells. Proc Natl Acad Sci U S A. 1964;51:29–36.
6. Ogawa M. Hemopoietic stem cells: stochastic differentiation and humoral control of proliferation. Environ Health Perspect. 1989;80:199–207.
7. Ogawa M, Pharr PN, Suda T. Stochastic nature of stem cell functions in culture. Prog Clin Biol Res. 1985;184:11–9.
8. Vogel H, Niewisch H, Matioli G. The self renewal probability of hemopoietic stem cells. J Cell Physiol. 1968;72(3):221–8.
9. Kestler HA, Wawra C, Kracher B, Kuhl M. Network modeling of signal transduction: establishing the global view. Bioessays. 2008;30(11–12):1110–25.
10. Kirouac DC, Madlambayan GJ, Yu M, Sykes EA, Ito C, Zandstra PW. Cell-cell interaction networks regulate blood stem and progenitor cell fate. Mol Syst Biol. 2009;5:293.
11. Orkin SH, Zon LI. Hematopoiesis: an evolving paradigm for stem cell biology. Cell. 2008;132(4):631–44.
12. McCulloch EA. Stem cells in normal and leukemic hemopoiesis (Henry Stratton Lecture, 1982). Blood. 1983;62(1):1–13.
13. McCulloch EA. Stem cell renewal and determination during clonal expansion in normal and leukaemic haemopoiesis. Cell Prolif. 1993;26(5):399–425.
14. Mehr R, Agur Z. Bone marrow regeneration under cytotoxic drug regimens: behaviour ranging from homeostasis to unpredictability in a model for hemopoietic differentiation. Biosystems. 1992;26(4):231–7.
15. Morrison SJ, Weissman IL. The long-term repopulating subset of hematopoietic stem cells is deterministic and isolatable by phenotype. Immunity. 1994;1(8):661–73.
16. Novak JP, Stewart CC. Stochastic versus deterministic in haemopoiesis: what is what? Br J Haematol. 1991;78(2):149–54.
17. Ogawa M. Stochastic model revisited. Int J Hematol. 1999;69(1):2–5.
18. Quesenberry P, Abedi M, Dooner M, et al. The marrow cell continuum: stochastic determinism. Folia Histochem Cytobiol. 2005;43(4):187–90.
19. Abkowitz JL, Catlin SN, Guttorp P. Evidence that hematopoiesis may be a stochastic process in vivo. Nat Med. 1996;2(2):190–7.
20. Roeder I, Glauche I. Towards an understanding of lineage specification in hematopoietic stem cells: a mathematical model for the interaction of transcription factors GATA-1 and PU.1. J Theor Biol. 2006;241(4):852–65.
21. Palani S, Sarkar C. Integrating extrinsic and intrinsic cues into a minimal model of lineage commitment for hematopoietic progenitors. PLoS Comput Biol. 2009;5:e1000518.
22. Chang HH, Hemberg M, Barahona M, Ingber DE, Huang S. Transcriptome-wide noise controls lineage choice in mammalian progenitor cells. Nature. 2008;453(7194):544–7.
23. Huang S, Guo YP, May G, Enver T. Bifurcation dynamics in lineage-commitment in bipotent progenitor cells. Dev Biol. 2007;305(2):695–713.
24. Becker AJ, Mc CE, Till JE. Cytological demonstration of the clonal nature of spleen colonies derived from transplanted mouse marrow cells. Nature. 1963;197:452–4.

25. Siminovitch L, McCulloch EA, Till JE. The distribution of colony-forming cells among spleen colonies. J Cell Physiol. 1963;62:327–36.
26. Fall C, Marland E, Wagner J, Tyson J. Computational cell biology. Vol. 20. New York: Springer; 2005.
27. Wilkinson D. Stochastic modelling for systems biology. New York: Chapman & Hall/CRC; 2006.
28. Palsson B. Systems biology: properties of reconstructed networks. Cambridge: Cambridge University Press; 2006.
29. Pearl J. Models, reasoning and inference. New York: Cambridge University Press; 2000.
30. Barabasi AL. Scale-free networks: a decade and beyond. Science. 2009;325(5939):412–3.
31. Amaral LA, Diaz-Guilera A, Moreira AA, Goldberger AL, Lipsitz LA. Emergence of complex dynamics in a simple model of signaling networks. Proc Natl Acad Sci U S A. 2004;101(44):15551–5.
32. Le Novere NH, Mi H, et al. The systems biology graphical notation. Nat Biotechnol. 2009;27(8):735–41.
33. HHMI/AAMC. Scientific foundations for the future physicians. 2009. https://www.aamc.org/download/271072/data/scientificfoundationsforfuturephysicians.pdf. Accessed 12 Oct 2014.
34. Council NR. BIO2010: transforming undergraduate education of future research biologists. Washington, DC: National Academies Press; 2003.
35. Wingreen N, Botstein D. Back to the future: education for systems-level biologists. Nat Rev Mol Cell Biol. 2006;7(11):829–32.
36. Killcoyne S, Carter GW, Smith J, Boyle J. Cytoscape: a community-based framework for network modeling. Methods Mol Biol. 2009;563:219–39.
37. Meier-Schellersheim M, Xu X, Angermann B, Kunkel EJ, Jin T, Germain RN. Key role of local regulation in chemosensing revealed by a new molecular interaction-based modeling method. PLoS Comput Biol. 2006;2(7):e82.
38. Mendes P, Hoops S, Sahle S, Gauges R, Dada J, Kummer U. Computational modeling of biochemical networks using COPASI. Methods Mol Biol. 2009;500:17–59.
39. Moore JH. Bioinformatics. J Cell Physiol. 2007;213(2):365–9.
40. Alon U. An introduction to systems biology: design principles of biological circuits. Boca Raton: Chapman & Hall/CRC; 2007.
41. Palsson BO. Systems biology, properties of reconstructed networks. New York: Cambridge University Press; 2006.
42. Kitano H. Foundations of systems biology. Cambridge: MIT Press; 2001.
43. Polanski A, Kimmel M. Bioinformatics. New York: Springer; 2007.

Chapter 2
Quantification and Modeling of Stem Cell–Niche Interaction

Axel Krinner and Ingo Roeder

Abstract Adult stem cells persist lifelong in the organism, where they are responsible for tissue homeostasis and repair. It is commonly assumed that their maintenance and function are facilitated in local environments called "stem cell niches." Although there is convincing evidence that a variety of niche components determine stem cell fate, the regulatory details of stem cell–niche interactions are widely unknown. To pave the way for a substantiated discussion of these interactions, we first focus on the stem cells themselves and describe the stem cell defining criteria and their implications. The fate of the cells that fulfill these criteria is regulated by a broad spectrum of factors and regulatory mechanisms. A summary of established components and their action is given exemplary for the hematopoietic system. The complexity resulting from the interplay of various cell types, signaling molecules, and extracellular structures can be boiled down to important key features as exemplified by the presented model of hematopoietic stem cell organization. Although neglecting many details, we show that this and similar models have the power to yield intriguing results as proven by the agreement of the presented model with experimental data and the predictions derived from model simulations. Finally, we will discuss the paradigm of systems biology and give a summary of the techniques that promise to unveil further details of the organization principles of stem cell niches at different levels. The synergistic effect of the described techniques together with the integration of their results into a unified model that allows quantitative evaluation and predictions may lead to a better and more systematic understanding of the most relevant niche elements and their interactions.

Keywords Stem cell niche · Hematopoiesis · Mathematical modeling · Systems biology

A. Krinner (✉) · I. Roeder
Faculty of Medicine Carl Gustav Carus, TU Dresden,
Institute for Medical Informatics and Biometry, Fetscherstr. 74, D-01307 Dresden, Germany
Tel.: +49-351-458-6233
e-mail: axel.krinner@tu-dresden.de

I. Roeder
Tel.: +49-351-458-6060
e-mail: ingo.roeder@tu-dresden.de

S. J. Corey et al. (eds.), *A Systems Biology Approach to Blood*,
Advances in Experimental Medicine and Biology 844, DOI 10.1007/978-1-4939-2095-2_2

Introduction

Although it is generally accepted that microenvironmental cues play a key role in regulating stem cell function, and although many individual regulatory mechanisms and pathways of cell–microenvironment interaction have been identified, a systemic understanding of stem cell–microenvironment interaction and its impact on stem cell fate regulation is still missing. This is also the case for hematopoietic stem cells (HSCs), which have been extensively studied for more than 40 years, starting, e.g., with the pioneering work of James Till and Ernest McCulloch in the early 1960s. The two scientists were able to demonstrate the existence of undifferentiated hematopoietic cells in the bone marrow (BM) that are capable of both, self-renewing and differentiating—two features that are classically used to define cells as *stem cells*. Based on serial transplantation experiments, Till and McCulloch showed that these (stem) cells are able to develop into spleen colonies of irradiated mice, which contain cells with an identical potential [1–3]. These were called colony-forming units in spleen (CFU-S cells) and regarded as stem cells. Later, they turned out to be progenitor cells, which are, in contrast to true stem cells, characterized by only a limited self-renewal and repopulation potential.

Clearly, the origin of CFU-S cells was the BM, but it was by no means clear, whether there are specific regions in the BM that functionally support stem and/or progenitor cells. Unlike other stem cell systems, such as the intestinal crypt [4], the BM is lacking an obviously structured spatial arrangement. This absence of clearly visible, stem-cell-supporting areas widely hampered the study of HSCs and their interactions with local microenvironmental components in the in vivo situation. Nevertheless, the perspective of an instructive local microenvironment of HSCs was introduced already in the early 1970s by John Trentin [5, 6] and Raymond Schofield [7]. Schofield proposed a concept that includes a context dependency of stem cell behavior. In this concept, stem cells live in a certain environment, the *niche*, where differentiation and maturation is prevented and thereby continuous proliferation and maintenance of stem cell potential is guaranteed. Therefore, stem cells lose their potential, if they lack this specific environment. This concept is consistent with the results of contemporary coculture experiments. For instance, Dexter and coworkers were able to maintain proliferative CFU-S cells over several months in vitro using a mixture of feeder cells from the BM, whereas these cells differentiated if cultured without feeder cells [8, 9].

Since these days, new ideas and experimental techniques have extended the list of cells and other microenvironmental factors that presumably act in combination to form the stem cell niche. Other factors, such as geometry and biomechanics, nutrient supply, signaling molecules, metabolic conditions, and contact dependent cues have been shown to contribute to the niche environment, too. Later in this chapter, we will give an overview of some important examples of these presumably stem-cell-regulating niche components with a particular focus on the hematopoietic system.

Defining Stem Cells

Before talking about stem cell regulatory components and effects of a niche environment, we need to precisely define what we mean by a stem cell or by stem cell potential. Because the term *stem cell* resulted from the conceptual aftermath of the discovery of a multipotent and self-renewing cell population, its definition almost exclusively contains functional criteria. Only in the case of embryonic stem (ES) cells [10, 11], the functional definition has its counterpart in a definition by origin. When the blastula is formed, this cell population emerges from the first differentiation step, the separation of trophoblast and inner cell mass. While the first forms only extraembryonic structures, all cell types of the embryo itself develop from the cells of the inner cell mass. Therefore, these cells are characterized as *pluripotent*. They are the source for the in vitro derivation of ES cell lines, which are usually denoted as *pluripotent ES cells*, as they preserve the potential to differentiate into cells of all tissue types. In vivo, the development of the embryo involves further differentiation steps beginning with the development of three germ layers. From those, the different tissues are derived and with this specification process the ability of the cells to generate cells from other tissues is lost. Pluripotency, therefore, turns into *multipotency*. Multipotent cells still have the potential to differentiate into various cell types of a particular tissue and are maintained as so-called *(adult) tissue stem cells* lifelong. They preserve their proliferation and self-renewal capacity as well as their multilineage potential in order to guarantee homeostasis and to repair damaged tissues, which represents the core of their functional definition [12].

Whereas the details of the definition of a tissue stem cell depend on its author, functional characteristics such as self-renewal, differentiation, and proliferative potential were always cornerstones of this definition. Tissue stem cells are defined by a number of qualities, which enable them to guarantee a lifelong maintenance and, in case of injury, to reconstitute a fully functional tissue. Over the years and with new experimental results, the definitions have been modified and a more flexible interpretation of this concept of a *functional* definition has been introduced. Flexibility has been included in the sense that stem cell fate decisions depend on the environment. This dependence results in some flexibility or even reversibility of stem cell properties and functionalities [13].

A general problem with the *functional* definition is the fact that it does not allow for a prospective selection of stem cells on an individual cell basis: Any particular assay (e.g., a colony formation assay) that is required for the examination of a particular cellular function (e.g., proliferative potential) will always alter the state of the cell. Therefore, the assessment of one function of a particular cell might impair the assessment of any other of its functions by another assay. In other words, the measurement process itself (to test for stem cell functionality) alters the object of measurement. This perception is the reason why Potten and Loeffler [14] compared this dilemma to Heisenberg's *uncertainty principle of quantum physics*. Although this analogy is certainly not perfect, it points to a very important aspect that applies to both areas: Any prospective statement about the function of a particular object (in our case a potential stem cell) can only be made in a probabilistic sense. This should be kept in mind if talking about stem cells; we will come back to this aspect later.

To meet this problem of characterization and selection of tissue stem cells, scientists have put large efforts in the development of purification protocols that enrich a cell population for functional stem cells. Fluorescence-activated cell sorting (FACS) applied simultaneously to a large set of cell surface markers has led and still leads to continuously refined selection protocols. Latest protocols allow for very high enrichment rates of HSCs with long-term repopulating ability (LTRA), which are considered as the *true* HSCs. As an example, the *Lin-Sca+c-Kit+* (*LSK*) *CD34-SLAM* (*CD244-CD48-CD150+*) marker combination allows to enrich mouse primary BM cells to a degree of up to one LTRA-HSC in two target cells [15, 16]. Surprisingly, for most of these markers, no functional, mechanistic link to LTRA has been found. However, it should be noted that despite the high enrichment, prospective statements about the purified cells are still only possible in a statistical, probabilistic sense.

Furthermore, there are two other flaws that are inherently connected with this characterization approach. First, for the application of sorting protocols, the cells have to be removed from their natural habitat. As mentioned above, such a treatment might alter cellular properties during this time of in vitro culture due to the dependence of stem cell properties on environment. Second, for assessment of in vivo functionality, the cells have to be reinjected into host animals. Usually lethally irradiated mice provide the environment that guarantees efficient engraftment. Unfortunately, irradiation does significantly damage the niche environment and the physiological structures in the BM [17–19]. Therefore, cellular and microenvironmental effects are inevitably confounded by the application of such assay protocols.

These two remarks bring us back to the role of the local microenvironment. In our opinion, it does not make sense to talk about HSCs without considering these cells as being embedded in a particular environmental context. This is most likely also true for any other tissue type. However, in the following, we will focus on the hematopoietic system and use HSCs as a model system to describe a general approach to systematically analyze the underlying mechanisms of microenvironment-based stem cell regulation. Herein, we will focus on a description of (potential) regulatory components of the stem cell niche and on mathematical modeling approaches to study the systems dynamics of stem cell–niche interactions. These two major paragraphs will be complemented by some thoughts about a potential road map for a more complete understanding of stem cell–niche interactions.

Components of the HSC Niche

Already decades ago, the BM has been identified as the natural environment and, therefore, a "niche" of HSCs. Basically, it is composed of a scaffold of extracellular matrix components, a cell population comprising cells of various lineages, and a fluid filling the rest of the space (Fig. 2.1). In the marrow, two main structures are

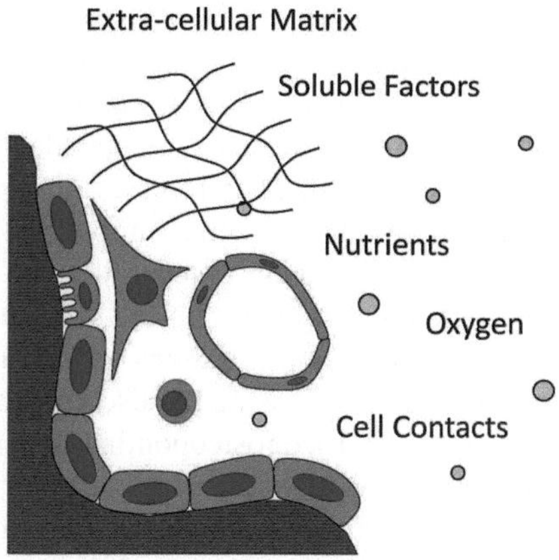

Fig. 2.1 Components of the niche. The niche environment comprises several factors. All of them are dynamically dependent on the cells of the niche environment. They provide growth factors, build and remodel the extracellular matrix and constitute endosteum and vascular network

obvious, mineralized bone and vascularization. Most of the cell types found within the marrow have been attributed to either of these basic structures. Directly associated with the bone is its lining, the endosteum. It is mainly composed of undifferentiated mesenchymal bone-lining cells and the two bone-remodeling cell types, osteoblasts (OBs) and osteoclasts. Also, there is the vascular system connecting the marrow to the rest of the organism by vessels and sinusoids. The walls of these tubular structures are formed by endothelial cells (ECs), which coexist with so-called perivascular cells in their direct vicinity.

There are a number of reports proposing that these two structures represent two distinct local environments in the BM: the endosteal and the perivascular environment, which form two distinguishable stem cell niches fostering different stem cell populations [20]. Whereas the so-called *endosteal niche* is associated with proliferative quiescence (low cell cycle activity), the *vascular niche* has been described to support stem cell proliferation [21, 22]. As a consequence of the two different environments, stem cells with LTRA are found preferentially in the endosteal niche, while the vascular niches hosts stem and progenitor cells with only short-term repopulating ability (STRA) [23]. In this view, the dormant cells form a reserve pool for emergencies, which can be repopulated after a potential emergency operation [23]. In contrast, a more recent study suggests a continuous and frequent exchange of cells between quiescent and proliferative states [24]. The hypothesis inevitably comes up that this exchange happens between the two niche environments. However, these studies only quantify the number of cell divisions in a certain time (using label retaining experiments), but their observations do not link transitions between dormant and proliferative states to translocations between the two niches.

A thorough identification of niche environment and function would require a separation of the two niches. As already shown by early histological studies, the interior vascular system of the bone is connected with the exterior system by a dense system of vessels through mineralized bone, which consequently indicates highly vascularized endosteum [25]. This difficulty of defining a spatial separation of vasculature and endosteum was recently confirmed by in vivo tracking experiments of HSCs in mice [26]. By fluorescence staining of blood, OBs and injected HSCs, relative positions of HSCs, vasculature, and OBs were measured. In this way, it was shown that sinusoids are abundant in the whole BM, though more dense in the BM cavities [26]. Therefore, a rigorous spatial separation of the two hypothesized niche environments seems impossible. Taking one step beyond, this might suggest integrating the signals emanating from the two presumed "niche environments" into one self-organizing system featuring one continuous niche. This view is supported by the fact that concentrations of various soluble molecules, e.g., chemokine (C-X-C motif) ligand 12 (CXCL12, also known as stromal cell-derived factor, SDF-1), stem cell factor (SCF), or osteopontin (OPN), seem to exhibit continuous rather than step-like gradients. Also, the supply of nutrients and oxygen continuously changes with distance from the bone surface. The latter observation was the origin of yet another idea, the metabolic niche [27, 28]. In such a continuous niche, all components may be present throughout the niche, although with certain tendencies or activities. We will now summarize these components of the niche by describing the cells themselves and their role within the BM, because they represent the active components in the BM that are motile, remodel the bone, and produce HSC-supporting factors.

Hematopoietic Cells

Hematopoietic Stem and Progenitor Cells

An important contribution to the niche organization is made by the HSCs themselves. It is their active migratory behavior that finally determines the niche by bringing them into particular environmental conditions and keeping them there. For example, most dormant HSCs are detected in an isolated position [23]. Also, it has been reported that dormancy and LTRA is associated with cells homing close to the endosteal surface [21]. Furthermore, several properties related to HSC migration, such as membrane fluctuations, cell adhesion, and cell motility, vary with distance to the bone [29]. A prominent cell-adhesion molecule that has been in the focus of discussion in recent years is N-cadherin. Intermediate levels of N-cadherin expression have been reported to indicate a quiescent state, while activated cells express low levels [30]. However, the conditional knockout of N-cadherin in mice illustrates the complexity of the niche system, since it caused no observable change in HSC frequency or repopulation potential [31]. An interesting link to the metabolic niche is given by the observation that reactive oxygen species downregulate N-cadherin in HSCs [32]. Further support of a hypoxic BM niche comes from Parmar and colleagues. They used a perfusion tracer to identify the location of most HSCs in an area of low perfusion [33]. Also consistent with the idea of a hypoxic in vivo niche is the analysis of HSCs in hypoxic culture. In vitro hypoxic conditions induce quiescence in hematopoietic cells [34] and support the Hoechst-stained side population in LSK cells that is commonly accepted as a typical HSC quality [35].

Many different factors have been identified in the context of the stem cell niche, including Angiopoietin-1 (Ang-1) [36], Kit-ligand (Kitl) [37], CXCL12 [38], thrombopoietin (TPO) [39], and OPN [40]. However, in most cases, the identity of their key cellular sources promoting this maintenance remains unclear. Just now conditional knockouts of known factors in hematopoietic cells begin to reveal the cell types most that are most important for a particular signaling route [41].

Macrophages and Monocytes

Recent studies suggest a key role for monocytes in maintenance of HSCs [42–44]. Chow et al. applied four different techniques to induce specific loss of defined subpopulations of monocytes and macrophages [42]. Loss of the addressed cells resulted in HSC mobilization into peripheral blood and spleen. It was accompanied by a 40 % reduction of CXCL12 that is known to critically regulate niche retention of HSCs via activation of its receptor CXCR4 [45, 46]. Addressing the transcription of CXCL12 and other HSC retention factors in stromal cells, it was shown that CXCL12, SCF, Ang-1, and vascular cellular adhesion molecule 1 (VCAM1) mRNAs were not reduced in OBs but in Nestin-positive osteoprogenitors/mesenchymal stem cells (MSCs). Interestingly, total cell numbers of both populations were not affected. These results indicate that the key factors themselves are regulated by further components as in this case the macrophage/monocyte cell numbers. In a similar approach, Winkler et al. [44] depleted phagocytes and also observed mobilization of HSCs. Transcripts of CXCL12, Ang-1, and SCF decreased in total BM and in endosteal stroma, too. Most striking was the simultaneous loss of osteomacs, a particular macrophage subpopulation specifically associated with the endosteal lining [44, 47]. Additionally, a significant reduction of bone remodeling activity was observed. In the depleted system, the proportion of bone surface lined with OBs and the amount of newly formed bone matrix decreased significantly. Thus, both studies nicely illustrate two aspects of the regulation of the stem cell niche: the tight interaction of different cell types, here HSCs, macrophages, and osteoprogenitors, and the complexity resulting from combination of various feedback mechanisms such as bone remodeling, cell numbers, and HSC mobilization.

Osteoclasts

Although osteoclasts take part in the process of bone remodeling, they do not belong to the mesenchymal lineage like OBs and osteocytes, but are derived from hematopoietic cells [48]. They are responsible for bone resorption and, therefore, for Ca2+ blood levels. The calcium-sensitive receptor (CaR) is expressed on various hematopoietic lineages and, in particular, on LSK cells [49, 50]. Ca signaling and its role in niche regulation were investigated by studying a CaR -/- mouse model [49]. In CaR -/- mice, BM cellularity and relative frequency of LSK cells among hematopoietic cells were clearly reduced. The function of fetal liver mononucleated CaR -/-

cells was tested by their transplantation into irradiated mice, and although 100 % survival was observed, homing of these cells in the BM was markedly reduced [49]. Despite no differences in surface expression of many homing related molecules (e.g., CD49d, CD62L and CXCR4) was found, they also showed a remarkably reduced adhesion to one of the main components of bone, collagen I. All together the osteoclasts represent another niche player that intimately connects signaling, extracellular matrix, cell migration, and control via differentiation.

Mesenchymal Cells

Mesenchymal Stem and Progenitor Cells

Like HSCs, MSCs are defined by their functional potential to self-renew, proliferate, and differentiate. As for HSCs, a strictly phenomenological characterization is limited. For MSCs, the multilineage potential comprises three main lineages: the chondrogenic, adipogenic, and osteogenic lineage [51]. In the BM, they directly participate in the regulation of hematopoiesis as adventitial reticular cells (ARCs) in humans [52] or in mice as CXCL12-abundant reticular (CAR) cells [38] or Nestin-positive cells [53]. Additionally, they differentiate into two other cell types that are involved in the control of a HSC niche: OBs [e.g., 54, 55] and adipocytes [56]. Within the BM, they are found in the reticular space as mural or subendothelial cells [57]. Definitely impressive is the variety of cytokines expressed by MSCs that are involved in niche regulation: SCF, leukemia inhibitory factor (LIF), SDF-1, Oncostatin M (OSM), bone morphogenetic protein-4 (BMP-4), Flt-3, and transforming growth factor-β (TGF-β) [57]. MSCs are also capable of producing a variety of interleukins [58], niche related adhesion molecules such as VCAM1 and N-cadherin [52, 53] or even the key hematopoietic growth factors G-CSF and GM-CSF [58]. However, since most of the related experiments have been carried out in vitro, their interpretation regarding the in vivo situation should be done with caution. The role of stromal cells for HSC fate was shown early by their coculture with HSCs where they support proliferation and differentiation in vitro [59]. Another indication of their role as niche keepers is given by subcutaneous transplantation of CD146 + MSCs into immunodeficient mice, where they are able to generate heterotopic BM, trigger its vacularization, and there eventually give rise to hematopoiesis [52].

Osteoblasts

Multiple studies have shown that OBs play a crucial role in supporting HSCs. Genetic data indicate that functional stem cells do need to interact with OBs [16, 20, 55]. In these studies that involved transgenic mice to address the effect of the factors BMP and parathyroid hormone, the number of the stromal pool of OBs was found to correlate with HSC number involving Notch-ligand and N-cadherin interactions

[54, 55]. Coculture with endosteal cells characterized by typical osteogenic markers (such as alkaline phosphatase and OPN) maintains the pluripotent state and hinders HSC proliferation [59] confirming the role of OBs in HSC regulation. Direct communication between HSCs and OBs is given, for example, by Ang-1/Tie2 signaling, which has been reported sustain HSC quiescence [36]. Thus, Ang-1/Tie2 signaling might directly correlate with the long-term repopulation ability of HSCs. However, two details that are mentioned rather rarely have to be considered: (1) OBs are a transient cell state in the osteoblastic lineage finally leading to osteocytes and (2) bone deposition by OBs is a dynamic process restricted to less than 10 % of the bone surface in adults [60]. This leads to the question on the influence of other cells in the osteoblastic lineage and the mechanisms of regulation. If only OBs would enable hematopoiesis and this regulation would act on a purely local scale, hematopoiesis would be limited to the sites of bone deposition. The solution for this conflict might be found in the role of pre- and post-osteoblastic stages. While the role of osteoprogenitors has already been confirmed, it remains elusive whether the abundant osteocytes contribute a regulatory function in the niche.

Adipocytes

The triple differentiation potential of MSCs includes both, osteogenic, and adipogenic lineages. Generally, lineage commitment is an exclusive choice and, therefore, the HSC-supporting OB population competes with the adipocytes for progenitor cells. Interestingly, a study evaluating the occurrence of HSCs in different body regions of wild-type mice and in fat-free transgenic mice has shown that the number of adipocytes in the BM correlates inversely with hematopoietic activity of the BM and suggests a negative regulation of hematopoiesis by adipocytes. Engraftment of HSCs in these fatless mice after irradiation is more efficient than in their wild-type litter mates [56]. Although this effect might be due to an apparent reciprocal correlation of adipocytes and OBs, the control of adipocyte/OB differentiation clearly represents a process that not only regulates HSC number and engraftment but also depends on biomechanics and, thus, introduces biomechanical stress to the set of regulatory mechanisms [61].

Endothelial Cells

Very early hints to a contribution of ECs to hematopoiesis were given in the 1970s when Knospe et al. [62] reported that hematopoietic regeneration in areas of curetted BM in adult mice corresponded with sites of BM sinusoidal vascular regeneration. Further evidence was given by coculture in vitro. Primary human BM ECs supported the proliferation and differentiation of human CD34 + cells (which represent a HSC-enriched subpopulation of BM cells) and produced several hematopoietic cytokines. This stem cell support by ECs is restricted to neither hematopoietic tissues nor HSCs, but is found in most stem cell systems [63]. Chute and coworkers, therefore, tested the effect of human ECs on self-renewal of human HSCs. Interestingly, noncontact culture of human BM or cord blood HSCs with primary human brain ECs induced a tenfold expansion of human HSCs with the potential to repopulate immunodeficient mice, suggesting that adult brain ECs produced soluble factors, which induce HSC self-renewal [64, 65]. Analysis of several candidate proteins revealed that concerted action of either angiopoietin-like 5, insulin-like growth-factor-binding protein-2 (IGFBP-2) or pleiotrophin together with early acting cytokines (SCF, TPO, Flt3-L) significantly supports the expansion of HSCs in vitro.

Adrenergic Neurons

Circulating HSCs and their progenitors exhibit robust circadian fluctuations in the peripheral blood [66]. They fluctuate in antiphase with the expression of the chemokine CXCL12 in the BM microenvironment. This cyclic release of HSCs follows the oscillations of the circadian clock and is transmitted by the sympathetic nervous system. BM adrenergic nerves secret noradrenaline and this signal leads to the rapid downregulation of CXCL12 via the β_3-adrenergic receptor and subsequent mobilization of HSCs. This interaction with the sympathetic nervous system adds a totally new aspect to the complex control mechanisms of the hematopoietic niche.

Already from the above given overview, it becomes clear that a mechanistic understanding of niche-driven HSC regulation is still a rather "white spot on the map of hematopoiesis." Although there is no doubt about the importance of the local environment in stem cell regulation, and although a number of important components of niche functionality have already been identified, a number of major ingredients for a systemic understanding of stem cell organization and its dependence on the local growth environment (GE) are still missing. These include (i) the spatial organization of niche components, (ii) the general rules of stem cell–niches "communication" (e.g., feedback mechanisms), as well as (iii) a quantification of the functional relationships between the individual components of the stem cell–niche complex.

One way to foster a comprehensive understanding of niche-mediated stem cell regulation is the application of systems biological methods. In particular, the application of mathematical models provides a means for quantitatively studying the effect of different regulatory rules (such as feedback loops or dose–response relations), can help to guide the experimental strategy and to foster a quantitative, mechanistic understanding. However, to be able to mathematically model the dynamics of stem cell systems, it is necessary (a) to derive adequate model assumptions, (b) to estimate model parameters, and (c) to experimentally test model predictions. In the following, we will give an overview on different strategies to measure and quantify stem cell–niche interactions and illustrate a modeling framework that is able to integrate these measurements and to quantitatively study emerging system properties.

Mathematical Modeling of Stem Cell–Niche Interactions

The Benefit of Models

As described in the previous sections, experimental research in the field of HSC biology has attracted a lot of attention in the past decades. In contrast, surprisingly, little theoretical work has been published. This lack of theoretical research may be partly due to the expectation that experimental approaches will be able to determine stem cells and stem cell functionality directly. However, the more we realize that this expectation is misleading (see, e.g., the unsuccessful search for *the* stem cell marker), theoretical concepts and mathematical models that allow to quantitatively test the concepts and to compare them directly to experimental data, are becoming more and more important to cope with the current lack of understanding.

A theoretical—or in other words *systems biological*—basis of (tissue) stem cell research that complements (but not replaces!) experimental approaches, will have several advantages. Theories and their formalizations in mathematical models:

- Provide presumptive mechanisms to explain and link a variety of observed phenomena and reveal how far data are consistent with one another and with the latent mechanisms.
- Help to direct experimentation due to predictions that can be investigated.
- Help to anticipate the impact of manipulations to a system and its response.
- May help to understand the similarities of construction principles between different tissues and/or species.

To guarantee that mathematical models indeed provide the above listed support for the experimental sciences, they have to meet a number of general requirements. In particular, to study and analyze stem cell–niche interactions, corresponding models have to:

- Be based on populations of individual cells to follow clonal development, to conform with the uncertainty principle and to enable considerations of population fluctuations.
- Consider growth environments and their interactions with cells.
- Be dynamic in time and space.
- Make assumptions on mechanisms that regulate proliferation, cellular differentiation, and cell–growth environment interactions (e.g., homing).
- Be comprehensive in the sense of being applicable to the normal, unperturbed in vivo homeostasis as well as to any in vivo or in vitro assay procedure. This means, the model has to adequately account for system–measurement interactions.

Of course, there are several strategies to use mathematical models to study different aspects of stem cell–niche interactions. One possibility is an elaborated modeling of particular (molecular) mechanisms, such as receptor–ligand interaction, signaling pathways or transcriptional regulations. Such an approach, which is commonly referred to as "bottom-up" approach, includes as many system components as possible, even at very small scales, to explain the resulting, higher-level system dynamics. It intends to describe almost the entire complexity of a certain (sub−)system and can certainly help to achieve a detailed quantitative understanding of specific regulatory processes. However, due to the complexity of the models and the corresponding huge number of model parameters, their application and their predictive power are usually restricted to particular (experimental) situations. Also, to be successful, this type of models requires detailed information on the molecular parameters, such as transcription-factor-binding rates, diffusion coefficients, or protein decay rates, which are very often not available.

An alternative modeling strategy is the formulation of rather simple models that explain higher order phenomena without considering the full complexity of the underlying system. In particular, such an approach intends to find the most simple, but still consistent explanation of a number of different phenomena (e.g., different assay systems, different cell types) in terms of general rules that are predicted to determine the system dynamics. A validation of such model can, e.g., be achieved by a (quantitative) comparison of modeling results with a defined list of critical phenomena. Although such a "top-down" approach is intentionally neglecting low-level (e.g., molecular) details, it is still very helpful to identify general regulatory principles and to predict the system behavior in different situations.

A Simple Model of HSCs–Niche Interaction

In the following subsection, we will describe a mathematical model that, building on the above-introduced top-down perspective, explains stem cell organization in the hematopoietic system. Specifically, the presented model focuses on the regulatory effect of microenvironmental cues, often called niche factors, and their interplay with (intrinsic) properties of individual HSCs. This model, which describes stem cell–niche interactions in a very simple and abstract way, has originally been proposed already in 2002 [13, 67]. Since then it has successfully been applied to a multitude of different experimental and clinical settings, e.g., by supporting the design of new experiments, by generating new biological hypotheses, or by predicting the effect of new therapeutic strategies [68–75]. For a detailed, technical description of the mathematical implementation and the parameters of the model, we refer the reader, e.g., to Roeder and Loeffler [67] or Roeder et al. [73].

The model is based upon three of general principles:

(a) *Functional flexibility*: The model assumes a separation of potential cellular properties and their actual use. Any cell can, but does not have to use its functional potentials in different situations.
(b) *Phenotypic flexibility*: Cellular properties (potentials and actual functions, see (a)) can, in principle, be gained and lost reversibly.
(c) *Context dependency*: Actual use of a certain cellular property or potential as well as the development and change of properties are context dependent.

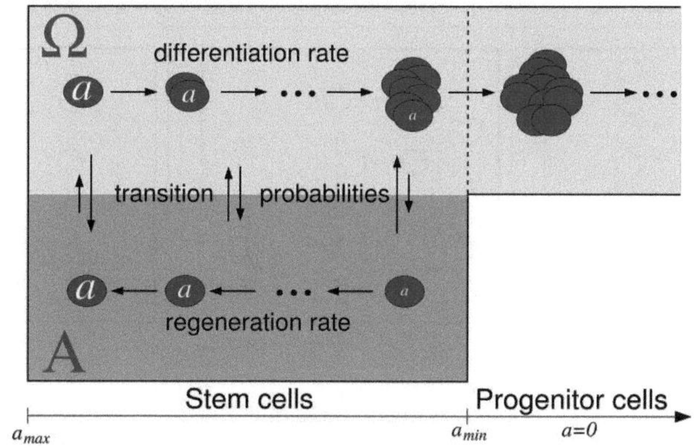

Fig. 2.2 Scheme of the HSC model. *HSCs* can reside in one of two functionally different signaling contexts, A and Ω. Their affinity a to reside in GE A reflected by the probabilities to change GEs as illustrated by vertical arrows. Evolution of affinity a and proliferation are context dependent. In A the cells are quiescent and regain affinity a, while in Ω they proliferate and gradually lose a. After passing the threshold $a_{\min}$, the cells eventually lose the capacity to switch into signaling context *A*

To implement these principles for the particular situation of the hematopoietic system in a mathematical framework, the following minimal set of assumptions is made:

1. To account for *context dependency*, we assume two different signaling contexts inside the BM, denoted as A and Ω, respectively. This assumption does in principle not exclude a higher numbers of signaling contexts (see discussion below), but is the simplest configuration that allows to model context dependency. All cells in the BM can reside in either A or Ω. The propensity of a particular cell to reside in A is modeled by a variable called context affinity a. Cells with large a tend to either stay in A (if residing in A) or to change to A (if residing in Ω), but these propensities shrink with decreasing value of a. Once a has fallen below a threshold $a_{\min}$, cells are no longer able to change to signaling context A.
2. *Phenotypic reversibility* is given by the context affinity a that can reversibly change within the interval $[a_{\min},1]$, with highest propensity for staying in/changing to A at $a = 1$.
3. Whereas cells in Ω do gradually lose context affinity a over time, cells in A can regain a up to its maximum value 1. Also, cells in A do not proliferate, although they have, in principle, the potential for proliferation. This potential is used, if the cell is found in context Ω. These assumptions introduce *functional flexibility* in the sense that both proliferation as well as losing/regaining of a are potential functions/behaviors of the cells that can be used under certain circumstances.

The design of this model (Fig. 2.2) translates the rather general assumptions inspired by biological observations (context dependence and flexibility) into a formal mathematical framework. Of course, as a first step, this model must comply with obvious requirements such as self-renewal, homeostasis, and in case of damage, regeneration. It turns out that the simple rules (1)–(3) result in the properties of a stem cell population: Simultaneously to the continuous production of cells with

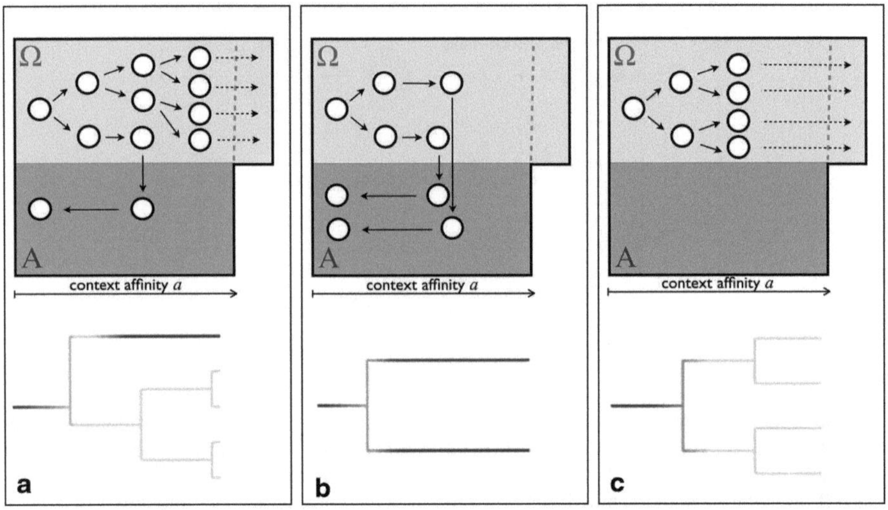

Fig. 2.3 Realization of symmetric and asymmetric cell fates in the HSC model. The proposed context-dependent, self-organizing stem cell model is capable to account for qualitatively different stem cell fates, namely asymmetric (**a**), symmetric self-renewing (**b**), and symmetric differentiating (**c**), without assuming different types of cell division. Depending on whether a stem cell (context affinity $a > a_{\min}$) is able to find the way into context A (the "niche"), it will/will not be able to regain its context affinity a and, therefore, to maintain/lose its stem cell potential. For detail, the reader is referred to Roeder and Lorenz [75]

$a < a_{\min}$ in Ω, the system maintains a stable population of cells with $a > a_{\min}$ due to proliferation in Ω and regeneration of a in A. It is also able to regenerate the above-described homeostatic situation after reducing cell numbers. In fact, it is in principle able to be "rebooted" from a single cell with context affinity $a > a_{\min}$. These system properties do exactly describe the functionality of a stem cell population: its cells are able to simultaneously self-renew their own population of undifferentiated cells ($a > a_{\min}$) and to produce differentiated cells ($a < a_{\min}$). Furthermore, the stem cell defining ability to repopulate a disturbed/depleted system is fully captured by the above-described model.

The model also provides useful predictions and generates (new) hypotheses that can guide further experimental strategies. Importantly, the predictions can be tested experimentally, because the model provides measurable quantities like time scales and cell numbers. In this way, it has been capable of addressing a wide range of biological phenomena regarding the stem cell population, such as age-related changes of HSC properties, proliferative heterogeneity of HSCs, or leukemia development and treatment, without changing the underlying basic assumptions [68–75].

For example, in contrast to other models of stem cell organization, this model does not explicitly consider different types of cell division, symmetric self-renewing, symmetric differentiating, and asymmetric (producing on stem and one differentiating cell). This decision process is replaced by the context dependency of cellular development, i.e., the loss and the regain of affinity a. Different developmental pathways of individual cells do generate symmetric and asymmetric cellular fates, without tying this process to the cell division event (Fig. 2.3). That means the model demonstrates that cell fate asymmetry, which is a necessary property of a regenerative tissue, can be induced by environmental effects rather than by a pre-determined cell-intrinsic program at the single cell level. However, on the cell population level, the assumption of an asymmetric average behavior of stem cells can still serve as an appropriate description.

There are a number of further model predictions that emphasize the interplay of cell-intrinsic potentials and cell-extrinsic (e.g., niche-induced) influences, which are able to challenge the functional potentials of the cells.

First, the model predicts the existence of two distinguishable functional states of HSCs, namely a rather quiescent and an actively proliferating one, between which these cells can reversibly change. This prediction, which is based on theoretical considerations on the effect of a heterogeneous local GE of HSCs (signaling contexts A and Ω), had already been formulated in 2002 [67]. This was 6 years before the same conclusion had been derived on the basis of experimental results: In 2008, Wilson et al. [23] demonstrated the existence of two HSC subpopulations: a deeply quiescent and a more proliferative one. Furthermore, these authors showed that in situations of stress or injury, quiescent cells can be activated into cell cycle, while most of them will return into proliferative quiescence once homeostasis has been reestablished. Our model, which predicts exactly this behavior, demonstrates that these reversible system dynamics are consistent with the concept of a niche-induced proliferation arrest. As shown by a quantitative comparison of model simulations with BrdU label dilution data, the heterogeneity observed in the HSC population (dormant vs. active cells) is naturally generated by a system that is driven by a particular niche environment, which induces a protective effect on HSCs [69]. Here, protective means to keep the cells in a rather inactive state while they maintain their full repopulation ability. From a conceptual point of view, we argue that proliferation and quiescence are just two sides of the same "stem cell coin," and that the dualism in the appearance of HSCs (dormant vs. proliferative) is an inherent system property. Moreover, this dualism and the reversibility of the actual cellular state make it highly questionable to consider these populations as being independent from each other. Thus, the model shows the consistency of a simple, self-organizing system, which uses a context- or niche-dependent cellular development as the driving force of stem cell regulation, with many in vitro and in vivo phenomena. In this sense, it provides theoretical evidence for the functional role of a stem cell niche in the hematopoietic system. Although this consistency check does not represent a formal proof for the (mechanistic) correctness of the model, it points out possible explanations for (biologically) unknown processes.

Another prediction, which results from the assumption of a context-dependent stem cell organization, and in particular from the potentially reversible change of HSCs between the functionally different signaling contexts, is the gradual loss of clonal heterogeneity of the HSC population over time [72]. Figure 2.4 illustrates this phenomenon, which is referred to as *clonal conversion*: If one assumes a certain degree of heterogeneity among HSCs with respect to their dynamics of changing between different signaling contexts, e.g., represented by their differentiation rate and their individual niche-binding propensity, one would predict a gradual loss of

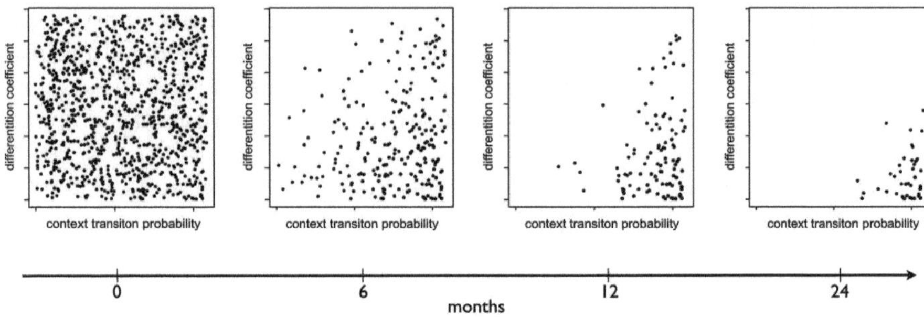

Fig. 2.4 Clonal conversion in the HSC model: in a simulation of a murine HSC population over 24 months, the initial heterogeneity in two cell-intrinsic parameters almost vanishes, if no *de novo* generation of heterogeneity is included. Each HSC is initialized with an individual combination of differentiation coefficient (quantifies the velocity of losing context affinity a) and context transition probability (controls transitions from context Ω to A). This cell-specific parameter configuration, which is inherited from mother to daughter cells, is illustrated by the *black dots* in the diagrams. Over time, the model predicts a selection of the fittest HSC clones with lower differentiation coefficient and higher probability per time step for changing into the ("niche") context. For details of the underlying simulation, we refer the reader to Roeder et al. [72]

this heterogeneity over time. An important assumption underlying this result is the strict inheritance of the individual cellular properties from mother to daughter cells. This is what we call a *clonal* property. As the cells compete for a common resource, the space in the stem cell-supporting niche environment, those clones with even slightly higher potential for homing to the niche, will ultimately outcompete the others. It is important to note, that only the number of clones (the cell "families" with identical properties), not the total number of HSCs in the system decreases. The latter remains constant for the dynamic regulation of the system. Clonal conversion is still a theoretical prediction in the hematopoietic system. To be experimentally validated, one needs to individually mark HSCs and to track their clonal progeny over time. Although this has been done in the past using random, retroviral integration sites [76, 77], a detailed, quantitative analysis has only recently become potentially feasible by the introduction of the method of cellular barcoding [78, 79]. A particular question to be addressed in this context is, whether the above-assumed heritable clonal properties indeed stay constant over time. Only if that would be true, the predicted clonal conversion and reduction of the diversity could be observed at the degree shown in Fig. 2.4. Alternatively, it is well possible that heterogeneity is constantly generated. A quantitative understanding of the interplay of clonal conversion and maintenance and generation of clonal heterogeneity is not only important from a basic science perspective but it is medically relevant, too. For example, it has recently been described that *polyclonality* (the simultaneous existence of many cell clones with different properties) in the T-cell population can inhibit the development of lymphomas [80]. Similar questions, such as the regulatory effect or the prognostic function of a polyclonal hematopoiesis for the potential outgrowth of leukemic (stem) cell clones, are discussed in the context of gene therapeutic applications [81–84].

Towards a Mechanistic Understanding of HSC–Niche Interactions

A Potential Road Map

Although the described model already explains a number of general regulatory principles and although it allows for (experimentally testable) model predictions, it is still very much restricted to a rather general level of description. Complex mechanisms are simplified in such a way that the general rules (e.g., the role of the context-dependent control of proliferative activity) can be quantitatively studied without considering underlying molecular details. The consistency of the model assumptions is tested using a "reverse-engineering" strategy. That means, the model assumptions are evaluated by comparing the system dynamics, predicted by the model under a number of different conditions, with experimental observations on the cellular and the tissue level. This way, the modeling can provide possible explanations of the system behavior in terms of the considered rules and properties. However, the model is not able to explain the lower level processes, such as signaling pathways that are triggering proliferative activity or receptor–ligand configurations that induce co-localization of certain cell types. To arrive at a detailed mechanistic explanation of cellular organization, with a focus on stem cell–niche interactions including the sub-cellular/molecular level, a number of prerequisites have to be achieved (Fig. 2.5):

I. Relevant regulatory components (e.g., genes or molecules) have to be identified and validated biologically.
II. Test systems, which allow for studying the effects of relevant regulatory components on cell behavior and function under controlled conditions, have to be established.
III. The spatial in vivo organization, including a quantification of the anatomical topology and/or geometry of regions of stem cell appearance, as well as the dynamics of stem cells within these regions has to be characterized.
IV. A theoretical framework that allows for a systematic analysis and for the integration of different data types needs to be implemented.

On the basis of these prerequisites, it will be possible to set up a combined experimental and theoretical program that will result in a much deeper understanding of stem cell–niche interactions. Partially, the above mentioned requests were already established. In the following, we will give a short summary of some of the corresponding results.

Coculture Systems and Molecular High-Throughput Screens

Cocultures with various BM stromal cell types have established the possibility of maintaining HSCs for longer periods of time in culture. Of course, this maintenance of HSCs in coculture depends on the cell type and involves the complexity of the

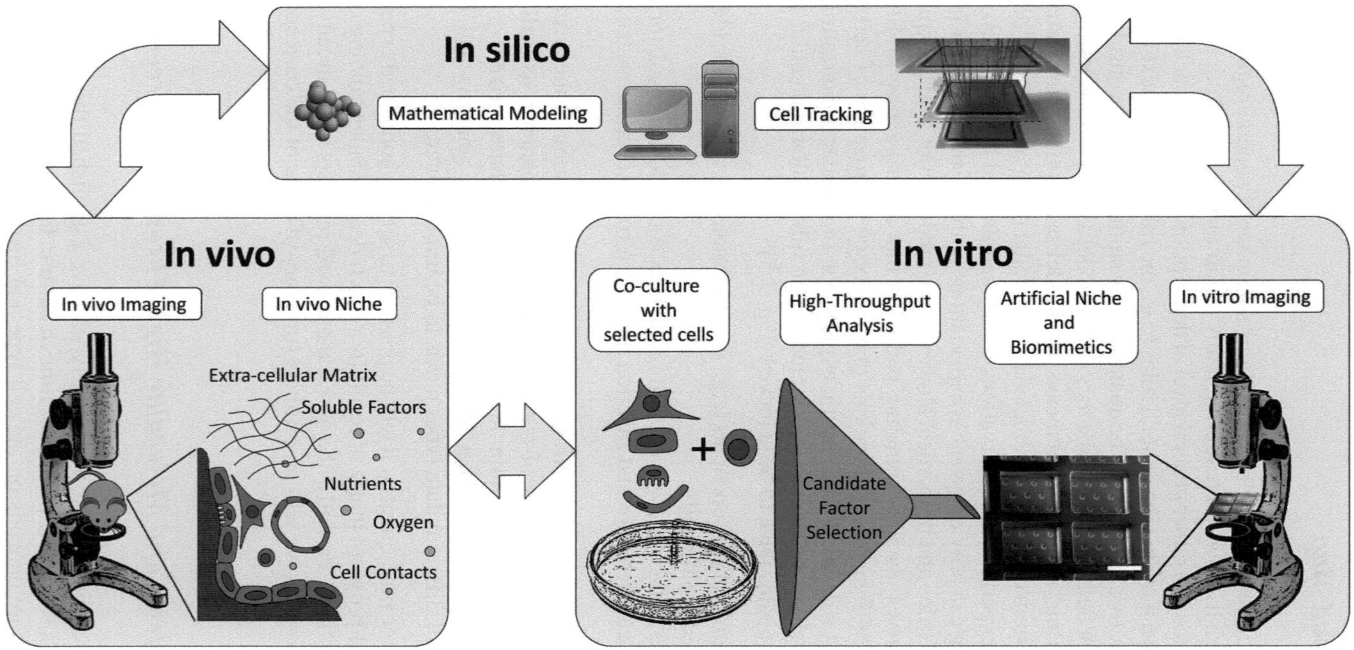

Fig. 2.5 Key elements for investigation of the niche. To understand the complex interplay of all factors in the niche both, studying the in vivo situation as well as the interplay of candidate factors in controlled in vitro settings is necessary. High-throughput analysis of cocultures acknowledges the central role of cellular interaction in the niche and permits to filter for the most important factors. Due to the heterogeneity of cell populations, single cell tracking is indispensable for a comprehensive picture. Finally, mathematical modeling and simulations integrate the insights into a complete picture and allow for predictions and model testing

wide spectrum of biological signals. Therefore, discriminating single factors is rather difficult, but molecularly assessing the transcriptional state of both HSCs and cocultured cells helps to identify important factors involved in the molecular crosstalk of HSCs and stromal cells. Specifically, high-throughput analysis of the transcriptome and proteome of the cocultured cells provides a means to select candidate factors to focus on in further studies.

Because heterogeneity not only is present in the stem cell niche but also seems to be a functional element of stem cell regulation, analysis of individual cells is almost mandatory. An approach to analyze individual cells in a specific environment, alone or in small colonies, is provided by advances in microfluidics and lab-on-a-chip techniques. Small cocultures of arbitrary composition can be cultured in individual capsules within an environment of tailored factor composition and biomechanical properties and analyzed at high throughput [85, 86]. Another technique to mention in this context is single-cell PCR that has the ability to unveil information on the heterogeneity of transcription states within cell populations or among many of the microfluidic cocultures [87].

Biomimetics/Artificial Niches

After identification of relevant niche components, their mode of action can be identified by an isolated analysis of these components in an in vitro setting, where all other factors can be controlled. Advances in material sciences permit to combine presumptive factors in artificial and observable environments. Different polymers, linkers, and proteins together with micro- and 3D-printing techniques can provide scaffolds of diverse biophysical properties and geometries [88]. Molecular modifications of these scaffolds by covalently attaching functional molecules permit to mimic biologically relevant cell–cell and cell–matrix contacts and to change scaffold degradability. Finally, perfusion and fed-batch approaches can control the availability of soluble factors and nutrients and the drainage of metabolites [89]. In principle, all the above factors can now be combined in a single high-throughput experiment. Automation in the laboratory facilitates combinatorial composition of various presumable niche components [90]. Taken together, the advances in controlling an increasing number of niche components in vitro pave the way to constructing artificial niches as valuable experimental setups.

Image Analysis and Single Cell Tracking (In Vitro)

The screening of niche components in vitro is an essential tool for studying their contributions to the stem cell niche. The results arising from these high-throughput experiments will generate a huge amount of data that mainly consist of micrographs. Automation, therefore, is not only necessary for performing the experiments but also for their analysis, where reliable segmentation algorithms for image analysis are necessary. This allows identification of individual cells and, therefore, quantification of cellular heterogeneity. This way, we can also assess the role of the latter for the self-organization of the niche. Because self-organization of the stem cell niche implies a dynamically changing system, time-lapse microscopy is necessary to follow this process. Single-cell-tracking tools have been identified as a key element for understanding the dynamics process of cell regulation [91, 92]. Increasing computational power as well as the development of new algorithms that allow for automatic cell recognition (segmentation) and tracking (registration and mapping) have already produced intriguing results on the genealogies of cells in vitro [93].

In Vivo Imaging

After describing possibilities to address the stem cell niche in vitro in an analytical or artificial manner, one still has to recognize that the true stem cell niche environment and, thus, the ultimate standard for stem cells is the in vivo situation. If we really want to understand the niche and its regulation, direct comparison of experimental results with the in vivo scenario is necessary. Taking into account the role of single cells in the maintenance of the population, this includes the knowledge about detailed locations and dynamics of individual cells. Only this level of understanding will lead to an unambiguous and comprehensive picture of the stem cell niche. In the past decade, considerable advances in vivo imaging of the BM could be achieved, e.g., by taking advantage of the thin bone of the calvarium in mice. Simultaneous visualization of transplanted HSCs, OBs, bone, and vasculature facilitates the in vivo tracking of individual HSCs and analysis of the relative positions of the visualized niche components [26]. This way, information on the migratory behavior of single HSCs becomes accessible and its analysis promises to shed light on the spatiotemporal organization of the stem cell niche.

Spatiotemporal Modeling

Each of the approaches described above provides information related to different levels of the organization hierarchy in the stem cell niche. A crucial step towards the understanding of the niche is now to integrate it into a formulated model. Paralleling the generation of information on different levels, various models can be applied according to the sort of experimental input. Besides the multitude of possible modeling techniques, there are some common characteristics for models of the spatiotemporal niche organization. Cell heterogeneity requires representation of the internal state of individual cells. The degree of detail, e.g., differentiation/cell cycle position or levels of protein/transcription factors is determined by the experimental context. To pursue understanding of the spatiotemporal organization implies to relate the internal states to dynamic spatial entities. Finally, the interaction of cells with other cells and local environment is crucial for the self-organization of the niche. Essentially, there are three aspects that have to be considered regarding this interaction: (1) The cells perceive information from the environment. Therefore, an interface between the environment (e.g., factors, nutrient, cell contacts) and the internal state of the cell has to be given by the model. (2) In order to process this external information and to change the internal state, the model needs a set of hypothesized or identified rules for this information processing. (3) Cellular actions that allow the cell to alter its environment according to its internal states, e.g., by motion, growth or production/degradation of nutrients, extra cellular matrix or signaling molecules constitute the last element of the model and close the circle of self-organization.

These general ingredients permit to focus on particular aspects of regulation as required by the experimental design and its input to the model. For example, if the aim of the model is to describe transcription factor networks, the internal state will be the most important aspect. In the other direction, the model has to provide predictions that are accessible by the experiments. Of course, the complexity of the stem cell niche will hamper modeling all aspects of niche regulation at once, but as exemplified above, also simplified models can be quite useful and produce intriguing results. As another example, a model of mesenchymal cells has addressed this HSC niche relevant cell type by a phenomenological, noise-driven model of cell differentiation [94]. Its stochastic approach succeeded well in describing the oxygen dependence of cell proliferation and differentiation. An adequate description of the population heterogeneity is featured by this phenomenological noise-driven model. The spatial representation allowed to match the differentiation dynamics in the oxygen gradient within a pellet culture and to capture the importance of cell–cell contacts for the production of cartilage. In a similar way, key qualities of the niche may be identified by comparison of hidden scenarios. An example may be given by combinations of different mechanisms of transport for positive and negative signals that cover different ranges like information transfer via adhesion, membrane-bound transport, and diffusion. Similar simplifications enable mathematical modeling to contribute substantially to the disentangling of the niche despite its complexity.

Concluding Remarks

Generally, there is no doubt about the importance of extrinsic, microenvironmental effects on the regulation and on fate decisions of tissue stem cells in general and of HSCs in particular. As described above by means of some examples, a number of theories about the localization, the components, and the key regulatory pathways are discussed in the scientific community. Partially, these theories are substantiated by different experimental results. However, a complete picture of stem cell–niche interaction and of the underlying regulatory and organizational principles is still lacking.

In this chapter, we tried to summarize (without claiming completeness) important facts about the stem cell niche in the hematopoietic system. In particular, we discussed the possibility to use mathematical models, as one important means of a systems biological approach, to quantitatively study the rules that drive the dynamics of stem cell systems. As shown above, even rather simple models can considerably help to identify regulatory principles, which are consistent with multiple in vitro and in vivo results, and, therefore, to guide further experimental strategies.

Nevertheless, we also pointed out that from our perspective a complete or, more realistic, a better understanding of stem cell organization and of the particular role of niche components in this process will require a concerted experimental and theoretical research program. In our opinion, any promising approach towards a systemic understanding of stem cell organization (and that is what *systems biology* is all about) should consist of: (i) experimental systems that allow for controlled quantification of the effect of individual regulatory components including the possibility for targeted system perturbations (may be in an in vitro setting), (ii) a theoretical framework that allows for interpretation and for quantitative comparison of theoretical (model) predictions with experimental data, and last but not least, (iii) an experimental system that allows to validate the obtained result for the in vivo situation. This also makes clear that systems biology is always a combination of experiment and theory. However, the theoretical component must not be forgotten; it can indeed considerably sharpen our scientific view. Along these lines, we would like to finish with a citation that is commonly attributed to Charles R. Darwin: "*Mathematics seems to endow one with something like a new sense*."

Acknowledgments This work has been supported by the German Ministry for Education and Research, BMBF-grant on Medical Systems Biology HaematoSys (BMBF-FKZ 0315452), and the Human Frontier Science Program, HFSP-grant RGP0051/2011.

References

1. Becker AJ, McCulloch EA, Till JE. Cytological demonstration of the clonal nature of spleen colonies derived from transplanted mouse marrow cells. Nature. 1963;197:452–4.
2. Siminovitch L, Till JE, McCulloch EA. Decline in colony-forming ability of marrow cells subjected to serial transplantation into irradiated mice. J Cell Physiol. 1964;64:23–31.
3. Till JE, McCulloch EA. A direct measurement of the radiation sensitivity of normal mouse bone marrow cells. Radiat Res. 1961;14:213–22.
4. Barker N, van de Wetering M, Clevers H. The intestinal stem cell. Genes Dev. 2008;22(14):1856–64.
5. Trentin J. Influence of hematopoietic organ stroma (hematopoieticinductive microenvironment) on stem cell differentiation. New York: Appleton-Century-Crofts; 1970. pp. 161–8.
6. Trentin JJ. Determination of bone marrow stem cell differentiation by stromal hemopoietic inductive microenvironments (HIM). Am J Pathol. 1971;65(3):621–8.
7. Schofield R. The relationship between the spleen colony-forming cell and the haemopoietic stem cell. Blood Cells. 1978;4(1–2):7–25.
8. Allen TD, Dexter TM. The essential cells of the hemopoietic microenvironment. Exp Hematol. 1984;12(7):517–21.

9. Dexter TM, Wright EG, Krizsa F, Lajtha LG. Regulation of haemopoietic stem cell proliferation in long term bone marrow cultures. Biomedicine. 1977;27(9–10):344–9.
10. Evans MJ, Kaufman MH. Establishment in culture of pluripotential cells from mouse embryos. Nature. 1981;292(5819):154–6.
11. Thomson JA, Itskovitz-Eldor J, Shapiro SS, Waknitz MA, Swiergiel JJ, Marshall VS, et al. Embryonic stem cell lines derived from human blastocysts. Science. 1998;282(5391):1145–7.
12. Lanza R, Blau H, Gearhart J, Hogan B, Melton D, Moore M, et al., editors. Handbook of stem cells. Burlington: Elsevier; 2004.
13. Loeffler M, Roeder I. Tissue stem cells: definition, plasticity, heterogeneity, self-organization and models—a conceptual approach. Cells Tissues Organs. 2002;171(1):8–26.
14. Potten CS, Loeffler M. Stem cells: attributes, cycles, spirals, pitfalls and uncertainties. Lessons for and from the crypt. Development. 1990;110(4):1001–20.
15. Yilmaz OH, Kiel MJ, Morrison SJ. SLAM family markers are conserved among hematopoietic stem cells from old and reconstituted mice and markedly increase their purity. Blood. 2006;107(3):924–30.
16. Kiel MJ, Yilmaz OH, Iwashita T, Terhorst C, Morrison SJ. SLAM family receptors distinguish hematopoietic stem and progenitor cells and reveal endothelial niches for stem cells. Cell. 2005;121(7):1109–21.
17. Shirota T, Tavassoli M. Alterations of bone marrow sinus endothelium induced by ionizing irradiation: implications in the homing of intravenously transplanted marrow cells. Blood Cells. 1992;18(2):197–214.
18. Fliedner TM, Bond VP, Cronkite EP. Structural, cytologic and autoradiographic (H3-thymidine) changes in the bone marrow following total body irradiation. Am J Pathol. 1961;38:599–623.
19. Chute JP, Muramoto GG, Salter AB, Meadows SK, Rickman DW, Chen B, et al. Transplantation of vascular endothelial cells mediates the hematopoietic recovery and survival of lethally irradiated mice. Blood. 2007;109(6):2365–72.
20. Wilson A, Trumpp A. Bone-marrow haematopoietic-stem-cell niches. Nat Rev Immunol. 2006;6(2):93–106.
21. Arai F, Suda T. Maintenance of quiescent hematopoietic stem cells in the osteoblastic niche. Ann N Y Acad Sci. 2007;1106:41–53.
22. Doan PL, Chute JP. The vascular niche: home for normal and malignant hematopoietic stem cells. Leukemia. 2012;26(1):54–62.
23. Wilson A, Laurenti E, Oser G, van der Wath RC, Blanco-Bose W, Jaworski M, et al. Hematopoietic stem cells reversibly switch from dormancy to self-renewal during homeostasis and repair. Cell. 2008;135(6):1118–29.
24. Takizawa H, Regoes RR, Boddupalli CS, Bonhoeffer S, Manz MG. Dynamic variation in cycling of hematopoietic stem cells in steady state and inflammation. J Exp Med. 2011;208(2):273–84.
25. De Bruyn PP, Breen PC, Thomas TB. The microcirculation of the bone marrow. Anat Rec. 1970;168(1):55–68.
26. Lo Celso C, Fleming HE, Wu JW, Zhao CX, Miake-Lye S, Fujisaki J, et al. Live-animal tracking of individual haematopoietic stem/progenitor cells in their niche. Nature. 2009;457(7225):92–6.
27. Cross M, Alt R, Niederwieser D. The case for a metabolic stem cell niche. Cells Tissues Organs. 2008;188(1–2):150–9.
28. Suda T, Takubo K, Semenza GL. Metabolic regulation of hematopoietic stem cells in the hypoxic niche. Cell Stem Cell. 2011;9(4):298–310.
29. Kohler A, Schmithorst V, Filippi MD, Ryan MA, Daria D, Gunzer M, et al. Altered cellular dynamics and endosteal location of aged early hematopoietic progenitor cells revealed by time-lapse intravital imaging in long bones. Blood. 2009;114(2):290–8.
30. Haug JS, He XC, Grindley JC, Wunderlich JP, Gaudenz K, Ross JT, et al. N-cadherin expression level distinguishes reserved versus primed states of hematopoietic stem cells. Cell Stem Cell. 2008;2(4):367–79.

31. Kiel MJ, Acar M, Radice GL, Morrison SJ. Hematopoietic stem cells do not depend on N-cadherin to regulate their maintenance. Cell Stem Cell. 2009;4(2):170–9.
32. Hosokawa K, Arai F, Yoshihara H, Nakamura Y, Gomei Y, Iwasaki H, et al. Function of oxidative stress in the regulation of hematopoietic stem cell-niche interaction. Biochem Biophys Res Commun. 2007;363(3):578–83.
33. Parmar K, Mauch P, Vergilio JA, Sackstein R, Down JD. Distribution of hematopoietic stem cells in the bone marrow according to regional hypoxia. Proc Natl Acad Sci U S A. 2007;104(13):5431–6.
34. Shima H, Takubo K, Tago N, Iwasaki H, Arai F, Takahashi T, et al. Acquisition of G(0) state by CD34-positive cord blood cells after bone marrow transplantation. Exp Hematol. 2010;38(12):1231–40.
35. Krishnamurthy P, Ross DD, Nakanishi T, Bailey-Dell K, Zhou S, Mercer KE, et al. The stem cell marker Bcrp/ABCG2 enhances hypoxic cell survival through interactions with heme. J Biol Chem. 2004;279(23):24218–25.
36. Arai F, Hirao A, Ohmura M, Sato H, Matsuoka S, Takubo K, et al. Tie2/angiopoietin- 1 signaling regulates hematopoietic stem cell quiescence in the bone marrow niche. Cell. 2004;118(2): 149–61.
37. Thoren LA, Liuba K, Bryder D, Nygren JM, Jensen CT, Qian H, et al. Kit regulates maintenance of quiescent hematopoietic stem cells. J Immunol. 2008;180(4):2045–53.
38. Sugiyama T, Kohara H, Noda M, Nagasawa T. Maintenance of the hematopoietic stem cell pool by CXCL12-CXCR4 chemokine signaling in bone marrow stromal cell niches. Immunity. 2006;25(6):977–88.
39. Qian H, Buza-Vidas N, Hyland CD, Jensen CT, Antonchuk J, Mansson R, et al. Critical role of thrombopoietin in maintaining adult quiescent hematopoietic stem cells. Cell Stem Cell. 2007;1(6):671–84.
40. Nilsson SK, Johnston HM, Whitty GA, Williams B, Webb RJ, Denhardt DT, et al. Osteopontin, a key component of the hematopoietic stem cell niche and regulator of primitive hematopoietic progenitor cells. Blood. 2005;106(4):1232–9.
41. Ding L, Saunders TL, Enikolopov G, Morrison SJ. Endothelial and perivascular cells maintain haematopoietic stem cells. Nature. 2012;481(7382):457–62.
42. Chow A, Lucas D, Hidalgo A, Mendez-Ferrer S, Hashimoto D, Scheiermann C, et al. Bone marrow CD169 + macrophages promote the retention of hematopoietic stem and progenitor cells in the mesenchymal stem cell niche. J Exp Med. 2011;208(2):261–71.
43. Christopher MJ, Rao M, Liu F, Woloszynek JR, Link DC. Expression of the G-CSF receptor in monocytic cells is sufficient to mediate hematopoietic progenitor mobilization by G-CSF in mice. J Exp Med. 2011;208(2):251–60.
44. Winkler IG, Sims NA, Pettit AR, Barbier V, Nowlan B, Helwani F, et al. Bone marrow macrophages maintain hematopoietic stem cell (HSC) niches and their depletion mobilizes HSCs. Blood. 2010;116(23):4815–28.
45. Broxmeyer HE. Chemokines in hematopoiesis. Curr Opin Hematol. 2008;15(1):49–58.
46. Lapidot T, Petit I. Current understanding of stem cell mobilization: the roles of chemokines, proteolytic enzymes, adhesion molecules, cytokines, and stromal cells. Exp Hematol. 2002;30(9):973–81.
47. Chang MK, Raggatt LJ, Alexander KA, Kuliwaba JS, Fazzalari NL, Schroder K, et al. Osteal tissue macrophages are intercalated throughout human and mouse bone lining tissues and regulate osteoblast function in vitro and in vivo. J Immunol. 2008;181(2):1232–44.
48. Suda T, Takahashi N, Martin TJ. Modulation of osteoclast differentiation. Endocr Rev. 1992;13(1):66–80.
49. Adams GB, Chabner KT, Alley IR, Olson DP, Szczepiorkowski ZM, Poznansky MC, et al. Stem cell engraftment at the endosteal niche is specified by the calcium-sensing receptor. Nature. 2006;439(7076):599–603.
50. House MG, Kohlmeier L, Chattopadhyay N, Kifor O, Yamaguchi T, Leboff MS, et al. Expression of an extracellular calcium-sensing receptor in human and mouse bone marrow cells. J Bone Miner Res. 1997;12(12):1959–70.

51. Charbord P. Bone marrow mesenchymal stem cells: historical overview and concepts. Hum Gene Ther. 2010;21(9):1045–56.
52. Sacchetti B, Funari A, Michienzi S, Di Cesare S, Piersanti S, Saggio I, et al. Self-renewing osteoprogenitors in bone marrow sinusoids can organize a hematopoietic microenvironment. Cell. 2007;131(2):324–36.
53. Mendez-Ferrer S, Michurina TV, Ferraro F, Mazloom AR, Macarthur BD, Lira SA, et al. Mesenchymal and haematopoietic stem cells form a unique bone marrow niche. Nature. 2010;466(7308):829–34.
54. Calvi LM, Adams GB, Weibrecht KW, Weber JM, Olson DP, Knight MC, et al. Osteoblastic cells regulate the haematopoietic stem cell niche. Nature. 2003;425(6960):841–6.
55. Zhang J, Niu C, Ye L, Huang H, He X, Tong WG, et al. Identification of the haematopoietic stem cell niche and control of the niche size. Nature. 2003;425(6960):836–41.
56. Naveiras O, Nardi V, Wenzel PL, Hauschka PV, Fahey F, Daley GQ. Bone-marrow adipocytes as negative regulators of the haematopoietic microenvironment. Nature. 2009;460(7252):259–63.
57. Dazzi F, Ramasamy R, Glennie S, Jones SP, Roberts I. The role of mesenchymal stem cells in haemopoiesis. Blood Rev. 2006;20(3):161–71.
58. Majumdar MK, Thiede MA, Mosca JD, Moorman M, Gerson SL. Phenotypic and functional comparison of cultures of marrow-derived mesenchymal stem cells (MSCs) and stromal cells. J Cell Physiol. 1998;176(1):57–66.
59. Balduino A, Hurtado SP, Frazao P, Takiya CM, Alves LM, Nasciutti LE, et al. Bone marrow subendosteal microenvironment harbours functionally distinct haemosupportive stromal cell populations. Cell Tissue Res. 2005;319(2):255–66.
60. Weinstein R. The clinical use of bone biopsy. In: Coe F, Favus M, editors. Disorders of bone and mineral metabolism. Philadelphia: Lippincott Williams & Wilkins; 2002. pp. 469–85.
61. McBeath R, Pirone DM, Nelson CM, Bhadriraju K, Chen CS. Cell shape, cytoskeletal tension, and RhoA regulate stem cell lineage commitment. Dev Cell. 2004;6(4):483–95.
62. Knospe WH, Gregory SA, Husseini SG, Fried W, Trobaugh FE Jr. Origin and recovery of colony-forming units in locally curetted bone marrow of mice. Blood. 1972;39(3):331–40.
63. Bautch VL. Stem cells and the vasculature. Nat Med. 2011;17(11):1437–43.
64. Chute JP, Saini AA, Chute DJ, Wells MR, Clark WB, Harlan DM, et al. Ex vivo culture with human brain endothelial cells increases the SCID-repopulating capacity of adult human bone marrow. Blood. 2002;100(13):4433–9.
65. Rosler ES, Brandt JE, Chute J, Hoffman R. An in vivo competitive repopulation assay for various sources of human hematopoietic stem cells. Blood. 2000;96(10):3414–21.
66. Mendez-Ferrer S, Lucas D, Battista M, Frenette PS. Haematopoietic stem cell release is regulated by circadian oscillations. Nature. 2008;452(7186):442–7.
67. Roeder I, Loeffler M. A novel dynamic model of hematopoietic stem cell organization based on the concept of within-tissue plasticity. Exp Hematol. 2002;30(8):853–61.
68. Glauche I, Cross M, Loeffler M, Roeder I. Lineage specification of hematopoietic stem cells: mathematical modeling and biological implications. Stem Cells. 2007;25(7):1791–9.
69. Glauche I, Moore K, Thielecke L, Horn K, Loeffler M, Roeder I. Stem cell proliferation and quiescence-two sides of the same coin. PLoS Comput Biol. 2009;5(7):e1000447.
70. Glauche I, Thielecke L, Roeder I. Cellular aging leads to functional heterogeneity of hematopoietic stem cells: a modeling perspective. Aging Cell. 2011;10(3):457–65.
71. Roeder I, Braesel K, Lorenz R, Loeffler M. Stem cell fate analysis revisited: interpretation of individual clone dynamics in the light of a new paradigm of stem cell organization. J Biomed Biotechnol. 2007;2007(3):84656.
72. Roeder I, Horn K, Sieburg HB, Cho R, Muller-Sieburg C, Loeffler M. Characterization and quantification of clonal heterogeneity among hematopoietic stem cells: a model-based approach. Blood. 2008;112(13):4874–83.
73. Roeder I, Horn M, Glauche I, Hochhaus A, Mueller MC, Loeffler M. Dynamic modeling of imatinib-treated chronic myeloid leukemia: functional insights and clinical implications. Nat Med. 2006;12(10):1181–4.

74. Roeder I, Kamminga LM, Braesel K, Dontje B, de Haan G, Loeffler M. Competitive clonal hematopoiesis in mouse chimeras explained by a stochastic model of stem cell organization. Blood. 2005;105(2):609–16.
75. Roeder I, Lorenz R. Asymmetry of stem cell fate and the potential impact of the niche: observations, simulations, and interpretations. Stem Cell Rev. 2006;2(3):171–80.
76. Drize NJ, Keller JR, Chertkov JL. Local clonal analysis of the hematopoietic system shows that multiple small short-living clones maintain life-long hematopoiesis in reconstituted mice. Blood. 1996;88(8):2927–38.
77. Jordan CT, Lemischka IR. Clonal and systemic analysis of long-term hematopoiesis in the mouse. Genes Dev. 1990;4(2):220–32.
78. Bystrykh LV, Verovskaya E, Zwart E, Broekhuis M, de Haan G. Counting stem cells: methodological constraints. Nat Methods. 2012;9(6):567–74.
79. Gerrits A, Dykstra B, Kalmykowa OJ, Klauke K, Verovskaya E, Broekhuis MJ, et al. Cellular barcoding tool for clonal analysis in the hematopoietic system. Blood. 2010;115(13):2610–8.
80. Newrzela S, Al-Ghaili N, Heinrich T, Petkova M, Hartmann S, Rengstl B, et al. T-cell receptor diversity prevents T-cell lymphoma development. Leukemia. 2012;26(12):2499-507.
81. Kustikova O, Brugman M, Baum C. The genomic risk of somatic gene therapy. Semin Cancer Biol. 2010;20(4):269–78.
82. Maetzig T, Brugman MH, Bartels S, Heinz N, Kustikova OS, Modlich U, et al. Polyclonal fluctuation of lentiviral vector-transduced and expanded murine hematopoietic stem cells. Blood. 2011;117(11):3053–64.
83. Kim YJ, Kim YS, Larochelle A, Renaud G, Wolfsberg TG, Adler R, et al. Sustained high-level polyclonal hematopoietic marking and transgene expression 4 years after autologous transplantation of rhesus macaques with SIV lentiviral vector-transduced CD34 + cells. Blood. 2009;113(22):5434–43.
84. Neff T, Beard BC, Peterson LJ, Anandakumar P, Thompson J, Kiem HP. Polyclonal chemoprotection against temozolomide in a large-animal model of drug resistance gene therapy. Blood. 2005;105(3):997–1002.
85. Kumachev A, Greener J, Tumarkin E, Eiser E, Zandstra PW, Kumacheva E. High-throughput generation of hydrogel microbeads with varying elasticity for cell encapsulation. Biomaterials. 2011;32(6):1477–83.
86. Tumarkin E, Tzadu L, Csaszar E, Seo M, Zhang H, Lee A, et al. High-throughput combinatorial cell co-culture using microfluidics. Integr Biol (Camb). 2011;3(6):653–62.
87. Sanchez-Freire V, Ebert AD, Kalisky T, Quake SR, Wu JC. Microfluidic single-cell real-time PCR for comparative analysis of gene expression patterns. Nat Protoc. 2012;7(5):829–38.
88. Lehmann K, Herklotz M, Espig M, Paumer T, Nitschke M, Werner C, et al. A new approach to biofunctionalisation and micropatterning of multi-well plates. Biomaterials. 2010;31(33):8802–9.
89. Csaszar E, Kirouac DC, Yu M, Wang W, Qiao W, Cooke MP, et al. Rapid expansion of human hematopoietic stem cells by automated control of inhibitory feedback signaling. Cell Stem Cell. 2012;10(2):218–29.
90. Ranga A, Lutolf MP. High-throughput approaches for the analysis of extrinsic regulators of stem cell fate. Curr Opin Cell Biol. 2012;24(2):236–44.
91. Kokkaliaris KD, Loeffler D, Schroeder T. Advances in tracking hematopoiesis at the single-cell level. Curr Opin Hematol. 2012;19(4):243–9.
92. Schroeder T. Long-term single-cell imaging of mammalian stem cells. Nat Methods. 2011;8(4 Suppl):30–5.
93. Scherf N, Franke K, Glauche I, Kurth I, Bornhauser M, Werner C, et al. On the symmetry of siblings: automated single-cell tracking to quantify the behavior of hematopoietic stem cells in a biomimetic setup. Exp Hematol. 2012;40(2):119–30 e9.
94. Krinner A, Zscharnack M, Bader A, Drasdo D, Galle J. Impact of oxygen environment on mesenchymal stem cell expansion and chondrogenic differentiation. Cell Prolif. 2009;42(4):471–84.

Chapter 3
Erythropoiesis: From Molecular Pathways to System Properties

Miroslav Koulnis, Ermelinda Porpiglia, Daniel Hidalgo and Merav Socolovsky

Abstract Erythropoiesis is regulated through a long-range negative feedback loop, whereby tissue hypoxia stimulates erythropoietin (Epo) secretion, which promotes an increase in erythropoietic rate. However, this long-range feedback loop, by itself, cannot account for the observed system properties of erythropoiesis, namely, a wide dynamic range, stability in the face of random perturbations, and a rapid stress response. Here, we show that three Epo-regulated erythroblast survival pathways each give rise to distinct system properties. The induction of Bcl-x_L by signal transducer and activator of transcription 5 (Stat5) is responsive to the rate of change in Epo levels, rather than to its absolute level, and is therefore maximally but transiently activated in acute stress. By contrast, Epo-mediated suppression of the pro-survival Fas and Bim pathways is proportional to the levels of stress/Epo and persists throughout chronic stress. Together, these elements operate in a manner reminiscent of a "proportional-integral-derivative (PID)" feedback controller frequently found in engineering applications. A short-range negative autoregulatory loop within the early erythroblast compartment, operated by Fas/FasL, filters out random noise and controls a reserve pool of early erythroblasts that is poised to accelerate the response to acute stress. Both these properties have previously been identified as inherent to

M. Socolovsky (✉)
Department of Cancer Biology, and Department of Pediatrics, University of Massachusetts Medical School, 364 Plantation Street, Lazare Research Building (LRB) Room 440A, Worcester, MA 01605, USA
Tel.: 508-856-3743
e-mail: Merav.Socolovsky@umassmed.edu

M. Koulnis · D. Hidalgo · E. Porpiglia
Department of Cancer Biology, University of Massachusetts Medical School, 364 Plantation Street, Lazare Research
Building (LRB) Room 440A, Worcester, MA 01605, USA
Tel.: 508-856-3704
e-mail: miroslav.koulnis@umassmed.edu

D. Hidalgo
e-mail: daniel.hidalgo@umassmed.edu

E. Porpiglia
e-mail: eporpigl@stanford.edu

S. J. Corey et al. (eds.), *A Systems Biology Approach to Blood*,
Advances in Experimental Medicine and Biology 844, DOI 10.1007/978-1-4939-2095-2_3

negative regulatory motifs. Finally, we show that signal transduction by Stat5 combines binary and graded modalities, thereby increasing signaling fidelity over the wide dynamic range of Epo found in health and disease.

Keywords Erythropoiesis · Erythroblast · Stat5 · Epo · EpoR · Fas · Apoptosis, Bim, Bcl-x_L, negative autoregulation · Negative feedback · Stability · Stress · PID controller · Flow cytometry · Signal transduction · Binary/digital signaling · Graded/analog signaling

Overview

Imagine receiving a box in the mail, containing a disassembled bicycle. You might examine and even admire the individual parts, but the manner in which they provide transportation does not become apparent until you work out how they fit together into an assembled bike. Our current status investigating the erythropoietic system is similar: Decades of research have delivered an inventory of well-studied components; the challenge is to understand how they interact to generate the remarkable behavior of the erythropoietic system. To do so, we need to devise ways of measuring and describing interactions between the various system parts. Some simple engineering concepts are useful, particularly those pertaining to feedback control and signal transduction.

It has long been known that erythropoiesis is regulated via a well-established oxygen-dependent negative feedback loop mediated by the hormone erythropoietin (Epo). However, quantitative experimental measurements of multiple parts of the system in mice show that its stability and rapid stress response require much more than a simple negative feedback loop; they depend on the joint action of both long- and short-range feedbacks, persistent and adapting signaling responses, and binary as well as analog signaling modalities.

A number of groups have proposed mathematical models of the erythropoietic system (e.g., [1–4]). The account below does not propose a specific model; instead, we focus on understanding how specific molecular pathways in erythroid progenitors contribute to system properties at the level of the whole organism [5–9]. We conclude by suggesting that the early erythroblast compartment carries a computational task similar to that of a "proportional-integral-derivative (PID)" controller, a feedback controller frequently used in engineering applications.

First Indications of Negative Feedback

In 1878, Paul Bert proposed that at high altitude, lower oxygen tension would drive a compensatory increase in red blood cells number [10]. This may have been one of the earliest biological "feedback" control mechanisms to be recognized, where the regulated parameter, namely, tissue oxygen tension, maintains its own level between narrow bounds by suppressing the production of the cells that supply it to tissues. In 1906, Carnot and DeFlandre proposed that negative feedback was mediated via a humoral factor, a hypothesis that was confirmed by Erslev in 1953 [11]. The hormonal mediator, Epo, was purified in 1971 [12] and cloned in 1985 [13, 14]. Its receptor, EpoR, was cloned by D'Andrea and Lodish in 1989 [15].

Principal Components and Operation of the Epo/pO_2 Negative Feedback Loop

Epo Production and Erythropoietic Rate

The production of red cells is absolutely dependent on both Epo and EpoR [16–18]. In the adult, Epo secretion by the kidney [10, 19] is induced by hypoxia via the hypoxia-inducible factors, HIF1 and HIF2, which are direct Epo transcriptional activators [10, 19]. Epo activates the EpoR on the surface of erythroid progenitors and precursors in the adult bone marrow and spleen or in the fetal liver, resulting in expansion of the erythroid progenitor pool and a consequent increase in erythropoietic rate [20–22]. Erythropoietic rate may be assessed by the number of blood reticulocytes, which are immature red cells that were released into the circulation over the prior 24-h period.

The total amount of oxygen delivered to tissues is dependent on the total number of red cells in the circulation (the red cell mass). The hematocrit, which is the fraction of blood volume made up of red cells, is a useful proxy for monitoring changes in the red cell mass during anemia, though less so in polycythemia [23]. In healthy humans, red cells circulate for 120 days, and the rate of red cell production equals the rate of removal of senescent red cells from the circulation. The remarkably large dynamic range of the Epo/pO_2 negative feedback loop becomes apparent when examining patients with anemia. Epo levels are found to increase exponentially with decreasing hematocrit, up to 1000-fold their basal level [24], in turn leading up to a tenfold increase in red cell production rate. Increased red cell production rate is the result of a process known as the "erythropoietic stress response," triggered by anemia and other conditions that lead to lower tissue oxygen tension, such as high altitude or cardiorespiratory disease [25, 26].

Epo Cellular Targets in Hematopoietic Tissue

Epo has evolved to regulate red cell production without affecting the kinetics of other blood lineages [27], limiting its action to committed erythroid progenitors. Of interest, recent hierarchical clustering of single-cell transcriptomes [28] supports the suggestion [29] that megakaryocitic–erythrocytic progenitors arise directly from hematopoietic stem cells, rather than by the longer commitment route of the "classical" hematopoietic hierarchy [30, 31]. This early commitment to the erythroid lineage presumably contributes to a fast erythropoietic response to acute stress. Epo targets were traditionally identified by their colony-forming potential in vitro: The "burst-forming unit-erythroid" (BFU-e), which give rise to large, $\sim$500-cell colonies [32, 33], and a later, colony-forming unit-erythroid (CFU-e), which gives rise to colonies of 8–32 cells [34]. CFU-e cells become entirely Epo dependent for survival during cell cycle S-phase of the last CFU-e generation [35, 36], an event that coincides with several other key commitment decisions including activation of the erythroid master transcriptional regulator GATA-1 and a consequent increase in EpoR expression [36]. This stage also marks maximal Epo responsiveness, as indicated by the peak level of activation of the intracellular signal transducer and activator of transcription 5 (Stat5) [8].

The erythroblast progeny of CFU-e undergo erythroid gene induction and mature into red cells in the space of three to five "differentiation divisions." Morphological criteria that include gradually decreasing cell size, condensing nucleus, and increased expression of hemoglobin, are used to classify erythroblasts into proerythroblasts (ProE), early and late basophilic, polychromatic, and orthochromatic erythroblasts. In recent years, several groups, including our own, have developed flow-cytometric approaches that make use of cell surface markers to distinguish increasingly mature erythroblasts and CFU-e cells [7, 36–41] (Fig. 3.1). We use the cell surface markers CD71 and Ter119 together with the flow-cytometric forward scatter (FCS) parameter to subdivide adult erythroblasts into ProE, EryA, EryB, and EryC subsets that form a developmental sequence (Fig. 3.1) [5–7, 37]. This approach allows us to investigate how the frequency, number, and biochemistry of differentiation stage-specific erythroid progenitors and precursors alter in hematopoietic tissue in vivo in response to Epo, hypoxia, and other forms of erythropoietic stress.

What Should a Model of the Erythropoietic System Explain?

A good model should allow us to explain the erythropoietic system's principal properties, which include:

1. *Stability*. In a healthy individual at sea level, the hematocrit, oxygen tension, and Epo remain within a narrow range for many years. This stability is maintained in spite of inevitable biochemical, cellular, and environmental " noise" that might be expected to generate fluctuations in the rates of red cell production and destruction. The Epo/pO_2 negative feedback loop cannot account for the observed stability, due to its inherent delay: Epo's target cells require three to five cell divisions before generating mature red cells. This delay might be expected to promote oscillations and hinder stability. The system's stability is therefore likely the result of additional mechanisms.
2. *A rapid response to stress*. Whatever the mechanisms that stabilize steady-state erythroid parameters in the face of unwanted fluctuations, they must nevertheless

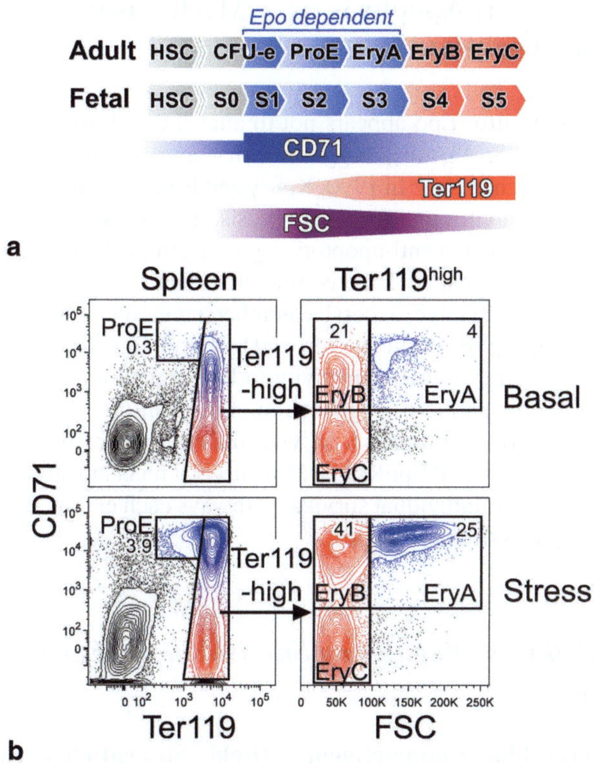

Fig. 3.1 Flow-cytometric identification of tissue erythroblasts. **a** Classification of adult and fetal erythropoietic tissue into increasingly differentiated erythroid precursors using flow-cytometric parameters *CD71*, *Ter119*, and forward scatter (*FSC*). Epo-dependent progenitors are marked by high expression of *CD71*. **b** Flow-cytometric *CD71*/*Ter119*/*FSC* profiles of mouse spleen either in the *basal* state (*top panels*) or 48 h following injection of erythropoietin (Epo; 300U/25 g mouse, "*stress*," *lower panels*). The principal expansion is seen in the Epo-dependent erythroblast subsets labeled "*ProE*" and "*EryA*" (in *blue*). Epo-independent late erythroblast subsets are colored in *red*. *HSC* hematopoietic stem cells, *CFU-e* colony-forming unit-erythroid, *ProE* proerythroblasts. (Published in Ref. [5])

allow a rapid increase in erythropoietic rate in response to physiologically relevant perturbations such as bleeding, anemia, or high altitude. Indeed, in the mouse, CFU-e, ProE, and EryA erythroblasts may increase 50–100-fold within 48–72 h in response to a single maximal Epo dose [5, 7, 42–44].

3. *A graded erythropoietic response to a graded increase in Epo levels, over a wide dynamic range*. EpoR signal transduction needs to reflect the wide dynamic range of extracellular Epo concentrations in basal and stress conditions.

Below, we will consider some recent insights into these questions.

Epo-Mediated Anti-Apoptosis as a Mechanism of Erythroid Expansion

Both in vivo and in vitro, Epo appears not to alter the cell cycle status of primary erythroid cells [20, 35, 45, 46], suggesting that erythroid expansion in stress is mediated by other mechanisms. In 1990, Koury and Bondurant determined that CFU-e -like cells cultured in vitro underwent apoptosis, unless rescued by Epo [47]. They suggested that Epo-mediated anti-apoptotic signaling might be the major mechanism regulating erythroid expansion. This requires first, that early CFU-e progenitors be continuously generated in large, excessive numbers, regardless of Epo concentration, and that they undergo apoptosis unless rescued by EpoR signaling. Second, erythroid progenitors should differ in their sensitivity to Epo signaling, so that an increasing Epo dose would rescue an increasing fraction of progenitors.

Our examination of erythroblasts in vivo during their response to stress supports this hypothesis. Several EpoR-activated survival mechanisms participate in the stress response. Further, individual survival pathways each endow the erythropoietic process with unique system properties.

Molecular Pathways Regulating Survival in the Early Erythroblast Compartment

The Early Erythroblast Compartment Is Highly Susceptible to Apoptosis

The last generation of CFU-e and their early erythroblast progeny together comprise the early erythroblast compartment. Cells in this compartment are highly susceptible to apoptosis, requiring Epo for survival in vitro and in vivo. By contrast, late erythroblasts are apoptosis resistant and Epo independent. Thus, in the basal state in healthy mice at sea level, 40–60 % of early erythroblasts in spleen and 10–20 % in bone marrow or fetal liver are undergoing apoptosis, as judged by Annexin V binding [5–7, 9]. By contrast, there is little apoptosis in late erythroblasts. With the acute onset of stress, such as an injection of Epo or low atmospheric oxygen, the number of apoptotic early erythroblasts declines rapidly, correlating with a rapid expansion in the early erythroblast pool and increased erythropoietic rate [5–7]. The highest rates of early erythroblast apoptosis are found in the mouse spleen, reflecting the low erythropoietic rate of this tissue in the basal state, and its massive potential for expansion during stress when apoptosis is inhibited. The susceptibility of the early erythroblast compartment to apoptosis is also apparent in its higher sensitivity to radiation injury, compared with later erythroblasts [48].

A Balance in Favor of Pro-Apoptotic Regulators in Early, but Not Late, Erythroblasts

The difference in apoptosis resistance between late and early erythroblasts is a result of their distinct expression pattern of apoptotic regulators. Early erythroblasts express high levels of pro-apoptotic regulators and only low levels of pro-survival proteins; this pattern inverts with differentiation, culminating in late erythroblasts that express high levels of pro-survival regulators, and only low levels of pro-apoptosis proteins. Thus, early erythroblasts have fourfold higher levels of the pro-apoptotic regulator Bim mRNA and protein, compared with late erythroblasts [6]; they also express higher mRNAs for the pro-apoptotic Bax and Bid [48]. Further, 30–50 % of early erythroblasts in the mouse spleen and fetal liver express the death receptor Fas on their cell surface, declining to $< 5\,\%$ in mature erythroblasts [5, 7, 9]. Conversely, levels of the pro-survival Bcl-x_L protein are low in early erythroblasts, increasing sixfold with differentiation into late erythroblasts.

Differentiation-Dependent Expression of Apoptosis Regulators Is Mediated by GATA-1 and Modified by EpoR Signaling During Stress

The differentiation-dependent shift in expression of apoptotic regulators is the result of GATA-1-mediated transcriptional regulation. GATA-1 directly induces Bcl-x_L [49]. It also suppresses Bim via its transcriptional target lymphoma-related factor (LRF) [50]. EpoR signaling enhances these effects as a function of Epo levels and stress. Epo preferentially targets the early erythroblast compartment, where it induces Bcl-x_L and suppresses both Bim and Fas, events that would not otherwise take place until later in differentiation [5, 7, 51]. The resulting pro-survival effect allows a larger number of early erythroblasts to give rise to viable progeny. Taken together, the dependence of the early erythroblast compartment on Epo for survival makes the early erythroblast pool, and consequently, erythropoietic rate, highly responsive to the prevailing Epo concentration.

A Graded Input/Output Relationship in the Regulation of Erythropoietic Rate by Epo Requires Erythroblast Heterogeneity

The regulation of erythropoietic rate requires a graded input, in the form of Epo concentration, to generate a graded output, measured as an appropriate increase in the number of CFU-e, ProE, and EryA early erythroblasts [5–7, 20]. However, the response to EpoR survival signaling at the single-cell level is binary, either life or death. A graded output at the population level therefore requires heterogeneity in the susceptibility of early erythroblasts to apoptosis and/or their sensitivity to EpoR's anti-apoptotic signaling. Indeed, CFU-e-like cells in vitro were found to be heterogeneous in their survival response to Epo, an effect that was shown to be independent of the number of cell-surface EpoR [52]. The molecular basis of this heterogeneity is currently unclear. A question for future investigation is whether it is the result of stochastic variation in the expression of apoptotic regulators and/or EpoR signal transduction components, or whether it is generated by distinct subpopulations of early erythroblasts, each with a deterministically different response to EpoR survival signaling.

Erythroblast Survival Pathways Each Generate Unique System-Level Properties

A number of EpoR-activated survival pathways have now been identified [51, 53–56], but relevance to erythropoiesis in vivo was documented for only a few [38, 57–60]. The study of these pathways in vivo became possible with the advent of flow-cytometric techniques [7, 37, 48]. Below we discuss three EpoR-activated survival pathways, which appear to be similar in vitro, in that they each rescue cells from apoptosis. However, their unique properties are revealed when they are studied in the context of the whole organism in vivo.

Short-Range Negative Autoregulation Through Fas and FasL Provides Stability and a Fast Stress Response at the Level of the Whole Organism

Early Erythroblasts Co-Express Fas and FasL, Suppressed by High Epo and Stress

Erythropoiesis takes place in the context of the erythroblastic island tissue niche, where a central macrophage is surrounded by one or more concentric layers of closely apposed erythroblasts [61, 62]. Intercellular interactions between cells in the island are likely contributors to erythropoietic regulation. One key interaction between erythroblasts is mediated via the death receptor, Fas, and its ligand, FasL [7, 9, 25, 37]. Using multiparameter flow cytometry in fresh mouse tissue we found that, surprisingly, both Fas and FasL are co-expressed on the surface of early, but not late, erythroblasts. As discussed below, Fas- and FasL-mediated intercellular interaction within the early erythroblast pool mediates negative autoregulation [5, 9]. Further, EpoR-mediated suppression of Fas and FasL contributes to the expansion of the early erythroblast pool during the stress response.

Fas as a Negative Regulator of Erythropoiesis

Several lines of evidence support the hypothesis that Fas and FasL are negative regulators of the early erythroblast pool and consequently, of erythropoietic rate:

(1) The likelihood that a given early erythroblast is undergoing apoptosis is highly correlated with its cell-surface Fas expression [7]. (2) With the onset of acute stress, Fas expression declines at the same time that apoptosis decreases and the number of early erythroblasts (ProE and EryA) increases [5, 7]. Conversely, early erythroblast Fas expression and apoptosis both increase following the transition from hypoxia back to atmospheric oxygen [5]. (3) During stress, the decline in the fraction of early erythroblasts expressing Fas is a function of the degree of stress and Epo levels (Fig. 3.2a) [5, 7]. Further, there is an inverse relationship between the total number of early erythroblasts and the early erythroblast Fas expression, both during the response to stress and during development (Fig. 3.2a, left panel) [5, 7, 9]. (4) An acute decrease in erythroblast FasL by transient administration of the decoy receptor Fas:Fc resulted in an acute increase in erythropoietic rate, reducing ProE and EryA apoptosis and doubling reticulocyte number by 48–72 h [5]. (5) We generated mice that lack either functional Fas or functional FasL, in which we also prevented the autoimmune syndrome that is usually associated with these mutations. We found that the absence of Fas or FasL resulted in a two- to fourfold increase in the number of early erythroblasts (ProE and EryA), confirming that Fas and FasL exert a negative regulatory effect on these cells. Further, the mice also had either an increased hematocrit, or blood Epo concentration that was lower than in normal mice [5]. Therefore, negative regulation of the early erythroblast pool by Fas and FasL lowers erythropoietic rate at the level of the whole animal.

Fas-Mediated Apoptosis Is a Component of a Short-Range Negative Autoregulatory Loop Within the Erythroblastic Island

Although Fas expression is regulated by Epo, whether or not it leads to apoptosis is not cell-autonomous, but instead depends on the probability that Fas-expressing and FasL-expressing early erythroblasts interact within the erythroblastic island (Fig. 3.2b). It is expected that the probability of such an encounter would be proportional to the square of the frequency of early erythroblasts within the island [5, 9], a relationship that we confirmed experimentally in both adult and fetal tissue [5, 7, 9] (Fig. 3.2a), strongly supporting the concept of Fas- and FasL-mediated negative autoregulation of the early erythroblast pool. Thus, Fas-mediated apoptosis is responsive to both the prevailing Epo levels and the actual frequency of early erythroblasts. Unexpected perturbations in early erythroblast frequency would immediately trigger correction, through altered rates of Fas-mediated apoptosis. Early erythroblast frequencies that are inappropriately high would result in a higher chance of intercellular interaction and apoptosis, quickly returning early erythroblast frequency to its appropriate set point. Conversely, a lower than required early erythroblast frequency would be corrected through the lower probability of an intercellular encounter and therefore a slower apoptotic loss. Its responsiveness to local changes means that Fas-mediated autoregulation has the potential to prevent propagation of any perturbation or "noise" in the early erythroblast pool into the mature red blood cell compartment, obviating the need for activating a correction via the long-range Epo/pO_2 negative feedback loop, and preventing unnecessary fluctuations in hematocrit and pO_2 (Fig. 3.2b).

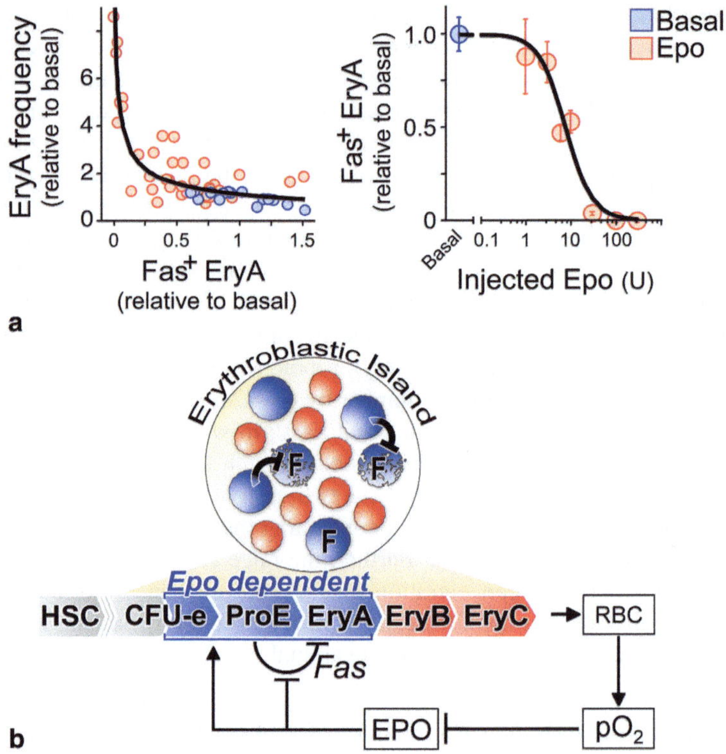

Fig. 3.2 Short-range negative autoregulation of early erythroblasts by Fas and FasL. **a** Interdependence of erythropoietin (*Epo*) concentration, early erythroblast (*EryA*) frequency and EryA expression of Fas in vivo. Mice were injected with varying doses of Epo, and spleen EryA frequency and Fas expression were examined on day 3 post injection. *Left panel* shows EryA cell frequency relative to basal frequency, plotted against the number of EryA cells that express Fas (*Fas*$^+$ *EryA*, expressed as a ratio to basal levels). Data points represent individual mice. *Blue* = mice in the basal state, *red* = mice injected with Epo. Data are fitted with a curve derived from the mathematical model described in [5]. *Right panel* shows the dependence of *Fas*$^+$ *EryA* on the dose of injected Epo, in the same dataset as in the *left panel*; all mice injected with a given Epo dose were pooled into one data point, mean ± sem. **b** Adjusted feedback diagram of the erythropoietic system. In addition to the long-range Epo/pO_2 negative feedback loop, there is a short-range Fas/FasL-mediated negative autoregulatory loop local to the early erythroblast compartment. Epo signaling inhibits local autoregulation by suppressing Fas expression. The illustrated erythroblastic island shows that the frequency of Fas/FasL-expressing early erythroblasts (in *blue*) within the island determines the probability of their interaction and potential apoptosis. Later erythroblasts (*EryB*, *EryC*) are illustrated in *red*. *F* = Fas. *HSC* hematopoietic stem cells, *CFU-e* colony-forming unit-erythroid, *ProE* proerythroblasts, *RBC* red blood cell. (Adapted from Ref. [5])

An independent line of evidence for negative autoregulation via Fas and FasL comes from a computational modeling approach in which we used erythroblast frequency data during early fetal liver development to predict potential regulatory intercellular interactions within the erythroblastic island [9]. We divided the erythroid differentiation sequence within the island into four increasingly mature subsets. We made dynamical measurements of the frequency of each of these subsets in vivo on successive embryonic days during the onset of definitive erythropoiesis in the mouse fetal liver, between embryonic days 12 and 14. This period is characterized by large changes in the erythroblastic island, which is initially dominated by early erythroblasts, rapidly giving way to higher frequencies of late erythroblasts [9]. To build the model, we generated a system of ordinary differential equations, in which the differentiation of early to late erythroblasts was modulated by intercellular interactions. We considered all potential feedforward and feedback interactions between the four erythroblast subsets, and altogether screened 298 networks, or model topologies, in each of which there were at least three such interactions. Once each of the models was fitted to the experimental data, potential intercellular interactions were ranked based on a number of metrics, including their ability to bestow robustness to the erythropoietic networks in which they were present. The highest ranking interaction in this analysis was a negative autoregulatory interaction within an early erythroblast compartment, equivalent to the experimentally identified Fas/FasL-mediated autoregulation.

Local Negative Autoregulation by Fas Stabilizes the Erythropoietic System

Fas-mediated negative autoregulation provides short-range correction of noise or perturbations of the early erythroblast pool (see above). This was made apparent when we examined mutant mice that lacked either Fas or FasL function. We found that, in addition to being significantly larger, the size of their early erythroblast pool was more variable between individual mice in the same population, compared with matched control mice, evident from the increased variance and increased coefficients of variation in early erythroblast subsets in both fetal and adult tissue [5, 9]. There was also increased variability in overall erythropoietic rate, as reflected by increased variability in the number of blood reticulocyte. These experimental findings suggest that Fas and FasL-mediated negative autoregulation contributes to the stability of the erythropoietic system.

Fas-Mediated Apoptosis Controls a Reserve Pool of Early Erythroblasts That Accelerates the Stress Response

Perhaps counter intuitively, mice lacking Fas or FasL showed a significantly delayed response to stress at multiple levels of the erythropoietic system, in spite of the absence of a negative regulator and of their already larger basal erythroblast pool [5]. Whereas control mice responded to a single high dose of Epo with a 30-fold expansion in the early erythroblast pool by day 2, the increase in Fas-deficient mice was smaller by 30 %, a shortfall equivalent to tenfold the size of the normal basal erythroblast pool. The Fas-deficient mice caught up with control mice by day 3, but the early shortfall caused a delay in the expansion of mature erythroblasts and a slower increase in hematocrit. Mice with impaired Fas function also took longer to elevate erythropoietic rate when placed in low atmospheric oxygen [5].

These findings suggest that Fas-mediated apoptosis nonredundantly controls approximately 30 % of the erythropoietic reserve available for the acute stress response (Fig. 3.3a). This reserve consists of cells that are continuously generated as part of the very large "input" into the early erythroblast compartment from earlier progenitor stages. In the basal state, these cells undergo apoptosis through a number of distinct pathways and do not contribute to the basal erythropoietic rate. With the onset of stress and the consequent increase in Epo, Fas expression is suppressed, and cells in the Fas-regulated reserve survive and contribute to the rapid stress response. In Fas-deficient mice, the absence of the Fas-regulated reserve slows down the response to stress.

The control of erythropoietic reserve through apoptosis appears at first glance to be a highly wasteful mechanism, since during health most of the CFU-e/early erythroblasts are continuously generated only to be lost. The finding that mice lacking Fas-mediated apoptosis are slower to respond to stress suggests that this "wasteful" mechanism may have evolved due to its fast stress response: It allows the fast recruitment of a large number of preexisting progenitors in response to the sudden onset of stress, such as might occur following injury and bleeding. An equivalent 30-fold expansion in erythroblasts through cell cycling would presumably add a considerable time lag to the Epo response.

The Role of the Negative Autoregulatory Motif

Negative autoregulation is a frequent motif in biological networks. Computational and experimental approaches in simple transcriptional networks in *E. coli* suggested that it has two principal effects: conferring resistance to random fluctuations, and accelerating the response to a stimulus [63–68]. The roles we outlined for the Fas/FasL autoregulatory interaction in early erythroblasts suggest that the negative autoregulatory motif may exert similar "logic" in higher-level intercellular networks, helping to maintain both stability and a fast stress response of tissue progenitors.

Contrasting Dynamic Responses of Bcl-X$_L$ and Bim Highlight Distinct Anti-Apoptotic Mechanisms for Acute and Chronic Phases of Stress

Bcl-X$_L$ Induction Is an Adapting Response to a Change in the Level of Epo/Stress

Bim and Bcl-x$_L$ are pro- and anti-apoptotic regulators of the Bcl-2 family [69–72]. EpoR activation in early erythroblasts leads to Bim suppression [6, 51] and to induction of Bcl-x$_L$ [6, 38, 57, 73]. We found that the time course of these two anti-apoptotic pathways differed markedly. In response to acute stress such as low atmospheric oxygen or Epo injection, early erythroblast Bcl-x$_L$ increased rapidly

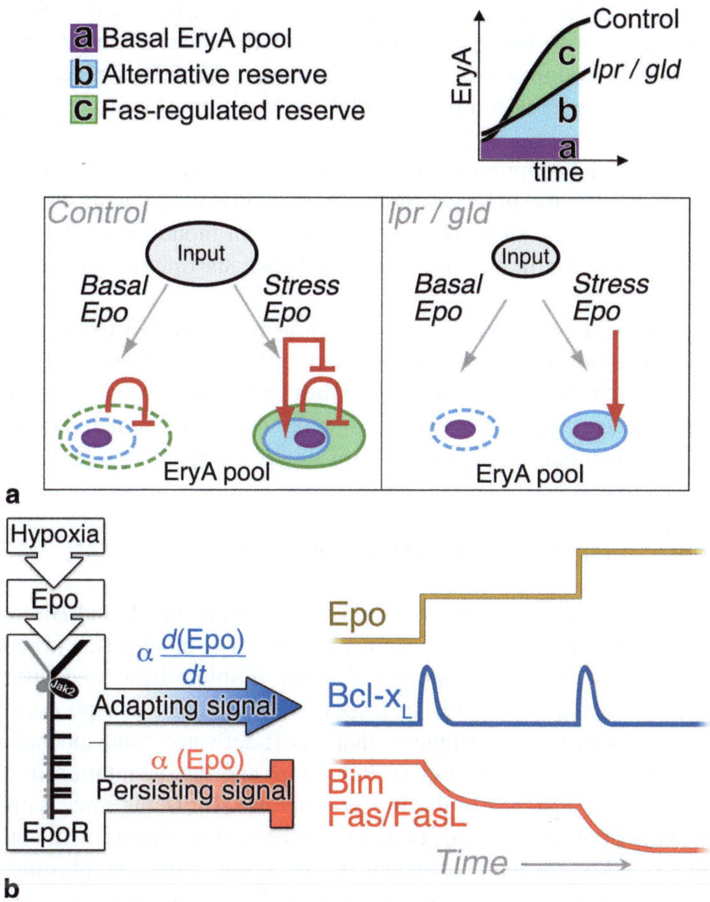

Fig. 3.3 Mechanisms that accelerate the acute stress response. **a** Fas/FasL negative autoregulation accelerates the response to stress. *Top panel* shows the slower response of *lpr/gld* mice, mutant for *Fas/FasL*, respectively. *Lower panels* suggest the mechanism of this delay. In control mice (*left panel*), *EryA* cells are continuously formed from earlier precursors ("*input*"). In the basal state, when erythropoietin (*Epo*) concentration is low, only a small fraction of these cells survives (*arrow* pointing left), forming the "*basal EryA pool*" (in *purple*). The EryA that undergo apoptosis (*dashed lines*) do so either through Fas ("Fas-regulated reserve," *green dashed line*) or alternative mechanisms ("alternative reserve," *blue dashed line*). Together, the EryA reserve is 30 to 60-fold the size of the basal pool [6]. During the acute response to stress, high Epo rescues the EryA reserve pools from apoptosis (*arrow* pointing to right), resulting in an immediate increase in the size of the surviving *EryA pool* (*solid green* and *blue* colors) and an increase in erythropoietic rate. In the basal state, the *lpr* and *gld* mice partially compensate for the chronic absence of Fas-mediated negative regulation by generating fewer EryA cells (a smaller input, right panel). In this way, the absence of Fas-mediated apoptosis does not excessively increase the basal EryA pool. However, the absence of the Fas-regulated reserve in *lpr* and *gld* mice reduces the number of EryA that may be immediately recruited into the surviving EryA pool during stress, delaying the stress response. (Published in Ref. [6]). **b** Contrasting dynamic EpoR signaling responses. Hypoxic stress and high Epo levels result in two distinct EpoR signals: A persistent inhibitory signal that is proportional to the prevailing Epo concentration, suppressing Bim and Fas expression; and an adapting signal, inducing $Bcl\text{-}x_L$, that

and peaked by 12–18 h. However, the peak was short-lived, with Bcl-x_L rapidly declining to basal levels by 48 h, in spite of the persistence of stress. By contrast, Bim suppression was slower, reaching its maximum by 48–72 h. Unlike Bcl-x_L, both Bim and Fas/FasL suppression persisted for the duration of stress/high Epo levels [5–7].

Consistent with its adapting response to acute stress, Bcl-x_L was not elevated in chronic stress conditions such as β-thalassemia. However, an acute stress stimulus superimposed on chronic stress reactivated a transient induction of Bcl-x_L that was very similar to that seen in an acute stress challenge starting from basal conditions [6].

These findings suggest that Bcl-x_L induction is responsive to changes in Epo concentration with respect to time, rather than to the absolute level of Epo or stress (Fig. 3.3b). In this way, it provides a rapid, though transient survival signal to early erythroblasts, accelerating the response to acute stress and remaining active until slower but persistent pathways, such as Bim suppression, are activated.

Mechanism of Adaptation in the Bcl-X_L Response

The contrasting dynamic responses of Bcl-x_L on the one hand, and Bim or Fas on the other, suggest that the EpoR generates at least two signal forms in response to high Epo: a persistent signal proportional to the absolute Epo concentration; and an adapting signal that is elicited in response to a change in Epo concentration (Fig. 3.3b). Our recent work suggests that EpoR-activated Stat5 phosphorylation (p-Stat5) is the key stimulus to Bcl-x_L induction, and that adaptation in p-Stat5 is responsible for the transience of the Bcl-x_L response [6]. Using mice with knockin mutations of the EpoR cytoplasmic domain, we found that adaptation in both p-Stat5 and the Bcl-x_L response was dependent on the EpoR distal cytoplasmic domain. We suggest it is the result of p-Stat5-mediated induction of suppressor of cytokine signaling (SOCS) family proteins, negative regulators that bind EpoR cytoplasmic domain phosphotyrosines and inhibit further p-Stat5 docking and activation [74–76].

The Early Erythroblast Compartment as a "PID" Controller

Why does high Epo elicit both persistent and adaptive responses? The speed of response to acute stress is paramount. However, it may be that fast-response pathways like Bcl-x_L are intrinsically riskier for the cell, and are therefore replaced in the chronic phase with slower but safer pathways. Indeed, persistently

is proportional to the rate of change in Epo concentration, and therefore, specifically accelerates erythropoietic rate in acute stress. The responses of each of these pathways to consecutive stepwise increases in Epo concentration is shown on the right. (Published in Ref. [6])

high Bcl-x_L is associated with polycythemia vera and other myeloproliferative disorders [77–80]. High levels of its close relative, Bcl-2, are associated with B cell lymphomas.

More generally, the levels of Bcl-x_L reflect the rate of change of Epo with respect to time. By contrast, the Bim and Fas responses are proportional to the actual Epo concentration over a wide range [5, 6]. The early erythroblast compartment is therefore processing both proportional and derivative functions of Epo.

These findings are reminiscent of a feedback controller frequently used in engineering applications and known as the PID controller [81]. A controller is a computational device that uses feedback to generate the output of a dynamical system, so that it matches a desired reference value (Fig. 3.4a). The controller is continuously fed an input, known as the "error," generated by the comparison of the actual system's output with the system's required set point. The error is processed by three types of elements within the controller, each with its own specific gain, tuned to match the system's requirements. These are first, "P," which generates an output that is proportional to the error signal. Second, "I," generating an output that is the integral of the system's errors over a recent, specified period of time. This element helps to predict the output based on the system's recent error history. Last, "D" generates an output that is a function of the error's derivative with respect to time. This element predicts the likely future error based on the current error's rate of change. All three elements are combined to generate the controller output that drives the system (Fig. 3.4a). PID controllers are often combined with additional elements such as noise filters to improve the system's output.

We propose that the early erythroblast compartment behaves as a controller of erythropoietic rate and contains elements similar to those of a PID controller (Fig. 3.4b). The error signal that feeds into this controller is Epo concentration, signaling the mismatch between the actual and desired oxygen availability, sensed by HIF1 and HIF2 (see above). Epo then regulates the size of the early erythroblast compartment via pathways that use the same "logic" as those of the PID controller elements. Thus, regulation of Fas, FasL, and Bim expression are similar to the "P" element since they are proportional to Epo concentration, while regulation of Bcl-x_L expression corresponds to the "D" element since it is a function of the derivative of Epo concentration. At the current time we have not identified an "I" element within the early erythroblast compartment in vivo. However, we propose that it may be represented by Epo-dependent CFU-e self-renewal, a process that Epo achieves in cooperation with glucocorticoids during erythropoietic stress [22, 82, 83]. The early erythroblast compartment controller also includes a noise-reducing filter operated by the Fas/FasL negative autoregulatory interaction. The final output from the controller, namely the number of surviving erythroblasts, is the sum total of all survival and self-renewal pathways, and directly drives erythropoietic rate, which in turn increases red cell mass and delivery of oxygen to tissues.

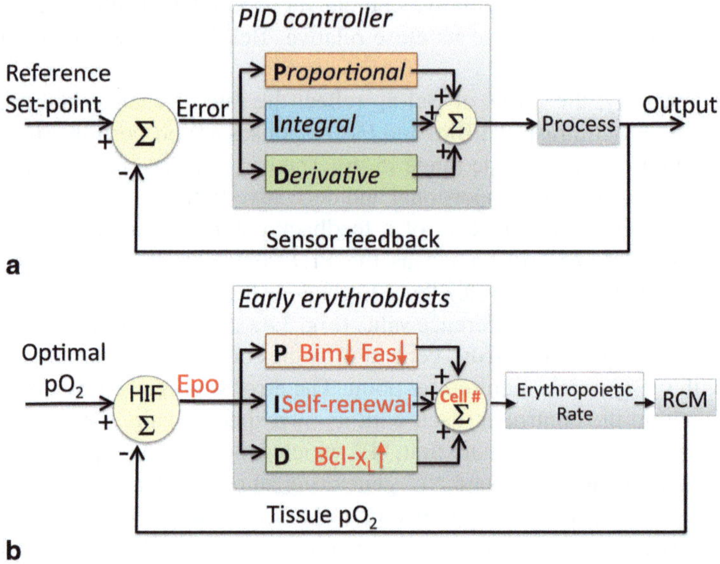

Fig. 3.4 The early erythroblast compartment is analogous to a proportional-integral-derivative (*PID*) controller. **a** Block diagram of a *PID controller*, a device that uses feedback to continuously generate the output of a dynamical system so as to match a desired reference value [81]. An "*error*" is continuously generated by the comparison of the system's required set point with the actual value of the parameter controlled by the system. Three elements within the controller generate output that is either proportional ("*P*"), integral ("*I*"), or a derivative ("*D*") function of the error signal. The output of the three elements is then combined $(\sum)$ to generate the system's output. **b** The *early erythroblast* compartment as a PID controller of erythropoietic rate. Erythropoietin (*Epo*) concentration represents the error signal, generated by the mismatch between the actual and desired oxygen availability, sensed by HIF1 and HIF2. Epo regulates the size of the early erythroblast compartment (the controller "*output*") via pathways that use the same "logic" as those of the PID controller elements. Fas and Bim suppression is proportional ("*P*") to Epo concentration while regulation of Bcl-x_L expression corresponds to the "*D*" element since it is a function of the derivative of Epo concentration. The "*I*" element may represent Epo-dependent CFU-e self-renewal. *CFU-e* colony-forming unit-erythroid, *RCM* red cell mass

Stat5 Signaling Combines Binary and Graded Dynamic Modalities

The Dimmer Switch Model of Stat5 Signaling

Mouse genetic models suggest that EpoR-mediated Stat5 activation by phosphorylation is essential for both basal erythropoiesis and for its acceleration during stress [38, 57, 84, 85]. A key challenge lies in understanding how Stat5 signaling encodes both basal and stress Epo signals, in turn eliciting distinct downstream responses. We investigated the dynamic features of the Stat5 phosphorylation signal (p-Stat5) in erythroblasts within freshly isolated tissue using flow cytometry [8]. We found

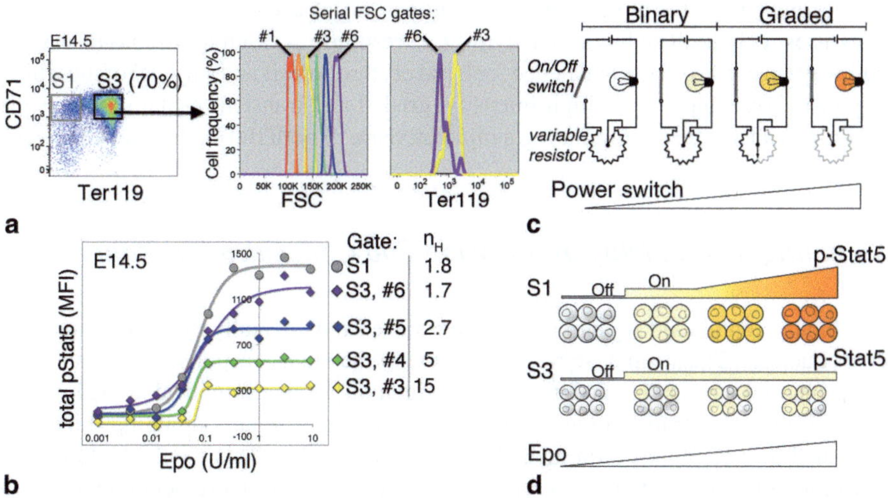

Fig. 3.5 Binary and graded modalities of signal transducer and activator of transcription 5 (Stat5) signaling in erythroblasts. **a, b** The steepness of the erythropoietin (Epo)/p-Stat5 dose/response curves increases, and the maximal p-Stat5 intensity decreases, with increasing erythroblast maturation. *Panel "A"* shows the gating strategy of freshly isolated embryonic day 14.5 fetal liver cells, based on *Ter119*, *CD71*, and forward scatter (*FSC*). The S3 subset is subdivided into increasingly mature (expressing increasing Ter119) smaller subsets, labeled FSC gates #1–#6. The dose/response curves for each subset (*panel B*) shows decreasing maximal intensity, and increasing steepness and Hill coefficient, with increasing erythroblast maturation [8]. **c, d** The dimmer-switch model of Stat5 signaling. Stat5 signaling in erythroblasts resembles a dimmer switch, which combines binary and graded components (*panel C*). The closing of an "*on/off*" switch completes an electric circuit, turning on a dim light. The further turning of the power-switch dial permits a gradual decrease of the circuit's resistance, with a consequent graded increase in the electric current and light intensity. *Panel* D shows a binary, decisive but low-intensity p-Stat5 signal in more mature S3 erythroblasts. In S1 early erythroblasts, this signal can increase further with a gradual further increase in Epo concentration. No further increase takes place in mature S3 erythroblasts, though the number of signaling erythroblasts increases with increasing Epo concentration [8]

two modalities of p-Stat5 signaling, dependent both on the level of Epo/stress and on erythroblast maturation. The Epo/pStat5 dose/response curve in early erythroblasts is graded and reaches the highest signal intensities in response to high Epo. By contrast, in later erythroblasts the Epo/pStat5 dose/response curve is much steeper, with a high Hill coefficient, but a low maximal intensity. This results in a low intensity but decisive, or binary, pStat5 response (Fig. 3.5a, 3.5b).

The transition from a high-intensity graded p-Stat5 signal in the earliest erythroblasts to a low-intensity binary response in later erythroblasts is the result of a decline in Stat5 protein [8]. We propose that the p-Stat5 signal is analogous to a light bulb operated through a dimmer switch (Fig. 3.5c, 3.5d). A binary switch controls the circuit and is responsible for activating a binary, though dim, light. A gradual further turning of the power switch dial permits a gradual further increase in the electric current and light intensity.

The intensity of the p-Stat5 signal determines the intracellular gene transcriptional responses it elicits. The low-intensity p-Stat5 response is essential for basal survival pathways in erythroblasts and hence for basal erythropoiesis and survival of the whole organism. By contrast, the high-intensity, graded and transient p-Stat5 response is specific to acute stress, stimulating a rapid increase in both Bcl-x_L [6] and CD71 [8].

Signaling with Fidelity over a Wide Epo Concentration Range

The binary and graded signaling modalities have fundamentally different functional consequences. The steep dose/response curve of the binary response filters out noise and generates a clear signal. This mode of signaling is therefore ideal at the low end of the Epo concentration range. A key disadvantage of binary signaling, however, is its inability to encode incremental changes in stimulus. This would exclude it as a useful signaling modality in erythropoietic stress, where incremental changes in Epo concentration signal the required incremental increases in erythropoietic rate. On the other hand, relying on a graded signaling response over the three log orders of magnitude of the Epo concentration range would make the response to basal Epo levels difficult to distinguish from noise. Stat5 bridges this conundrum by combining the binary and graded signaling modalities, so that binary signaling dominates and provides signaling fidelity in the low-intensity signaling range of basal erythropoiesis, and graded signaling is reserved for the high-intensity stress range.

Conclusions and Outlook

Computational models of the erythropoietic system have treated the erythroid progenitor compartment as a "black box," disregarding the biochemical pathways that control progenitor functions. Our experimental approach simultaneously measures dynamical change at multiple levels of the erythropoietic system in vivo, including the molecular responses of defined sequential differentiation stages of erythroid progenitors and precursors. This approach shows that specific biochemical pathways at the single-cell level are directly responsible for the large-scale system properties of erythropoiesis. It suggests that the early erythroblast compartment is a complex and sophisticated feedback controller where molecular and cellular pathways compute the optimal response to the pO_2 "error signal," namely the prevalent Epo concentration. To date, we identified molecular mechanisms that enhance stability in the face of unwanted perturbations using short-range feedback and an accelerated stress response. Further, the logic "tools" used to achieve these large-scale functions are similar to those used in engineering applications, and include binary and graded signal transduction modalities, negative autoregulation, and elements of the PID feedback controller. A better and detailed dynamical understanding of molecular erythropoiesis, combined with knowledge of system engineering, should allow biologists to refine a molecular model of erythropoiesis that would allow a mechanistic assessment of pathology and improved design of therapies.

Acknowledgments This work was supported by R01DK099281, R21HL113978, the American Society of Hematology Bridge Grant award, and the Leukemia and Lymphoma Scholar award. The authors thank Ziv Scully (MIT) for discussion.

References

1. Loeffler M, Pantel K, Wulff H, Wichmann HE. A mathematical model of erythropoiesis in mice and rats Part 1: structure of the model. Cell Prolif. 1989;22(1):13.
2. Crauste F, Pujo-Menjouet L, Genieys S, Molina C, Gandrillon O. Adding self-renewal in committed erythroid progenitors improves the biological relevance of a mathematical model of erythropoiesis. J Theor Biol. 2008;250(2):322–38.
3. Belair J, Mackey MC, Mahaffy JM. Age-structured and two-delay models for erythropoiesis. Math Biosci. 1995;128(1–2):317–46.
4. Crauste F, Demin I, Gandrillon O, Volpert V. Mathematical study of feedback control roles and relevance in stress erythropoiesis. J Theor Biol. 2010;263(3):303–16.
5. Koulnis M, Liu Y, Hallstrom K, Socolovsky M. Negative autoregulation by Fas stabilizes adult erythropoiesis and accelerates its stress response. PLoS ONE. 2011;6(7):e21192.
6. Koulnis M, Porpiglia E, Porpiglia PA, et al. Contrasting dynamic responses in vivo of the Bcl-xL and Bim erythropoietic survival pathways. Blood. 2012;119(5):1228–39.
7. Liu Y, Pop R, Sadegh C, Brugnara C, Haase VH, Socolovsky M. Suppression of Fas-FasL coexpression by erythropoietin mediates erythroblast expansion during the erythropoietic stress response in vivo. Blood. 2006;108(1):123–33.
8. Porpiglia E, Hidalgo D, Koulnis M, Tzafriri AR, Socolovsky M. Stat5 signaling specifies basal versus stress erythropoietic responses through distinct binary and graded dynamic modalities. PLoS Biol. 2012;10(8):e1001383.
9. Socolovsky M, Murrell M, Liu Y, Pop R, Porpiglia E, Levchenko A. Negative autoregulation by FAS mediates robust fetal erythropoiesis. PLoS Biol. 2007;5(10):e252.
10. Haase VH. Hypoxic regulation of erythropoiesis and iron metabolism. Am J Physiol Renal Physiol. 2010;299(1):F1–13.
11. Erslev A. Humoral regulation of red cell production. Blood. 1953;8(4):349–57.
12. Goldwasser E, Kung CK. Purification of erythropoietin. Proc Natl Acad Sci U S A. 1971;68(4):697–8.
13. Lin FK, Suggs S, Lin CH, et al. Cloning and expression of the human erythropoietin gene. Proc Natl Acad Sci U S A. 1985;82(22):7580–4.
14. Jacobs K, Shoemaker C, Rudersdorf R, et al. Isolation and characterization of genomic and cDNA clones of human erythropoietin. Nature. 1985;313(6005):806–10.
15. D'Andrea AD, Lodish HF, Wong GG. Expression cloning of the murine erythropoietin receptor. Cell. 1989;57(2):277–85.
16. Wu H, Liu X, Jaenisch R, Lodish HF. Generation of committed erythroid BFU-E and CFU-E progenitors does not require erythropoietin or the erythropoietin receptor. Cell. 1995;83:59–67.
17. Kieran MW, Perkins A, Orkin S, Zon L. Thrombopoietin rescues in vitro erythroid colony formation from mouse embryos lacking the erythropoietin receptor. Proc Natl Acad Sci U S A. 1996;93:9126–31.
18. Lin CS, Lim SK, D'Agati V, Costantini F. Differential effects of an erythropoietin receptor gene disruption on primitive and definitive erythropoiesis. Genes Dev. 1996;10:154–64.

19. Yoon D, Ponka P, Prchal JT. Hypoxia. 5. Hypoxia and hematopoiesis. Am J Physiol Cell Physiol. 2011;300(6):C1215–22.
20. Papayannopoulou T, Finch CA. On the in vivo action of erythropoietin: a quantitative analysis. J Clin Invest. 1972;51(5):1179–85.
21. Lodish HF, Ghaffari S, Socolovsky M, Tong W, Zhang J. Intracellular signaling by the erythropoietin receptor. In: Elliott SG, Foote M, Molineux G, editors. Erythropoietins, erythropoietic factors, and erythropoiesis: molecular, cellular, preclinical, and clinical biology. 2nd ed. Basel:Birkhäuser; 2009. pp. 155–74.
22. Hattangadi SM, Wong P, Zhang L, Flygare J, Lodish HF. From stem cell to red cell: regulation of erythropoiesis at multiple levels by multiple proteins, RNAs, and chromatin modifications. Blood. 2011;118(24):6258–68.
23. Huber H, Lewis SM, Szur L. The influence of anaemia, polycythaemia and splenomegaly on the relationship between venous haematocrit and red-cell volume. Br J Haematol. 1964;10:567–75.
24. Erslev AJ. Clinical erythrokinetics: a critical review. Blood Rev. 1997;11(3):160–7.
25. Socolovsky M. Molecular insights into stress erythropoiesis. Curr Opin Hematol. 2007;14(3):215–24.
26. Paulson RF, Shi L, Wu DC. Stress erythropoiesis: new signals and new stress progenitor cells. Curr Opin Hematol. 2011;18(3):139–45.
27. Jelkmann W. Erythropoietin: structure, control of production, and function. Physiol Rev. 1992;72(2):449–89.
28. Guo G, Luc S, Marco E, et al. Mapping cellular hierarchy by single-cell analysis of the cell surface repertoire. Cell Stem Cell. 2013;13(4):492–505.
29. Adolfsson J, Mansson R, Buza-Vidas N, et al. Identification of Flt3+ lympho-myeloid stem cells lacking erythro-megakaryocytic potential a revised road map for adult blood lineage commitment. Cell. 2005;121(2):295–306.
30. Akashi K, Traver D, Miyamoto T, Weissman IL. A clonogenic common myeloid progenitor that gives rise to all myeloid lineages. Nature. 2000;404(6774):193–7.
31. Akashi K, Kondo M, von Freeden-Jeffry U, Murray R, Weissman IL. Bcl-2 rescues T lymphopoiesis in interleukin-7 receptor-deficient mice. Cell. 1997;89:1033–41.
32. Iscove NN, Sieber F. Erythroid progenitors in mouse bone-marrow detected by macroscopic colony formation in culture. Exp Hematol. 1974;3:32.
33. Axelrad AA, McLeod DL, Shreeve MM, Heath DS. Properties of cells that produce erythrocytic colonies in vitro. In: Robinson WA, editor. Hemopoiesis in culture. Washington: U.S. Government Printing Office; 1974.
34. Stephenson JR, Axelrad AA, McLeod DL, Shreeve MM. Induction of colonies of hemoglobin-synthesizing cells by erythropoietin in vitro. Proc Natl Acad Sci U S A. 1971;68(7):1542–6.
35. Landschulz KT, Boyer SH, Noyes AN, Rogers OC, Frelin LP. Onset of erythropoietin response in murine erythroid colony-forming units: assignment to early S-phase in a specific cell generation. Blood. 1992;79(10):2749–58.
36. Pop R, Shearstone JR, Shen Q, et al. A key commitment step in erythropoiesis is synchronized with the cell cycle clock through mutual inhibition between PU.1 and S-phase progression. PLoS Biol. 2010;8(9):e1000484.
37. Koulnis M, Pop R, Porpiglia E, Shearstone JR, Hidalgo D, Socolovsky M. Identification and analysis of mouse erythroid progenitors using the CD71/TER119 flow-cytometric assay. J Vis Exp. 2011(54):2809.
38. Socolovsky M, Nam H, Fleming MD, Haase VH, Brugnara C, Lodish HF. Ineffective erythropoiesis in Stat5a(−/−)5b(−/−) mice due to decreased survival of early erythroblasts. Blood. 2001;98(12):3261–73.
39. Pronk CJ, Rossi DJ, Mansson R, et al. Elucidation of the phenotypic, functional, and molecular topography of a myeloerythroid progenitor cell hierarchy. Cell Stem Cell. 2007;1(4):428–42.
40. McGrath KE, Bushnell TP, Palis J. Multispectral imaging of hematopoietic cells: where flow meets morphology. J Immunol Methods. 2008;336(2):91–7.

41. Chen K, Liu J, Heck S, Chasis JA, An X, Mohandas N. Resolving the distinct stages in erythroid differentiation based on dynamic changes in membrane protein expression during erythropoiesis. Proc Natl Acad Sci U S A. 2009;106(41):17413–18.
42. Wickrema A, Bondurant MC, Krantz SB. Abundance and stability of erythropoietin receptor mRNA in mouse erythroid progenitor cells. Blood. 1991;78(9):2269–75.
43. Wickrema A, Krantz SB, Winkelmann JC, Bondurant MC. Differentiation and erythropoietin receptor gene expression in human erythroid progenitor cells. Blood. 1992;80(8):1940–9.
44. Broudy VC, Lin N, Brice M, Nakamoto B, Papayannopoulou T. Erythropoietin receptor characteristics on primary human erythroid cells. Blood. 1991;77(12):2583–90.
45. Nijhof W, Wierenga PK, Pietens J, Bloem R. Cell kinetic behaviour of a synchronized population of erythroid precursor cells in vitro. Cell Tissue Kinet. 1984;17(6):629–39.
46. Nijhof W, de Haan G, Pietens J, Dontje B. Mechanistic options of erythropoietin-stimulated erythropoiesis. Exp Hematol. 1995;23(4):369–75.
47. Koury MJ, Bondurant MC. Erythropoietin retards DNA breakdown and prevents programmed death in erythroid progenitor cells. Science. 1990;248:378–81.
48. Peslak SA, Wenger J, Bemis JC, et al. Sublethal radiation injury uncovers a functional transition during erythroid maturation. Exp Hematol. 2011;39(4):434–45.
49. Gregory T, Yu C, Ma A, Orkin SH, Blobel GA, Weiss MJ. GATA-1 and Erythropoietin cooperate to promote erythroid cell survival by regulating bcl-x_L expression. Blood. 1999;94(1):87–96.
50. Maeda T, Ito K, Merghoub T, et al. LRF is an essential downstream target of GATA1 in erythroid development and regulates BIM-dependent apoptosis. Dev Cell. 2009;17(4):527–40.
51. Abutin RM, Chen J, Lung TK, Lloyd JA, Sawyer ST, Harada H. Erythropoietin-induced phosphorylation/degradation of BIM contributes to survival of erythroid cells. Exp Hematol. 2009;37(2):151–8.
52. Kelley LL, Koury MJ, Bondurant MC, Koury ST, Sawyer ST, Wickrema A. Survival or death of individual proerythroblasts results from differing erythropoietin sensitivities: a mechanism for controlled rates of erythrocyte production. Blood. 1993;82(8):2340–52.
53. Menon MP, Karur V, Bogacheva O, Bogachev O, Cuetara B, Wojchowski DM. Signals for stress erythropoiesis are integrated via an erythropoietin receptor-phosphotyrosine-343-Stat5 axis. J Clin Invest. 2006;116(3):683–94.
54. Wood AD, Chen E, Donaldson IJ, et al. ID1 promotes expansion and survival of primary erythroid cells and is a target of JAK2V617 F-STAT5 signaling. Blood. 2009;114(9):1820–30.
55. Sathyanarayana P, Dev A, Fang J, et al. EPO receptor circuits for primary erythroblast survival. Blood. 2008;111(11):5390–9.
56. Bouscary D, Pene F, Claessens YE, et al. Critical role for PI 3-kinase in the control of erythropoietin-induced erythroid progenitor proliferation. Blood. 2003;101(9):3436–43.
57. Socolovsky M, Fallon AEJ, Wang S, Brugnara C, Lodish HF. Fetal anemia and apoptosis of red cell progenitors in Stat5a$^{-/-}$5b$^{-/-}$ mice: a direct role for Stat5 in Bcl-X_L induction. Cell. 1999;98:181–91.
58. Halupa A, Bailey ML, Huang K, Iscove NN, Levy DE, Barber DL. A novel role for STAT1 in regulating murine erythropoiesis: deletion of STAT1 results in overall reduction of erythroid progenitors and alters their distribution. Blood. 2005;105(2):552–61.
59. Wagner KU, Claudio E, Rucker EB 3rd, et al. Conditional deletion of the Bcl-x gene from erythroid cells results in hemolytic anemia and profound splenomegaly. Development. 2000;127(22):4949–58.
60. Guihard S, Clay D, Cocault L, et al. The MAPK ERK1 is a negative regulator of the adult steady-state splenic erythropoiesis. Blood. 2010;115(18):3686–94.
61. Mohandas N, Chasis JA. The erythroid niche: molecular processes occurring within erythroblastic islands. Transfus Clin Biol. 2010;17(3):110–1.
62. Socolovsky M. Exploring the erythroblastic island. Nat Med. 2013;19(4):399–401.
63. Thieffry D, Huerta AM, Perez-Rueda E, Collado-Vides J. From specific gene regulation to genomic networks: a global analysis of transcriptional regulation in *Escherichia coli*. Bioessays. 1998;20(5):433–40.

64. Alon U. Network motifs: theory and experimental approaches. Nat Rev Genet. 2007;8(6):450–61.
65. Savageau MA. Comparison of classical and autogenous systems of regulation in inducible operons. Nature. 1974;252(5484):546–9.
66. Rosenfeld N, Elowitz MB, Alon U. Negative autoregulation speeds the response times of transcription networks. J Mol Biol. 2002;323(5):785–93.
67. Becskei A, Serrano L. Engineering stability in gene networks by autoregulation. Nature. 2000;405(6786):590–3.
68. Camas FM, Blazquez J, Poyatos JF. Autogenous and nonautogenous control of response in a genetic network. Proc Natl Acad Sci U S A. 2006;103(34):12718–23.
69. Green DR. At the gates of death. Cancer Cell. 2006;9(5):328–30.
70. Chao DT, Korsmeyer SJ. BCL-2 family: regulators of cell death. Annu Rev Immunol. 1998;16:395–419.
71. O'Connor L, Strasser A, O'Reilly LA, et al. Bim: a novel member of the Bcl-2 family that promotes apoptosis. EMBO J. 1998;17(2):384–95.
72. Inaba T. Cytokine-mediated cell survival. Int J Hematol. 2004;80(3):210–14.
73. Silva M, Grillot D, Benito A, Richard C, Nunez G, Fernandez-Luna JL. Erythropoietin can promote erythroid progenitor survival by repressing apoptosis through Bcl-XL and Bcl-2. Blood. 1996;88:1576–82.
74. Kile BT, Schulman BA, Alexander WS, Nicola NA, Martin HM, Hilton DJ. The SOCS box: a tale of destruction and degradation. Trends Biochem Sci. 2002;27(5):235–41.
75. Larsen L, Ropke C. Suppressors of cytokine signalling: SOCS. APMIS. 2002;110(12):833–44.
76. Sarna MK, Ingley E, Busfield SJ, et al. Differential regulation of SOCS genes in normal and transformed erythroid cells. Oncogene. 2003;22(21):3221–30.
77. Silva M, Richard C, Benito A, Sanz C, Olalla I, Fernandez-Luna JL. Expression of Bcl-x in erythroid precursors from patients with polycythemia vera. N Engl J Med. 1998;338(9):564–71.
78. Garcon L, Rivat C, James C, et al. Constitutive activation of STAT5 and Bcl-xL overexpression can induce endogenous erythroid colony formation in human primary cells. Blood. 2006;108(5):1551–4.
79. Hsieh PP, Olsen RJ, O'Malley DP, et al. The role of Janus Kinase 2 V617 F mutation in extramedullary hematopoiesis of the spleen in neoplastic myeloid disorders. Mod Pathol. 2007;20(9):929–35.
80. Rubert J, Qian Z, Andraos R, Guthy DA, Radimerski T. Bim and Mcl-1 exert key roles in regulating JAK2V617 F cell survival. BMC Cancer. 2011;11:24.
81. Astrom KJ, Murray RM. Feedback systems: an introduction for scientists and engineers. Princeton:Princeton University Press; 2008.
82. Bauer A, Tronche F, Wessely O, et al. The glucocorticoid receptor is required for stress erythropoiesis. Genes Dev. 1999;13(22):2996–3002.
83. von Lindern MZ, Mellitzer G, et al. The glucocorticoid receptor cooperates with the erythropoietin receptor and c-Kit to enhance and sustain proliferation of erythroid progenitors in vitro. Blood. 1999;94(2):550–9.
84. Longmore GD. A unique role for Stat5 in recovery from acute anemia. J Clin Invest. 2006;116(3):626–8.
85. Zhu BM, McLaughlin SK, Na R, et al. Hematopoietic-specific Stat5-null mice display microcytic hypochromic anemia associated with reduced transferrin receptor gene expression. Blood. 2008;112(5):2071–80.

Chapter 4
Systems Biology of Megakaryocytes

Alexis Kaushansky and Kenneth Kaushansky

Abstract The molecular pathways that regulate megakaryocyte production have historically been identified through multiple candidate gene approaches. Several transcription factors critical for generating megakaryocytes were identified by promoter analysis of megakaryocyte-specific genes, and their biological roles then verified by gene knockout studies; for example, GATA-1, NF-E2, and RUNX1 were identified in this way. In contrast, other transcription factors important for megakaryopoiesis were discovered through a systems approach; for example, c-Myb was found to be critical for the erythroid versus megakaryocyte lineage decision by genome-wide loss-of-function studies. The regulation of the levels of these transcription factors is, for the most part, cell intrinsic, although that assumption has recently been challenged. Epigenetics also impacts megakaryocyte gene expression, mediated by histone acetylation and methylation. Several cytokines have been identified to regulate megakaryocyte survival, proliferation, and differentiation, most prominent of which is thrombopoietin. Upon binding to its receptor, the product of the c-Mpl proto-oncogene, thrombopoietin induces a conformational change that activates a number of secondary messengers that promote cell survival, proliferation, and differentiation, and down-modulate receptor signaling. Among the best studied are the signal transducers and activators of transcription (STAT) proteins; phosphoinositol-3-kinase; mitogen-activated protein kinases; the phosphatases PTEN, SHP1, SHP2, and SHIP1; and the suppressors of cytokine signaling (SOCS) proteins. Additional signals activated by these secondary mediators include mammalian target of rapamycin; β(beta)-catenin; the G proteins Rac1, Rho, and CDC42; several transcription factors, including hypoxia-inducible factor 1α(alpha), the homeobox-containing proteins HOXB4 and HOXA9, and a number of signaling mediators that are reduced,

K. Kaushansky (✉)
Office of the Sr. Vice President, Health Sciences, Stony Brook Medicine,
Health Sciences Tower, Level 4, Room 255, Stony Brook, NY 11794-8430, USA
Tel.: 631-444-2121
e-mail: kenneth.kaushansky@stonybrook.edu

A. Kaushansky
Malaria Program, Seattle Biomedical Research Institute,
307 Westlake Ave., North, Suite 500, Seattle, WA 98109, USA
Tel.: 206-256-7126
e-mail: alexis.kaushansky@seattlebiomed.org

S. J. Corey et al. (eds.), *A Systems Biology Approach to Blood*,
Advances in Experimental Medicine and Biology 844, DOI 10.1007/978-1-4939-2095-2_4

including glycogen synthase kinase 3α(alpha) and the FOXO3 family of forkhead proteins. More recently, systematic interrogation of several aspects of megakaryocyte formation have been conducted, employing genomics, proteomics, and chromatin immunoprecipitation (ChIP) analyses, among others, and have yielded many previously unappreciated signaling mechanisms that regulate megakaryocyte lineage determination, proliferation, and differentiation. This chapter focuses on these pathways in normal and neoplastic megakaryopoiesis, and suggests areas that are ripe for further study.

Keywords Megakaryocytes · Platelets · Systems biology · Protein array · Genomics · Signal transduction · Transcription factors · Thrombopoietin · Cell proliferation · Endomitosis

Background

An adequate supply of platelets is essential to repair the minor vascular damage that occurs with daily life, and to initiate thrombus formation in the event of overt vascular injury. Platelets also play a critical role in cardiovascular disease [1], wound repair [2], the innate immune response [3], and the biology of metastatic cancer [4]. The average platelet count in humans ranges from 150 to 350 × 10^9/L, with some diurnal variation, although the level for any individual is maintained within fairly narrow limits when adjusted for time of day. The range of tolerable platelet counts is broad, but once platelet counts fall below 50 × 10^9/L, the risk of pathological bleeding rises substantially.

Once derived from a bi-potent erythroid–megakaryocyte progenitor cell, the megakaryoblast undergoes a series of divisions, during which time the cytoplasm begins to express platelet-specific proteins (e.g., β(beta)1-tubulin), the cell surface membrane becomes decorated by a number of platelet-specific adhesive proteins (e.g., integrin α(alpha)IIb/β3, integrin αII/β1, glycoprotein GpIb, GpVI), cytoplasmic granules and their constituents (e.g., platelet factor 4, transforming growth factor β1, von Willebrand factor, P-selectin) assemble, and internal membranes (specialized for rapid calcium flux or for proplatelet formation) begin to form. After four to six cycles of cell division, mitosis begins to abort in anaphase. As DNA synthesis continues despite aborted mitosis, a process termed endomitosis, the megakaryocyte becomes highly polyploid. During the endomitotic phase of the megakaryocyte life cycle, gene transcription becomes synchronized on all copies of the platelet-specific structural genes, resulting in massive translation of critical platelet proteins, required for the impressive growth in cell volume during this phase of megakaryocyte development. The result is an extremely large mature megakaryocyte that contains 64, 128, or even 256 times the normal chromosome complement. At this point, exvaginations of internal membranes form, driven by a breakdown of the circumferential actin cytoskeleton and projection of long filaments of β1-tubulin; these proplatelet processes then branch and fragment into platelets.

A wide range of pathological conditions lead to both reduced (thrombocytopenia) and increased (thrombocytosis) platelet counts. Hence, the mechanisms that regulate the production of megakaryocytes and platelets are of keen interest for both health and disease.

Components of the Gene Expression Apparatus that Affect Megakaryocyte Lineage Development

Candidate Gene Approaches

Once a hematopoietic stem cell (HSC) commits to the myeloid lineage, the stochastic rise and fall of lineage-specific transcription factors influences its precise hematopoietic lineage fate. Using reporter gene analyses of the 5′ untranslated region of megakaryocyte and mixed erythroid/megakaryocyte-specific genes to probe for its regulatory sites, at least six transcription factors, GATA1, GATA2, RUNX1, Fli-1, FOG1, and NF-E2, have been identified as critical for megakaryocyte gene expression and differentiation into mature platelets. More recently, using a variety of methods, a number of genes that drive megakaryocyte-specific epigenetic changes have also been identified.

GATA-1, termed for its core DNA-binding sequence (AGATAG), is vital for committing primitive multipotent progenitors to the erythroid/megakaryocyte pathway. However, the transcription factor also is critical later in megakaryopoiesis, for cytoplasmic development. For example, reduction in GATA-1 expression impairs granule and demarcation membrane formation in murine megakaryocytes [5]. The promoters for integrin α_{IIb}, glycoprotein (Gp)Ib, GpVI, GpIX, and platelet factor-4 genes display consensus sequences for both GATA-1 and members of the Ets family of transcription factors (e.g., FLI-1), deletion of which reduces or eliminates reporter gene expression [6–8], at least in mature hematopoietic cells; these were among the first studies demonstrating the need for multiple lineage-specific transcription factors to drive lineage-specific gene expression. Additional megakaryocyte targets of GATA-1 include the inositol 1,4,5-triphosphate receptor 2 (*InsP*^{3}-*R2*), which encodes endoplasmic reticulum channels that control calcium efflux [9] enzymes of arachidonic acid metabolism, and the G-protein-coupled P2Y receptors, 1 and 12 [10].

Another transcription factor that plays a critical role in the final stages of megakaryocyte maturation is NF-E2. Initially described as an erythroid-specific, basic leucine zipper family transcription factor [11], NF-E2 is composed of a ubiquitously expressed p18 subunit and a p45 protein also present in megakaryocytes. NF-E2 binds to tandem AP-1-like motifs, such as those seen in the second DNAse hypersensitive site of the β-globin locus control region, and is required for β-globin expression [12]. However, genetic elimination of p45 failed to significantly affect erythropoiesis. Rather, p45-deficient mice display prominent alterations in megakaryocyte development and severe thrombocytopenia [13] leading to death from widespread hemorrhage. Examination of the marrow reveals modest expansion of megakaryocytes but failure of the cells to produce platelets because of defects in cytoplasmic maturation, including substantial reductions in platelet granules and membranes. The genetic targets of NF-E2 include β1-tubulin [14], thromboxane synthesis [15], both expressed relatively late in megakaryocyte maturation, consistent with the phenotype of NF-E2-deficient marrow megakaryocytes, highly polyploid cells that express integrin αIIbβ3 and platelet factor-4, but lack platelet granules and the internal membranes required for proplatelet formation.

FLI-1 is an Ets family transcription factor first identified as the site of insertion of Friend murine leukemia virus. Its oncogenic potential in humans was established with the identification of a fusion of the Ewing's sarcoma gene (EWS) and the Ets domain of Fli1 in pediatric Ewing's sarcoma, and the interaction of Fli-1 with Tel in acute myeloid leukemia (AML). A role in normal hematopoiesis was suggested when overexpression of Fli-1 was shown to lead to massive overproduction of erythrocytes, and established when hemizygous loss of Fli-1 was shown to be causative of the thrombocytopenia of Paris–Trousseau syndrome [16]. Consistent with this conclusion is the finding that loss of the C-terminal region of Fli1 causes significant thrombocytopenia in mice, [17] and loss of both Fli-1 and another Ets factor, Erg, is required for both HSC and megakaryocyte development [18].

RUNX1, also termed AML1 and core-binding factor (CBF)α, is a member of the Runt (Rnt) family of transcription factors, which stabilize the interaction of CBFβ with DNA [19]. The gene for RUNX1 is located on chromosome 21 (21q22.12), and encodes a protein of 453 amino acids, including the DNA-binding Rnt domain and the transactivating domain (TAD). Translocation between the RUNX1 site on chromosome 21 and the ETO gene on chromosome 8 produces the AML–ETO oncogene, found in approximately half of patients with M2 acute myelogenous leukemia [20]. The activity of RUNX1 is modulated by a number of posttranslational modifications, including phosphorylation by mitogen-activated protein kinases (MAPKs) and cyclin-dependent kinases, and by interaction with several binding partners. Once activated, the RUNX1/CBFβ complex binds to the consensus sequence TGTGNNN and enhances transcription of a myriad of megakaryocytic genes, including platelet factor-4 and integrin αIIb, and a number of genes that promote cell proliferation [19].

FOG1, or Friend of GATA, was initially identified as a protein that interacts with GATA-1 on erythroid promoters without interacting directly with DNA [21]. Its direct role in megakaryocyte development has been determined through promoter reporter analyses; for example, FOG interacts with GATA in the regulation of megakaryocyte-specific genes [22]. FOG1 is a member of the zinc-finger domain proteins that employ two zinc fingers to bind DNA. In the case of FOG1, the amino terminal finger binds the protein to GATA transcription factors, and mutations of this domain are responsible for human dyserythropoietic anemia and thrombocytopenia [21, 23]. In addition to the transcriptional activity of FOG that regulates megakaryopoiesis, a number of epigenetic factors have been identified that mediate its action. For example, mutations in FOG1 are associated with a gray platelet syndrome-like macrothrombocytopenia. The causative mutation in FOG1 blocks the interaction of the transcription factor with the nucleosome remodeling and deacetylase complex (NuRD) [24].

In addition to transcription factors that direct megakaryocyte-specific gene expression, several genes or gene products in megakaryocytes are regulated by microRNAs (miRNAs). miRNAs are a large group of ~22-nucleotide noncoding RNAs that are evolutionarily conserved and modulate protein expression by blocking target gene transcription or degrading the corresponding mRNA [25]. There are over 400 human miRNAs and several are implicated in megakaryopoiesis. In megakaryocytes, miRNA-150 expression is reduced as cells differentiate towards megakaryocytes, and overexpression of miRNA-150 inhibits megakaryopoiesis in in vitro and mouse models, potentially by targeting Meis1 and Ets1 transcripts [26]. In contrast, miRNA-150 levels increase during megakaryocyte differentiation, in a thrombopoietin-dependent manner, but not in erythroid cells [27]. miRNA-150 targets *c-Myb* expression, reducing its expression, and as downregulation of *c-Myb* promotes megakaryocytic differentiation at the expense of erythroid cells [28], the pathway of miRNA-150 to c-Myb helps to explain the effects of the hormone on enhancing megakaryocyte production. Additional regulatory miRNAs with potential roles in megakaryopoiesis include miRNA-34a, which when overexpressed in primitive hematopoietic cells increased megakaryocyte colony formation and modulates c-Myb expression [29], and miRNA-27a which targets the transcription factor RUNX1 [30].

Systems Approaches to Identify Novel Members of the Megakaryocytic Gene Expression Apparatus

Several investigators have used gene arrays to identify previously undescribed transcription factors that regulate the megakaryocyte lineage. For example, Fuhrken and colleagues induced CD34 + human marrow cells into the granulocytic or megakaryocytic lineage, and then focused upon the differentially expressed genes that bear a transcription factor signature [31]. Nearly 200 differentially expressed transcription factor motif-containing genes were identified, including several known to be expressed in the lineage, such as GATA-1, Fli-1, and MafG, lending credence to the veracity of the approach. Novel genes identified that were subsequently shown to be differentially expressed in megakaryocytes at the protein level included FHL2, MXD1, E2F3, and RFX5. However, other than the LIM domain protein FHL2, for which genetic elimination results in a failure of erythropoiesis and megakaryopoiesis [32], there are no subsequent studies that have determined if these genes play an important biological role in megakaryopoiesis.

With the availability of a growing database of transcription factors that influence hematopoiesis, and a fairly robust understanding of the factors that influence their expression and biological activities, many believe that this information can be used to model the networks that regulate lineage fate determination, ultimately utilizing such network models to reprogram progenitor cells, predict the effects of specific loss or gain of a specific transcription factor, or to drive specific differentiation patterns in primitive cells. One such effort that has met with some success is a Boolean model developed to predict the pathways taken, as a common myeloid progenitor cell gives rise to neutrophils, macrophages, erythroid cells, and megakaryocytes [33]. The success of the modeling effort was confirmed; when the model predictions of the levels of 11 distinct transcription factors in each population of cells was compared to the values measured by quantitative reverse transcription polymerase chain reation (RT-PCR), a remarkable concurrence was found.

Like systems approaches to identifying novel transcription factors that regulate megakaryocyte gene expression programs, a number of investigators have systematically sampled megakaryocyte miRNAs and tested their function. To discover novel regulatory pathways during megakaryocytic differentiation, soon after miRNA expression arrays were developed, Garzon and colleagues performed miRNA expression profiling of in vitro-differentiated megakaryocytes derived from primitive hematopoietic progenitors [34]. They found that *miRNA-10a, miRNA-126, miRNA-106, miRNA-10b, miRNA-17,* and *miRNA-20* were down-modulated as megakaryocytes differentiated from their immediate progenitors. Furthermore, using ectopic expression of various miRNAs found in megakaryocytes, they went on to show the functional consequences of several of these regulatory elements. A more comprehensive analysis of miRNA expression in differentiating megakaryocytes was more recently published [35]. A comprehensive assessment of miRNAs differentially expressed between normal cells and those derived from patients with the myeloproliferative neoplasms (MPNs), essential thrombocythemia (ET), and primary myelofibrosis (PMF) has begun to reveal some of the complex changes that can accompany malignant transformation [36]. Along similar lines, Girardot and colleagues sampled miRNAs in patients with MPNs in an attempt to help explain the down-modulation seen in the receptor, c-Mpl, in megakaryocytes and platelets from such patients. They found that miRNA-28 bound to and degraded c-Mpl mRNA, potentially accounting for the observed phenomenon [37].

Genes Regulated by the Megakaryocytic Transcriptional Apparatus

Candidate Gene Approaches

The transcription factors that drive the expression of a number of megakaryocyte-specific genes have been determined by classic promoter reporter analysis, and have included GATA-1, GATA-2, RUNX1, Fli-1, SCL, SP-1, NF-E2, EGR1, FOG1, among others (e.g. [22, 38]). Once the role(s) of individual transcription factors was identified, effective transcriptional complexes were then identified [39]. Many of the genes regulated by these transcriptionally active proteins have been studied in detail, and include the genes encoding integrin α_{IIb}, glycoprotein GpIb, GpVI, GpIX, and platelet factor-4, *InsP*^{3}*-R2*, enzymes of arachadonic acid metabolism, several G-protein-coupled P2Y receptors, β1-tubulin, thromboxane synthesis, as detailed above.

Novel Findings Due to Systems Approaches

One of the earliest systematic studies of the genes differentially expressed in the three major hematopoietic lineages—erythroid, myeloid, and megakaryocytic—was reported in 2005 [40]. Megakaryocytes derived from thrombopoietin-induced CD34 + marrow cells were sampled by gene array analysis at 4, 7, and 11 days following culture initiation. Genes expected to be differentially expressed included those encoding integrin αIIb, integrin α6, integrin β3, GpIb, GpV, GpIX, FOG2, IRS1, Thromboxane A2 receptor, and vascular endothelial cell growth factor, as well as many previously undescribed genes, lending much credence to the general approach of employing arrays to discovering new pathways that are important for megakaryocyte development. In a similar approach, novel genes potentially involved in megakaryocyte apoptosis were identified by Chen and colleagues [41].

Based on the repeated identification of binding sites for multiple transcription factors in the regulatory regions of several dozen well-described megakaryocyte-specific genes, an innovative approach to identify additional, previously undescribed genes important for megakaryocyte development was launched using chromatin immunoprecipitation (ChIP)–Seq technology. Tijssen and colleagues performed ChIP–Seq for GATA-1/2, RUNX1, Fli-1, and SCL in megakaryocytes derived from cord blood CD34 + cells, identifying 4722 GATA-1, 2475 GATA-2, 7345 RUNX1, 8688 Fli-1, and 3085 SCL binding sites in the genome [42]. Approximately 150 genes were identified that bound all five transcription factors, including many well-described megakaryocyte-specific proteins (e.g., integrin αIIB and GpIbβ), and using bioinformatics approaches, nearly one half of these genes had been previously reported expressed in megakaryocytes. Eight genes that were also expressed in zebrafish were subsequently randomly chosen for loss of function studies, to determine whether they might be functionally important; morpholino antisense-induced knockdown of each of these genes significantly impaired thrombocyte and/or erythrocyte formation in the resultant fish. In this way, the genes *march2, max, smox, pttg1lp, emilin1,* and *sufu* were identified as critical for stem cell and/or megakaryocyte formation.

An approach related to ChIP-Seq is termed formaldehyde-assisted isolation of regulatory elements (FAIRE), a method that identifies regions of open (likely transcriptionally active) chromatin, and hence, functionally important for a potentially important gene that had been previously identified from genome-wide association studies, quantitative trait locus analyses, or any other genetic systems approach. Paul and colleagues used FAIRE to confirm that rs342293, a noncoding polymorphism that correlates with platelet volume and function, is open in megakaryocytes and closed in erythroid cells, and identifies an Evi1-binding site in the *PIK3CG* gene, known to be involved in platelet membrane biogenesis [43].

Polyphosphate-4-phosphatase (P4P) was first identified by subtraction cloning between normal and GATA-1 knockdown megakaryocytes [44]. Consistent with its mode of discovery, the P4P gene was subsequently shown to display a functional GATA-1 site. One of the initially unexplained features of GATA-1 knockdown mice is their (dysmorphic) megakaryoblasts, which are highly abundant and proliferate in vitro far greater than control cells [5]. P4P catalyzes hydrolysis of the D-4-position phosphate of $PI_{3,4}P$ and $PI_{3,4,5}P$. These membrane phospholipids are products of thrombopoietin-stimulated phosphoinositol-3-kinase (PI3K) action on membrane phospholipids, which then play an important role in the proliferation and survival of the cells. When reintroduced into the knockdown mice, P4P diminishes the exuberant growth characteristic of the knockdown cells [44], findings similar to the phenotype of cells from PTEN or SHIP knockout mice, enzymes that hydrolyze the D-3 and D-5 positions of $PI_{3,4,5}P$.

Cytokines

Candidate Gene Approaches

Thrombopoietin

The primary regulator of platelet production is thrombopoietin, an acidic glycoprotein produced in many organs but primarily in the liver, kidney, and bone marrow. Most paths to discovery of thrombopoietin were dependent on identification of the *c-Mpl* proto-oncogene. Based on the presence of two copies of a characteristic structural motif, an 200 amino acid module containing four spatially conserved Cys residues, 14 β-sheets, a juxtamembrane Trp–Ser–Xaa–Trp–Ser sequence, a 20–25-residue transmembrane domain, and a 70–500-amino-acid intracellular domain containing short sequences that bind intracellular kinases, it was postulated that the c-Mpl product was an orphan cytokine receptor. Using three different approaches, four separate groups identified the ligand for the c-Mpl proto-oncogene; the use of the recombinant protein in marrow cultures allowed the outgrowth of megakaryocytes, and its administration to mice lead to massive megakaryopoiesis and thrombocytosis. Genetic elimination of thrombopoietin in mice leads to severe thrombocytopenia, and while only a single pedigree of humans null for thrombopoietin has been described, genetic elimination of c-Mpl leads to severe neonatal thrombocytopenia, and ultimately, HSC failure. Once available in large amounts, crystallization of thrombopoietin was accomplished, and its tertiary fold was found remarkably similar to that of erythropoietin and many members of the growth hormone/prolactin family of cytokines.

Stromal Cell-Derived Factor-1

Chemokines, defined by their ability to support chemotaxis, play multiple roles in blood cell physiology [45]. Four classes of the 8- to 12-kDa polypeptides have been recognized, based on the spacing of cysteine residues close to the amino terminus of the proteins (so called CC and CXC chemokines). All chemokine receptors are members of the seven-transmembrane family of receptors that signal through heterotrimeric G proteins, and are subtyped by the family of chemokines they serve.

The CXC chemokine CXCL12 (also termed stromal cell-derived factor (SDF)-1 and its receptor CXCR4 are critical for normal hematopoietic stem-cell trafficking and function. Most chemokines and chemokine receptors are promiscuous, they can bind to, or bind multiple counter-receptors, respectively; in contrast, the phenotype of genetic elimination of both CXCR4 and CXCL12 are almost identical [46, 47], a severe defect in stem cell homing to and retention in the marrow, resulting in hematopoietic failure. In addition, megakaryocytes display CXCR4 [48] and migrate in response to an CXCL12 concentration gradient [49]. Several groups have shown that CXCL12 augments thrombopoietin-induced megakaryocyte growth in suspension culture [48, 50]. Later studies have shown that the synergy between CXCL12 and other stimuli on megakaryocyte growth extends to cell surface adhesion [51].

Other Cytokines

A number of other cytokines have been described to affect megakaryocyte growth in vitro, almost always in concert with thrombopoietin or in plasma- or serum (a source of thrombopoietin)-containing cultures. Included in this group are other members of the cytokine receptor family, including interleukin (IL)-3, granulocyte–macrophage colony-stimulating factor, leukemia inhibitory factor, erythropoietin, IL-6, and IL-11, and members of the receptor tyrosine kinase family, including stem cell factor, Flt3 ligand and fibroblast growth factor. Of note, except for stem cell factor, genetic elimination of any of these cytokines fails to substantially alter resting platelet levels of the resultant animal.

Membranes/Receptors

Candidate Gene (Hypothesis-Driven) Approaches

c-Mpl Receptor

The myeloproliferative leukemia virus was first described in 1986 as causing a murine leukemia. The viral oncogene was identified in 1990, and the cellular proto-oncogene in 1992. Upon inspection of its predicted primary structure, it was immediately apparent that c-Mpl encoded a cytokine receptor. The cloning and characterization of thrombopoietin as the c-Mpl ligand 2 years later confirmed this hypothesis. The c-Mpl receptor is expressed primarily in hematopoietic tissues, specifically in megakaryocytes, their precursors (e.g., HSC), and their progeny (platelets). For the most part, c-Mpl is constitutively expressed in these tissues, although the level of receptor display is modulated by thrombopoietin-induced receptor internalization and degradation. In the absence of stimulation, hematopoietic cytokine receptors such as c-Mpl exist in a homodimeric state, in a conformation that holds the cytoplasmic domains far apart (e.g., 73 Å in the unliganded erythropoietin receptor). Upon binding the cognate ligand, the receptor conformation shifts, bringing the two cytoplasmic domains into close juxtaposition (39 Å in the liganded erythropoietin receptor). Additional studies indicate that the intracytoplasmic domain proximal to box1 and box2 cytoplasmic domains constitutively bind JAK family kinases, even in an inactive state. Upon ligand binding, the closer juxtaposition of the two tethered kinases is thought to allow their cross-activation, initiating signal transduction.

Mutational studies of the thrombopoietin receptor have revealed a great deal of signaling subtlety. By introducing a series of progressively longer helical segments into the transmembrane domain of the receptor, to introduce progressive degrees of cytoplasmic domain "orientation twist," Staerk and colleagues have found that depending on how the cytoplasmic domain homodimer is oriented, the receptor can be "off," "inducibly on" (i.e., responsive to thrombopoietin, the normal state), or "constitutively on," associated with a myeloproliferative phenotype in vivo [52]. As the signals sent from each of these receptor forms are different, an increasing understanding of the tertiary structure of the thrombopoietin receptor could lead to an enhanced ability to intervene disorders of c-Mpl.

Interferon Receptors

The most common cause of thrombocytosis in humans is inflammation. Much of the enhanced megakaryopoiesis due to inflammation has been traced to IL6-induced enhanced expression of hepatic thrombopoietin [53]; however, interferons could play a role as well. Interferon receptors are expressed on the megakaryocyte surface membrane, but not on platelets [54]. Culture of megakaryocytes or megakaryocyte cell lines has been shown to activate STAT1, and its downstream signals, leading to megakaryocyte maturation [55].

Adhesive Receptors

A large number of adhesive proteins are present on the surface of megakaryocytes and platelets. Included in this class are integrins and non-integrins, both of which have the capacity to enter into adhesive reactions that tether cells together in a regulable fashion, colloquially referred to as "molecular velcro." Less well appreciated is that both types of adhesive cell surface proteins also send signals that affect the activation state of the cell, and, in some cases, have an impact on megakaryopoiesis.

Integrins are heterodimeric proteins composed of 1 of 14 or more α chains and 1 of 8 or more β chains. The most important megakaryocyte and platelet integrins are αIIb/β3, which serve as a fibrinogen receptor, integrin α2/β1, one of two collagen receptors, and integrin α4β1, a fibronectin receptor. The effects of collagen on platelet activation is well known—it modestly activates aggregation and allows adhesion to exposed subendothelial collagen. Likewise, the interaction of fibrinogen with integrin αIIb/β3 is critical for platelet aggregation, directly cross-linking adjacent platelets in the growing platelet plug. Less well appreciated is the effect of integrins and other adhesive receptors on megakaryopoiesis. Genetic elimination of GpIb leads to modest thrombocytopenia, as seen in patients with Bernard–Soulier syndrome [56]. Engagement of integrin α4/β1 with specific fragments of fibronectin (that cannot interact with integrin α5/β1) stimulates megakaryocyte growth in vitro [57]. And integrin αIIbβ3 is critical for proplatelet formation and platelet release [58]. Megakaryocytes and platelets display two distinct collagen receptors, integrin α2β1 and GpVI [59]. The latter collagen receptor is thought to be the predominant signaling receptor, engaging cAMP and cGMP, the Fcγ receptor; the Src kinases, Fyn and Lyn; and the G protein, CDC42, among others [60–62]. Thus, while the platelet and megakaryocyte adhesion receptors are predominantly thought of in terms of platelet function, i.e., molecular velcro, they also play critical roles in megakaryocyte development and platelet production.

Novel Findings Due to Systems Biology

Given the relatively similar ontological background of erythroblasts and megakaryoblasts (the bi-potent "Meg-Erythroid" progenitor), a systematic search for megakaryocyte-specific cell surface receptors might best be conducted by finding differentially expressed genes predicting to encode transmembrane-containing proteins. Such a differential erythroid/megakaryocytic gene array was conducted by Macaulay and colleagues [63], who found many of the expected genes (integrin αIIb, integrin β3, PECAM1, CD9, platelet factor-4), but several previously unidentified or poorly characterized transcripts. Among the latter, of some interest were G6b, G6f, and the succinate receptor, SUCNR1. G6b has also been identified by others as a novel immunoreceptor tyrosine-based inhibitory motif (ITIM)-containing protein [64], and later found to modulate collagen-induced platelet activation mediated by GpVI [65], and the finding of succinate receptors on megakaryocytes and platelets provides an explanation for the enhancing effect of succinate on a variety of platelet agonists.

Signaling Molecules/Adaptors

Candidate Gene (Hypothesis-Driven) Approaches

Prior to the cloning of thrombopoietin, extensive studies using numerous hematopoietic family cytokines, especially growth hormone and erythropoietin, provided a roadmap for candidate-signaling molecules and adapters that might be utilized by megakaryopoietic cytokines. With the cloning and characterization of thrombopoietin and naturally occurring and engineered c-Mpl-bearing cells, the tools required to confirm or refute the role of these candidate genes in megakaryopoiesis was at hand.

The signaling pathways that promote cell survival, proliferation, and differentiation in megakaryopoiesis have been widely studied. Like that for erythropoietin and the erythropoietin receptor, c-Mpl binding of thrombopoietin results in the juxtaposition of two molecules of the signaling kinase Jak2, with subsequent phosphorylation of the contralateral Jak2, and several tyrosine residues of the receptor (Y_{591}, Y_{625}, Y_{630}). Once Jak2 is active and Y_{625} and Y_{630} are phosphorylated, signal transducer and activator of transcription (STAT)3 and STAT5, and in some settings STAT1 are recruited and phosphorylated, leading to their dimerization and translocation into the nucleus. STAT5 is indispensable for the normal megakaryocyte development, as *Stat5*-deficient mice show impaired platelet production [66]. Moreover, transgenic mice with megakaryocytic lineage-specific overexpression of a dominant negative form of STAT3 display reduced platelet recovery following 5-FU-induced myelosuppression [67]. One target of HGF-induced STAT5 that mediates these effects on cell survival is the anti-apoptotic molecule BclXL [68].

Coincident with STAT activation, the adapter proteins SHC and LNK bind to P-Y_{625}, recruiting son-of-sevenless (SOS) and then Ras, and recruiting p85 phosphoinositol-3-kinase, and its kinase domain, p110. Activation of the MAPK pathway in megakaryocytes by Ras is required for maturation of megakaryocytic progenitor cells and the generation of highly polyploid cells [69]. One MAPK-dependent pathway that contributes to megakaryocyte differentiation is its activation of RUNX1 by phosphorylation (see below, and [70]). Phosphorylation of AKT by thrombopoietin-activated PI3K controls cell-cycle progression and cell survival [71] through silencing of FOXO family of transcription factor [72], which if left unchecked would stimulate expression of the cell-cycle inhibitor p27 and the proapoptotic molecule fas ligand.

The STAT proteins are also affected by additional c-Mpl-related adaptor molecules. For example, overexpression of LNK, a known inhibitor of cytokine signaling, inhibits thrombopoietin-induced STAT5 and MAPK activation in 32D-mpl cells, introduction of LNK into bone marrow Lin-cells reduced thrombopoietin-dependent growth, and ploidy of megakaryocytes and increased megakaryocyte ploidy was found in LNK-deficient mice [73, 74]. The calcium- and integrin-binding (CIB) protein appears to play a similar role, as genetic elimination of CIB1 is associated with thrombocytosis, increased megakaryocyte ploidy, and enhanced thrombopoietin-induced activation of PI3K/Akt and MAPK, although proplatelet formation was impaired, perhaps due to a megakaryocyte adhesion and migration defect [75].

A number of small G proteins are also activated during megakaryopoiesis. RhoA, Rac1, and CDC42 are critical for megakaryocyte maturation. For example, like most/all diploid cells, RhoA localizes to the cleavage furrow in diploid megakaryocytes undergoing late anaphase. However, this subcellular localization is disturbed during megakaryocytic endomitosis, although the molecular explanation is unclear. Likewise, CDC42 interacts with WASP, the protein product of the gene mutated in Wiscott–Aldrich syndrome, a disorder characterized by modest thrombocytopenia. Rac1, RhoA, and CDC42 are also critical for megakaryocyte proplatelet formation. Maturing megakaryocytes display reduced phosphorylation of the myosin light chain

(MLC), levels of which are mediated by these G proteins. Of note, inhibition of histone deacetylases with newly emerging cancer chemotherapeutic agents leads to reduced levels of Rac1, RhoA, and CDC42, enhanced levels of phopho-MLC, and reduced proplatelet formation, thought to account for the thrombocytopenia due to these new drugs [76].

Once a megakaryocyte is stimulated by thrombopoietin and other pro-proliferative stimuli, the induced signals must be extinguished, lest uncontrolled proliferation ensue. At least three mechanisms down-modulate the sensitivity of cells to further growth factor signaling: (1) induction of suppressors of cytokine signaling (SOCS) proteins, (2) activation of phosphatases that remove P-Tyr sites from Jak family kinases and cytokine receptors, and (3) a number of adaptor proteins that negatively regulate signaling. Removal of activated c-Mpl from the cell surface by endocytosis will be discussed below. SOCS proteins are induced by STAT-mediated transcription, and once translated bind to P-Y residues in c-Mpl and Jak2, precluding binding of additional signaling molecules and triggering proteolytic destruction [77]. Hematopoietic cells contain a number of phosphatases that eliminate P-Y from receptors and signaling adaptors (e.g., SHP1); their physiological importance is illustrated by disorders of macrophage activation or profound erythropoiesis. Recent evidence suggests that the dual-function phosphatase, PTEN, is also important for hematopoiesis. Moreover, thrombopoietin has been shown to induce the expression of SOCS1 and SOCS3 [78], implicating these STAT–induced proteins in the regulation of thrombopoietin signaling. Finally, several proteins that bind to c-Mpl, either directly or indirectly, initiate signals that dampen the proliferative signals that emanate from c-Mpl. In addition to LNK, mentioned above, the Src family kinase Lyn down-modulates thrombopoietin-induced proliferation [79], likely acting downstream of the focal adhesion kinase, FAK [80].

Once exposed to thrombopoietin, c-Mpl is rapidly removed from the cell surface, in a clathrin-dependent process [81]. The molecular components required for the removal of surface proteins are clathrin triskeletons and adaptor protein (AP)2 complexes [82]. Target proteins for clathrin-mediated endocytosis bear recognition sequences NPXY or YXXθ (theta; $\theta =$ bulky hydrophobic), LL, or an acidic cluster; of note, human c-Mpl bears two YRRL sequences (Y_{521}RRL and Y_{591}RRL), highly conserved in the murine receptor. Previously, we showed that the Y_{591}F mutant c-Mpl molecule displays enhanced thrombopoietin signaling [83]. Moreover, reduced receptor internalization and enhanced signaling could explain why truncation of c-Mpl beyond S_{574} [84] or L_{582} [83] signals so well, despite elimination of the P-Y residues that activate STATs and MAPKs. We found that cell surface clearance of c-Mpl is greatly diminished when Y_{591} is mutated to F, an effect associated with intense and prolonged signaling, and that Y_{591} is part of the clathrin/AP2 complex recognition site. Moreover, Y_{521} is responsible for the trafficking of internalized c-Mpl to the lysosome, as its mutation to F allows enhanced recycling of the internalized receptor to the cell surface [81]. c-Mpl contains two intracellular lysine residues that are potential targets for ubiquitination (K_{553}, K_{573}) and hence might mediate its degradation through the proteasome. The use of mutant forms of the c-Mpl receptor demonstrated that c-Cbl is an E3 ligase for the ubiquitinated c-Mpl receptor, and is responsible for receptor-signaling intensity.

In addition to cytokine and chemokine binding, the engagement of other megakaryocyte cell surface proteins induces activation of adaptors and signaling molecules. Upon binding cognate ligand, the cytoplasmic domain of the integrin receptors undergo conformational changes, eliminating binding sites for some, and revealing binding sites for other adaptors and signal transducers. Among the adaptors that bind once platelets are activated, talin, Rap1, and RIAM are critical for the conformational change that allows fibrinogen engagement during aggregation [85].

Novel Findings Due to Systems Biology

As noted above, STAT1 activation is critical for megakaryocyte maturation. Recently, an exciting corollary to the levels of megakaryocyte expression of STAT1 was identified in patients with MPNs.

When an acquired, activating mutation of Jak2 (Jak2V_{617}F) was identified in virtually all patients with polycythemia vera (PV), and half of patients with ET and PMF, a dilemma arose: How can one mutation lead to three distinct (albeit related) diseases [86]. One ready explanation would be that additional genetic alterations could explain the phenotypic diversity of the same primary, proliferative stimulus (an activated Jak2). By performing gene array analysis of the clonal erythroid and megakaryocytic colonies derived from 20 patients with ET, and 16 with PV, Chen and colleagues concluded that the inherent level of STAT1 expression and activation, which varies considerably from patient to patient, contributes to whether a patient that acquires Jak2V_{617} F develops PV or ET [63] 63]. Several conclusions from this work deserve comment. First, the gene expression profiles of progenitor-derived colonies from different patients display far more variation between patients than between normal and malignant progenitors of the same individual. Second, normal megakaryocytic colonies display more STAT1 than normal erythroid colonies. Third, expression of a constitutively active STAT1 in normal human CD34+ progenitor cells skews the balance of erythroid and megakaryocyte outgrowth towards the latter. Fourth, since patients with either of these MPNs circulate both normal and Jak2V_{617} F-positive progenitor cells, the levels of STAT1 in both neoplastic and normal colonies could be independently assessed, allowing the pre-mutant expression profile to be determined; patients with heterozygous expression of Jak2V_{617}F ET display more STAT1 in their *normal* hematopoietic colonies than those with PV. These investigators conclude that the inherent variability in STAT1 expression, prior to the acquisition of Jak2V_{617}F, determines whether PV or ET develops in any given patient with that kinase mutation.

The marrow of patients with PMF display altered megakaryopoiesis, with densely clustered arrays of megakaryocytes with maturation defects and ineffective thrombopoiesis. In an attempt to understand the genetic basis for this phenotype, Theophile and colleagues performed low-density gene array analysis, focusing on genes that influence cell survival, on laser-dissected megakaryocytes from patients with PMF and normal individuals. They found that many anti-apoptotic genes tended to be overexpressed in PMF megakaryocytes, especially BNIP3 [87]. However, given some recent results from Chen and colleagues [88], indicating that the premorbid level of STAT1 expression could be responsible for the disease phenotype (PV vs. ET, see above), it would need to be explored whether patients with PMF might display enhanced anti-apoptotic genes at baseline, which could then account for their PMF phenotype upon acquiring Jak2V_{617} F. In keeping with this notion, Olthaf and colleagues reported that another STAT, STAT5, plays a similar lineage-switching function [89]. When they reduced expressed expression of STAT5 in CD34+ using an RNA interference (RNAi) approach, they found enhanced megakaryocytic output of the cultures; in contrast, when an activated form of STAT5 was introduced into the same starting cell population, erythropoiesis was enhanced.

As noted above, engagement of megakaryocyte and platelet adhesion receptors leads to both platelet activation and megakaryopoiesis. Thus, a systems approach to identifying new adaptor proteins for these receptors is of great interest. The search for platelet activation adaptors and signaling molecules will be discussed in Chap. 14, but several studies should be commented upon here due to the effects of integrin α4β1, integrin αIIbβ3, and GpIb on megakaryopoiesis. Using the global proteome machine database to create an in silico protein interaction tool, Zhang and colleagues queried which proteins might interact with the cytoplasmic domains of the integrin αIIb and integrin β3 subunits [90]. In addition to talin1, Rap1, the partnering integrin subunit, tubulin, vinculin, and plekstrin, expected from prior experimental studies, they also identified myosin, GpIb, α-actinin, profiling, and the adaptor protein 14-3-3. Likewise, for integrin β1, using the in silico tool, they found integrin α2, integrin α5, integrin α6, integrin αIIb, γ (gamma)actin, multimerin1, thrombospondin1, talin1, filamin A, vinculin, myosin, vimentin—only some of which were previously known. To confirm their predictions, they then performed formaldehyde cross-linking of activated platelets, added an anti-integrin β1 antibody and performed mass spectroscopy on the immunoprecipitate; over 80 % of the predicted proteins were confirmed to interact with integrin β1 experimentally.

Genome-wide association studies (GWAS) hold the promise of identifying in an unbiased way a large number of genes that were previously unsuspected to play a role in a simple or complex trait. Much has been written about the advantages and potential pitfalls of GWAS, but it is clear that its power to generate testable hypotheses is impressive. Recently, a GWAS was conducted on 68,000 individuals of European descent, focusing on platelet size and number [91]. From the analysis, 68 loci were identified as contributing to one or both of these traits; 43 with platelet count, 25 with platelet volume, and 16 with both. Several of the genes were previously described as impacting megakaryocyte or platelet formation, including those encoding thrombopoietin, GpIb, Wiskott–Aldrich-like protein, LNK, Cbl, β1-tubulin, NF-E2, and cyclin A. Several other genes were identified that were previously known to play important functional roles in megakaryocytes or platelets, including CD9 (a tetraspanin), integrin αllb, protease-activated receptor (PAR)1, the major thrombin receptor, and the adhesion receptor CD226. When loss-of-function analyses were conducted in Drosophila on genes not currently known to affect hematopoiesis or platelet function, several hematopoietic phenotypes were identified, including ablation of all primitive hematopoiesis (*arhgef3, ak3, rnf145, and jmjd1c*), or thrombocytes alone (*tpma*; the homologue of human TPM1). As this GWAS accounted for less than 10 % of the genetic variability in human platelet number and size, it is safe to assume that there are many additional genetic determinants affecting megakaryopoiesis.

Genes Targeted by Signaling Apparatus

Candidate Gene Approaches

Once the primary regulators of megakaryopoiesis were identified, a number of signaling pathways that affect megakaryocyte development were identified. For example, MafB, which enhances GATA-1 and Ets activity during megakaryoblast differentiation [25], is induced by activation of ERK1/2, one of the primary downstream events of thrombopoietin stimulation [26]. The transcription factor RUNX1 (also termed CBFα2 and AML1) was identified from reporter gene analyses of several genes that regulate hematopoiesis. Its critical role in megakaryopoiesis was established by gene mapping studies in patients with familial platelet disorder/predisposition to AML; such patients, but not their normal family members, display haploinsufficiency of RUNX1, linking the gene to thrombocytopenia [92]. This conclusion was subsequently confirmed by its genetic elimination in mice [93]. In addition to its upregulation in cytokine-stimulated megakaryocytes, in response to phosphorylation by thrombopoietin-induced activation of ERK1/2, RUNX1 forms a complex with CBF-β and together with GATA-1, induces integrin α_{IIb} and integrin α_2 expression in megakaryocyte-like cells. Moreover, phosphorylated RUNX1 then induces expression of p19 INK4, which leads to endomitosis arrest and megakaryocyte maturation [94].

A number of studies have demonstrated an important role for thrombopoietin in HSC biology. For example, the hormone supports cell survival in cultures of highly purified murine HSCs [95], and genetic elimination of thrombopoietin or its receptor leads to an approximate tenfold reduction in the ability of HSCs to repopulate lethally irradiated mice. The molecular targets that support this function include at least three transcription factors, each pathway identified by prior work identifying a critical role of each transcription factor in HSC biology. Three- to fourfold overexpression of the homeobox-containing transcription factor HoxB4 results in a tenfold expansion of HSCs following transplantation [96]. Thrombopoietin-induced activation of p38 MAPK enhances expression of HoxB4, in a USF1/2-dependent fashion [97]. Likewise, modest overexpression of HoxA9 leads to greatly enhanced HSC numbers [98]; thrombopoietin-induced activation of PI-3 K induces expression of the heterodimeric partner of HoxA9, MEIS1 [99], and once phosphorylated by activated MAPK, MEIS-1 acts to translocate HoxA9 to the nucleus. Finally, based on the supportive role of vascular endothelial growth factor (VEGF) on HSCs [100], the capacity of thrombopoietin to induce hypoxia-inducible factor (HIF)1α to drive expression of VEGF [101] was demonstrated.

Novel Findings Due to Systems Biology

One of the most unique features of megakaryocytes is endomitosis, the process by which mitosis is aborted in mid-anaphase, coupled to reenter into DNA synthesis, producing cells that become polyploid, containing 64, 128, and even 256 times the normal chromosomal complement in a single lobated nucleus. Polyploidization is felt to be necessary for the cell enlargement required for efficient platelet formation, allowing amplification of the genome to support the massive protein synthesis that fuels the impressive cell growth that characterizes megakaryocyte development. Early studies demonstrated that gene expression occurs in concert in polyploidy megakaryocytes; that is, for example, in a 32N cell, all 32 copies of a gene are undergoing transcription simultaneously, and transcriptionally silent genes remain silent at all the multiple loci [102]. Another important question is whether endomitosis might specifically drive certain transcriptional programs. To address this question, Raslova and colleagues sorted human megakaryocytes into various ploidy classes, and performed gene chip analysis on each class of cells [103]. Surprisingly, only 105 genes were modulated during endomitosis, comparing 2N cells to 16N cells. Of the genes down-modulated in the more mature cells, nearly all were members of classes that regulate DNA replication, transcription and repair, and most of the upregulated genes were involved in cytoskeletal maturation (e.g., actin, β1-tubulin, MYH9), platelet adhesion/aggregation (e.g., integrins, GpIb, PECAM), and signal transduction (e.g., MAPK, FYN, RAC1, RAB1B).

Systematic Approaches to Addressing Major Issues in Megakaryopoiesis

Candidate gene approaches have advanced our understanding of the molecular machinery that mediates megakaryocyte differentiation, development, and platelet production. Despite this, a number of key questions remain. For instance, what signals are essential to transition megakaryocytes from normal diploid mitosis to polyploid endomitosis, or from a large intact megakaryocyte into one that fragments into platelets. What factors other than thrombopoietin are essential for "normal" megakaryocyte and platelet levels? Are platelets formed from megakaryocytes of lower ploidy functionally different from platelets formed from high-ploidy cells?

With several proteins critical for basic megakaryocyte development now functionally defined, the identification of subsequent factors, very likely to display more subtle effects, is certain to be increasingly challenging. Additionally, genes critical to answering the next set of major questions in hematopoiesis, in general, and megakaryopoiesis, in particular, might not be "usual players." For these reasons, in order to continue to advance our understanding, the field must undertake a series of experiments that utilize unbiased methodology, to further advance our understanding of megakaryopoiesis. Although far from comprehensive, what follows is a summary of techniques that will likely be part of the investigation of megakaryopoiesis in the decades to come.

Uncovering Critical Protein–Protein Interactions

As discussed above, megakaryocytes display cell surface receptors that are shared with other cell types (i.e., integrins) and receptors are relatively unique to these cells (i.e., c-Mpl is expressed primarily on megakaryocytes, their precursors, and their progeny). Thus, the wiring diagram that describes potential signaling events in megakaryocytes cannot be accurately inferred from models in other cell types. For this reason, mapping the protein–protein interactions that occur for megakaryocyte-specific receptors is critical. Fortunately, proteomic techniques have been established to identify interacting partners.

One critical technology that has already been utilized in the field is protein domain microarrays. Protein microarray technology provides a means to study protein–ligand interactions in vitro in a noncompetitive format. Importantly, it can be used to obtain quantitative information on binding affinities. In a typical protein microarray experiment, the target proteins are spotted in a regular pattern at high spatial density on a solid support, usually a chemically derivatized glass substrate or a glass-supported nitrocellulose membrane. The spotted proteins become immobilized on the surface and, following a blocking step, are incubated with a labeled probe (e.g., a protein, peptide, nucleic acid, or small molecule). After a brief washing step, protein–ligand interactions are identified using the label on the probe. For example, if the probe has been labeled with a fluorophore, the array is simply scanned for fluorescence.

Most eukaryotic proteins are modular in nature. For example, they might comprise both catalytic domains and interaction domains that, to a first approximation, can be extracted from their host protein without loss of function. Thus, a domain-oriented strategy that provides an effective way to circumvent the difficulty associated with producing full-length recombinant proteins can be used to study the recognition properties of entire families of related proteins in high throughput. Recently, this approach was taken to study the Src-homology-2 (SH2) and phosphotyrosine binding (PTB) domain-containing proteins that interact with pY_{630} of c-Mpl. This study identified a new binding partner, Tensin2, and demonstrated that it is a key activator of the pro-survival Akt cascade in megakaryocytes [104].

It is impractical to express and purify all protein domains in the human genome, so broader approaches to the study of protein–protein interaction are often advantageous. For this purpose, immunoprecipitation followed by mass spectrometry (IP-MS) has been extensively used in a number of fields. In this type of experiment, the protein of interest is immunoprecpitated under native conditions, then the subsequent precipitant lyzed, digested, and peptides subjected to tandem mass spectrometry. This allows for the identification of binding partners to the protein of interest. More extensive use of this technique will enable a more complete picture of what proteins interact with critical megakaryocyte signaling molecules.

Comparison of Intracellular Signaling in Different Cell Populations

In addition to a more comprehensive understanding of the molecular basis of direct interactions that contribute to signaling in megakaryocytes, a broader understanding of signaling pathways involved in each of the phenotypic changes described above is necessary. To date, this has been challenging due to the difficulty in collecting large number of primary megakaryocytes undergoing specific changes. However, with the development of a number of new technologies, understanding the signaling changes that occur in cell populations with only a small number of cells is rapidly becoming possible.

Antibody-Based Approaches

A number of new technologies have made analyses that were previously only approachable by traditional western blotting, possible on a smaller scale. A new technology, Luminex xMAP, based on the principles of flow cytometry, uses colored microspheres linked to antibodies to quantify protein abundance or posttranslation modification [105]. The advantage over traditional western blotting is the ability to obtain this measurement with as few as ten cells. xMAP technology is a medium-throughput approach, and is available commercially. Additionally, antibody microarray technology similarly uses a very small amount of total cell lysate per measurement. In an antibody array, a wide variety of antibodies are attached to a glass slide, and lysates, which have been modified to be labeled on either the N- or C-terminus with a fluorescent tag, are incubated with the slide. These arrays are commercially available and able to monitor differences between a wide range of signaling proteins. Alternatively, arrays can be fabricated from lysates. In a lysate microarray, each sample is printed on a glass slide coated with nitrocellulose. The slide is then treated as a membrane would be in the process of performing a western blot. Unfortunately, it is critical to validate each antibody that is used for a particular cell line for use on protein microarrays, by comparing data obtained on control arrays to that obtained by western blotting [106]. Although antibody validation will always be a challenge in these approaches, the ability to obtain western-blot-like information from only thousands of cells is a major advantage.

Perhaps the most promising antibody-based technique is a fusion between flow cytometry and fluorescence microscopy, termed ImageStream. ImageStream technology allows for a cellular sample to be stained with multiple antibodies, just as one might do in analyzing or isolating subpopulations in peripheral blood mononuclear cells or other samples. However, in addition to providing levels of total protein present in the given cell, by taking an image of each cell that is analyzed, subcellular localization can also be determined. This combination of single-cell analysis, and the ability to mark a given cell with up to tens of markers, is extremely powerful, and will surely advance our understanding of signaling in hard-to-isolate cell populations such as megakaryocytes [107, 108].

Non-Antibody-Restricted Approaches

For experiments that require truly unbiased experimentation that antibody-based approaches cannot provide, a number of quantitative mass-spectrometry-based approaches have been developed. The two most common approaches now used routinely are termed stable isotope labeling with amino acids in cell culture (SILAC) and isobaric tags for relative and absolute quantization (iTRAQ). In both types of experiments, proteins from samples to be compared are labeled with a given tag, digested, then mixed. Finally, samples are then analyzed by liquid chromatography and tandem mass spectrometry and quantities of peptides containing one tag are compared with peptides containing another tag. This allows for quantification of each peptide between multiple samples.

Analysis of Transcriptional Changes

In the past decades, the development of DNA microarray and RNA–seq technologies has greatly enhanced our ability to probe the entire complement of RNA sequences for information. Although multiple technologies exist for this purpose, the two most commonly used approaches are transcription arrays and RNA–seq. In a transcriptional array, a large number of short nucleotide sequences are printed on a glass slide, and RNA extracted from a particular experimental condition, labeled with a fluorescent dye, and hybridized to the array. If RNA from a particular gene is expressed, it will bind the corresponding printed spot, and produce a signal. Microarrays have now been successfully used for decades, in > 40,000 journal articles. Although this approach has now become the "gold standard" in transcriptome analyses, it requires that probes predict transcribed sequences, since each spot on an array must correspond to a given sequence, and arrays typically cannot probe the entirety of sequences contained within the human genome. Thus, sequences that are thought to encode noncoding sequences are typically excluded.

In an atypical cell type such as the megakaryocyte, transcription of sequences that are typically believed to be silent is not outside of the realm of possibility. Next-generation sequences approaches have allowed for the development of RNA–seq-based approaches. In RNA–seq, RNA generated in a particular experimental condition is directly sequenced, so assumptions do not need to be made regarding which sequences are transcribed. Moving forward, this approach will likely be invaluable, not only in comparing the response of megakaryocytes to different environmental conditions and identifying molecular differences in cell populations at different developmental time points, but also potentially for identifying novel transcripts unique to megakaryocytes.

Conclusions, Challenges, and Future Directions

Much has been learned since megakaryocytes were first recognized as the origin of blood platelets in 1906. With the cloning and characterization of thrombopoietin, the field has advanced significantly, because of the capacity to grow the cells in vitro and analyze their biochemistry and cell biology. But much remains to be discovered, in order that a more complete understanding of these cells translate into both a better appreciation of the unique biological properties of endomitosis and proplatelet formation and fragmentation, and the ability to manipulate the cells for therapeutic benefit be realized. In addition to the genes already identified and detailed in this chapter, the systems approaches described hold great potential to expand our inventory of potentially important determinants of megakaryocyte biology. In this realm, systems approaches will be of great benefit. The function of potentially novel regulators of the megakaryocyte must undergo a thorough functional analysis. RNAi hold great promise as a function-defining tool, as extensive libraries of RNAi are now available to systematically inhibit virtually all human and murine genes. Thus, we are on the verge of a systematically determined global description of the wiring diagram of the megakaryocyte and its immediate product, the blood platelet. Given the vital role these blood cells play in hemostasis, inflammation, cancer, and cardiovascular disease, the future holds great promise to greatly impact many of the most prevalent diseases of man.

References

1. Yousuf O, Bhatt DL. The evolution of antiplatelet therapy in cardiovascular disease. Nat Rev Cardiol. 2011;8:547–59.
2. Nurden AT. Platelets, inflammation and tissue regeneration. Thromb Haemost. 2011;105 (Suppl 1):S13–33.
3. Semple JW, Italiano JE, Jr., Freedman J. Platelets and the immune continuum. Nat Rev Immunol. 2011;11:264–74.
4. Gay LJ, Felding-Habermann B. Contribution of platelets to tumour metastasis. Nat Rev Cancer. 2011;11:123–34.
5. Shivdasani RA, Fujiwara Y, McDevitt MA, Orkin SH. A lineage-selective knockout establishes the critical role of transcription factor GATA-1 in megakaryocyte growth and platelet development. EMBO J. 1997;16:3965–73.
6. Lemarchandel V, Ghysdael J, Mignotte V, Rahuel C, Romeo PH. GATA and Ets cis-acting sequences mediate megakaryocyte-specific expression. Mol Cell Biol. 1993;13:668–76.
7. Bastian LS, Kwiatkowski BA, Breininger J, Danner S, Roth G. Regulation of the megakaryocytic glycoprotein IX promoter by the oncogenic Ets transcription factor Fli-1. Blood. 1999;93:2637–44.
8. Furihata K, Kunicki TJ. Characterization of human glycoprotein VI gene 5′ regulatory and promoter regions. Arterioscler Thromb Vasc Biol. 2002;22:1733–9.
9. Lacabaratz-Porret C, et al. Biogenesis of endoplasmic reticulum proteins involved in Ca^{2+} signalling during megakaryocytic differentiation: an in vitro study. Biochem J. 2000;350(Pt 3):723–34.

10. Muntean AG, Crispino JD. Differential requirements for the activation domain and FOG-interaction surface of GATA-1 in megakaryocyte gene expression and development. Blood. 2005;106:1223–31.
11. Andrews NC, Erdjument-Bromage H, Davidson MB, Tempst P, Orkin SH. Erythroid transcription factor NF-E2 is a haematopoietic-specific basic-leucine zipper protein. Nature. 1993;362:722–8.
12. Bean TL, Ney PA. Multiple regions of p45 NF-E2 are required for beta-globin gene expression in erythroid cells. Nucleic Acids Res. 1997;25:2509–15.
13. Shivdasani RA, et al. Transcription factor NF-E2 is required for platelet formation independent of the actions of thrombopoietin/MGDF in megakaryocyte development. Cell. 1995;81: 695–704.
14. Lecine P, Italiano JE Jr, Kim SW, Villeval JL, Shivdasani RA. Hematopoietic-specific beta 1 tubulin participates in a pathway of platelet biogenesis dependent on the transcription factor NF-E2. Blood. 2000;96:1366–73.
15. Deveaux S, et al. p45 NF-E2 regulates expression of thromboxane synthase in megakaryocytes. EMBO J. 1997;16:5654–61.
16. Raslova H, et al. FLI-1 monoallelic expression combined with its hemizygous loss underlies Paris-Trousseau/Jacobsen thrombopenia. J Clin Invest. 2004;114:77–84.
17. Moussa O, et al. Thrombocytopenia in mice lacking the carboxy-terminal regulatory domain of the Ets transcription factor Fli1. Mol Cell Biol. 2012;30:5194–206.
18. Kruse EA, et al. Dual requirement for the ETS transcription factors Fli-1 and Erg in hematopoietic stem cells and the megakaryocyte lineage. Proc Natl Acad Sci U S A. 2009;106:13814–9.
19. Friedman AD. Cell cycle and developmental control of hematopoiesis by Runx1. J Cell Physiol. 2009;219:520–4.
20. Wei Q, Paterson BM. Regulation of MyoD function in the dividing myoblast. FEBS Lett. 2001;490:171–8.
21. Freson K, et al. Molecular cloning and characterization of the GATA1 cofactor human FOG1 and assessment of its binding to GATA1 proteins carrying D218 substitutions. Hum Genet. 2003;112:42–9.
22. Gaines P, Geiger JN, Knudsen G, Seshasayee D, Wojchowski DM. GATA-1- and FOG-dependent activation of megakaryocytic alpha IIB gene expression. J Biol Chem. 2000;275:34114–21.
23. Freson K, et al. Platelet characteristics in patients with X-linked macrothrombocytopenia because of a novel GATA1 mutation. Blood. 2001;98:85–92.
24. Wang Y, et al. Pleiotropic platelet defects in mice with disrupted FOG1-NuRD interaction. Blood. 2011;118:6183–91.
25. Thomas M, Lieberman J, Lal A. Desperately seeking microRNA targets. Nat Struct Mol Biol. 2010;17:1169–74.
26. O'Connell RM, et al. Sustained expression of microRNA-155 in hematopoietic stem cells causes a myeloproliferative disorder. J Exp Med. 2008;205:585–94.
27. Lu J, et al. MicroRNA-mediated control of cell fate in megakaryocyte-erythrocyte progenitors. Dev Cell. 2008;14:843–53.
28. Emambokus N, et al. Progression through key stages of haemopoiesis is dependent on distinct threshold levels of c-Myb. EMBO J. 2003;22:4478–88.
29. Navarro F, et al. miR-34a contributes to megakaryocytic differentiation of K562 cells independently of p53. Blood. 2009;114:2181–92.
30. Ben-Ami O, Pencovich N, Lotem J, Levanon D, Groner Y. A regulatory interplay between miR-27a and Runx1 during megakaryopoiesis. Proc Natl Acad Sci U S A. 2009;106:238–43.
31. Fuhrken PG, et al. Gene Ontology-driven transcriptional analysis of CD34+ cell-initiated megakaryocytic cultures identifies new transcriptional regulators of megakaryopoiesis. Physiol Genomics. 2008;33:159–69.
32. Li L, et al. A requirement for Lim domain binding protein 1 in erythropoiesis. J Exp Med. 2010;207:2543–50.

33. Krumsiek J, Marr C, Schroeder T, Theis FJ. Hierarchical differentiation of myeloid progenitors is encoded in the transcription factor network. PLoS ONE. 2011;6:e22649.
34. Garzon R, et al. MicroRNA fingerprints during human megakaryocytopoiesis. Proc Natl Acad Sci U S A. 2006;103:5078–83.
35. Opalinska JB, et al. MicroRNA expression in maturing murine megakaryocytes. Blood. 2010;116:e128–38.
36. Hussein K, et al. MicroRNA expression profiling of megakaryocytes in primary myelofibrosis and essential thrombocythemia. Platelets. 2009;20:391–400.
37. Girardot M, et al. miR-28 is a thrombopoietin receptor targeting microRNA detected in a fraction of myeloproliferative neoplasm patient platelets. Blood. 2010;116:437–45.
38. Gannon AM, Kinsella BT. Regulation of the human thromboxane A2 receptor gene by Sp1, Egr1, NF-E2, GATA-1, and Ets-1 in megakaryocytes. J Lipid Res. 2008;49:2590–604.
39. Eisbacher M, et al. Protein-protein interaction between Fli-1 and GATA-1 mediates synergistic expression of megakaryocyte-specific genes through cooperative DNA binding. Mol Cell Biol. 2003;23:3427–41.
40. Komor M, et al. Transcriptional profiling of human hematopoiesis during in vitro lineage-specific differentiation. Stem Cells. 2005;23:1154–69.
41. Chen C, et al. A systems-biology analysis of isogenic megakaryocytic and granulocytic cultures identifies new molecular components of megakaryocytic apoptosis. BMC Genomics. 2007;8:384.
42. Tijssen MR, et al. Genome-wide analysis of simultaneous GATA1/2, RUNX1, FLI1, and SCL binding in megakaryocytes identifies hematopoietic regulators. Dev Cell. 2011;20:597–609.
43. Paul DS, et al. Maps of open chromatin guide the functional follow-up of genome-wide association signals: application to hematological traits. PLoS Genet. 2011;7:e1002139.
44. Vyas P, Norris FA, Joseph R, Majerus PW, Orkin SH. Inositol polyphosphate 4-phosphatase type I regulates cell growth downstream of transcription factor GATA-1. Proc Natl Acad Sci U S A. 2000;97:13696–701.
45. Rollins BJ. Chemokines. Blood. 1997;90:909–28.
46. Nagasawa T, et al. Defects of B-cell lymphopoiesis and bone-marrow myelopoiesis in mice lacking the CXC chemokine PBSF/SDF-1. Nature. 1996;382:635–8.
47. Ma Q, et al. Impaired B-lymphopoiesis, myelopoiesis, and derailed cerebellar neuron migration in CXCR4- and SDF-1-deficient mice. Proc Natl Acad Sci U S A. 1998;95:9448–53.
48. Wang JF, Liu ZY, Groopman JE. The alpha-chemokine receptor CXCR4 is expressed on the megakaryocytic lineage from progenitor to platelets and modulates migration and adhesion. Blood. 1998;92:756–64.
49. Hamada T, et al. Transendothelial migration of megakaryocytes in response to stromal cell-derived factor 1 (SDF-1) enhances platelet formation. J Exp Med. 1998;188:539–48.
50. Hodohara K, Fujii N, Yamamoto N, Kaushansky K. Stromal cell-derived factor-1 (SDF-1) acts together with thrombopoietin to enhance the development of megakaryocytic progenitor cells (CFU-MK). Blood. 2000;95:769–75.
51. Avecilla ST, et al. Chemokine-mediated interaction of hematopoietic progenitors with the bone marrow vascular niche is required for thrombopoiesis. Nat Med. 2004;10:64–71.
52. Staerk J, et al. Orientation-specific signalling by thrombopoietin receptor dimers. EMBO J. 2011;30:4398–413.
53. Kaser A, et al. Interleukin-6 stimulates thrombopoiesis through thrombopoietin: role in inflammatory thrombocytosis. Blood. 2001;98:2720–25.
54. Negrotto S, et al. Expression and functionality of type I interferon receptor in the megakaryocytic lineage. J Thromb Haemost. 2011;9:2477–85.
55. Huang Z, et al. STAT1 promotes megakaryopoiesis downstream of GATA-1 in mice. J Clin Invest. 2007;117:3890–9.
56. Nurden P, Nurden AT. Congenital disorders associated with platelet dysfunctions. Thromb Haemost. 2008;99:253–63.
57. Fox NE, Kaushansky K. Engagement of integrin alpha4beta1 enhances thrombopoietin-induced megakaryopoiesis. Exp Hematol. 2005;33:94–9.

58. Larson MK, Watson SP. Regulation of proplatelet formation and platelet release by integrin alpha IIb beta3. Blood. 2006;108:1509–14.
59. Nuyttens BP, Thijs T, Deckmyn H, Broos K. Platelet adhesion to collagen. Thromb Res. 2011;127(Suppl 2):S26–9.
60. Margarucci L, et al. Collagen stimulation of platelets induces a rapid spatial response of cAMP and cGMP signaling scaffolds. Mol Biosyst. 2011;7:2311–9.
61. Jung SM, Moroi M. Platelet glycoprotein VI. Adv Exp Med Biol. 2008;640:53–63.
62. Akbar H, et al. Gene targeting implicates Cdc42 GTPase in GPVI and non-GPVI mediated platelet filopodia formation, secretion and aggregation. PLoS ONE. 2011;6:e22117.
63. Macaulay IC, et al. Comparative gene expression profiling of in vitro differentiated megakaryocytes and erythroblasts identifies novel activatory and inhibitory platelet membrane proteins. Blood. 2007;109:3260–9.
64. Senis YA, et al. A comprehensive proteomics and genomics analysis reveals novel transmembrane proteins in human platelets and mouse megakaryocytes including G6b-B, a novel immunoreceptor tyrosine-based inhibitory motif protein. Mol Cell Proteomics. 2007;6:548–64.
65. Mori J, et al. G6b-B inhibits constitutive and agonist-induced signaling by glycoprotein VI and CLEC-2. J Biol Chem. 2008;283:35419–27.
66. Kirito K, Kaushansky K. Transcriptional regulation of megakaryopoiesis: thrombopoietin signaling and nuclear factors. Curr Opin Hematol. 2006;13:151–6.
67. Kirito K, et al. A functional role of Stat3 in in vivo megakaryopoiesis. Blood. 2002;99: 3220–27.
68. Silva M, et al. Erythropoietin can promote erythroid progenitor survival by repressing apoptosis through Bcl-XL and Bcl-2. Blood. 1996;88:1576–82.
69. Rojnuckarin P, Drachman JG. Kaushansky K. Thrombopoietin-induced activation of the mitogen-activated protein kinase (MAPK) pathway in normal megakaryocytes: role in endomitosis. Blood. 1999;94:1273–82.
70. Hamelin V, Letourneux C, Romeo PH, Porteu F, Gaudry M. Thrombopoietin regulates IEX-1 gene expression through ERK-induced AML1 phosphorylation. Blood. 2006;107:3106–13.
71. Geddis AE, Fox NE, Kaushansky K. Phosphatidylinositol 3-kinase is necessary but not sufficient for thrombopoietin-induced proliferation in engineered Mpl-bearing cell lines as well as in primary megakaryocytic progenitors. J Biol Chem. 2001;276:34473–9.
72. Nakao T, Geddis AE, Fox NE, Kaushansky K. PI3K/Akt/FOXO3a pathway contributes to thrombopoietin-induced proliferation of primary megakaryocytes in vitro and in vivo via modulation of p27(Kip1). Cell Cycle. 2008;7:257–66.
73. Tong W. Lodish HF. Lnk inhibits Tpo-mpl signaling and Tpo-mediated megakaryocytopoiesis. J Exp Med. 2004;200:569–80.
74. Takizawa H, et al. Growth and maturation of megakaryocytes is regulated by Lnk/Sh2b3 adaptor protein through crosstalk between cytokine- and integrin-mediated signals. Exp Hematol. 2008;36:897–906.
75. Kostyak JC, Naik MU, Naik UP. Calcium- and integrin-binding protein 1 regulates megakaryocyte ploidy, adhesion, and migration. Blood. 2012;119:838–46.
76. Bishton MJ, et al. Deciphering the molecular and biologic processes that mediate histone deacetylase inhibitor-induced thrombocytopenia. Blood. 2011;117:3658–68.
77. Wormald S, Hilton DJ. Inhibitors of cytokine signal transduction. J Biol Chem. 2004;279: 821–4.
78. Wang Q, Miyakawa Y, Fox N, Kaushansky K. Interferon-alpha directly represses megakaryopoiesis by inhibiting thrombopoietin-induced signaling through induction of SOCS-1. Blood. 2000;96:2093–9.
79. Lannutti BJ, Minear J, Blake N. Drachman JG. Increased megakaryocytopoiesis in Lyn-deficient mice. Oncogene. 2006;25:3316–24.
80. Hitchcock IS, et al. Roles of focal adhesion kinase (FAK) in megakaryopoiesis and platelet function: studies using a megakaryocyte lineage specific FAK knockout. Blood. 2008;111:596–604.

81. Hitchcock IS, Chen MM, King JR, Kaushansky K. YRRL motifs in the cytoplasmic domain of the thrombopoietin receptor regulate receptor internalization and degradation. Blood. 2008;112:2222–31.
82. Collins BM, McCoy AJ, Kent HM, Evans PR, Owen DJ. Molecular architecture and functional model of the endocytic AP2 complex. Cell. 2002;109:523–35.
83. Drachman JG, Kaushansky K. Dissecting the thrombopoietin receptor: functional elements of the Mpl cytoplasmic domain. Proc Natl Acad Sci U S A. 1997;94:2350–5.
84. Luoh SM, et al. Role of the distal half of the c-Mpl intracellular domain in control of platelet production by thrombopoietin in vivo. Mol Cell Biol. 2000;20:507–15.
85. Banno A, Ginsberg MH. Integrin activation. Biochem Soc Trans. 2008;36:229–34.
86. Kaushansky K. On the molecular origins of the chronic myeloproliferative disorders: it all makes sense. Blood. 2005;105:4187–90.
87. Theophile K, Hussein K, Kreipe H, Bock O. Expression profiling of apoptosis-related genes in megakaryocytes: BNIP3 is downregulated in primary myelofibrosis. Exp Hematol. 2008;36:1728–38.
88. Chen E, et al. Distinct clinical phenotypes associated with JAK2V617F reflect differential STAT1 signaling. Cancer Cell. 2010;18:524–35.
89. Olthof SG, et al. Downregulation of signal transducer and activator of transcription 5 (STAT5) in CD34 + cells promotes megakaryocytic development, whereas activation of STAT5 drives erythropoiesis. Stem Cells. 2008;26:1732–42.
90. Zhang CC, et al. In silico protein interaction analysis using the global proteome machine database. J Proteome Res. 2011;10:656–68.
91. Gieger C, et al. New gene functions in megakaryopoiesis and platelet formation. Nature. 2011;480:201–8.
92. Song WJ, et al. Haploinsufficiency of CBFA2 causes familial thrombocytopenia with propensity to develop acute myelogenous leukaemia. Nat Genet. 1999;23:166–75.
93. Ichikawa M, et al. AML-1 is required for megakaryocytic maturation and lymphocytic differentiation, but not for maintenance of hematopoietic stem cells in adult hematopoiesis. Nat Med. 2004;10:299–304.
94. Gilles L, et al. P19INK4D links endomitotic arrest and megakaryocyte maturation and is regulated by AML-1. Blood. 2008;111:4081–91.
95. Sitnicka E, et al. The effect of thrombopoietin on the proliferation and differentiation of murine hematopoietic stem cells. Blood. 1996;87:4998–5005.
96. Thorsteinsdottir U, Sauvageau G, Humphries RK. Enhanced in vivo regenerative potential of HOXB4-transduced hematopoietic stem cells with regulation of their pool size. Blood. 1999;94:2605–12.
97. Kirito K, Fox N, Kaushansky K. Thrombopoietin stimulates Hoxb4 expression: an explanation for the favorable effects of TPO on hematopoietic stem cells. Blood. 2003;102:3172–8.
98. Thorsteinsdottir U, et al. Overexpression of the myeloid leukemia-associated Hoxa9 gene in bone marrow cells induces stem cell expansion. Blood. 2002;99:121–9.
99. Kirito K, Fox N, Kaushansky K. Thrombopoietin induces HOXA9 nuclear transport in immature hematopoietic cells: potential mechanism by which the hormone favorably affects hematopoietic stem cells. Mol Cell Biol. 2004;24:6751–62.
100. Gerber HP, et al. VEGF regulates haematopoietic stem cell survival by an internal autocrine loop mechanism. Nature. 2002;417:954–8.
101. Kirito K, Kaushansky K. Thrombopoietin stimulates vascular endothelial cell growth factor (VEGF) production in hematopoietic stem cells. Cell Cycle. 2005;4:1729–31.
102. Raslova H, et al. Megakaryocyte polyploidization is associated with a functional gene amplification. Blood. 2003;101:541–4.
103. Raslova H, et al. Interrelation between polyploidization and megakaryocyte differentiation: a gene profiling approach. Blood. 2007;109:3225–34.

104. Jung AS, Kaushansky A, Macbeath G, Kaushansky K. Tensin2 is a novel mediator in thrombopoietin (TPO)-induced cellular proliferation by promoting Akt signaling. Cell Cycle. 2011;10:1838–44.
105. Skogstrand K, et al. Simultaneous measurement of 25 inflammatory markers and neurotrophins in neonatal dried blood spots by immunoassay with xMAP technology. Clin Chem. 2005;51:1854–66.
106. Wolf-Yadlin A, Sevecka M, MacBeath G. Dissecting protein function and signaling using protein microarrays. Curr Opin Chem Biol. 2009;13:398–405.
107. Basiji DA, Ortyn WE, Liang L, Venkatachalam V, Morrissey P. Cellular image analysis and imaging by flow cytometry. Clin Lab Med. 2007;27:653–70.
108. George TC, et al. Quantitative measurement of nuclear translocation events using similarity analysis of multispectral cellular images obtained in flow. J Immunol Methods. 2006;311:117–29.

Chapter 5
Systems Biology of Platelet–Vessel Wall Interactions

Yolande Chen, Seth Joel Corey, Oleg V. Kim and Mark S. Alber

Abstract Platelets are small, anucleated cells that participate in primary hemostasis by forming a hemostatic plug at the site of a blood vessel's breach, preventing blood loss. However, hemostatic events can lead to excessive thrombosis, resulting in life-threatening strokes, emboli, or infarction. Development of multi-scale models coupling processes at several scales and running predictive model simulations on powerful computer clusters can help interdisciplinary groups of researchers to suggest and test new patient-specific treatment strategies.

Keywords Platelets · Blood vessels · Hemostasis · Thrombosis

Multiple Functions of Platelets in Blood Clot Development

First observed through simple microscopy in the nineteenth century, platelets and leukocytes interact with blood vessel walls [1, 2]. Since those first observations, an impressive body of information has accumulated to establish the centrality of

Y. Chen (✉)
RH Lurie Comprehensive Cancer Center, Northwestern University Feinberg School of Medicine, Rm 5-220, 303 E. Superior St., Chicago, IL 60611, USA
Tel: 312-503-9359
e-mail: yolande-chen@northwestern.edu

S. J. Corey
Department of Pediatrics and Cell & Molecular Biology, Northwestern University Feinberg School of Medicine and Lurie Children's Hospital of Chicago, Chicago, IL, USA

O. V. Kim
Department of Applied and Computational Mathematics and Statistics, University of Notre Dame, Notre Dame, IN, USA

M. S. Alber
Department of Applied and Computational Mathematics and Statistics, University of Notre Dame, Notre Dame, IN, USA

Department of Medicine, Indiana University School of Medicine, Indianapolis, IN, USA

S. J. Corey et al. (eds.), *A Systems Biology Approach to Blood*, Advances in Experimental Medicine and Biology 844, DOI 10.1007/978-1-4939-2095-2_5

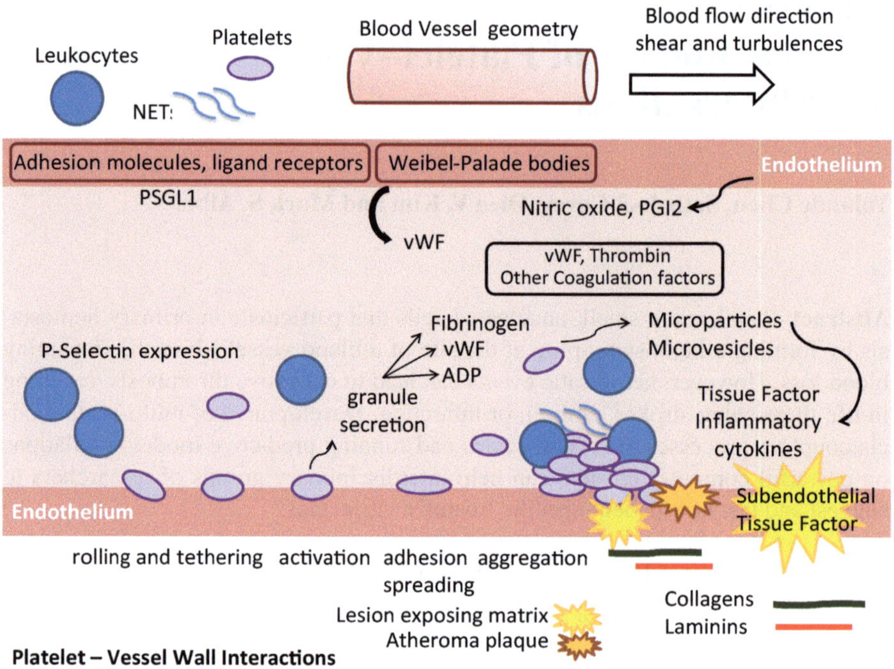

Fig. 5.1 Platelets and vessel walls interact in multiple ways. Platelets and vessel wall cells express adhesion molecules, surface receptors, and release substances that initiate or regulate cascading thrombosis events. Matrix components of the vessel wall, when blood flow is exposed through a damaged endothelium, also elicit clotting reactions from blood cells. Moreover, blood cells and endothelial cells also produce microparticles and microvesicles that carry procoagulant activity

platelets in vascular thrombosis. The main steps in thrombosis formation classically include the tethering, rolling/translocation, and adhesion of leukocytes and platelets to the exposed matrix at the damaged blood vessel site. Arterial thrombosis remains the most common cause of myocardial infarction and stroke, resulting in significant morbidity and mortality. To prevent thrombosis, many patients are treated with antiplatelet agents [3]. By producing chemical messengers, microparticles, and vascular changes, platelets also promote cancer and inflammation [4–6]. Therefore, there is a great need to understand more platelet–leukocyte–endothelial interactions (Fig. 5.1) and translate that into more effective, less toxic therapies.

A major challenge in quantitative understanding of hemostasis/thrombosis is to integrate various processes occurring during thrombus development. To predict how variations of multiple factors associated with platelet activity affect thrombus development is of great biomedical importance. However, there are significant challenges in developing such understandings. For instance, platelet–vessel wall receptor–ligand interactions occur at nanometer scale, whereas blood flow dynamics in the vicinity of a thrombus is a macroscopic event developing over the scale of hundreds of micrometers to millimeters. Coupling various processes is a complex, challenging task.

By reviewing main platelet functions, their roles in hemostasis/thrombosis, and computational approaches to simulate clotting events, this chapter establishes a rationale for a systems approach to platelet physiology. These include modeling of coagulation reactions, platelet activation, platelet dynamics, platelet–platelet interactions, and blood flow. The integrative modeling approaches are described to provide the basis for multi-scale computational models of thrombus development. First, biological background on platelet functioning is given, including platelet adherence and activation, intracellular and extracellular signaling, relation to tumor metastasis, and global approaches to study platelet–vessel wall interactions. Then, several recent integrative modeling methods of thrombus development involving cellular signaling, platelet–platelet, platelet–flow, and platelet–wall interactions are highlighted. Finally, a concluding perspective is offered on the role of platelets in hemostasis/thrombosis and tumor progression as well as the role of system biology in testing new therapeutic targets.

Platelet Adherence and Activation

Platelets interact with their environment through specialized receptors, many of which are integrins [7]. One principle mediator is platelet P-selectin (CD62P), a cell adhesion molecule stored in platelet alpha granules that interacts with P-selectin ligand 1 (PSGL-1, or CD162, found on leukocytes and endothelial cells). Endothelial cells possess granules called Weibel–Palade bodies, which release von Willebrand Factor (vWF) and P-selectin when activated. Interaction between P-selectin and its ligand occurs during tethering and rolling on the endothelium [1]. Importantly, the platelet surface receptor for vWF is glycoprotein Ib (GPIb; CD42). Lack of cleavage of high molecular weight vWF multimers results in thrombotic thrombocytopenic purpura, a catastrophic disorder. On the other hand, defective GPIb expression or activity results in excessive bleeding, the Bernard–Soulier disorder. vWF binding to GPIb induces downstream cytoskeletal actin rearrangement via Filamin A. Filamin A which regulates intracellular signaling [8] is responsible for a solid anchorage of GPIb [9]. Filamin A mutations have recently been found as a cause for thrombocytopenia [10] and abnormal platelet function [11].

Another critical mediator for platelet interactions is integrin αIIbβ3 (CD41), a cell surface receptor for fibrinogen. This integrin is defective in individuals suffering from Glanzmann's thrombasthenia, a bleeding disorder. When fibrinogen binds to the integrin, a cascade of downstream events, which include cytoskeletal rearrangements via talin and other signaling involving rous sarcoma (SRC) and focal adhesion kinase (FAK)-tyrosine kinases [12], is set in motion. Equally important is the presence of collagen receptors at the surface of the platelet, the better known of which are GPVI and α2β1. GPVI in particular seems to have a prominent role during collagen exposure to platelet surface and subsequent activation [12]. Platelets also express the C-type lectin receptor (CLEC-2, which triggers a downstream signaling cascade similar to that of GPVI, including activation of the spleen tyrosine kinase (Syk)) [12].

G-protein-coupled receptors include the thrombin receptor, adenosine diphosphate (ADP) receptors (P2Y1, P2Y12 which is a target for several "antiplatelet" drugs in thrombotic disease treatment) [3, 13], and thromboxane A2 receptor. They are activated in response to soluble factors from the blood. Exposure of the subendothelial matrix provides additional potent activators of platelet activation and coagulation. Platelet receptors recognize matrix components, collagen (via α2β1, GPVI), and laminin (via α6β1). Targeting of the laminin receptor, like the integrin α6β1, has been shown to be a promising strategy in the treatment of arterial thrombosis [14]. Recently, platelets were found to express CXCR6, the receptor specific to chemokine CXCL16. CXCL16 present on atherosclerotic lesions was found to enhance platelet adhesion to the endothelium after high arterial shear stress and to injured vascular wall [15]. Other cell surface proteins on platelets, such as Eph kinases and EphrinB1 [16], semaphorins [17] or Gas6 receptors [18], may promote thrombus formation. Gap junction channels, such as connexin 37 [19], may also enhance thrombus formation.

The list of platelet cell surface receptors or adhesion molecules will likely continue to grow, adding to complexity of platelet interactions and function. For instance, by expressing CD40-L, Fc receptors to immunoglobulins, and toll-like receptors [5], platelets serve as effectors in the immune system. Platelet-derived microparticles containing pro-inflammatory cytokines such as IL-1β contribute to inflammation [6].

Intracellular Cell Signaling and Inter-platelet Signaling

Resulting from those diverse ligand–receptor stimuli (Fig. 5.2), downstream signaling events involve phospholipid metabolism, generation of cAMP and cGMP second messengers, and Ca^{2+} release from the dense tubular system. These pathways lead to cytoskeletal reorganization, calpain activation, and signal amplification and diversification due to protein and lipid kinases. Phosphatase activity increases too [20]. Many of these pathways intersect or interact with each other in a complex fashion. Besides outside–inside signaling, there is also inside–outside signaling, as in the case of αIIbβ3 integrin activation [12, 21].

Extraplatelet Signaling

Upon platelet stimulation, alpha or dense granule release occurs, leading to platelet activation amplification, since the released substances (fibrinogen, vWF, ADP, thromboxane A2, thrombospondin-1) will further activate platelets and contribute to hemostasis. One adaptive measure is to induce blood vessel constriction, which is achieved through thromboxane A2. However, alpha granules release not only prothrombotic but also pro-angiogenic (VEGF) and anti-angiogenic (endostatin, thrombospondin) factors [22–25]. Intensively studied are platelet microparticles and exosomes. Microparticles are shed from membranes and their size varies from 100 to 1000 nm [26], while exosomes are secreted and overall smaller (30–100 nm) [27].

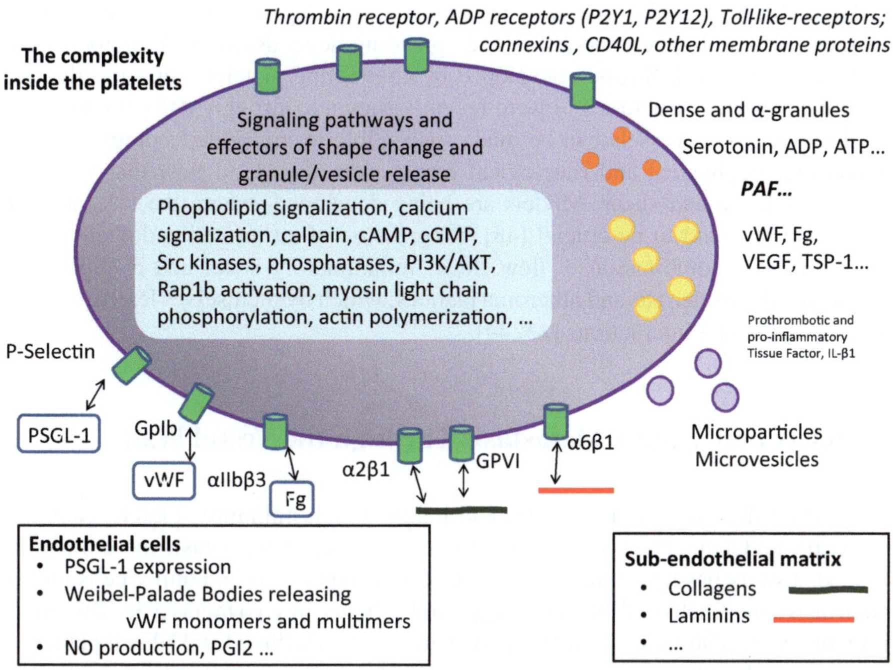

Fig. 5.2 Platelets express a diversity of cell surface receptors and chemical substances that interact with the vessel wall components during hemostasis. Downstream signaling events take place that involve phospholipid signalization, Ca^{2+} flux, calpain activation, cAMP- and cGMP-level modulation, cytoskeletal players and their modulators and diverse kinases. Many of these pathways intersect or interact with each other in a complex manner

Microparticles are shed from the platelet membrane and carry tissue factor activity, thus being procoagulant [12, 28]. Exosome secretion from platelets has also been described, but their role is less well known [29]. Inflammation contributes to thrombus formation through the interaction of platelets with leukocytes (neutrophils and monocytes) [30, 31]. Indeed, thrombosis could more generally be seen as an effector of innate immunity [32]. Moreover, neutrophils also contribute to thrombus formation, including via formation of neutrophil extracellular traps (NET) [33–37].

When a blood vessel is injured, the subendothelial matrix is exposed, and matrix components such as the collagens or laminins serve as potent activators of platelet activation. Furthermore, endothelial cells release prothrombotic factors such as vWF from their Weibel–Palade bodies. Interestingly, vWF release from the endothelial Weibel–Palade bodies is dependent on essential autophagy genes Atg5 or Atg7, and pharmacological inhibitors of autophagic flux lead to increased bleeding time [38]. On the other hand, the endothelium also releases substances that are inhibitory to platelet activation, by secreting nitric oxide (NO) and the downstream modulation of cGMP levels [39–43], or by secreting prostacyclin(PGI2) [44].

On a different scale, platelet contraction forces depend on microenvironment stiffness [45] and platelet adhesion and spreading depends on local microenvironment geometry [46]. Stresses developed by contracting platelets significantly alter thrombus internal structure and were recently shown to strongly deform embedded erythrocytes [47]. Vessels can be made susceptible to injury depending on vessel geometry, biophysical and rheological forces from the blood flow that will produce turbulence and shear. Models are being developed, where the role of vWF is still being found to be crucial [48]. Progression of an injured endothelium, resulting from a combination of flow shear, inflammatory state, and dyslipidemia, results in atherosclerosis and atheroma plaques, which by themselves lead to specific atheroma–platelet interactions [15, 44].

Platelets and Tumor Metastasis Through the Vessel Wall

Apart from their role in thrombosis and inflammation/immunity, platelets are also currently under intense scrutiny for their role in cancer metastasis dissemination. Indeed, a study has found that interactions between platelets and tumor cells induce an invasive mesenchymal-like phenotype and enhance in vivo metastasis. This effect is mediated by platelet-derived TGFβ, which activates the Smad and NF-κB pathways in cancer cells [4]. Specific to vessel wall effects, tumor cell-activated platelets secreted ADP, which facilitates cancer cell penetration past blood vessels. The P2Y2 receptor on blood vessel cells is necessary for this effect [49].

Global Approaches in Studying Platelet–Vessel Wall Interactions

To discover molecular mechanisms of platelet interactions with the vessel wall, researchers have been relying on narrowed and focused approaches (Fig. 5.3). One example is observing human disease and trying to pin down the molecular defect. Then one might want to search for partners of known actors. Subsequently, to confirm the relevance, one might pursue on knocking down or knocking out an element and observing the cell or organism for modified behaviors. More recently, researchers have been adding new global approaches in the field of biology, enriching knowledge through global and systemic approaches [50]. These new global approaches include:

1. Mutagenesis screening by N-ethyl-N-nitrosourea (ENU). For example, BcL-xL was discovered to regulate platelet half-life or platelet number [51].
2. Genome-wide association studies (GWAS). Human genomic variations are associated with cardiovascular outcomes, platelet size, or number, as reported by European study consortiums [52]. Jones et al. found that single nucleotide variants in platelet endothelial aggregation receptor 1 (PEAR1*)*, guanine nucleotide exchange factors (GEFs) for Rho family GTPases (VAV3*), and* IP3 receptor *(*ITPR1*) were associated with modified* platelet response to platelet agonists [53].

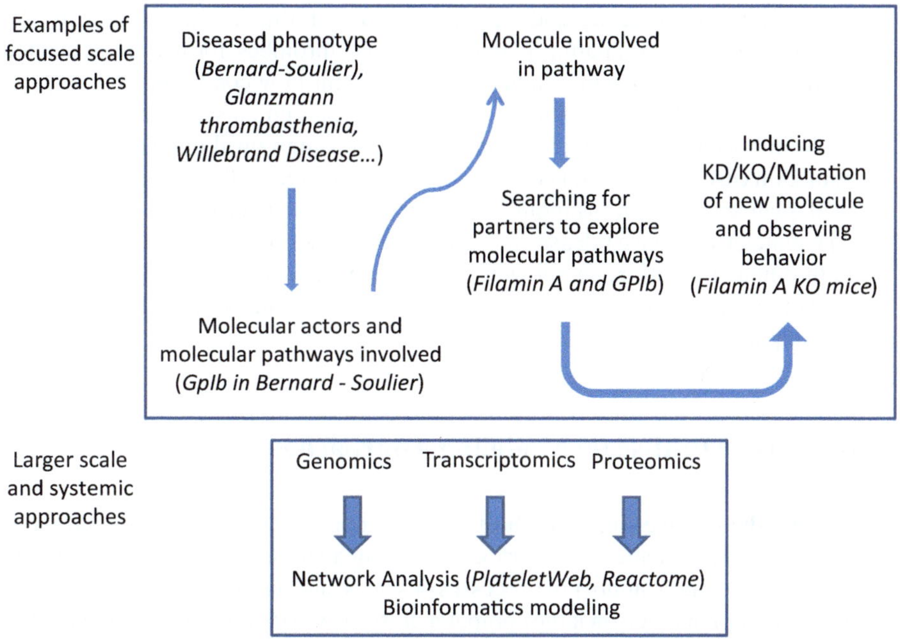

Fig. 5.3 Systems approach to hemostasis control by platelet–blood vessel interaction. A highly informative approach is based on multi-scale analysis of human bleeding disorders at biochemical, molecular, genetic, and organismal levels

3. Transcriptomics. Rowley et al. recently reported a comprehensive transcriptome study of human and mouse platelets [54]. This approach led to the discovery of connexin 37 in platelet aggregation [19].
4. Proteomics. Comprehensive platelet proteomics have been performed [55] and can be further narrowed down to subfields as "secretome" or "phosphoproteome" [56]. These proteomic studies revealed that secretogranin III, cyclophilin A, and calumenin were secreted by platelets after thrombin stimulation and found in atherosclerotic plaques [57] and nitrous oxide-treatment abrogated platelet activation by thrombin and prevented thrombin-induced translocation of gelsolin, filamin, 14–3-3 ζ, phosphatidylinositol 3-kinase-gamma isoform, and growth factor receptor-bound protein 2 (Grb2) [58].
5. Network analysis. Network analysis tools are being developed and databases are being made available online. For example, cPlateletWeb (http://plateletweb.bioapps.biozentrum.uni-wuerzburg.de) is an Internet-based platform organizing signaling network [59, 60]. A study based on PlateletWeb found a novel interaction between vasodilator-stimulated phosphoprotein and Abelson interactor 1 in human platelets [61]. Another database is Reactome (http://www.reactome.org), where extensive data have been collected, analyzed, and grouped in different pathways [62].

6. Informatics for modeling and simulations. Bioinformatics can be used to develop models for simulating platelet function, thrombus formation associated with pro- or anticoagulant gradients, and different flow conditions [63–69]. Indeed, thrombus formation studies have mostly focused on separate components of thrombogenesis, which can be numerous and subgrouped into categories: coagulation cascades of blood coagulation factors, platelet adhesion to the vascular wall, platelet aggregation among themselves, internal platelet activation phenomena, platelet substance release and amplification reaction, white blood cells' roles in thrombus formation, vessel wall product release, vessel wall injury and exposure of thrombogenic elements, atheroma genesis, and blood flow shear variation and impact.

Integrative Multi-scale Modeling Approaches

Integrating the relative role of each of the mentioned elements to model thrombus formation has been challenging. Various modeling approaches have been proposed, that integrate a certain number of processes or scales, and tested with the assistance of simulations implemented on large computer clusters (see [63, 65, 70, 83] for review). One target goal for future research would be to improve modeling ability to predict platelet/vessel wall behavior and thrombus formation by integrating simulations of the molecular signature characteristics, mechanical properties of agonist/antagonist, blood flow and viscoelastic properties of a blood vessel. Below we outline several existing modeling approaches that combine several scales.

Explicit incorporation of single platelet dynamics into a three-dimensional thrombus formation model has been described in Pivkin et al. [71] where each platelet was represented in a simplified way as a spherical object, while red blood cells were treated using continuum submodel describing their density. The model also included an ADP-induced platelet activation mechanism. Model simulations accurately reproduced the thrombus growth rate as a function of blood velocity obtained in experiments [72].

Fogelson and Guy developed a microscale platelet aggregation model in which individual platelets were modeled as fluid-filled closed membranes immersed in a viscous liquid [73]. This model allowed for simulations of individual platelet motion and their interactions with each other and with surrounding medium. In the model by Mody and King [74], the hydrodynamic effects of the oblate spheroidal shape of platelets and the proximity of a wall on cell–cell collisions were investigated. Collision time and contact area and collision frequency were compared between spherical shape vs. platelet-type oblate shape on one hand and presence or not of a proximal bounding wall. The approach used calculation of forces and torques acting on each particle in the fluid system (gravity, bond forces, and repulsion between two surfaces in close proximity). The study showed that the contact time between two platelets during collision close to the wall was greater than the contact time during a collision far from the wall. The wall proximity had a greater influence on platelet–platelet collisions than on sphere–sphere collisions. The method also used Huang and Hellums' mathematical model of the shear-induced platelet aggregation [75–77], where high shear resulted in increased platelet aggregation.

Leiderman and Fogelson [78] described a model of blood coagulation under flow that included coagulation biochemistry, chemical activation, and deposition of platelets, and a two-way interaction between fluid dynamics and growth of platelet mass. Expanding on a previously published Kuharsky and Fogelson model [79], this approach now described how tissue factor threshold triggered production of thrombin and how the wall shear rate and near-wall increased platelet concentration affected thrombus growth. The porous nature of the thrombus, allowing for advective and diffusive transport within itself, was also accounted for in the extended model. Xu et al. [63, 80–83] proposed a multi-scale model-coupling submodels of coagulation reactions, platelet dynamics, and blood flow, where platelets were represented as extended objects with fluctuating boundaries based on the cellular Potts model [84].

The importance of quantifying transport of coagulation factors within thrombus microenvironment was emphasized in Ref. [69]. By combining in vitro experiments and continuum-modeling approach of thrombus hydrodynamics, authors showed that both diffusivity and advection of blood proteins through the porous thrombus structure affect platelet–thrombus interaction and play an essential role in blood clot growth dynamics. Following this work, Stalker et al. demonstrated, using mouse injury model, that regional platelet-packing density emerged in parallel with differences in intrathrombus molecular transport and predicted that these differences affect thrombus growth and stability [85].

Wu et al. [86] presented a three-dimensional multi-scale platelet–blood flow–vessel wall interaction model, which combined three biological scales crucial for the early platelet aggregation. The model included hybrid cell membrane submodel of platelet elasticity, stochastic receptor–ligand binding submodel of cell adhesion kinetics and Lattice Boltzmann submodel of blood flow. Adhesion kinetics involved specific receptor–ligand pairs, namely vWF-GpIb complexes. At subcellular level, to simulate vWF–GpIb and GpIb-vWF–GpIbαbinding, individual molecules were represented by elastic springs. This was justified by the demonstration that the receptor–ligand binding is probabilistic in nature. Individual filaments of the cytoskeleton network of platelet membrane were modeled as coarse-grained harmonic potentials. At cellular level, a novel continuum description of the cell membrane was used. The subcellular and cellular components were integrated by distributing GpIbα receptors over verticies of the cytoskeleton network and by superimposing the lipid bilayer and the network. The model allowed investigation of how platelet stiffness, GPIb receptor expression, and platelet–platelet interaction affect platelet–wall adhesion quantified in terms of platelet pause time. To reduce the computational time cost, the model was implemented on graphical processing units (GPUs) computer cluster. Predictive simulations revealed that platelet deformation, interactions between platelets in the vicinity of the vessel wall, as well as the number of functional GPIbα platelet receptors played significant roles in the platelet adhesion to the injury site (Fig. 5.4, from Ref. [86]).

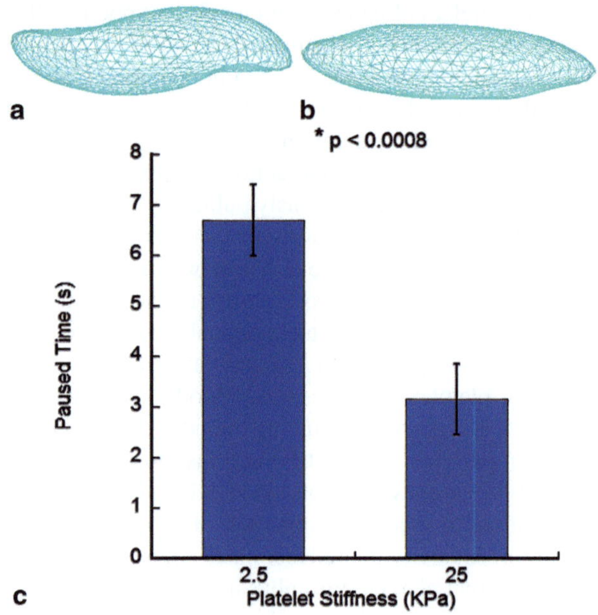

Fig. 5.4 Simulated deformations of platelet structure during adhesion to the vessel wall for platelets stiffnesses of **a** 2.5 kPa and **b** 25 kPa. **c** The effect of platelet membrane stiffness on the platelet pause time. (Originally published in Ref [86] Open Access: http://rsta.royalsocietypublishing.org/content/372/2021/20130380.long)

Recently, a multi-scale model was presented in Flamm, Diamond et al. [64, 65, 87], which included four components: the fluid flow (using a lattice Boltzmann method), the transport of soluble substances (using convection–diffusion–reaction equations), motion and binding of platelets leading to their deposition (using a lattice kinetic Monte Carlo algorithm), and the activation state of each platelet (using a neural network for cellular signaling). A pairwise agonist scanning approach had been found to allow handling of large datasets of measured calcium mobilization to predict an individual's platelet responses to pairwise combinations of ADP (which activates P2Y1/P2Y12 receptors), U46619 (which has properties similar to Thromboxane A2), and convulxin (which activates GPVI receptor to collagen). A neural network for cellular signaling was used to predict patient-specific responses to drugs. The resulting simulations were compared with experimental results in a system using blood flowing on collagen in microfluidic devices at different shear rates. The simulations were used to predict the individual's drug sensitivity to cyclooxygenase (COX) inhibitors and P2Y1 receptor antagonists in three different blood donors.

Perspective and Conclusions

Increasing and organizing our knowledge on platelet–vessel wall interactions and combining it with novel multi-scale computational models to test new biological hypothesis will help devise treatments for human disease where excessive thrombosis occurs, while attempting to minimize the risk of bleeding. Moreover, platelets and vessels are not only implicated in "pure" thrombotic states and inflammatory disease but also in other pathological processes such as cancer and metastases. Furthermore, how both platelets and blood vessels may contribute to tumor growth and tumor dissemination can be by itself a part of a systems approach to cancer biology. Relying only on in vivo studies is impractical and time consuming. Genetic knockout animal models have their own limits, as nonhuman animals' biology differs from human biology in many aspects, and a complete functional knockout might not yield information on certain conditions, where a dosage effect or a mutated state might be the actual determinant pathogenesis. While hypothesis-based experiments are being performed, concomitantly developing and refining multi-scale models and running simulations on powerful computer clusters will enable biomedical community to accelerate testing of new therapeutic targets. Systems biology is thus becoming a novel empowering tool to devise new less toxic treatments more efficiently and economically.

Acknowledgments Research of Mark Alber and Oleg Kim reported in this publication was supported by NIH U01HL116330, Yolande Chen by an American Heart Association Post-Doctoral Fellowship, and Seth Corey by NIH R21HL106462.

References

1. Wagner DD, Frenette PS. The vessel wall and its interactions. Blood. 2008;111(11):5271–81.
2. Jackson SP. Arterial thrombosis-insidious, unpredictable and deadly. Nat Med. 2011;17(11):1423–36.
3. Michelson AD. Antiplatelet therapies for the treatment of cardiovascular disease. Nat Rev Drug Discov. 2010;9(2):154–69.
4. Labelle M, Begum S, Hynes RO. Direct signaling between platelets and cancer cells induces an epithelial-mesenchymal-like transition and promotes metastasis. Cancer Cell. 2011;20(5):576–90.
5. Rondina MT, Weyrich AS, Zimmerman GA. Platelets as cellular effectors of inflammation in vascular diseases. Circ Res. 2013;112(11):1506–19.
6. Boilard E, et al. Platelets amplify inflammation in arthritis via collagen-dependent microparticle production. Science. 2010;327(5965):580–3.
7. Kauskot A, Hoylaerts MF. Platelet receptors. Handb Exp Pharmacol. 2012;210:23–57.
8. Falet H, et al. A novel interaction between FlnA and Syk regulates platelet ITAM-mediated receptor signaling and function. J Exp Med. 2010;207(9):1967–79.
9. Feng S, et al. Filamin A binding to the cytoplasmic tail of glycoprotein Ibalpha regulates von Willebrand factor-induced platelet activation. Blood. 2003;102(6):2122–9.
10. Nurden P, et al. Thrombocytopenia resulting from mutations in filamin A can be expressed as an isolated syndrome. Blood. 2011;118(22):5928–37.
11. Berrou E, et al. Heterogeneity of platelet functional alterations in patients with filamin A mutations. Arterioscler Thromb Vasc Biol. 2013;33(1):e11–8.
12. Versteeg HH, et al. New fundamentals in hemostasis. Physiol Rev. 2013;93(1):327–58.
13. Gachet C P2Y(12) receptors in platelets and other hematopoietic and non-hematopoietic cells. Purinergic Signal. 2012;8(3):609–19.
14. Schaff M, et al. Integrin alpha6beta1 is the main receptor for vascular laminins and plays a role in platelet adhesion, activation, and arterial thrombosis. Circulation. 2013;128(5):541–52.
15. Borst O, et al. The inflammatory chemokine CXC motif ligand 16 triggers platelet activation and adhesion via CXC motif receptor 6-dependent phosphatidylinositide 3-kinase/Akt signaling. Circ Res. 2012;111(10):1297–307.

16. Prevost N, et al. Signaling by ephrinB1 and Eph kinases in platelets promotes Rap1 activation, platelet adhesion, and aggregation via effector pathways that do not require phosphorylation of ephrinB1. Blood. 2004;103(4):1348–55.
17. Zhu L, et al. Regulated surface expression and shedding support a dual role for semaphorin 4D in platelet responses to vascular injury. Proc Natl Acad Sci U S A. 2007;104(5):1621–6.
18. Angelillo-Scherrer A, et al. Role of Gas6 receptors in platelet signaling during thrombus stabilization and implications for antithrombotic therapy. J Clin Invest. 2005;115(2):237–46.
19. Vaiyapuri S, et al. Gap junctions and connexin hemichannels underpin hemostasis and thrombosis. Circulation. 2012;125(20):2479–91.
20. Senis YA. Protein-tyrosine phosphatases: a new frontier in platelet signal transduction. J Thromb Haemost. 2013;11(10):1800–13.
21. Stalker TJ, et al. Platelet signaling. Handb Exp Pharmacol. 2012;210:59–85.
22. Italiano JE Jr, et al. Angiogenesis is regulated by a novel mechanism: pro- and antiangiogenic proteins are organized into separate platelet alpha granules and differentially released. Blood. 2008;111(3):1227–33.
23. Italiano JE Jr, Battinelli EM. Selective sorting of alpha-granule proteins. J Thromb Haemost. 2009;7(Suppl 1):173–6.
24. Battinelli EM, Markens BA, Italiano JE Jr. Release of angiogenesis regulatory proteins from platelet alpha granules: modulation of physiologic and pathologic angiogenesis. Blood. 2011;118(5):1359–69.
25. Battinelli EM, et al. Anticoagulation inhibits tumor cell-mediated release of platelet angiogenic proteins and diminishes platelet angiogenic response. Blood. 2014, 123(1):101–12.
26. Thery C, Ostrowski M, Segura E. Membrane vesicles as conveyors of immune responses. Nat Rev Immunol. 2009;9(8):581–93.
27. Raposo G, Stoorvogel W. Extracellular vesicles: exosomes, microvesicles, and friends. J Cell Biol. 2013;200(4):373–83.
28. Mause SF. Platelet microparticles: reinforcing the hegemony of platelets in atherothrombosis. Thromb Haemost. 2013;109(1):5–6.
29. Heijnen HF, et al. Activated platelets release two types of membrane vesicles: microvesicles by surface shedding and exosomes derived from exocytosis of multivesicular bodies and alpha-granules. Blood. 1999;94(11):3791–9.
30. von Bruhl ML, et al. Monocytes, neutrophils, and platelets cooperate to initiate and propagate venous thrombosis in mice in vivo. J Exp Med. 2012;209(4):819–35.
31. Darbousset R, et al. Tissue factor-positive neutrophils bind to injured endothelial wall and initiate thrombus formation. Blood. 2012;120(10):2133–43.
32. Engelmann B, Massberg S. Thrombosis as an intravascular effector of innate immunity. Nat Rev Immunol. 2013;13(1):34–45.
33. Fuchs TA, et al. Extracellular DNA traps promote thrombosis. Proc Natl Acad Sci U S A. 2010;107(36):15880–5.
34. Demers M, et al. Cancers predispose neutrophils to release extracellular DNA traps that contribute to cancer-associated thrombosis. Proc Natl Acad Sci U S A. 2012;109(32):13076–81.
35. Chen K, et al. Endocytosis of soluble immune complexes leads to their clearance by FcgammaRIIIB but induces neutrophil extracellular traps via FcgammaRIIA in vivo. Blood. 2012;120(22):4421–31.
36. Duerschmied D, et al. Platelet serotonin promotes the recruitment of neutrophils to sites of acute inflammation in mice. Blood. 2013;121(6):1008–15.
37. Martinod K, et al. Neutrophil histone modification by peptidylarginine deiminase 4 is critical for deep vein thrombosis in mice. Proc Natl Acad Sci U S A. 2013;110(21):8674–9.
38. Torisu T, et al. Autophagy regulates endothelial cell processing, maturation and secretion of von Willebrand factor. Nat Med. 2013;19(10):1281–7.
39. Li Z, et al. A stimulatory role for cGMP-dependent protein kinase in platelet activation. Cell. 2003;112(1):77–86.
40. Zhang G, et al. Biphasic roles for soluble guanylyl cyclase (sGC) in platelet activation. Blood. 2011;118(13):3670–9.

41. Gambaryan S, Friebe A, Walter U Does the NO/sGC/cGMP/PKG pathway play a stimulatory role in platelets? Blood. 2012;119(22):5335–6; author reply 5336–7.
42. Tsikas D, et al. Extra-platelet NO and NO(+)-containing drugs are potent inhibitors of platelet aggregation in humans by cGMP-dependent and cGMP-independent mechanisms. Blood. 2012;119(22):5337–9; author reply 5339.
43. Sylman JL, et al. Transport limitations of nitric oxide inhibition of platelet aggregation under flow. Ann Biomed Eng. 2013;41(10):2193–205.
44. Schulz C, Massberg S Platelets in atherosclerosis and thrombosis. Handb Exp Pharmacol. 2012;210:111–33.
45. Lam WA, et al. Mechanics and contraction dynamics of single platelets and implications for clot stiffening. Nat Mater. 2011;10(1):61–6.
46. Kita A, et al. Microenvironmental geometry guides platelet adhesion and spreading: a quantitative analysis at the single cell level. PLoS ONE. 2011;6(10):e26437.
47. Cines DB, et al. Clot contraction: compression of erythrocytes into tightly packed polyhedra and redistribution of platelets and fibrin. Blood. 2014;123(10):1596–603.
48. Westein E, et al. Atherosclerotic geometries exacerbate pathological thrombus formation poststenosis in a von Willebrand factor-dependent manner. Proc Natl Acad Sci U S A. 2013;110(4):1357–62.
49. Schumacher D, et al. Platelet-derived nucleotides promote tumor-cell transendothelial migration and metastasis via P2Y2 receptor. Cancer Cell. 2013;24(1):130–7.
50. Macaulay IC, et al. Platelet genomics and proteomics in human health and disease. J Clin Invest. 2005;115(12):3370–7.
51. Mason KD, et al. Programmed anuclear cell death delimits platelet life span. Cell. 2007;128(6):1173–86.
52. Gieger C, et al. New gene functions in megakaryopoiesis and platelet formation. Nature. 2011;480(7376):201–8.
53. Jones CI, et al. A functional genomics approach reveals novel quantitative trait loci associated with platelet signaling pathways. Blood. 2009;114(7):1405–16.
54. Rowley JW, et al. Genome-wide RNA-seq analysis of human and mouse platelet transcriptomes. Blood. 2011;118(14):e101–11.
55. Burkhart JM, et al. The first comprehensive and quantitative analysis of human platelet protein composition allows the comparative analysis of structural and functional pathways. Blood. 2012;120(15):e73–e82.
56. Dittrich M, et al. Platelet protein interactions: map, signaling components, and phosphorylation groundstate. Arterioscler Thromb Vasc Biol. 2008;28(7):1326–31.
57. Coppinger JA, et al. Characterization of the proteins released from activated platelets leads to localization of novel platelet proteins in human atherosclerotic lesions. Blood. 2004;103(6):2096–104.
58. Pena E, et al. Proteomic signature of thrombin-activated platelets after in vivo nitric oxide-donor treatment: coordinated inhibition of signaling (phosphatidylinositol 3-kinase-gamma, 14-3-3zeta, and growth factor receptor-bound protein 2) and cytoskeleton protein translocation. Arterioscler Thromb Vasc Biol. 2011;31(11):2560–9.
59. Dittrich M, et al. Understanding platelets. Lessons from proteomics, genomics and promises from network analysis. Thromb Haemost. 2005;94(5):916–25.
60. Boyanova D, et al. PlateletWeb: a systems biologic analysis of signaling networks in human platelets. Blood. 2012;119(3):e22–e34.
61. Dittrich M, et al. Characterization of a novel interaction between vasodilator-stimulated phosphoprotein and Abelson interactor 1 in human platelets: a concerted computational and experimental approach. Arterioscler Thromb Vasc Biol. 2010;30(4):843–50.
62. Jupe S, et al. Reactome-a curated knowledgebase of biological pathways: megakaryocytes and platelets. J Thromb Haemost. 2012;10(11):2399–402
63. Xu Z, et al. Computational approaches to studying thrombus development. Arterioscler Thromb Vasc Biol. 2011;31(3):500–5.

64. Flamm MH, et al. Multiscale prediction of patient-specific platelet function under flow. Blood. 2012;120(1):190–8.
65. Flamm MH, Diamond SL. Multiscale systems biology and physics of thrombosis under flow. Ann Biomed Eng. 2012;40(11):2355–64.
66. Stalker TJ, et al. Hierarchical organization in the hemostatic response and its relationship to the platelet-signaling network. Blood. 2013;121(10):1875–85.
67. Skorczewski T, Erickson LC, Fogelson AL. Platelet motion near a vessel wall or thrombus surface in two-dimensional whole blood simulations. Biophys J. 2013;104(8):1764–72.
68. Voronov RS, et al. Simulation of intrathrombus fluid and solute transport using in vivo clot structures with single platelet resolution. Ann Biomed Eng. 2013;41(6):1297–307.
69. Kim OV, et al. Fibrin networks regulate protein transport during thrombus development. PLoS Comput Biol. 2013;9(6):e1003095.
70. Leiderman K, Fogelson A. An overview of mathematical modeling of thrombus formation under flow. Thromb Res. 2014;133(Suppl 1):S12–4.
71. Pivkin IV, Richardson PD, Karniadakis G. Blood flow velocity effects and role of activation delay time on growth and form of platelet thrombi. Proc Natl Acad Sci U S A. 2006;103(46):17164–9.
72. Begent N, Born GV. Growth rate in vivo of platelet thrombi, produced by iontophoresis of ADP, as a function of mean blood flow velocity. Nature. 1970;227(5261):926–30.
73. Fogelson AL, Guy RD. Immersed-boundary-type models of intravascular platelet aggregation Comput Methods. Appl Mech Eng. 2008;197:2087–104.
74. Mody NA, King MR. Platelet adhesive dynamics. Part I: characterization of platelet hydrodynamic collisions and wall effects. Biophys J. 2008;95(5):2539–55.
75. Huang PY, Hellums JD. Aggregation and disaggregation kinetics of human blood platelets: part III. The disaggregation under shear stress of platelet aggregates. Biophys J. 1993;65(1):354–61.
76. Huang PY, Hellums JD. Aggregation and disaggregation kinetics of human blood platelets: part II. Shear-induced platelet aggregation. Biophys J. 1993;65(1):344–53.
77. Huang PY, Hellums JD. Aggregation and disaggregation kinetics of human blood platelets: part I. Development and validation of a population balance method. Biophys J. 1993;65(1):334–43.
78. Leiderman K, Fogelson AL. Grow with the flow: a spatial-temporal model of platelet deposition and blood coagulation under flow. Math Med Biol. 2011;28(1):47–84.
79. Kuharsky AL, Fogelson AL. Surface-mediated control of blood coagulation: the role of binding site densities and platelet deposition. Biophys J. 2001;80(3):1050–74.
80. Xu Z, et al. A multiscale model of thrombus development. J R Soc Interface. 2008;5(24):705–22.
81. Xu Z, et al. A multiscale model of venous thrombus formation with surface-mediated control of blood coagulation cascade. Biophys J. 2010;98(9):1723–32.
82. Xu Z, et al. Multiscale model of fibrin accumulation on the blood clot surface and platelet dynamics. Methods Cell Biol. 2012;110:367–88.
83. Xu Z, et al. Multiscale models of thrombogenesis. Wiley Interdiscip Rev Syst Biol Med. 2012;4(3):237–46.
84. Graner F, Glazier JA. Simulation of biological cell sorting using a two-dimensional extended Potts model. Phys Rev Lett. 1992;69(13):2013–6.
85. Stalker TJ, et al. A systems approach to hemostasis: 3. Thrombus consolidation regulates intrathrombus solute transport and local thrombin activity. Blood. 2014;124(11):1824–31.
86. Wu Z, et al. Three-dimensional multi-scale model of deformable platelets adhesion to vessel wall in blood flow. Phil Trans R Soc A. 2014;372(2021):1–23.
87. Diamond SL, et al. Systems biology of platelet-vessel wall interactions. Front Physiol. 2013;4:229.

Chapter 6
Systems Approach to Phagocyte Production and Activation: Neutrophils and Monocytes

Hrishikesh M. Mehta, Taly Glaubach and Seth Joel Corey

Abstract Granulocyte differentiation and immune response function is a dynamic process governed by a highly coordinated transcriptional program that regulates cellular fate and function, often in a context-dependent manner. Advances in high-throughput technologies and bioinformatics have allowed us to better understand complex biological processes at the genomic and proteomic levels. Components of the environmental milieu, along with the molecular mechanisms that drive the development, activation, and regulation of granulocytes, have since been elucidated. In this chapter, we present the intricate network in which these elements come together and influence one another. In particular, we describe the critical roles of transcription factors like PU.1, CCAAT/enhancer-binding protein (C/EBPα; alpha), C/EBPε (epsilon), and growth factor independent-1 (Gfi-1). We also review granulocyte colony-stimulating factor (G-CSF) receptor-induced signal transduction pathways, their influence on proliferation and differentiation, and the cooperativity of cytokines and chemokines in this process.

Keywords Systems biology · Phagocyte · Granulocyte · Neutrophil · Macrophage · Transcription factor · Granulopoiesis · Chemotaxis · Phagocytosis · Apoptosis

Phagocytes constitute the primary line of host defense through the highly coordinated process of chemotaxis; ingestion of microbes, particles, and cells; and production

S. J. Corey (✉)
Departments of Pediatrics and Cell & Molecular Biology, Northwestern University Feinberg School of Medicine and Lurie Children's Hospital of Chicago, 303 E. Superior Street, Lurie 5-107, Chicago, IL 60611, USA
e-mail: s-corey@northwestern.edu

H. M. Mehta
Feinberg School of Medicine, Northwestern University, 303 E. Superior Street, Lurie 5-220, Chicago, IL 60611, USA
e-mail: hrishi89@hotmail.com

T. Glaubach
Department of Pediatrics, Lurie Children's Hospital of Chicago, Chicago, IL 60611, USA
Tel: 312-227-4000
e-mail: tglaubach@luriechildrens.org

S. J. Corey et al. (eds.), *A Systems Biology Approach to Blood*,
Advances in Experimental Medicine and Biology 844, DOI 10.1007/978-1-4939-2095-2_6

and secretion of peptides, lipids, and reactive oxygen species (ROS). This biological function has been critical for the success of multicellular organisms and is found in elementary forms such as dichtyostelium. Phagocytes provide the cornerstone of the innate immune system. Unlike the adaptive immune response that requires prior microbe exposure and time to develop specific antigen recognition, phagocytes are critical for the rapid, nonspecific targeting and elimination of infectious pathogens. These events involve complex interactions between the host, pathogen recognition, and phagocytic effector cells that must be tightly regulated. Efficient innate immune responses must be balanced against prevention of unabated inflammation linked to autoimmune and inflammatory disease states. For example, the recognition and subsequent phagocytosis of apoptotic neutrophils by macrophages is a key homeostatic event in the resolution of inflammation. The components and molecular mechanisms governing the development, activation, and regulation of these cells have been established, and they present the basis for this chapter and future work with a systems analysis.

From a common myeloid progenitor (CMP) cell, phagocytic cells develop into highly specialized cells. Granulocytes, also known as polymorphonuclear leukocytes (PMNs), consist of neutrophils, basophils, and eosinophils. Of these, the neutrophil is the most predominant circulating leukocyte in humans, whereas lymphocytes predominate in mice. Peripheral blood monocytes undergo a process of activation and differentiation to become resident tissue macrophages. Both cell types are descended from a common hematopoietic progenitor cell, the colony-forming unit granulocyte/macrophage (CFU-GM).

Granulocyte production must be sufficient and dynamic to protect the host against infection, but not excessive as to cause chronic inflammation and tissue damage. Production begins at the earliest stage when the hematopoietic stem cell (HSC) is recruited from a pool of dormant HSCs (Figs. 6.1 and 6.2). A granulocyte requires 12 days in the bone marrow before it leaves to eventually reside in tissues. In peripheral blood, neutrophil levels are finely controlled ($2\text{–}8 \times 10^3/\text{mm}^3$) and their circulating half-life is brief (~ 6 h). Most of the bone marrow activity ($\sim 67\,\%$ of the cells belong to the myeloid, nonerythroid lineage) is directed toward a continuous, prodigious degree of neutrophil production ($\sim 5\text{–}10 \times 10^{10}$/day) [1]. Life-threatening sepsis occurs when absolute neutrophil counts are less than $0.5 \times 10^3/\text{mm}^3$. Granulocyte production can respond quickly to severe infection with a log-fold increase in circulating neutrophil counts ($\sim 3\text{–}5 \times 10^4/\text{mm}^3$) within 48–72 h. When healthy adult volunteers received a single dose of granulocyte colony-stimulating factor (GCSF), neutrophil counts increased rapidly, peaking at 12 h and returning to baseline by 48–72 h [2]. Bone marrow reserve is critical. The neonatal neutrophil bone marrow storage pool is decreased compared to the adult counterpart [3, 4]. Moreover, there is delayed neutrophil response to infection (3–4 h in the neonate compared to 30–90 min in the adult) [5]. Thus, neonates are especially susceptible to neutrophil exhaustion when stressed by severe infection (sepsis). Understanding neutrophil production and kinetics has led to improved survival in stressed neonates [6, 7].

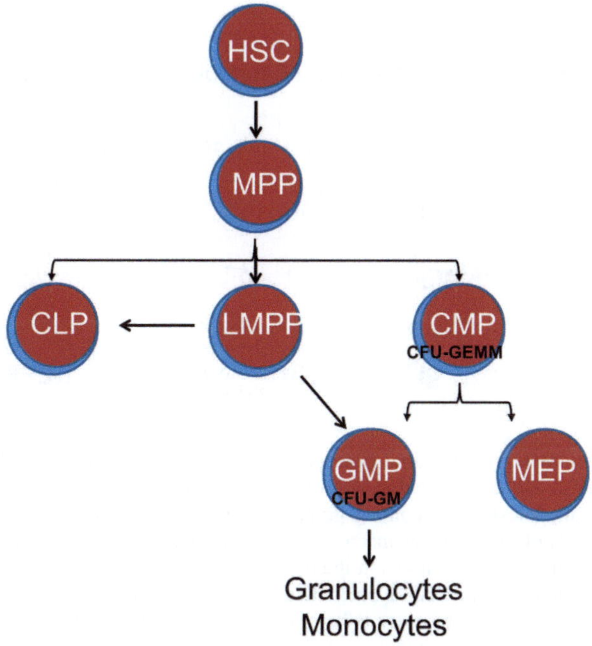

Fig. 6.1 Overview of derivation of granulocytes and monocytes from hematopoietic stem cells. Hematopoietic stem cells (*HSC*) give rise to multipotent progenitors (*MPP*) which can produce three lineage-committed progenitors, which are the classical common lymphoid progenitors (*CLP*) and common myeloid progenitor (*CMP*) and a nonclassical lymphoid/myeloid multipotent progenitor (*LMPP*). CMPs produce colony-forming units which consist of granulocytes erythrocytes, monocytes, and megakaryocytes (*CFU-GEMM*). Granulocyte and monocyte progenitors (*GMP*) and megakaryocyte and erythroid progenitors (*MEP*) descend from CMP. However, GMP can also be derived from LMPP. CFU-GM colony forming unit-granulocyte and monocyte

Granulopoiesis

Granulopoiesis is a complex process by which a CMP, under the stimulation of cytokines interleukin-3 (IL3), GCSF, and/or granulocyte macrophage colony-stimulating facto GMCSF, induces CFU-granulocytes erythrocytes, monocytes, and megakaryocytes (GEMM) to differentiate into CFU-GM, the common precursor for both neutrophils and monocytes. Myelopoiesis involves stem and progenitor cells that generate also megakaryocytes (and platelets) and erythrocytes (Fig 6.1). This hierarchy has been challenged and a newer paradigm has emerged to reflect evidence that the cells of both the innate (neutrophils and macrophages) and adaptive (T and B cells) immune system are derived from a lymphoid/myeloid multipotent progenitor (LMPP; Fig 6.1), which does not give rise to megakaryocytes or erythrocytes [8, 9]. Other cytokines involved in multipotential lineage progenitors include thrombopoietin and Flt3. The precise combination(s) of growth factors and stromal factors that lead to the production of a specific granulocyte remains poorly understood. The CFU-GM

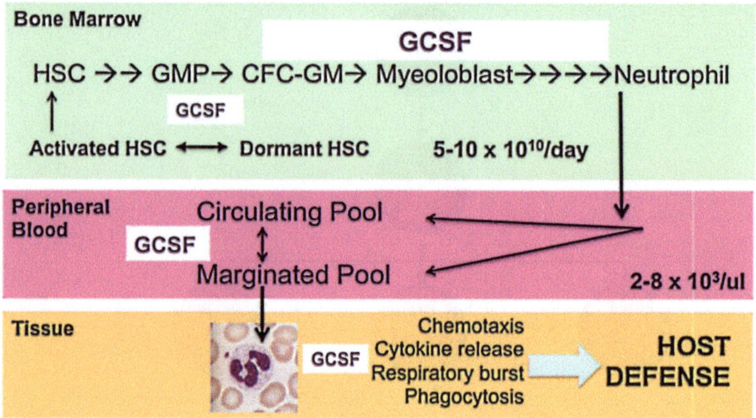

Fig. 6.2 Multiscale analysis of granulopoiesis. Granulocytes are produced in the bone marrow through self-renewal and differentiation of hematopoietic stem cells (*HSC*). These pluripotential HSC initiate granulopoiesis by becoming granulocyte/macrophage progenitor cells (*GMP*) and colony-forming cells of granulocytes/macrophages (*CFC-GM*). Soon, they become myeloblasts, which are easily identifiable precursors in the bone marrow. This prodigious amount of stem cell renewal and differentiation produces approximately 10 billion cells per day in each adult person. The number of circulating granulocytes is kept within a narrow range of 2000–8000 per μl. The bone marrow is able to respond quickly to infectious stimuli and amplify granulocyte numbers by several fold. The granulocytes leave the blood vessels and migrate (chemotaxis) to the tissues where they release cytokines, engulf microbes, and produce reactive oxygen species (*ROS*) through a respiratory burst involving the NADPH oxidase. Granulocyte colony-stimulating factor (*G-CSF*) is the primary cytokine responsible for stem cell expansion toward the granulocyte, inducer of differentiation of the myeloblasts, and enhancer of granulocyte function. *NADPH* nicotinamide adenine dinucleotide phosphate

stem cell differentiates into either a CFU-G or CFU-M stem cell. The development of mature granulocytes from hematopoietic precursor cells is controlled by a small number of transcription factors and complex gene regulatory networks, including those encoding growth factors and their receptors, enzymes, adhesion molecules, and transcription factors. In particular, PU.1, CCAAT/enhancer-binding protein (C/EBPα; alpha), C/EBPε (epsilon), and growth factor independent (Gfi)-1 have emerged as critical players, master regulators of myeloid development [10], and constitute a gene regulatory networking for granulopoiesis. A systematic study of the regulatory components and their complex interactions will enable higher-order understanding of how granulocytes are produced, how they modulate other immune responses and themselves, and how aberrations in these pathways lead to disease states.

GCSF is the most important hematopoietic growth factor that drives the production, proliferation, and differentiation of myeloid progenitor and precursor cells, beginning with the bone marrow HSC and terminating as a mature neutrophil released into the periphery. It also acts to enhance the survival and function of mature neutrophils by delaying apoptosis [1, 2] and acting cooperatively with other cytokines (e.g., IL-8 and tumor necrosis factor (TNF)) to activate or "prime" the neutrophil, in a dose-dependent manner [3, 4]. The clinical utility of GCSF is well evidenced by its use in the treatment and survival of patients with congenital and chemotherapy-induced neutropenias.

GCSF acts through its cognate receptor (GCSFR) via multiple signal transduction pathways, including the Janus kinase (JAK)/signal transducer and activator of transcription (STAT), Ras/mitogen-activated protein kinase (MAPK), and phosphatidylinositol 3-kinase (PI3K)/protein kinase B (Akt) pathways, described in detail below. Seven isoforms of GCSF receptor, termed class I through VII, have been identified by screening placental and myeloid leukemia complementary DNA (cDNA) libraries [11–14]. These isoforms result from the alternative spicing of the GCSFR mRNA. Although multiple isoforms of GCSFR were identified, only the class I and class IV isoforms appear to be important for granulopoiesis, as these were the only two isoforms validated to be expressed in normal and leukemic hematopoietic cells using quantitative PCR [15]. The class I isoform represents the wild type since it is the predominantly expressed GCSFR isoform. The class IV GCSFR is an alternatively spliced, truncated isoform lacking the distal 87 amino acids, which are replaced by a novel 34 amino acid sequence [13]. Quantitative analysis of class I and class IV isoform expression revealed very low relative levels of class IV in mature circulating neutrophils compared to class I [15]. However, in CD34+ cells from adult bone marrow, a higher ratio of class IV to I expression was observed, suggesting that class IV expression drops during differentiation and is therefore developmentally regulated [16]. Interestingly increased class IV to class I ratio was also observed in leukemic cell lines and patient samples from acute myeloid leukemia (AML) patients [15]. Functional studies of the class IV isoform demonstrated that the class IV GCSFR as differentiation defective, but able to promote proliferation [17]. The class IV isoform is similar in structure to a series of truncated GCSFR mutants resulting from somatic nonsense mutations identified in patients with severe congenital neutropenia that developed acute myelogenous leukemia [18–20]. Similar to the class IV GCSFR, these truncated forms of GCSFR also conferred a maturation arrest with enhanced proliferation [19, 20].

Ligand-induced dimerization of the GCSFR rapidly triggers downstream signal transduction pathways including JAK/STAT, Ras/MAPK, and Lyn-PI3K/Akt signal transduction pathways [21–29]. The proximal cytoplasmic domain of the GCSFR contains Box 1 and Box 2, which are conserved in the hematopoietic cytokine receptor superfamily. The distal domain also contains a di-leucine receptor internalization signal and four tyrosine residues (Y704, Y729, Y744, and Y764 in the human receptor sequence). The tyrosine residues can be phosphorylated and serve as docking sites for SH2-containing proteins. The truncated forms of GCSFR lack three of four tyrosine residues (Y729, Y744, Y769) in the distal domain, which strongly implicate them in promoting differentiation signaling. The distal domain also contains a di-leucine receptor internalization signal and is also a target of suppressor of cytokine signaling (SOCS) protein binding which serves to target the receptor for ubiquitination. Together, they contribute to signaling termination via receptor degradation [30]. Thus, the loss of the distal domain results in increased receptor signaling promoted by both reduced internalization and also increased recycling of internalized receptor to the surface. In the full-length class I GCSFR, cell signaling and receptor internalization synergize to promote granulopoiesis by coupling differentiation signaling with attenuation of long-term proliferation.

The GCSFR dimerization-induced JAK transautophosphorylation promotes activation of STAT5 and STAT3 protein phosphorylation which then translocate to the nucleus and promote gene expression. STAT5 is primarily implicated in promoting proliferation, whereas STAT3 is implicated in promoting both proliferation and differentiation. STAT3 can bind to multiple sites on the GCSFR, which is dependent on the GCSF dose, resulting in STAT3 activation. Activation of STAT3 promotes differentiation indirectly by inducing growth arrest, but does not induce the differentiation program, demonstrating the role of GCSFR signaling in maintenance of granulocyte precursors and cell fate determination. STAT3 induces cell cycle termination by inducing expression of feedback inhibitors such as SOCS3 that inhibits the JAK/STAT pathway and promotes signal termination by ubiquitin-mediated receptor degradation. STAT3 also promotes expression of $p27^{kip1}$, an inhibitor of cyclin-dependent kinases, and promoting cell cycle arrest and, perhaps, differentiation. Another important regulator of granulocyte maturation is SHP2. A protein tyrosine phosphatase, SHP2 favors granulopoiesis over monopoiesis by promoting the expression of the transcription factor C/EBPα [31–33]. One possible target of SHP2 is runt-related transcription factor 1 (RUNX1), a transcription factor that induces the expression of C/EBPα [34]. SHP2 also represents a nonconventional downstream target of JAK2 for the class IV isoform of GCSFR. Class IV-mediated proliferation was identified to be mediated by a nonclassical JAK2-SHP2 pathway as opposed to the classical JAK/STAT pathway [16].

The transcription factors PU.1 and globin transcription factor (GATA)-1 act to inhibit each other and are implicated in early determination of the MPP to either LMPP or megakaryocyte erythrocyte progenitor (MEP) [35, 36]. Higher PU.1 levels inhibit the activity of GATA-1, which inhibits the erythrocyte development and by default promotes LMPP generation [37]. Thus, PU.1 is deterministic during the early stage of granulocyte/monocyte formation, which depends on inactivation of GATA-1. Development along the LMPP lineage progresses to formation of CLP and granulocyte macrophage progenitor (GMP). At this stage of development, PU.1 and C/EBPα appear to act in concert to promote GMP formation over CLP. C/EBPα expression is observed in CMP and GMP but not in CLP and MEP [38] and thus its expression would direct the decision to generate GMP from LMPP. Along with C/EBPα expression, graded expression of PU.1 also determines lymphoid versus myeloid decision making in LMPP. High PU.1 expression is observed in macrophages and lower expression in B cells, which leads to PU.1 dose-dependent activation of signals that guide macrophage development whereas low PU.1 levels promote B cell development [39].

Determination of cell fate in GMP to form either macrophages or granulocytes is dependent on the interplay of C/EBPα and PU.1. High PU.1 expression promotes increased expression of Egr1,2 and Nab2, which collectively promote monopoiesis by both promoting expressing monocyte specific genes and also inhibit neutrophil-specific genes [10, 39]. An important determinant of neutrophil formation, Gfi-1, is inhibited by Egr2/Nab complex. However, Gfi-1 itself can repress Egrs in addition to repressing PU.1 expression [40]. Expression of Gfi-1 is promoted by C/EBPα. Thus, Gfi-1 and Egrs represent secondary regulators of late-stage granulocyte and monocyte lineage commitment by counteractive regulation of gene expression. The secondary regulators along with the master regulators help make cell fate decisions between granulocytes and monocyte lineages [40]. Gfi-1 and C/EBPε (epsilon) have been identified as regulators of terminal neutrophil differentiation.

Gene Expression Analysis of Neutrophil Development

System-level studies such as microarray and proteomic analysis using mass spectrometry have revealed distinct patterns of gene expression, defining the phagocytic precursor cell at different stages of differentiation. Microarray analysis indicated that neutrophil development can be divided into two segments: early and terminal differentiation. Gene expression analysis of early differentiation revealed expected upregulation of cell cycle proteins in the order HSC < MPP < CLP/CMP along with selective expression of myeloid genes in CMPs but not CLPs supporting existence of lineage specific genes and patterning [41]. Additionally, genes associated with HSCs were downregulated in MPP, CMP, and CLP. Multiple studies have been carried out in both microarray and proteomic scales to identify cellular differences during terminal differentiation. The milieu of genes and proteins identified has been characterized under functional groups to simplify their role during terminal differentiation.

Granule proteins are functional components of the terminally differentiated neutrophils that are released either into phagosomes or to the extracellular space. Granule proteins are stored in azurophilic, specific, and gelatinase granules, also classified as primary, secondary, and tertiary granules based on their sequential production [42]. Microarray-based analysis of morphologically defined stages in terminal neutrophil differentiation identified 16 new proteins which involved proteases, protease inhibitors, and signaling molecules [43]. Proteomic analysis of granule proteins was performed by subcellular fractionation of the three different types of granules and identified 286 proteins [44]. Additional proteomic studies have further identified proteins specific to granules or plasma membrane as well as their localization in lipid rafts [45–48].

Cell surface protein expression follows developmental patterns, which allows them to be used as cell surface markers for differentiation. Identification of cell surface proteins also provides the mechanisms by which phagocytic cells or their precursors can interact with the environment and define cell fate or help gauge changes in the extracellular environment. Microarray analysis showed low-level expression of receptors involved in inflammatory responses such as some IL, interferon, transforming growth factor, and chemokine receptors during early terminal differentiation stages. Expression of these receptors increase in terminally differentiated neutrophils [43]. An increase in GCSFR and GMCSFR was also observed along with a reduction in MCSFR expression. Thus, increased receptor expression profile shows priming of the cell to detect inflammatory responses.

Microarray gene expression patterns obtained from highly purified subsets of cells representing terminal differentiation stages of neutrophil differentiation demonstrated expected expression of cell cycle and apoptosis proteins [49]. A hallmark of differentiation is cell cycle arrest and downregulation of cell cycle proteins during an early terminal differentiation stage. The cell cycle promoting proteins cyclin-dependent kinases (cdk) 2, 4, and 6 and E2F target genes were upregulated. However, E2F expression was not downregulated, suggesting that other transcription factors were downregulating E2F targets. Inhibitor of cell cycle $p27^{kip1}$ was increased, a target of GCSF-mediated STAT3 and also under the control of C/EBPα (alpha) and C/EBPε (epsilon). Differential expression of apoptosis-related genes during terminal differentiation stages shows a difference in the mechanism of apoptosis at early and late stages, indicative of the cellular function. Early stages involve upregulation of p53-mediated apoptosis pathway, which surveys DNA damage, thus preventing proliferation of mutations. However, gene expression profile at the nonproliferative stage of differentiation showed an upregulation in ligand–receptor-mediated apoptosis pathway genes, with a concurrent reduction in p53-induced apoptotic genes. Thus, the profile suggests apoptosis in neutrophils is mediated upon activation of the neutrophil and in response to inflammatory cytokines.

Monocytopoiesis

Monocytes and macrophages are important effectors of the innate immune response and inflammation. Monopoiesis proceeds from the monoblast in the bone marrow to the circulating monocyte in the periphery, and eventually matures without proliferation to the tissue macrophage. However, monocyte development may be less linear than classically understood, and not necessarily a mere developmental intermediate between bone marrow precursors and tissue macrophages. Evidence suggests the possibility of a macrophage dendritic cell progenitor (MDP) and additionally shows that some subsets of both dendritic cells and tissue macrophages do not originate from monocytes in a steady state. Furthermore, monocytes may carry out specific effector functions during inflammation without further differentiation to macrophage or dendritic cell [50–53]. However, this dichotomy of thought between the classical hierarchy of monopoiesis and the novel existence of an MDP bone marrow progenitor has yet to be resolved.

GMCSF (or CSF2) is different from GCSF in that it acts on all granulocytes, monocytes, and macrophages. Because of this broad activity and the vast expression of the GMCSF receptor on hematopoietic cells, it was originally thought that the action of GMCSF was critical to the regulation and maintenance of the granulocyte and monocyte populations. However, deletion of neither the gene for GMCSF nor the GMCSF receptor had a significant impact on myelopoiesis but revealed an unexpected role for GMCSF in pulmonary homeostasis [54–56] 56. Additional evidence suggests GMCSF plays a vital role in stress or emergency myelopoiesis, with resultant increased production of granulocytes and monocytes in the bone marrow and stimulation of their survival and function in the tissues where they are recruited [57, 58].

Function of Neutrophils and Monocytes

Monocytes circulate in the bone marrow, blood, and spleen and are thought not to proliferate at steady state. In the setting of infection, monocytes are released from the bone marrow into the peripheral blood and migrate to sites of inflammation or injury where they mature to express distinct effector phenotypes [59]. A large portion of undifferentiated monocytes are also contained in the spleen, and serve as a storage reservoir for additional rapid deployment to sites of injury or infection [50]. Monocytes express chemokine receptors and pathogen associated pattern recognition receptors (e.g., toll-like receptors; TLRs) that mediate this process. Migration to tissues and further differentiation to inflammatory macrophages or dendritic cells is likely determined by the inflammatory milieu and the nature of the invading pathogen and TLR [59].

Macrophages are resident tissue phagocytes important for maintenance of tissue health via the clearance of apoptotic cells and other debris. Like their monocyte predecessors, macrophages also express a wide range of pattern recognition receptors that make them efficient effectors of the innate immune response in addition to their role in tissue homeostasis [60]. However, macrophages also play a vital role in initiating the adaptive immune response as antigen-presenting cells via MHC II. The developmental origin and more detailed function of tissue macrophage subsets remain poorly understood.

Neutrophils and monocytes released from the bone marrow can circulate for 24 h and 1–3 days respectively. The process of recruitment of neutrophils and monocytes involves recruitment by chemoattractants, IL-8, and bacterial proteins (N-formylmethionyl-leucyl-phenylalanine (fMLF), peptidoglycans). The delivery of neutrophils to the site starts off with "rolling" along the blood vessel walls and is mediated by low-affinity interactions between the selectin family of proteins [61, 62]. L-selectins are expressed on neutrophils which interact with transient and sequentially expressed P- and E-selectins on the inflammatory endothelial cells. Interaction between the selectins is followed by interaction of β2 (beta2)-integrins on neutrophils and intercellular adhesion molecule (ICAM) 1 and ICAM2 on endothelial cell wall (tethering). Integrin-binding affinity is increased in neutrophils upon activation by chemokines, which results in opening up of the integrin receptor conformation and of the ligand-binding pocket [63]. Increased affinity of the integrins brings the rolling of neutrophils to a stop, followed by transmigration across the vascular wall to the tissues. Microarray analysis of neutrophils exposed to fMLF showed an increase in pro-inflammatory molecules such as IL-8, TNF, IL1B, and both CXC and CC type chemokine [64]. Increased expression of pro-inflammatory cytokines contributes to delaying of apoptosis, which is essential for neutrophil function. In support, downregulation of apoptotic proteins was also observed. Additionally, in agreement with other studies, an upregulation of cytoskeletal reorganization proteins and adhesion-mediating molecules was observed [64, 65].

Priming of neutrophils is a process that activates the neutrophil and increases expression of proteins that are required for increased activity and also delay apoptosis.

Priming agents include GCSF, GMCSF, IL-8, bacterial lipopolysaccharides (LPS) and TNF-α (alpha) [66–69]. Priming with GCSF enhances chemotaxis and mobilization of neutrophils to the site of injury, whereas GMCSF is involved in promoting a more robust response that is involved in both delaying apoptosis and increasing the bactericidal activity of neutrophils by promoting expression of antiapoptotic proteins and cell surface receptors involved in recognition of antigens [70]. A role in antigen presentation was also evident from the increased expression of major histocompatibility complex II (MHC II). Priming with LPS enhances the bactericidal activity, with recruitment of components necessary for the assembly of nicotinamide adenine dinucleotide phosphate (NADPH) oxidase complex. In addition, priming with LPS also induced expression of proteins required for the NF-κB pathway [64, 71].

Phagocytosis is the penultimate step in neutrophil and macrophage function, wherein they ingest to get rid of the invading bacteria or apoptotic cells. The process of phagocytosis involves recognition of the bacteria or antigen via opsonization with antibodies or complement that is recognized by receptors on neutrophils. Opsonized bacteria are recognized by receptors against the Fc region of the antibody. Bacteria opsonized by complement bind are then able to bind to receptors like CD11b/CD18 on the neutrophil surface. Binding of opsonized bacteria to the activated neutrophil surface initiates changes in cytoskeleton and membrane to promote ingestion of the organism as a phagosome. Finally, the phagosome after sequential integration with the neutrophil granules turns into a phagolysosome and the invading organism is killed by exposure to products derived from ROS and antimicrobial granule proteins such as proteases, gelatinase, peroxidase, and other degradative enzymes.

Transcriptome analysis of phagocytosis identified expression of several hundred messenger RNAs (mRNAs) within 2 h of exposure [72, 73]. The changes in expression are divided into an early response and late response. Early response included cytokines and chemokines that act as pro-inflammatory molecules and aid in further recruitment of monocytes and neutrophils. Late-stage transcriptional changes involve upregulation of proapoptotic proteins of the receptor-mediated apoptotic pathway, such as TNFα, TNFR1, and tumor necrosis factor related apoptosis inducing ligand receptor (TRAILR). Other changes include proteins that are involved in the signal transduction pathway that involve TLRs. The downregulation of proteins that are involved in antibody- and complement-opsonized microbe recognition parallels the apoptotic expression.

ROS play a very important role in phagocytosis-mediated killing of bacteria by neutrophils. The production of ROS in neutrophils is mediated by an enzyme complex NADPH oxidase, which is composed of several components: oxidase specific ($p22^{phox}$, $p47^{phox}$, $p67^{phox}$, and $gp91^{phox}$) and guanosine triphosphate (GTPase; Rac1/2) [74]. Components of NADPH oxidase are present either in the cytoplasm or in either the plasma membrane or secretory vesicle membrane. Priming with GMCSF, TNFα, and LPS triggers phosphorylation of oxidase components and recruitment of the cytosolic components to the phagocyte membrane. Assembled NADPH oxidase mediates transfer of electrons from extracellular NADPH to oxygen within the phagosome, resulting in formation of a superoxide anion [75]. The superoxide anion dismutates to form hydrogen peroxide, which then oxidizes chloride anion to form hypochlorous acid. The reaction is catalyzed by myeloperoxidase, which resides in azurophilic granules and is released into the phagosome upon degranulation. Other products formed by hydrogen peroxide-mediated oxidation include hydroxyl radical. Together they have strong microbicidal activity.

Neutrophils contain highly toxic components used to kill microbes, but these molecules do not differentiate between host and pathogen. Neutrophils undergo apoptosis 24 h after they leave the bone marrow. The transcriptome analysis of neutrophils in the bone marrow and in peripheral blood show an upregulation of proapoptotic genes which indicates that neutrophils are destined to die as soon as they differentiate [43]. Priming of neutrophils delays the apoptotic response by upregulating antiapoptotic genes; however, upon phagocytosis, the transcriptional program now directs apoptosis of the neutrophils, termed as delayed apoptosis. The delayed apoptosis program is mediated by death receptors and is accompanied by a decreased inflammatory response. The decreased inflammatory response and mediation of apoptosis promotes resolution of the immune response and removal of the infection by macrophages. A downregulation of NF-κB expression is observed which halts the antiapoptotic response [76].

References

1. Summers C, Rankin SM, Condliffe AM, Singh N, Peters AM, Chilvers ER. Neutrophil kinetics in health and disease. Trends Immunol. 2010;31:318–24.
2. Kerst JM, de Haas M, van der Schoot CE, Slaper-Cortenbach IC, Kleijer M, von dem Borne AE, van Oers RH. Recombinant granulocyte colony-stimulating factor administration to healthy volunteers: induction of immunophenotypically and functionally altered neutrophils via an effect on myeloid progenitor cells. Blood. 1993;82:3265–72.
3. Ohls RK, Li Y, Abdel-Mageed A, Buchanan G Jr, Mandell L, Christensen RD. Neutrophil pool sizes and granulocyte colony-stimulating factor production in human mid-trimester fetuses. Pediatric Res. 1995;37:806–11.
4. Erdman SH, Christensen RD, Bradley PP, Rothstein G. Supply and release of storage neutrophils. A developmental study. Biol Neonate. 1982;41:132–7.
5. al-Mulla ZS, Christensen RD. Neutropenia in the neonate. Clin Perinatol. 1995;22:711–39.
6. Cairo MS, Christensen R, Sender LS, Ellis R, Rosenthal J, van de Ven C, Worcester C, Agosti JM. Results of a phase I/II trial of recombinant human granulocyte-macrophage colony-stimulating factor in very low birthweight neonates: significant induction of circulatory neutrophils, monocytes, platelets, and bone marrow neutrophils. Blood. 1995;86:2509–15.
7. Uguz A, Coskun M, Yuzbey S, Kizilors A, Karadogan I, Gura A, Yoldas B, Oygur N, Yegin O. Apoptosis of cord blood neutrophils and their response to colony-stimulating factors. Am J Perinatol. 2002;19:427–34.
8. McKercher SR, Torbett BE, Anderson KL, Henkel GW, Vestal DJ, Baribault H, Klemsz M, Feeney AJ, Wu GE, Paige CJ, et al. Targeted disruption of the PU.1 gene results in multiple hematopoietic abnormalities. EMBO J. 1996;15:5647–8.
9. Scott EW, Simon MC, Anastasi J, Singh H. Requirement of transcription factor PU.1 in the development of multiple hematopoietic lineages. Science. 1994;265:1573–7.
10. Dahl R, Walsh JC, Lancki D, Laslo P, Iyer SR, Singh H, Simon MC. Regulation of macrophage and neutrophil cell fates by the PU.1:C/EBP[alpha] ratio and granulocyte colony-stimulating factor. Nat Immunol. 2003;4:1029–36.
11. Fukunaga R, Seto Y, Mizushima S, Nagata S. Three different mRNAs encoding human granulocyte colony-stimulating factor receptor. Proc Natl Acad Sci U S A. 1990;87:8702–6.

12. Iwasaki H, Shimoda K, Okamura S, Otsuka T, Nagafuji K, Harada N, Ohno Y, Miyamoto T, Akashi K, Harada M, et al. Production of Soluble Granulocyte Colony-Stimulating Factor Receptors from Myelomonocytic Cells. J Immunol. 1999;163:6907–11.
13. Larsen A, Davis T, Curtis BM, Gimpel S, Sims JE, Cosman D, Park L, Sorensen E, March CJ, Smith CA. Expression cloning of a human granulocyte colony-stimulating factor receptor: a structural mosaic of hematopoietin receptor, immunoglobulin, and fibronectin domains. J Exp Med. 1990;172:1559–70.
14. Bernard T, Gale RE, Linch DC. Analysis of granulocyte colony stimulating factor receptor isoforms, polymorphisms and mutations in normal haemopoietic cells and acute myeloid leukaemia blasts. Br J Haematol. 1996;93:527–33.
15. White SM, Ball ED, Ehmann WC, Rao AS, Tweardy DJ. Increased expression of the differentiation-defective granulocyte colony-stimulating factor receptor mRNA isoform in acute myelogenous leukemia. Leukemia. 1998;12:899–906.
16. Mehta HM, Futami M, Glaubach T, Lee DW, Andolina JR, Yang Q, Whichard Z, Quinn M, Lu HF, Kao WM, et al. Alternatively spliced, truncated gcsf receptor promotes leukemogenic properties and sensitivity to jak inhibition. Leukemia. 2014;28:1041–51. doi:10.1038/leu.2013.321.
17. White SM, Alarcon MH, Tweardy DJ. Inhibition of granulocyte colony-stimulating factor-mediated myeloid maturation by low level expression of the differentiation-defective class IV granulocyte colony-stimulating factor receptor isoform. Blood. 2000;95:3335–40.
18. Dong F, Brynes RK, Tidow N, Welte K, Lowenberg B, Touw IP. Mutations in the gene for the granulocyte colony-stimulating-factor receptor in patients with acute myeloid leukemia preceded by severe congenital neutropenia. N Engl J Med. 1995;333:487–93.
19. Dong F, Hoefsloot LH, Schelen AM, Broeders CA, Meijer Y, Veerman AJ, Touw IP, Lowenberg B. Identification of a nonsense mutation in the granulocyte-colony-stimulating factor receptor in severe congenital neutropenia. Proc Natl Acad Sci U S A. 1994;91:4480–4.
20. Dong F, van Paassen M, van Buitenen C, Hoefsloot LH, Lowenberg B, Touw IP. A point mutation in the granulocyte colony-stimulating factor receptor (G-CSF-R) gene in a case of acute myeloid leukemia results in the overexpression of a novel G-CSF-R isoform. Blood. 1995;85:902–11.
21. Nicholson SE, Oates AC, Harpur AG, Ziemiecki A, Wilks AF, Layton JE. Tyrosine kinase JAK1 is associated with the granulocyte-colony-stimulating factor receptor and both become tyrosine-phosphorylated after receptor activation. Proc Natl Acad Sci U S A. 1994;91:2985–8.
22. Shimoda K, Iwasaki H, Okamura S, Ohno Y, Kubota A, Arima F, Otsuka T, Niho Y. G-CSF induces tyrosine phosphorylation of the JAK2 protein in the human myeloid G-CSF responsive and proliferative cells, but not in mature neutrophils. Biochem Biophysical Res Commun. 1994;203:922–8.
23. Tian SS, Tapley P, Sincich C, Stein RB, Rosen J, Lamb P. Multiple signaling pathways induced by granulocyte colony-stimulating factor involving activation of JAKs, STAT5, and/or STAT3 are required for regulation of three distinct classes of immediate early genes. Blood. 1996;88:4435–44.
24. Corey SJ, Burkhardt AL, Bolen JB, Geahlen RL, Tkatch LS, Tweardy DJ. Granulocyte colony-stimulating factor receptor signaling involves the formation of a three-component complex with Lyn and Syk protein-tyrosine kinases. Proc Natl Acad Sci U S A. 1994;91:4683–7.
25. Corey SJ, Dombrosky-Ferlan PM, Zuo S, Krohn E, Donnenberg AD, Zorich P, Romero G, Takata M, Kurosaki T. Requirement of Src kinase Lyn for induction of DNA synthesis by granulocyte colony-stimulating factor. J Biol Chem. 1998;273:3230–5.
26. Futami M, Zhu QS, Whichard ZL, Xia L, Ke Y, Neel BG, Feng GS, Corey SJ. G-CSF receptor activation of the Src kinase Lyn is mediated by Gab2 recruitment of the Shp2 phosphatase. Blood. 2011;118:1077–86.
27. Zhu QS, Robinson LJ, Roginskaya V, Corey SJ. G-CSF-induced tyrosine phosphorylation of Gab2 is Lyn kinase dependent and associated with enhanced Akt and differentiative, not proliferative, responses. Blood. 2004;103:3305–12.
28. de Koning JP, Soede-Bobok AA, Schelen AM, Smith L, van Leeuwen D, Santini V, Burgering BM, Bos JL, Lowenberg B, Touw IP. Proliferation signaling and activation of Shc, p21Ras,

and Myc via tyrosine 764 of human granulocyte colony-stimulating factor receptor. Blood. 1998;91:1924–33.
29. Wang L, Xue J, Zadorozny EV, Robinson LJ. G-CSF stimulates Jak2-dependent Gab2 phosphorylation leading to Erk1/2 activation and cell proliferation. Cell Signal. 2008;20:1890–9.
30. Wolfler A, Irandoust M, Meenhuis A, Gits J, Roovers O, Touw IP. Site-specific ubiquitination determines lysosomal sorting and signal attenuation of the granulocyte colony-stimulating factor receptor. Traffic. 2009;10:1168–79.
31. de Bruin AM, Libregts SF, Valkhof M, Boon L, Touw IP, Nolte MA. IFNgamma induces monopoiesis and inhibits neutrophil development during inflammation. Blood. 2012;119:1543–54.
32. Zhang L, Friedman AD. SHP2 tyrosine phosphatase stimulates CEBPA gene expression to mediate cytokine-dependent granulopoiesis. Blood. 2011;118:2266–74.
33. Huang G, Zhang P, Hirai H, Elf S, Yan X, Chen Z, Koschmieder S, Okuno Y, Dayaram T, Growney JD, et al. PU.1 is a major downstream target of AML1 (RUNX1) in adult mouse hematopoiesis. Nat Genet. 2008;40:51–60.
34. Guo H, Ma O, Speck NA, Friedman AD. Runx1 deletion or dominant inhibition reduces Cebpa transcription via conserved promoter and distal enhancer sites to favor monopoiesis over granulopoiesis. Blood. 2012;119:4408–18.
35. Orkin SH, Shivdasani RA, Fujiwara Y, McDevitt MA. Transcription factor GATA−1 in megakaryocyte development. Stem Cells. 1998;2(16 Suppl):79–83.
36. Arinobu Y, Mizuno S, Chong Y, Shigematsu H, Iino T, Iwasaki H, Graf T, Mayfield R, Chan S, Kastner P, et al. Reciprocal activation of GATA−1 and PU.1 marks initial specification of hematopoietic stem cells into myeloerythroid and myelolymphoid lineages. Cell Stem Cell. 2007;1:416–27.
37. Zhang P, Zhang X, Iwama A, Yu C, Smith KA, Mueller BU, Narravula S, Torbett BE, Orkin SH, Tenen DG. PU.1 inhibits GATA−1 function and erythroid differentiation by blocking GATA−1 DNA binding. Blood. 2000;96:2641–8.
38. Traver D, Miyamoto T, Christensen J, Iwasaki-Arai J, Akashi K, Weissman IL. Fetal liver myelopoiesis occurs through distinct, prospectively isolatable progenitor subsets. Blood. 2001;98:627–35.
39. Laslo P, Spooner CJ, Warmflash A, Lancki DW, Lee HJ, Sciammas R, Gantner BN, Dinner AR, Singh H. Multilineage transcriptional priming and determination of alternate hematopoietic cell fates. Cell. 2006;126:755–66.
40. Spooner CJ, Cheng JX, Pujadas E, Laslo P, Singh HA. A recurrent network involving the transcription factors PU.1 and Gfi1 orchestrates innate and adaptive immune cell fates. Immunity. 2009;31:576–86.
41. Akashi K, He X, Chen J, Iwasaki H, Niu C, Steenhard B, Zhang J, Haug J, Li L. Transcriptional accessibility for genes of multiple tissues and hematopoietic lineages is hierarchically controlled during early hematopoiesis. Blood. 2003;101:383–89.
42. Borregaard N, Cowland JB. Granules of the human neutrophilic polymorphonuclear leukocyte. Blood. 1997;89:3503–21.
43. Theilgaard-Monch K, Jacobsen LC, Borup R, Rasmussen T, Bjerregaard MD, Nielsen FC, Cowland JB, Borregaard N. The transcriptional program of terminal granulocytic differentiation. Blood. 2005;105:1785–96.
44. Lominadze G, Powell DW, Luerman GC, Link AJ, Ward RA, McLeish KR. Proteomic analysis of human neutrophil granules. Mol Cell Proteomics. 2005;4:1503–21.
45. Jethwaney D, Islam MR, Leidal KG, de Bernabe DB, Campbell KP, Nauseef WM, Gibson BW. Proteomic analysis of plasma membrane and secretory vesicles from human neutrophils. Proteome Sci. 2007;5:12.
46. Uriarte SM, Powell DW, Luerman GC, Merchant ML, Cummins TD, Jog NR, Ward RA, McLeish KR. Comparison of proteins expressed on secretory vesicle membranes and plasma membranes of human neutrophils. J Immunol. 2008;180:5575–81.
47. Feuk-Lagerstedt E, Movitz C, Pellme S, Dahlgren C, Karlsson A. Lipid raft proteome of the human neutrophil azurophil granule. Proteomics. 2007;7:194–205.

48. Nebl T, Pestonjamasp KN, Leszyk JD, Crowley JL, Oh SW, Luna EJ. Proteomic analysis of a detergent-resistant membrane skeleton from neutrophil plasma membranes. J Biol Chem. 2002;277:43399–409.
49. Rosmarin AG, Yang Z, Resendes KK. Transcriptional regulation in myelopoiesis: hematopoietic fate choice, myeloid differentiation, and leukemogenesis. Exp Hematol. 2005;33:131–43.
50. Swirski FK, Nahrendorf M, Etzrodt M, Wildgruber M, Cortez-Retamozo V, Panizzi P, Figueiredo JL, Kohler RH, Chudnovskiy A, Waterman P, et al. Identification of splenic reservoir monocytes and their deployment to inflammatory sites. Science. 2009;325:612–16.
51. Auffray C, Sieweke MH, Geissmann F. Blood monocytes: development, heterogeneity, and relationship with dendritic cells. Annu Rev Immunol. 2009;27:669–92.
52. Yona S, Jung S. Monocytes: subsets, origins, fates and functions. Curr Opin Hematol. 2010;17:53–9.
53. Gordon S, Taylor PR. Monocyte and macrophage heterogeneity. Nature reviews. Immunology. 2005;5:953–64.
54. Dranoff G, Mulligan RC. Activities of granulocyte-macrophage colony-stimulating factor revealed by gene transfer and gene knockout studies. Stem Cells. 1994;1(12 Suppl):173–82. (discussion 182–174).
55. Lieschke GJ, Stanley E, Grail D, Hodgson G, Sinickas V, Gall JA, Sinclair RA, Dunn AR. Mice lacking both macrophage- and granulocyte-macrophage colony-stimulating factor have macrophages and coexistent osteopetrosis and severe lung disease. Blood. 1994;84:27–35.
56. Stanley E, Lieschke GJ, Grail D, Metcalf D, Hodgson G, Gall JA, Maher DW, Cebon J, Sinickas V, Dunn AR. Granulocyte/macrophage colony-stimulating factor-deficient mice show no major perturbation of hematopoiesis but develop a characteristic pulmonary pathology. Proc Natl Acad Sci U S A. 1994;91:5592–96.
57. Zhan Y, Lieschke GJ, Grail D, Dunn AR, Cheers C. Essential roles for granulocyte-macrophage colony-stimulating factor (GM-CSF) and G-CSF in the sustained hematopoietic response of *Listeria* monocytogenes-infected mice. Blood. 1998;91:863–9.
58. Nakata K, Akagawa KS, Fukayama M, Hayashi Y, Kadokura M, Tokunaga T. Granulocyte-macrophage colony-stimulating factor promotes the proliferation of human alveolar macrophages in vitro. J Immunol. 1991;147:1266–72.
59. Serbina NV, Jia T, Hohl TM, Pamer EG. Monocyte-mediated defense against microbial pathogens. Annu Rev Immunol. 2008;26:421–52.
60. Gordon S. Pattern recognition receptors: doubling up for the innate immune response. Cell. 2002;111:927–30.
61. Lawrence MB, Springer TA. Leukocytes roll on a selectin at physiologic flow rates: distinction from and prerequisite for adhesion through integrins. Cell. 1991;65:859–73.
62. Lawrence MB, Springer TA. Neutrophils roll on E-selectin. J Immunol. 1993;151:6338–46.
63. Chigaev A, Zwartz G, Graves SW, Dwyer DC, Tsuji H, Foutz TD, Edwards BS, Prossnitz ER, Larson RS, Sklar LA. Alpha4beta1 integrin affinity changes govern cell adhesion. J Biol Chem. 2003;278:38174–82.
64. Zhang X, Kluger Y, Nakayama Y, Poddar R, Whitney C, DeTora A, Weissman SM, Newburger PE. Gene expression in mature neutrophils: early responses to inflammatory stimuli. J Leukoc Biol. 2004;75:358–72.
65. Boldt K, Rist W, Weiss SM, Weith A, Lenter MC. FPRL−1 induces modifications of migration-associated proteins in human neutrophils. Proteomics. 2006;6:4790–99.
66. DeLeo FR, Renee J, McCormick S, Nakamura M, Apicella M, Weiss JP, Nauseef WM. Neutrophils exposed to bacterial lipopolysaccharide upregulate NADPH oxidase assembly. J Clin Invest. 1998;101:455–63.
67. Guichard C, Pedruzzi E, Dewas C, Fay M, Pouzet C, Bens M, Vandewalle A, Ogier-Denis E, Gougerot-Pocidalo MA, Elbim C. Interleukin−8-induced priming of neutrophil oxidative burst requires sequential recruitment of NADPH oxidase components into lipid rafts. J Biol Chem. 2005;280:37021–32.
68. Dang PM, Dewas C, Gaudry M, Fay M, Pedruzzi E, Gougerot-Pocidalo MA, El Benna J. Priming of human neutrophil respiratory burst by granulocyte/macrophage colony-stimulating factor (GM-CSF) involves partial phosphorylation of p47(phox). J Biol Chem. 1999;274:20704–8.

69. Suzuki S, Kobayashi M, Chiba K, Horiuchi I, Wang J, Kondoh T, Hashino S, Tanaka J, Hosokawa M, Asaka M. Autocrine production of epithelial cell-derived neutrophil attractant−78 induced by granulocyte colony-stimulating factor in neutrophils. Blood. 2002;99:1863–5.
70. Kobayashi SD, Voyich JM, Whitney AR, DeLeo FR. Spontaneous neutrophil apoptosis and regulation of cell survival by granulocyte macrophage-colony stimulating factor. J Leukoc Biol. 2005;78:1408–18.
71. Tsukahara Y, Lian Z, Zhang X, Whitney C, Kluger Y, Tuck D, Yamaga S, Nakayama Y, Weissman SM, Newburger PE. Gene expression in human neutrophils during activation and priming by bacterial lipopolysaccharide. J Cell Biochem. 2003;89:848–61.
72. Subrahmanyam YV, Yamaga S, Prashar Y, Lee HH, Hoe NP, Kluger Y, Gerstein M, Goguen JD, Newburger PE, Weissman SM. RNA expression patterns change dramatically in human neutrophils exposed to bacteria. Blood. 2001;97:2457–68.
73. Kobayashi SD, Voyich JM, Buhl CL, Stahl RM, DeLeo FR. Global changes in gene expression by human polymorphonuclear leukocytes during receptor-mediated phagocytosis: cell fate is regulated at the level of gene expression. Proc Natl Acad Sci U S A. 2002;99:6901–6.
74. Quinn MT, Gauss KA. Structure and regulation of the neutrophil respiratory burst oxidase: comparison with nonphagocyte oxidases. J Leukoc Biol. 2004;76:760–81.
75. Babior BM. NADPH oxidase. Curr Opin Immunol. 2004;16:42–7.
76. Kobayashi SD, Braughton KR, Whitney AR, Voyich JM, Schwan TG, Musser JM, DeLeo FR. Bacterial pathogens modulate an apoptosis differentiation program in human neutrophils. Proc Natl Acad Sci U S A. 2003;100:10948–53.

Part II
Physiological Processes

This part contains not only reviews concerning basic systems biology of physiological processes, but also two chapters concerning modeling and data analysis; recent advances in experimental techniques produced a huge volume of data, which has to be organized before it is understood.

Two paradigms are presented. In Chap. 7 Marek Kimmel "Stochasticity and determinism in models of hematopoiesis" represents a novel view of modeling in hematopoiesis, synthesizing deterministic and stochastic approaches. Whereas the stochastic models work in situations where chance dominates, for example when the number of cells is small, or under random mutations, the deterministic models are more important for large-scale, normal hematopoiesis. It is also argued that distributed environments such as hematopoietic niches may have a major impact on dynamics. In Chap. 8 Rosemary Braun "Systems analysis of high–throughput data" focuses on multi–gene analysis methods and the integration of expression data with domain knowledge (such as biological pathways) and other gene-related information (e.g., sequence or methylation data) to identify novel functional modules in the complex cellular interaction network. Integrative approaches are presented to extract information about causal relationships, which is scrambled by the high dimensionality of the data and the complex nature of biological interaction networks.

The review of major physiological processes opens with Chap. 9 by Ka Tat Siu and Alex Minella "Developing a systems-based understanding of hematopoietic stem cell cycle control". The chapter is mostly concerned with determination of genes that regulate the major outcomes of hematopoietic stem cells (HSC) mitotic divisions: towards the generation of two new HSCs (symmetric, self-renewing), one HSC and one hematopoietic progenitor cell with multi-lineage potential (HPC) (asymmetric), or two HPCs (symmetric, progenitor pool-repleting). The objectives of these studies are to understand fundamental HSC biology and discover mechanistic underpinnings of bone marrow stem cell diseases, including the myelodysplastic syndromes (MDS) and aplastic anemias. In Chap. 10 Chifman, Laubenbacher, and Torti "A systems biology approach to iron metabolism" reports recent advances in the study of iron metabolism that have revealed multiple intricate pathways that are essential to the maintenance of iron homeostasis. This complexity makes a systems biology approach crucial, with its enabling technology of computational models based on a mathematical description of regulatory systems. Systems biology may represent a new strategy for understanding imbalances in iron metabolism and their underlying causes.

In Chap. 11 Nabil Azhar and Yoram Vodovotz "Innate immunity in Disease—Insights from Mathematical Modeling and Analysis" makes an argument that any rational efforts at modulating inflammation via the blood compartment must involve computational modeling. The acute inflammatory response is a complex defense mechanism that has evolved to respond rapidly to injury, infection, and other disruptions in homeostasis. The complex role of inflammation in health and disease has made this biological system difficult to understand comprehensively and modulate rationally for therapeutic purposes. Consequently, systems approaches have been applied in order to characterize dynamical properties and identify key control points in inflammation. Another modeling-oriented approach is represented by Chap. 12 by Lily Chylek, Bridget Wilson and William Hlavacek "Modeling biomolecular site dynamics in immunoreceptor signaling systems". It follows up on the topic of immunological modeling. The focus here is on the dynamic, site-specific, and context-dependent nature of interactions in immunoreceptor signaling (i.e., the biomolecular site dynamics of immunoreceptor signaling). The challenges associated with capturing these details in computational models have been met through use of rule-based modeling approaches.

The immune system plays a central role in human health. The activities of immune cells, whether defending an organism from disease or triggering a pathological condition such as autoimmunity, are driven by the molecular machinery of cellular signaling systems. Decades of experimentation have elucidated many of the biomolecules and interactions involved in immune signaling and regulation, and recently developed technologies have led to new types of quantitative, systems-level data. To integrate such information and develop non-trivial insights into the immune system, computational modeling is needed, and it is essential for modeling methods to keep pace with experimental advances. In this chapter, we focus on the dynamic, site-specific, and context-dependent nature of interactions in immunoreceptor signaling (i.e., the biomolecular site dynamics of immunoreceptor signaling), the challenges associated with capturing these details in computational models, and how these challenges have been met through use of rule-based modeling approaches.

Part II of the current volume is concluded with Chap. 13 by Elizabeth Gardiner and Robert Andrews "Structure and function of platelet receptors initiating blood clotting". The focus is the structure and function of key platelet receptors involved in thrombus formation and coagulation in health and disease, with a particular focus on platelet glycoprotein (GP)Iba. This latter is the major ligand-binding subunit of the platelet GPIb-IX-V complex, that binds the adhesive ligand, von Willebrand factor (VWF), and is co-associated with the platelet-specific collagen receptor, GPVI. Investigation of platelet receptors is so important because recent studies reveal the link between coagulation and other pathophysiological processes, including platelet activation, inflammation, cancer, the immune response and infectious diseases.

Seth Joel Corey, MD, MPH
Chicago, IL
Mark Kimmel, PhD
Houston, TX
Joshua N. Leonard, PhD
Evanston, IL

Chapter 7
Stochasticity and Determinism in Models of Hematopoiesis

Marek Kimmel

Abstract This chapter represents a novel view of modeling in hematopoiesis, synthesizing both deterministic and stochastic approaches. Whereas the stochastic models work in situations where chance dominates, for example when the number of cells is small, or under random mutations, the deterministic models are more important for large-scale, normal hematopoiesis. New types of models are on the horizon. These models attempt to account for distributed environments such as hematopoietic niches and their impact on dynamics. Mixed effects of such structures and chance events are largely unknown and constitute both a challenge and promise for modeling. Our discussion is presented under the separate headings of deterministic and stochastic modeling; however, the connections between both are frequently mentioned. Four case studies are included to elucidate important examples. We also include a primer of deterministic and stochastic dynamics for the reader's use.

Keywords Hematopoiesis · Leukemias · Stem cells · Dynamical systems · Stochastic processes · Molecular determinism · Driver and passenger mutations

Introduction

The role of stochastic events in hematopoiesis has been discussed for the past 60 years since the beginnings of experimental hematology by Till and McCulloch [1]. The two opposing paradigms, deterministic hematopoiesis based on the firm regulation of peripheral blood cell populations, and stochastic hematopoiesis based on variability observed in seeded bone marrow cells, are still awaiting a grand synthesis. This is in spite of the existence of substantial experimental findings, particularly those in the recent decade, using techniques of single-cell measurements. Disease-accompanying dynamics have been over the years variously modeled as deterministic or stochastic. Examples of stochastic phenomena observed in hematopoiesis include, but are not limited to:

M. Kimmel (✉)
Department of Statistics and Bioengineering, Rice University, 2102 Duncan Hall,
6100 Main St., Houston, TX 77005, USA
Tel.: (713) 348-5255
e-mail: kimmel@rice.edu

S. J. Corey et al. (eds.), *A Systems Biology Approach to Blood*,
Advances in Experimental Medicine and Biology 844, DOI 10.1007/978-1-4939-2095-2_7

- Stochastic fluctuations in the number of hematopoietic stem cell (HSC) making self-renewal versus commitment decisions result in high variability in the magnitude of the response to infection.
- The same stochastic fluctuations may lead to depletion of the HSC compartment when facing massive infections such as neonatal sepsis.
- Presence of variant proteins in molecular switches responding to hematopoietic growth factors such as granulocyte colony-stimulating factor (GCSF) leads to aberrant proliferation and leukemia, again with an important chance component.
- Molecular switches under stochastic fluctuations in molecular pathways and receptor noise may become reversible, which results in reversibility and plasticity at the level of the HSC and early committed cell level.

Recently, a third approach is emerging, which may be termed the molecular determinism (term coined based on ideas in [2, 3]). According to molecular determinism, stochastic variability of the proliferating bone marrow cells can be reduced to complicated series of deterministic events including molecular switches, which are multistable by nature and which trigger proliferation and/or maturation decisions. This is distinct from older proposals involving chaotic dynamics [4, 5].

Mathematical, and in particular stochastic, principles have been used to explain the balance of factors contributing to behavior of a cell population as a whole. However, new techniques for gathering data and probing biological processes at a molecule and cell level continuously provide unprecedented amounts of new information, which leads to reexamination of these models. This has led to a renewed skepticism concerning stochastic modeling as a paradigm. As argued by Snijder and Pelkmans [2], deterministic approach (or, what was called "molecular determinism" earlier in the current chapter) can resolve apparently stochastic phenomena with deterministic variability. They argue that cell-state parameters, such as cell size, growth rate, and cell cycle state, can be used to explain cell-to-cell variability, similarly as spatial cell population context parameters such as local cell density and location on cell colony edges. Tracing back cell-to-cell variability in time over multiple cell cycles may identify inherited, predetermining factors in cells of the same lineage. Snijder and Pelkmans [2] also advocate repeated stimulation of the same cells to help identify the presence of deterministic factors in seemingly stochastic cell-to-cell variability. Complicated dynamics leading to chaotic (and sometimes indistinguishable from stochastic) behavior has been appreciated for some time. For example, existing mathematical models of cell cycle regulation (cf. e.g., [6] and references therein) rely on nonlinear regulatory functions to control cell population distribution. However, these models also include a very real phenomenon of uneven allocation of constituents to progeny cells, which arguably is either truly stochastic or is indistinguishable from stochastic. Moreover, the idea of "backtracking" complicated (chaotic) trajectories seems to be doubtful from mathematical viewpoint. Schroeder [7] discussed the need for long-term continuous follow-up on individual cells in order to understand the specific rules of proliferation and differentiation. This chapter also touches upon issues such as influence of imaging techniques on cell behavior and difficulty in cell tracking using existing software.

Returning to molecular determinism, a very good example of this approach seems to be the paper by Takizawa et al. [8], concerning a purely deterministic and demand-driven integrated model of regulation of early hematopoiesis. This models is very complex and it involves "view of how cytokines, chemokines, as well as conserved pathogen structures, are sensed, leading to divisional activation, proliferation, differentiation, and migration of HSCs and progenitor cells, all aimed at efficient contribution to immune responses and rapid reestablishment of hematopoietic homeostasis." Takizava et al. [8] paper is too involved physiologically to be discussed at length here. Let us notice that it contrasts with the simpler (and stochastic) models of Ogawa [9] and Abkowitz et al. [10]. In these models, the branching process (bp) paradigm is used at its simplest, with cells depicted as independent individuals, splitting at random and possibly interacting with a limited number of smaller entities.

Another current concept is that of nongenetic variability as a substrate for natural section, as espoused by Huang's group [11]. For example, slow fluctuations in mammalian cells are the expression of heritability (memory) of protein abundance in successive generations of normal or cancer cells [12, 13]. One example is the noninherited form of drug resistance in cancer. Theoreticians have been suggesting this for several decades because of similar experimental evidence. The memories of protein abundance and dynamic homeostasis, which implied slow fluctuations in individual cells, were important constituents of many of the cell cycle regulation and unequal division models [14–16]. Development of resistance to chemotherapy by gene amplification (genetic, but nonmutation driven) has been pondered by theorists equally long ago [17, 18].

Questions about the dynamics of hematopoiesis are resurfacing due to new experimental studies concerning lineage-specific growth factors, morphogens, the microenvironment, and the plasticity of stem cells. These new findings allow a re-examination of two long-standing questions: whether hematopoiesis is stochastic or deterministic, and whether it is discrete or continuous. These issues exist for other non-HSC systems; however, hematopoiesis serves as the most informative and accessible mammalian tissue system to look for answers [1]. Since quantitative systems analysis based on multi-scale modeling is needed to understand the complexity and dynamics of hematopoiesis, determining the correct approach to this modeling is of more than academic interest. Much work has been recently published on this topic and some of it will be reviewed in the current chapter. We will first pose three key questions and then use a simple "toy" model to explain basic ideas and problems.

Question 1. Is Hematopoiesis Deterministic or Stochastic?

Experimental data suggest stochastic factors play a role in determining fate of daughter cells of a stem cell [19, 20]. However, it is not clear at which critical junction stochasticity operates in lineage-specific regulation (principal examples being erythropoietin (Epo)-driven erythropoiesis and GCSF-driven granulopoiesis. Recent systemic and modeling studies of dynamics of signaling pathways in cells at various stages of hematopoiesis, underscore the role of bistable (or multistable) switches, which can direct the cell towards "fates" such as differentiation in various directions, proliferation, or apoptosis [21]. These switches, as described and modeled, are essentially deterministic circuits, displaying a series of stable and unstable steady states [22]. The stable steady states correspond to distinct patterns of expression of target genes, characteristic of a given cell "fate." Small change in initial conditions at individual cell's level or in type or strength of receptor activation results in switching from one stable work regime to another [23]. Although this paradigm explains the interplay of positive and negative feedbacks in cells, it does not explain the intrinsic stochasticity, implied by both classical and more recent experiments on hematopoietic cells [9, 10]. Independently, there exists a sizeable body of evidence that eukaryotic cells may make individual decisions based on nondeterministic rules [24, 25]. The sources of intrinsic stochasticity in eukaryotic cells are related to processes in which a small number of interacting molecules may trigger a large-scale effect [26]. Stochastic effects may provide robust evolutionarily adaptive mechanisms [27]. A critical property of hematopoiesis is the ability to protect against environmental insults (e.g., infection), which may require a design incorporating stochastic dynamics.

Question 2. Do HSCs and Their Progeny Constitute Discrete Subsets or a Continuum?

The general question of stem cell plasticity and, in particular, the reversibility of the HSC has gained much interest due to stem cell engineering and induced pluripotent stem cells. A related question is to what extent the succession and timing and commitment and differentiation (maturation) processes in hematopoiesis can be altered or "stretched" within the bounds of normality. On an operational level, is it sufficient to model hematopoiesis in the terms of discrete stages or is it necessary to include continuous maturation?

Question 3. What Role Is Played in Hematopoiesis by Spatial Effects?

The usual approach has been to treat the process as spatially uniform in both the bone marrow and peripheral circulation. However, recent research on niches and environments in the bone marrow and the interaction of HSC and mesenchymal stem cells (see [28]), has led to a realization that spatial effects and interaction between spatial and stochastic effects cannot be ignored. Such interactions in mathematical models result in qualitatively new dynamics (as in Roeder's 2006 model [29]). The reason is that spatial separation provides opportunity for small colonies of cells to fix stochastic fluctuations despite the fact the total size of the population is large.

Primer in Deterministic and Stochastic Dynamics

This primer is intended for the wide audience who study systems biology, and depending on your expertise can be skipped or used as a loose reference.

Deterministic Dynamical Systems

To understand deterministic models of hematopoiesis, we will need definitions and theorems from the mathematical theory of *dynamical systems*. In general, these are objects that evolve in time and, when "stopped" as a result of either a physical or a "thought" intervention, can be restarted and continued "as if nothing happened." This is known as the "continuation principle." Mathematically, let us denote by $x(t; t_0, x_0)$ the state of the system at time t, if at time t_0 the state was x_0, i.e., $x(t_0; t_0, x_0) = x_0$. In these terms, the continuation principle can be stated as $x(t + s; t_0, x_0) = x(t; t_0 + s; x(s; t_0, x_0))$.

The most commonly employed dynamical systems have the form of *differential equations* (*DEs*; including multidimensional or infinitely dimensional DEs). In this setup, $x(t; t_0, x_0)$ is the solution of the DE of the form $\mathrm{d}x/\mathrm{d}t = f(x)$, with solution $x(t)$ satisfying the $x(t_0) = x_0$. Such solution can be denoted by $x(t; t_0, x_0)$, and it can be proved that it satisfies the continuation principle. Solutions of DEs have been extensively studied and therefore constitute a convenient tool.

Solution $x(t; t_0, x_0)$ of a DE $\mathrm{d}x/\mathrm{d}t = f(x)$ is *stable* if there can be found a disc of radius δ in the space of initial conditions, such that the solution stays forever in a "pipe" of a desired radius ε. In mathematical terms, for each ε there exists a δ such that if $|x_0' - x_0| < \delta$, then $|x(t; t_0, x_0') - x(t; t_0, x_0)| < \varepsilon$, for all $t \geq t_0$. Solution $x(t; t_0, x_0)$ is *asymptotically stable* if it is stable and moreover $|x(t; t_0, x_0') - x(t; t_0, x_0)|$ converges to 0 with t converging to infinity. In many cases, it is interesting to investigate stability of the equilibrium solution, i.e., the solution along which the time derivative $\mathrm{d}x/\mathrm{d}t = 0$. This solution is a constant function solving the equation $0 = f(x)$. Useful mathematical tools for investigating stability include the *Lyapunov functions* and *characteristic equations*.

Bifurcation Most commonly applied to the mathematical study of dynamical systems, a bifurcation occurs when a small smooth change made to the parameter values of a system causes a sudden qualitative or topological change in its behavior. The name "bifurcation" was first introduced by the mathematician Henri Poincaré in 1885. The two best-known types of bifurcation are *exchange of stability* and *Hopf bifurcation*. In exchange of stability, change of parameter causes a new equilibrium to appear, which becomes stable, while the old one remains but becomes unstable. In Hopf bifurcation, a stable equilibrium becomes unstable, with oscillations around it appearing at the same time.

Chaos theory is a field of study in mathematics, with applications in disciplines including physics and biology. In a chaotic system, small differences in initial conditions (even those due to rounding errors) yield widely diverging trajectories, rendering long-term prediction impossible in general. Chaotic systems are deterministic, so their trajectories are mathematically determined by their initial conditions, with no random elements involved. However, the deterministic nature of these systems does not make them predictable. In many ways, chaos can mimic randomness (stochasticity).

Stochastic Processes

Proliferation of cells is frequently stochastic. Therefore, it is useful to introduce definitions and theorems from the theory of probabilities and the theory of stochastic processes. The following account is a brief intuitive introduction. The definitions will be highlighted by italics.

Random variable (rv) X is, intuitively, a numerical result of observation, which displays random variation. The notation $X(\omega)$ highlights the dependence of the rv on the "chance" element ω of the sample space Ω (wherever it is superfluous, the ω is omitted. *Stochastic process* (or *random function*) $X(t;\omega)$ is, intuitively, a function of time t with a random component. Mathematically, it is a family of rv's parameterized by time. Function of time $X(t;\omega)$, with ω fixed is called the *realization* or the *sample path* of the process. *Self-recurrence* is an important property of the stochastic process $X(t;\omega)$. Suppose that a process such that at $X(0;\omega) = x_0$ is stopped at some time t_0. Then, if also $X(t_0;\omega) = x_0$, the continuation process restarted from time t_0 is identical (it has the same distributions) as the original process, except that it is shifted by t_0. A process with such property is called self-recurrent. Self-recurrence may be considered a rephrasing of a causality principle. It leads to recurrent relationships for a wide class of stochastic processes, including Markov processes and bp's.

Markov process is a process with a limited memory (the *Markov property*). Intuitively, given the state of the process at time s, the future of the process (at some t such that $s < t$) depends only on this state and not on its states at times before s. Mathematically, $\Pr[X(t) \in A|X(u), u \leq s < t] = \Pr[X(t) \in A|X(s), s < t]$, where A is a subset of the state space of the process (space of values assumed by the process). The probability listed above is the transition probability from state $x = X(s)$ to the set of states A, in time t s. If the states of the process form a finite or *denumerable* set, then the process is called a *Markov chain*. In this case, it is possible to define a matrix (finite or infinite) of transition probabilities between states $P(t) = [P_{ij}(t)]$, where $P_{ij}(t) = \Pr[X(s+t) = j|X(s) = i]$.

Branching process (bp) is a random collection of individuals (such as particles, objects, or cells), proliferating according to rules involving various degrees of randomness of the individual's life length and the number of progeny of an individual. The unifying principle is the so-called branching property, which states that the longevity and type of progeny of a newborn particle, conditional on the current state of the process, are independent of any characteristics of other particles present at this time or in the future. The branching property is a form of self-recurrence, as defined earlier on. *Galton–Watson bp (G–W bp)* is the simplest bp. It evolves in discrete time measured by nonnegative integers. At time 0, an *ancestor individual* (a particle, cell, or object) is born. At time 1, the ancestor dies, producing a random number of progeny. Each of these becomes an ancestor of an independent *subprocess*, distributed identically as the whole process. This definition implies that the numbers of progeny produced by each particle ever existing in the process are independent identically distributed rv's and that all particles live for one time unit. Discrete time moments coincide with generations of particles. The number of particles existing in the G–W bp, as a function of time, constitutes a time-discrete Markov chain.

Bp's occur frequently in biological systems. They serve as models for proliferating cells, amplified genes, and shortening telomeres. Bp is *critical* if the expected (mean) count of progeny of a particle is equal to 1. It is *supercritical*, if the mean count of progeny of a particle is greater than 1 and *subcritical* if it is less than 1. This classification leads to profound differences in *asymptotic properties* (properties after sufficiently long time) of the process. In particular, critical bp's behave in a paradoxic way since they become *extinct* (i.e., all particles die out) with probability 1, while the expected number of particles stays constant. Asymptotic properties of subcritical bp's are summarized by the *Yaglom's theorem*, which states that for a subcritical bp, which also becomes extinct with probability 1, there exists a *quasi-stationary* distribution, conditional on nonextinction. This means that the sample paths that do not become extinct will be, for times sufficiently long, distributed according to a law (distribution) that does not vary with time (i.e., *stationary*). All bp's share the property of *instability*, which means that, as time tends to infinity, the bp becomes either extinct or indefinitely large. Instability is due to the independent assumptions inherent in the definition of a bp (see earlier on).

Supercritical bp's have the property called the *exponential steady state*, which characterizes populations growing without spatial or selective constraints, the condition in which the number of individuals increases or decreases exponentially, while the proportions of individuals in distinct age classes and any other identifiable categories remain constant. Related notion is that of the *Malthusian parameter*, i.e., a parameter α such that the number $Z(t)$ of particles present in the process, normalized by dividing it by $\exp(\alpha t)$, converges to a limit rv, as time tends to infinity.

Markov bp is a type of time-continuous bp. At time 0, an ancestor individual (a particle, cell, or object) is born. The ancestor lives for time τ, which is an exponentially distributed rv, and then the ancestor dies, producing a random number of progeny. Each of these becomes an ancestor of an independent subprocess, distributed identically as the whole process. The number of particles existing in the Markov bp, as a function of time, is a Markov chain.

Type space is a collection of possible particle (cell) types existing in a bp. If there is more than one but finitely many types, the process is called multitype. *Multitype G–W bp* is a generalization of the usual (single-type) G–W bp. In the multitype process, each individual belongs to one of a finite number of types. At time 0, an ancestor individual (a particle, cell, or object), of some type, is born. Processes started by individuals of different types are generally different. At time 1, the ancestor dies, producing a random number of progeny of various types. The distribution of progeny counts depends on the type of parent. Each of the first-generation progeny becomes an ancestor of an independent subprocess, distributed identically as the whole process (modulo ancestor's type). In the multitype process, asymptotic behavior depends on the matrix of expected progeny count. Rows of this matrix correspond to the parental types, and columns to the progeny types. The largest positive eigenvalue of this matrix (the *Perron–Frobenius* eigenvalue), is the Malthusian parameter (see earlier on) of the process, provided the process is supercritical (the Perron–Frobenius eigenvalue larger than 1) and *positive regular*. The latter means that parent of any given type will have among its (not necessarily direct) descendants individuals of all possible types, with nonzero probability.

Probability generating function (*pgf*) is a function $f_X(s)$ of a symbolic argument s, which is an equivalent of the distribution of a nonnegative integer-valued rv X. If numbers $p_0, p_1, p_2, \ldots$ constitute the distribution of rv X, i.e., $\Pr[X = k] = p_k$, then the pgf of rv X is defined as $f_X(s) = E(s^X) = \sum_{i=0}^{\infty} p_i$, for $s \in [0,1]$. Use of the pgf simplifies mathematical derivations involving nonnegative integer rv's and processes, among them the bp's.

Poisson process is one of the most important stochastic processes, since it is frequently used as a model for mutation dynamics. It can be intuitively defined as a random collection of time points having the properties of complete randomness (the counts of events in any two disjoint time intervals are independent), and stationarity (the probability of an event occurring in a short time interval $(t, t + \Delta t)$ is equal to $\lambda\ \Delta t + o(\Delta t)$, where a small $o(\Delta t)$ with respect to Δt has the property it converges to 0 faster than Δt itself, i.e., $o(\Delta t)/\Delta t \to 0$, as $\Delta t \to 0$. Constant λ is called the *intensity* of the process. The number N of epochs of the Poisson process in an interval of length t has Poisson distribution of the form $p_n = \Pr[N = n] = \exp(-\lambda t)\lambda^n/n!$, for $n = 0,1,2,\ldots$, and the time intervals T between any two epochs have exponential distribution with the same parameter λ, i.e., the density of distribution of T is equal to $f_T(t) = \lambda \exp(-\lambda t)$.

The *Wright–Fisher* (*W–F*) model is a stochastic construct that is frequently used in genetics to explain loss of variants in a finite population. In brief, it is assumed that there exist N individuals, which produce progeny from one synchronous generation to another. Individuals in generation $n + 1$ are copies of individuals randomly and independently chosen from generation n, so that the probability that individuals $1,2,\ldots,i,\ldots,N$ are represented respectively $k_1, k_2, \ldots, k_i, \ldots, k_N$ times in the succeeding generation is equal to $p_{k_1,k_2,\ldots,k_i,\ldots,k_N} = (1/N)^{k_1+k_2+\ldots+k_i+\ldots+k_N} N!/(k_1!k_2!\ldots,k_i!\ldots k_N!)$, where $k_1 + k_2 + \ldots + k_i + \ldots + k_N = N$, i.e., the rv's $k_1, k_2, \ldots, k_i, \ldots, k_N$ are *multinomially* distributed. One consequence is that in each generation there exists a nonzero probability of one (or more) individuals leaving no progeny. This leads eventually (after a finite number of generations) to loss of copies of all the individuals except one. This individual (or genetic variant) is called *fixed*. The W–F model differs from the G–W bp, in that the former has a fixed total number of individuals, while in the latter the number of individuals is fluctuating from one generation to another. The W–F model applies in situations in which the environment pressure makes population size change unlikely.

Turing pattern formation One of the most interesting mathematical phenomena arising in models structured by spatial coordinates is *pattern formation* via *diffusion-driven instability (DDI)*. Discovered by Alan Turing, this effect has been used to explain emergence of biological, physical, and chemical patterns, such as patterns in colonies of microorganisms, embryo segmentation, or dynamics of the Belousov–Zhabotinsky reactions. The usual mathematical framework is that of the system of at least two reaction-diffusion equations, i.e., *partial differential equations (PDEs)* of the form

$$\partial u/\partial t = D_1 \Delta_x u + f(u, v),$$
$$\partial v/\partial t = D_2 \Delta_x v + g(u, v),$$

where $u(x,t)$and $v(x,t)$ are defined as functions of spatial coordinates x and time t, $\partial(\,\cdot\,)/\partial t$ id partial differentiation with respect to time, $\Delta_x(\,\cdot\,)$ is the second order partial differentiation operator with respect to the spatial coordinates x (*diffusion operator* or *Laplacian*), and nonlinear functions $f(u,v)$ and $g(u,v)$ are reaction terms. Spatial pattern is a stable spatially heterogeneous equilibrium solution (DDI or Turing pattern), which arises in the reaction-diffusion system, but which does not exist for the corresponding pure reaction system

$$\partial u/\partial t = f(u, v),$$
$$\partial v/\partial t = g(u, v),$$

for which only spatially homogeneous (constant in x) solution exist.

Deterministic Models of Hematopoiesis

Deterministic Models of Regulatory Feedbacks

Simplest Model of Hematopoiesis

Case study: Simple Hierarchical Model Some basic concepts on which hemat–opoiesis models are based can be explained using a simplified deterministic model of granulopoiesis. Let us consider a sequence of populations of bone marrow cells, where:

- Population $i = 0$ consists of the HSC
- Populations $i = 1, \dots, J$ include several stages of cells committed to granulopoiesis
- Populations $i = J + 1, \dots, I$ include differentiated granulopoietic precursors
- Population $i = I + 1$ includes the mature blood granulocytes

Let us denote by $N_i(t)$, $i = 0, \dots, I+1$, $t = 0, \Delta t, \dots, k\Delta t$ the respective numbers of cells in population i (or, in other words, of type i) at times $k\Delta t$ being the integer multiples of the duration of the cell cycle, this latter for this example assumed equal for all cell populations. Let us further assume that cells of type $i = 1, \ldots, I$ proliferate without losses, i.e., each produces two viable progeny cells upon division. Mature granulocytes do not proliferate but at each time point $k\Delta t$ they die with probability $\alpha\Delta t$. As a result, the expected lifetime of a granulocyte is equal to $1/(\alpha\Delta t)$.

The rules of maturation and differentiation are described as follows:

- Each of the HSC (type $i = 0$) progeny remains a HSC with probability $1 - d$ and differentiates into a cell committed to the granulocyte lineage with probability $d_0 < d$ such that $d - d_0$ is the probability of commitment to the remaining lineages.
- Each of the progeny of the committed and progenitor type i cells may either remain type i (with probability $1 - d_i$) or become type $i + 1$ (with probability d_i).

These rules lead to the following system of difference equations for the expected numbers of cells of all types:

$$N_0(t + \Delta t) = qN_0(t) \tag{7.1}$$

$$N_i(t + \Delta t) = q_i N_i(t) + p_{i-1}N_{i-1}(t),\ i = 0, \ldots, I \tag{7.2}$$

$$N_{I+1}(t + \Delta t) = N_{I+1}(t) + p_I N_I(t) - (\alpha\Delta t)N_{I+1}(t), \tag{7.3}$$

where for brevity we denote $p = 2d$, $p_i = 2d_i$, and $q_i = 2(1-d_i)$, so that $p+q_0 < 2$ and $p_i + q_i = 2$.

Equations (7.1–7.3) can be solved recursively. In particular, we are able to compute steady-state (equilibrium) values assuming $N_i(t) = N_i = \text{const}$. This results in the following expressions:

$$N_i = N_0 \left(\prod\nolimits_{j=0}^{i-1} p_j\right)\left(\prod\nolimits_{j=1}^{i} (1 - q_j)\right),\ i = 1, \ldots, I \tag{7.4}$$

$$N_{I+1} = N_0 \left(\prod\nolimits_{j=0}^{I} p_j\right)\left(\prod\nolimits_{j=1}^{I} (1 - q_j)\right)(\alpha\Delta t)^{-1}. \tag{7.5}$$

Let us note that biologically feasible steady state exists if $d_i \geq 1/2$. Further, for simplicity, we will assume that the probabilities of commitment to the subsequent stage are all equal to $d_i = \delta$ for $i = 1, \ldots, J$, hence all $p_i = 2\delta = \pi$ and $q_i = 2(1-\delta) = \psi$ for $i = 1, \ldots, J$. We may also assume that all differentiated precursors differentiate further at the subsequent division, i.e., $d_i = 1$ for $i = J + 1, \ldots, I$, hence $p_i = 2$ and $q_i = 0$ for $i = J + 1, \ldots, I$. This yields

$$N_{I+1} = N_0 p_0[\pi(1 - \psi)]^J 2^{I-J}(\alpha\Delta t)^{-1} = N_0 p_0[2\delta(2\delta - 1)]^J 2^{I-J}(\alpha\Delta t)^{-1}. \tag{7.6}$$

Nonlinear (polynomial) action of GCSF feedback Equation (7.6) demonstrates that the number of peripheral granulocytes at equilibrium depends polynomially on the commitment probability δ. Therefore, if at the normal equilibrium the value of δ is below 1 (i.e., not all committed cells commit further), then if GCSF action increases δ

so that in each of the J stages of committed cells more cells commit further upon each division, this causes the equilibrium number of peripheral granulocytes to increase (neglecting transients) with the Jth power of δ, i.e., nonlinearly. Therefore, small deviations of δ may cause large changes of the equilibrium number of $N_{I+1}(t)$, the number of mature cells.

Need for a negative internal feedback of HSC If the commitment probability p_0 of HSC is kept unchanged and probability d is equal to ½, then (see Eq. (8.1)) the steady state $N_0(t) = N_0 = \text{const}$ is maintained. However, if the GCSF feedback causes d to exceed ½, then Equation (7.1) shows that the HSC will be geometrically (exponentially) depleted with time. To prevent this from happening in long term, a protective mechanism is needed. Interplay between the GCSF feedback and the internal negative feedback was first considered by Arino and Kimmel [15]. It has been showed there that some forms of the feedback may not be sufficient for a complete return to equilibrium, a possibility observed in some disease states such as neonatal sepsis [30].

What Does the Simplified Model Fail to Explain?

I. *Interindividual and temporal fluctuations in the number of granulocytes.* The simplified model is a mean value model, so it does not account for fluctuations caused by stochastic events at the level of HSC and further amplified in the commitment/differentiation cascade.
II. *Feedbacks.* The model is also missing the explicit form of the GCSF and internal feedbacks, although it helps realizing these are needed.
III. *Differentiation arrest and dedifferentiation.* The model does not involve the chance state of molecular circuitry guiding the cell to commitment/differentiation, instead it uses aggregate coefficients d_i the simplification it shares with many published models [9, 10].

Deterministic Feedbacks in Hematopoiesis

Deterministic mathematical theory of cell production systems primarily relies on systems of nonlinear DEs. These models perform differently depending on the configuration of regulation feedbacks. Cell production systems are self-renewing cell populations which maintain the continuous supply of differentiated functional cells to various parts of a living organism. The dynamics of cell production systems attracted the attention of biologists and mathematicians a long time ago in the context of blood cell production [31]. Despite differences depending on the type of cells considered, certain common elements can be found in all the cell production systems and their models. First, there exists a self-renewing subpopulation of stem cells. Stem cell divisions can produce both stem cells and cells of greater lineage commitment, called the precursor cells. The precursor cells, in turn, produce cells with an even greater degree of maturity. After a certain number of maturation (differentiation) stages, the mature (differentiated) cells are produced. They usually do not have the ability to divide, and after fulfilling their specific tasks, are removed from the organism.

In normal conditions, the cell production system maintains a constant supply of differentiated mature cells. In the emergency cases, when for some reasons the organism suffers from the loss of certain mature cells (such as loss of erythrocytes in a hemorrhage) the system reacts, providing an increased supply of cells. These two postulates imply that the system is regulated through a long-range feedback mechanism detecting the perturbations in the number of mature cells and accordingly adjusting the production rate of the stem and precursor cells.

It seems logical to suppose at least one more regulation feedback exists. Indeed, the long-range feedback would have a tendency towards "draining" the stem cell population to compensate for the loss. Then, if all the stem cells were committed towards maturation, the whole system might collapse, since only the stem cells are truly self-renewing. Therefore, another feedback should "cut off" the supply of precursor cells if the number of stem cells decreases, preventing the system from extinction. This will be called a short-range feedback.

Based on ideas similar to those presented above, mathematical models of cell production systems were constructed, mainly for various lines of the blood-forming system in man and in experimental animals. For example, Mackey's periodic neutropenia model (as cited in Haurie et al. [32]) described the effects of a short-range feedback of the stem cell cycle, while the Wazewska and Lasota model included the long-range feedback only [33]. Recently, various possible configurations of short-, mid-, and long-range feedbacks have been discussed and analyzed mathematically [34].

Case study: Configuration of Feedbacks in a Deterministic Model of Hemat–opoiesis We will use as a case study the series of models devised by Arino and Kimmel [35]. Considering these models will explain the modeling paradigm, which has been later on perfected in various ways. The models are based on the following assumptions (Fig. 7.1):

1. Stem cell proliferation dynamics is represented by a cell cycle model consisting of two phases: active and passive. A stem cell leaving mitosis enters the passive phase and then it may either transform into a more mature precursor cell or enter the active phase (and then divide and enter the passive phase again). It is assumed that the cell residence time in the resting phase has the exponential distribution with parameter $\alpha(t)$ (the reciprocal of the mean residence time in this phase). Such a hypothesis is consistent with the Smith–Martin model of the cell cycle. The probability of stem cell differentiation (transformation) is denoted by $d(t)$. The residence time in the active phase is equal to T. We understand that our "active phase" is $S + G1 + M$, where S stands for the deoxyribonucleic acid (DNA) synthesis. G2 denotes the premitotic phase and M the cell division (mitosis). Our "passive phase" is assumed to be $G0 + G1$, where G0 is the resting (quiescent or "storage") phase, while G1 is the initial growth phase.

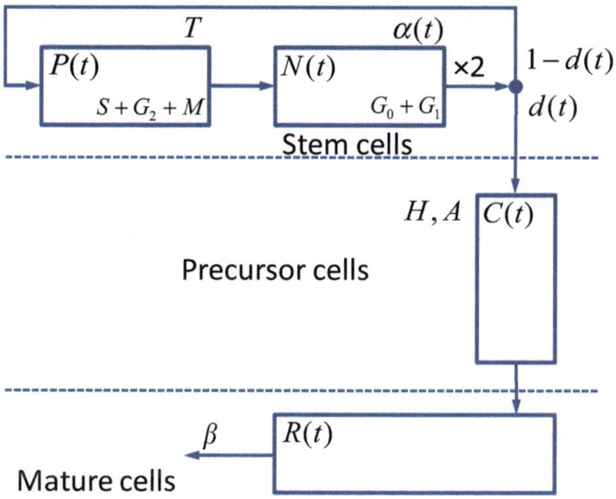

Fig. 7.1 Structural model of the hematopoietic system. $P(t)$, $N(t)$, $C(t)$, and $R(t)$ are the number of cycling stem cells, dormant stem cells, precursor cells, and mature cells, respectively. $\alpha(t)$ is the exit rate from the dormant stem cell compartment: T is the residence time in the active stem cell compartment, $d(t)$ is the fraction of differentiating stem cells. H and A are the transit time and amplification coefficient of the precursor cell compartment and β is the mature cell death rate. (Adapted from Ref. [35])

2. Regulated factors are $d(t)$ probability of stem cell differentiation and/or $\alpha(t)$ reciprocal of the mean residence time in the passive phase.
3. Each stem cell, once differentiated, produces after time H an average number of A mature (completely differentiated) cells. Quantities A and H represent all the stages of the precursor cells maturation, division, and so forth.
4. Mature cell life length is a rv with exponential distribution with expected value $1/\beta$.

Model structure implied by the assumptions (1)–(4) is depicted in Fig. 7.1. The equation for the stem cell number $N(t)$ in $G0 + G1$, takes the following form:

$$\dot{N}(t) = -\alpha(t)N(t) + 2(1 - d(t))\alpha(t)N(t).$$

The equation for the number $R(t)$ of mature cells is:

$$\dot{R}(t) = -\beta R(t) + r(t),$$

where $r(t)$ is the rate of cell flow into the mature cell compartment. Assumption (3) implies
that:

$$r(t) = Ad(t - H)\alpha(t - H)N(t - H),$$

so that

$$\dot{R}(t) = -\beta R(t) + r(t)Ad(t - H)\alpha(t - H)N(t - H).$$

We may also compute the number $P(t)$ of cells present at time t in the active phase of the stem cell cycle:

$$P(t) = \int_{t-T}^{t} (1 - d(\tau))\alpha(\tau)N(\tau)d\tau.$$

Equations above provide a complete description of the cell production system dynamics, if the regulated factors $\alpha(t)$ and $d(t)$ are specified.

Depletion and Nonunique Equilibria We will make the case for the possibility of depletion of HSC and nonunique equilibria, by considering the deterministic model of erythropoietic regulation [35].

Model 1: The fraction $d(t)$ of differentiating stem cells is an increasing function of the number of dormant stem cells: $d(t) = g[N(t)]$. The rate $\alpha(t)$ of the outflow from the dormant stem cell compartment is a decreasing function of the number of mature cells: $\alpha(t) = h[R(t)]$. Intuitively, the mature cell number is influencing the production rate of stem cells, while the contents of the "storage" dormant compartment controls the proportion of differentiating stem cells.
Model 2: In this variant, both $\alpha(t)$ and $d(t)$ depend on the mature cell number: $d(t) = g[R(t)]$, $\alpha(t) = h[R(t)]$ with $g(\cdot)$ and $h(\cdot)$ being decreasing functions. The assumption that both feedbacks here are designed to "exploit" the stem cell population causes system instability.
Model 3: This is, in a sense, a reversal of model 1. The long-range feedback controls the differentiating stem cell fraction, while the "defensive" one, the exit rate from the dormant compartment: $d(t) = g[R(t)]$, $\alpha(t) = h[N(t)]$, where $g(\cdot)$ and $h(\cdot)$ are decreasing.
Model 4: This is a special case of model 1, with $d(t) = 1/2$. In this case, model equations assume the form, $\dot{N}(t) = -h[R(t)]N(t) + h[R(t-T)]N(t-T)$ and $\dot{R}(t) = -\beta R(t) + (A/2)h[R(t-H)]N(t-H)$.

Importance of the Internal Feedback Models 1, 2, and 3 have (under additional hypotheses; see the exhaustive discussion in Arino and Kimmel [35]), two equilibria, the trivial one $(N, R) = (0{,}0)$ and the nontrivial one $(N, R) = (\tilde{N}, \tilde{R})$, which is a solution of nonlinear algebraic equations involving functions $g(\cdot)$ and $h(\cdot)$. Without getting into mathematical details, we can state, that in models 1 and 3, which involve autonomic internal feedbacks of the dormant HSC, the trivial equilibrium usually (i.e., for a region of parameter values) repels solutions, while the nontrivial one attracts them. Hence, the system is resistant to shocks. In model 2, which does not include an internal feedback, the situation is reversed, given a deviation from the nontrivial equilibrium, the system decays to the trivial one. This supports the assertion that without an internal feedback, the hematopoietic system may be unstable.

Nonunique equilibria of model 4 display an unusual behavior. Function $V(t) = N(t) + 2P(t)$ equal to the number of stem cells in the dormant phase plus twice the number of stem cells in the proliferative phase ("potential" number of HSC) stays constant along the trajectories of the system. Therefore, also at the equilibrium it will be the same value it had at time 0. Simple calculations show that in model 4, initial conditions dictate the equilibrium value: If the system undergoes a shock such as depletion of bone marrow HSC, it will forever linger near that low value. This situation only concerns an impaired internal feedback, but it may correspond to a specific biological defect such as one caused by a defective cytokine receptor (see further on).

Topics concerning configuration and functional forms of deterministic models of feedbacks have been further developed in more recent works of other authors. As an example, Marciniak-Czochra, Stiehl, and coworkers consider a range of general issues related to the question of hierarchy in the cell production systems, such as the granulopoietic system, using rigorous mathematical approaches [36]. HSCs are characterized by their ability of self-renewal to replenish the stem cell pool and differentiation to more mature cells. The subsequent stages of progenitor cells also share some of this dual ability. It is yet unknown whether external signals modulate proliferation rate or rather the fraction of self-renewal. They propose three multicompartment models, which rely on a single external feedback mechanism. In model 1, the signal enhances proliferation, whereas the self-renewal rates in all compartments are fixed. In model 2, the signal regulates the rate of self-renewal, whereas the proliferation rate is unchanged. In model 3, the signal regulates both proliferation and self-renewal rates. The study demonstrates that a unique strictly positive stable steady state can only be achieved by regulation of the rate of self-renewal. Furthermore, it requires a lower number of effective cell doublings. To maintain the stem cell pool, the self-renewal ratio of the HSC has to be greater than or equal to 50 % and it has to be higher than the self-renewal ratios of all downstream compartments. Interestingly, the equilibrium level of mature cells depends only on the parameters of self-renewal of HSC and it is independent of the parameters of dynamics of all upstream compartments. The model is compatible with the increase of leukocyte numbers following HSC transplantation. A more theoretical analysis of feedbacks has been published in ref. [37, 38]. In another paper, Marciniak-Czochra and Stiehl [39] find that that certain conditions have to be met for proliferative parameters of stem cells relative to those of the committed cells. Otherwise, the stem cells die out and their function is fulfilled by cells of one of the committed stages, the one that satisfies these conditions. Their other contributions concern replicative senescence of HSCs [40]. Logic of control, in a more intuitive framework, but considering competing feedbacks, has been considered in the works of Lander and coworkers [41].

Much of classical deterministic analysis has been developed over past 35 years by the school of Mackey and his coworkers. Initially he collaborated with Lasota and Wazewska, who had developed the first mathematical model of erythroid production [33]. They suggested that decreasing the rate of erythroid precursor maturation increases the steady-state level of nonproliferating erythroid cells. A patient would quickly recover red blood cell levels following treatment-induced anemia. Since erythroid precursor maturation rate increases with Epo levels, which are negatively correlated with blood oxygen content, the model suggests that by increasing a patient's blood oxygen level one can accelerate the rate of erythrocyte recovery following chemotherapeutic insult or radiation therapy. This conclusion was successfully validated in patients, showing the insight mathematical modeling can provide into disease. Another major topic was cytopenias and leukemias. Some neutropenias and anemias are characterized by periodic oscillations in blood cell counts [42]. The dynamics of these disorders has attracted model making, with the goal that the model will illuminate their pathophysiology as well as normal hematopoiesis. The oscillatory behavior seen in these diseases is thought by some to be due to irregular feedback control [42]. Other, more sophisticated models suggest that the abnormality lies not in the feedback loop but in an elevated neutrophil apoptotic rate that perturbs the normal regulation of stem cell dynamics [43].

There exists a category of recent deterministic models, which account for variability among disease cases or different types of disease. For example, Stiehl et al. [44] consider heterogeneity of responses to bone marrow transplants; they developed a model-based methodology for using averaged clinical trial data to estimate responses of individual patients. Other papers address the role of multiple cell lineages, including evolution of leukemia, competition between healthy and leukemic cells and dynamics of multiclonal structure of acute myeloid leukemia (AML) [45, 46]. One implication is that enhanced self-renewal may be a key mechanism in the clonal selection process. Simulations suggest that fast proliferating and highly self-renewing cells dominate at primary diagnosis, while relapse following therapy-induced remission is triggered mostly by highly self-renewing but slowly proliferating cells. A similar framework was applied to myelodysplastic syndrome (MDS) which is an important example of a malignant disease with a hypothetical stem cell origin [47]. The results stress the importance of self-renewal in cancer dynamics and allow concluding that invoking slowly proliferating cancer cells helps explain clinical dynamics and observations such as treatment resistance.

Models with Structure

Case Study: Structured Roeder Model and Competing Feedback In Roeder's model, HSCs are assumed to exist in two growth compartments: quiescent (denoted by A) and proliferating (denoted by Ω). At the beginning of every time step (representing 1 h), a stem cell may transfer from A to Ω with probability ω or from Ω to A with probability α. Each stem cell has a time-dependent affinity, denoted by $a(t)$, and the affinity ranges between $a_{\min}$ and $a_{\max}$ (which are estimated to be 0.002 and 1.0, respectively [29]). A cell with a high affinity has a high chance of remaining in the A environment or transferring to it. Likewise, a cell with a low affinity is more likely to remain in the Ω environment or transfer to it, where it starts proliferating. The transition probabilities can be variously defined.

Proliferating cells in the Ω compartment progress through various stages of the cell cycle: G1, S, G2, and M. The G1 phase is the longest period of growth during which the cell generates new organelles. The S phase is the period when DNA synthesis and replication occurs. The G2 phase is the short period of growth when the cell prepares for mitosis, and the M phase, or mitosis, is when the cell divides into two daughter cells. Only Ω cells in the G1 phase of the cell cycle can transfer to A. The Ω cells spend about two thirds of their time in the G1 phase. For each cell that remains in the A compartment, its affinity increases by a factor of r (estimated as 1.1). Similarly, cells that remain in Ω, decrease their affinity by a factor of d (estimated as 1.05). The affinity of a cell stops increasing once it reaches the maximal value, a_{max}. Stem cells whose affinity reaches the minimum affinity, a_{min}, differentiate into a proliferating precursor and then into a nonproliferating mature cell. Each cell in Ω has an internal time counter, $c(t)$, that indicates its position in the cell cycle (measured in hours). Each time step is equivalent to 1 h. Consequently, at each time step, $c(t)$ increases by 1. After $c(t)$ reaches its maximal value of 48, it recycles back to 0 at the next time step, resulting in a 49-h cell cycle. Cells entering Ω start with a counter that is set at $c(t) = 32$ corresponding to the beginning of the S phase. For the first 17 h, the cell progresses through the S, G2, and M phases and divides into two cells once $c(t) = 48$. Then for the next 32 h, ($c(t) = 0, \ldots, 31$), the cell remains in the G1 phase. If at the end of this period the cell has not transferred to A, it reenters the S, G2, and M phases and the cycle repeats.

In the original work of Roeder et al. [29], an agent-based model has been used to follow the dynamics of stem cell counts in the bone marrow. In a subsequent work by Kim et al. [48], a quasi-stochastic (operating on mean values) model has been derived, for the purpose of elucidating the progression of chronic myelogenous leukemia (CML). As it can be noticed, the mechanism of A↔ Ω transitions amounts to an elaborate protective system for the dormant HSC (the "internal feedback" of Arino and Kimmel [35]). This is one of the places in the model, where stochastic effects are likely to play a major role, since HSC may be organized in relatively small clonal colonies [28].

Continuous Maturation

Cell differentiation is a process by which dividing hematopoietic precursor cells become specialized, less proliferative, and equipped to perform specific functions. More generally, differentiation occurs many times during the development of a multicellular organism as the organism changes from a single zygote to a complex system with cells of different types. During tissue repair and normal cell turnover, a steady supply of somatic cells is ensured by proliferation of corresponding adult stem cells (such as the hemopoietic stem cells), which retain the capability for self-renewal. Nonhematologic cancers are likely to originate from a population of cancer stem cells that have properties comparable to those of stem cells [49]. Stem cell state and fate depend on the environment, which ensures that the critical stem cell character and activity in homeostasis is conserved, and that repair and development are accomplished [50]. While different genetic and epigenetic processes are involved in formation and maintenance of different tissues, the dynamics of population depends on the relative importance of symmetric and asymmetric cell divisions, cell differentiation, and death. One established view of the differentiation process is that of a series of discrete compartments, which can be modeled by a system of ordinary DEs describing dynamics of cells at different maturation stages and transition between the stages. This corresponds to the assumption [51, 52] that in each lineage of cell precursors there exists a discrete chain of maturation stages, which are sequentially traversed. However, differentiated precursors form such a clear sequence only under homeostatic (steady-state) conditions. Committed cells generally form a continuous sequence, which may involve incremental stages, part of which may be reversible. These observations invoke not only the fundamental biological questions of whether cell differentiation is a discrete or a continuous process and what the measure of cell differentiation is but also the question of how to choose an appropriate modeling approach. Is the pace of maturation (commitment) dictated by successive divisions, or is maturation a continuous process decoupled from proliferation? Early continuous maturation models have been conceived by Mackey group [43], but the question of correspondence of the discrete compartment and continuous maturation models is still open.

A recent paper that mathematically compares these two types of models is Doumic et al. [53]. The continuous maturation model has the form of a partial DE of transport type—a structured population equation with a nonlinear feedback loop. This models the signaling process due to cytokines, which regulate the differentiation and proliferation process. The dynamics of the model is compared to that of its discrete counterpart. Without an attempt to describe the details, let us note that the continuous and discrete maturation model (which assume the same mode of regulation) differ with respect to dynamics. One of the manifestations is that the discrete model has a richer set of stable equilibria including some with the upper stages of the hierarchy depleted, but the lower ones still capable of maintaining proliferation. Another mathematical approach is to unify the continuous and discrete maturation in a single model [54]. An earlier attempt includes mathematical proof of equivalence between a transport type approach and a bp [55].

Stochastic Models of Hematopoiesis

Stochastic processes (in particular the bp's) can be used to model biological phenomena of some complexity, at cellular or subcellular levels [56]. Probabilistic population dynamics arise from the interplay of the population growth pattern with probability. Thus, the classical G–W bp defines the pattern of population growth using sums of independent and identically distributed (iid) rv's; the population evolves from generation to generation by the individuals getting iid numbers of children. This mode of proliferation is frequently referred to as "free growth" or "free reproduction." The "simple deterministic model of hematopoiesis" considered earlier on is in fact describing the expected (mean) trajectories of a multitype G–W process.

The formalism of the G–W process provides insight into one of the fundamental problems of cell populations, the extinction problem and its complement, the question of size stabilization: If a freely reproducing population does not die out, can it stabilize, or does it have to grow beyond bounds? The answer is that there are no freely reproducing populations with stable sizes. Population size stability, if it exists in the real world, is the result of forces other than individual reproduction, of the interplay between populations, and their environment. This is true for processes much more general than the G–W process. A fusion between branching and environment pressure constitutes a challenge for stochastic models (if one does not wish to resort to simulation only); the same way, selection constitutes a challenge to population genetics models.

If unlimited growth models make more sense in the context of proliferation of cancer cells, then what the rate of the unlimited growth is? It can be answered not only within the generation counting framework of G–W type processes but also in more general branching models. In all these frameworks, in the supercritical case, when the average number of progeny of an individual is greater than 1, the growth pattern is asymptotically exponential. The parameter of this exponential growth is the famous Malthusian parameter. In the supercritical case, we can answer not only questions about the rate of growth but also questions about the asymptotic composition of nonextinct populations. What will the age distribution tend to be? What is the probability of being firstborn? What is the average number of second cousins? Importantly for biological applications, many of these questions do not have natural counterparts in deterministic models of unlimited growth.

Case Study: Model of SNC → MDS → AML Transition [57]

Severe congenital neutropenia (SCN) is a life-threatening infection in children which can be avoided through the use of recombinant GCSF. However, SCN often transforms into secondary MDS (sMDS) and then into secondary AML (sAML). A great unresolved clinical question is whether chronic, pharmacological doses of GCSF contribute to this transformation [58]. A number of epidemiological clinical trials have demonstrated a strong association between exposure to GCSF and sMDS/sAML [59–62] Mutations in the distal domain of the GCSF receptor (GCSFR) have been isolated from patients with SCN who developed sMDS/sAML or patients with de novo MDS [63]. Recently, clonal evolution over approximately 20 years was documented in a patient with SCN who developed sMDS/sAML [64]. Clonal evolution of a sick hematopoietic progenitor cell in SCN involves perturbations in proximal and distal signaling networks triggered by a mutant GCSFR. A summary of signaling pathways taking part in the response to GCSF in normal and mutant cells is presented in Fig. 7.2. Transition from SCN → sMDS → sAML involves chance mechanisms, such as mutations, drift, and transcription and receptor noise, which require that stochastic models to be used [1].

The model of this process in Kimmel and Corey [57] assumes that a limited mutation load at the SCN phase causes neutropenia and fluctuations of cell population size. With time, accumulation of driver mutations causes expansion of mutant clones, which however are not yet expanding at a dramatic rate. At some point in time,

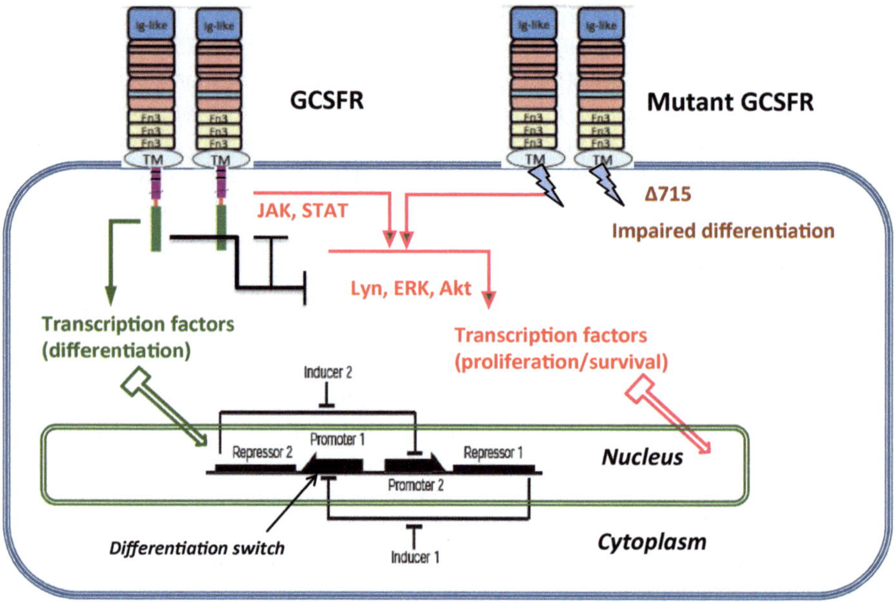

Fig. 7.2 Dynamic stochastic model of *impaired differentiation* in granulocyte precursors. Granulocyte colony-stimulating factor (*GCSF*) signaling occurs through its cognate receptor, granulocyte colony-stimulating factor receptor (*GCSFR*). It involves both proximal signaling networks consisting of signaling molecules such as *Lyn, JAK2, Akt*, and *ERK1/2*, and distal gene regulatory networks consisting of transcription factors. Together, these signaling networks promote proliferation, survival, and differentiation. In patients with severe congenital neutropenia, the earliest known mutations to contribute to transformation to secondary MDS or AML are nonsense mutations in the *GCSFR* gene. This mutation leads to a truncated receptor, one of the more common being *GCSFR* delta 715. (From Ref. [57]. See more at: http://journal.frontiersin.org/Journal/10.3389/fonc.2013.00089/full#sthash.ASCNmDwj.dpuf. Copyright: ©2013 Kimmel and Corey)

mutations accumulate sufficiently to cause a major change in the proliferation law and the now malignant cell population starts rapidly expanding. In summary: (i) At the time of diagnosis of SCN, GCSF therapy is initiated, which induces an initial series of X driver mutations, occurring at random times. (ii) The X-th mutation causes transition to the MDS, during which further Y mutations occur. (iii) After $X+Y$ mutations, the AML stage begins, during which the subsequent mutant clones grow at increasing rate, which in turn shortens the times at which still new mutations appear. In the model, the increasing proliferation rate of successive mutant clones causes acceleration of growth of the malignant bone marrow stem cell population, which shortens the time interval to appearance of new clones, which in turn increases proliferation rate, and so forth; this results in a positive feedback (Fig. 7.3). Stochastic nature of the process (the times to appearance of each next mutant are random) causes a spread of the timing of the subsequent mutations, particularly the first X mutations during the SCN phase. This may result in the transition to MDS not manifesting itself for a very long time in a fraction of cases.

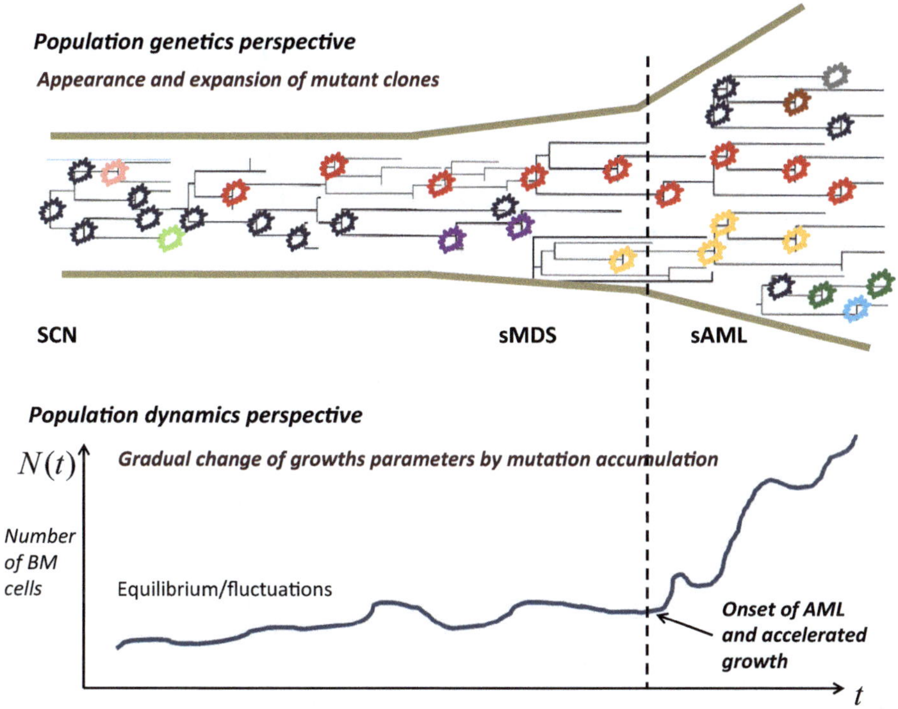

Fig. 7.3 Proliferating healthy cells in the bone marrow mutate at random times, possibly influenced by superpharmacological doses of *GCSF*. As long as the cell population size is kept in check, genetic drift and selection remove many of the mutants, whereas some mutants persist. When the population expands, new mutant clones become more easily established. At some point, a qualitative change in the proliferation rate occurs and the now malignant cell population starts rapidly expanding. *GCSF* granulocyte colony-stimulating factor, *sMDS* secondary myelodysplastic syndrome, *sAML* secondary acute myeloid leukemia, *BM* bone marrow. (From Ref. [57]. See more at: http://journal.frontiersin.org/Journal/10.3389/fonc.2013.00089/full#sthash.ASCNmDwj.dpuf. Copyright: ©2013 Kimmel and Corey)

Stochastic dynamics plays a major role in the model. For a new subclone, stochastic theory is used to estimate extinction probability, with extinction after more than a few cell generations being negligible in view of the growth advantage of the new clone. However, the time at which the next mutation occurs in a cell clone is also stochastic and it is as a rule more dispersed for the slower-growing clones. Therefore, the time to reach the threshold number of bone marrow stem cells (which in our model defines the time at sAML diagnosis) is an rv. One of the questions asked is whether dispersion of this time matches the wide distribution of the times at diagnosis [65].

Mathematically, the population-genetic effect of population size-dependent accumulation of mutations occurs as a natural consequence of the proliferation law in the form of a multitype G–W bp [66]. (i) Consecutively arising surviving mutant clones are numbered with the index k, ranging from 1 to k; time interval between the appearance of the k-th and $k+1$-st surviving mutant clones is denoted by τ_k. k-th mutant

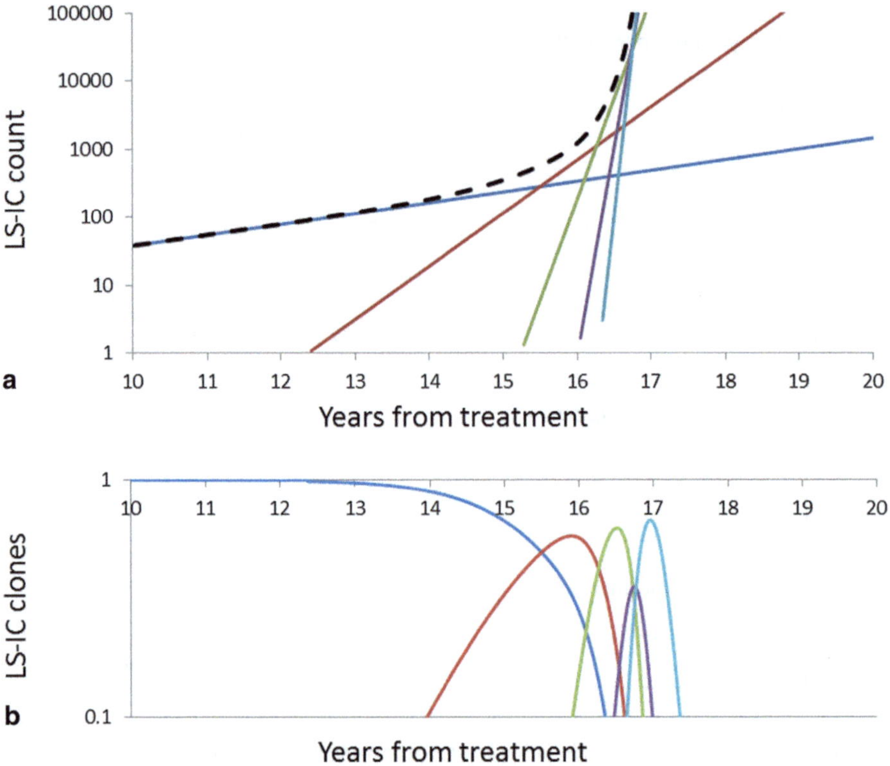

Fig. 7.4 Summary of successive driver mutations in the natural course of the SCN → sMDS → sAML transition. **a** Cell counts in successive mutant clones. *Straight lines* with increasing slopes: cell counts in successive mutant clones. *Thick dashed line*: total mutant cell count. **b** Relative proportions of cells belonging to successive mutant clones. *SCN* severe congenital neutropenia, *sMDS* secondary myelodysplastic syndrome, *sAML* secondary acute myeloid leukemia, *BM* bone marrow. (From Ref. [57]. See more at: http://journal.frontiersin.org/Journal/10.3389/fonc.2013.00089/full#sthash.ASCNmDwj.dpuf. Copyright: ©2013 Kimmel and Corey)

cells have accumulated k driver mutations. (ii) All clones expand as G–W bp's. Cell life length is constant and equal to T, and at that time the cell either produces two progeny with probability b_k (cell type k) or dies (or becomes quiescent or differentiated, which does not make a difference for disease dynamics) with probability $1 - b_k$. (iii) A cell of type k can mutate upon its birth (for definiteness) to type $k + 1$ with probability u. These three rules allow one to derive the probability distributions of time intervals τ_k, probabilities of survival of each clone, and expected growth laws of each clone. Fig. 7.4 depicts the impact of successive driver mutations on the natural course of the SCN → sMDS → sAML transition. Fig. 7.4a depicts counts $N_i(t)$ of cells in successive mutant clones as a function of time. Straight lines with increasing slopes are counts of cells in successive mutant clones. We observe that the time intervals separating the origins of successive clones are decreasing with each mutation event. Thick dashed line represents the total mutant cell count. Fig. 7.4b depicts relative proportions $n_i(t) = N_i(t)/\sum_j N_j(t)$ of cells belonging to successive mutant clones. It is interesting to observe that clones with increasing numbers of mutations dominate transiently, until they are replaced by other clones with higher proliferative capacity (selective value).

It is somewhat surprising that under any combination of coefficients of the model, the range of simulated times at which sAML arises is rather narrow [57]. This outcome is in contrast to the wide spread of times at diagnosis summarized in Rosenberg et al. [65]. The observed distributions of times (and ages) at diagnosis can be matched if a large interindividual variability is assumed. This illustrates how important it is to take into account potential different sources of stochasticity when modeling human disease.

Underlying stochastic effects Most likely mechanisms creating stochastic behavior in hematopoiesis are: (i) asymmetric division of progeny cells, with resulting difference in their fates, and (ii) on–off switching of differentiation status of cells accomplished by hormonal controls such as GCSF or Epo.

Asymmetric division is a possible mechanism by which randomness is inserted into stem cells' decision making. In the past, a prevailing hypothesis concerning stem cell decisions was that at each stem cell division, one of the progeny becomes a committed cell whereas the other remains a stem cell, in this manner providing a perfect balance between commitment and self-renewal. There exist at least two problems with this simple paradigm: first, that this does not seem satisfactory when the demand for committed cells is greater than average, and second, that it has been observed that stem cells can divide both asymmetrically and symmetrically. Observations are consistent with stochastic decisions as to which mode of division to choose. Consequences for population dynamics are different for different stochastic scenarios of asymmetric division, even if on the average these scenarios produce 50–50 committed and stem cells.

An interesting discussion of symmetry and asymmetry in stem cell division has been proposed by Schroeder [67]. Discussing findings in an experimental paper by Wu et al. [68], Schroeder [67] considers a catalogue of versions of symmetric and asymmetric divisions. Symmetric division: undifferentiated hemopoietic precursor cells (HPCs) produce two undifferentiated progeny, whose later fate decisions are not linked to the parent's mitosis. Hypothetical mechanisms of asymmetric divisions in HPCs include: (i) orientation of the division plane that leads to positioning of only one of the progeny close enough to localized extrinsic signals provided by a self-renewal or differentiation niche, (ii) generation of two identical undifferentiated progeny, which being in close spatial contact immediately after mitosis engage in reciprocal feedback signaling, leading to differentiation of only one of them, and (iii) intrinsic cell fate determinants segregate asymmetrically between daughter cells, instructing either self-renewal or differentiation of the receiving daughter. Let us notice that these distinct scenarios do not produce distinctions in deterministic models, where only averages matter, but they lead to possibly widely divergent scenarios in stochastic dynamics. As an example of mechanism (iii), Wu [68] found that Numb, a negative modulator of Notch signaling, which is known to asymmetrically segregate to one progeny during asymmetric division is indeed frequently enriched in one of the two emerging progeny cells. This has been accomplished by analyzing Numb localization in HPCs that had been fixed during mitosis and visualized.

Switches have been proposed to effectively translate the hormonal signals into decision about commitment or further progression. An archetypical molecular bistable switch (Gardner, Collins, and Cantor genetic toggle switch [69]) is deterministic and involves a system of two genes, the products of which are mutual cross-repressors. The system also involves two activators, which momentarily annul the action of the repressors and allow the system to switch. From mathematical point of view, the switch is a dynamical system with two stable equilibria separated by an unstable one. A more sophisticated switch, which moreover is based on confirmed molecular mechanism, is the Laslo switch [21].

However, a molecular switch may involve stochastic mechanisms, which make its action less predictable. This may mean that, if the level of fluctuations is sufficiently high, the switch is oscillating between the two stable equilibria, before or instead of being absorbed by one of them. Such behavior has been observed. In disease state, we may have to do with an aberrant switch with dynamics altered by mutation in one of the important molecular circuits.

Several transcription factors play key roles in regulating myelopoiesis and granulopoiesis. These include the ETS protein PU.1 and the cytosine-cytosine-adenosine-adenosine-thymidine (CCAAT)-enhancer-binding protein-α (CEBP-α), and are often referred to as the "master regulators" of myeloid development [21]. Although PU.1 is sometimes considered to induce myeloid versus lymphoid and monocyte versus granulocyte differentiation, the data suggest that the effects of PU.1 are more complex than this. Similarly, CEBP-α has been considered to direct granulocyte versus monocyte differentiation. PU.1 and CEBP-α constitute a gene regulatory network with bistable properties. Gene regulatory networks may be modified by protein abundance and posttranslational modification, both of which were shown to be induced by activation of cytokine receptors such as GCSFR via kinases. First, PU.1 and CEBP-α undergo serine/threonine phosphorylated triggered by GCSFR activation. Second, GCSFR activation modulates CEBP-α expression, which influences PU.1 function via unidentified mechanisms. Third, GCSFR also influences Gfi-1 expression and activity.

For the granulocyte lineage, the most essential growth factor is GCSF. Its cognate receptor, GCSFR is a member of the hematopoietin cytokine receptor superfamily, which includes receptors for many of the interleukins, colony-stimulating factors (e.g., Epo), cytokines (e.g., leptin), and hormones (e.g., prolactin). As a drug, recombinant human GCSF is used widely to reduce the duration of chemotherapy-induced neutropenia and mobilize into the periphery hematopoietic progenitor cells for transplant [70]. A number of clinical disorders demonstrate importance of GCSF/GCSFR (see below). Mutations in the GCSFR have been found in patients with SCN, MDS, and AML [59, 71].

Laslo et al. [21] describe a regulatory network demonstrating bistability based on a feedback loop between two transcriptional repressors (Egr/Nab-2 and Gfi-1) of PU.1 and GATA-1 genes that drive a common myeloid progenitor cell toward either granulocyte or macrophage fate. As mentioned above, deterministic toggle, or bistable, switch is a circuit which has two stable equilibria, usually separated by an unstable one [69]. Stochastic toggle switches have much richer behavior [72]. Instead of a monotonous approach to the stable equilibrium, the absorbing state is reached via a "saw-like" trajectory. If the time before absorption extends over more than a single cell cycle, the cell remains uncommitted, or in one of the the "intermediate" states, as for example in the paper by Laslo et al. [73] where existence of graded states of cells was experimentally observed and theoretically predicted (albeit using a deterministic switch). On the theoretical side, state space methods have been used by Michaels et al. [74] to find the range of dynamical behaviors exhibited by Laslo-type switch.

Jaruszewicz et al. [75] demonstrate that in a system of bistable genetic switch, the randomness characteristics control in which of the two epigenetic attractors the cell population will settle. They focus on two types of randomness: the one related to gene switching and the one related to protein dimerization. Change of relative magnitudes of these random components for one of the two competing genes introduces a large asymmetry of the protein stationary probability distribution and changes the relative probability of individual gene activation. Increase of randomness associated with a given gene can both promote and suppress activation of the gene. Each gene is repressed by an increase of gene switching randomness and activated by an increase of protein dimerization randomness. In summary, the authors demonstrated that randomness may determine the relative strength of the epigenetic attractors, which may provide a unique mode of control of cell fate decisions.

Traulsen et al. [76] concentrate on the role of hierarchy of the hematopoietic system, discussing the influence of mutations in the hematopoietic system. Although mutations can occur in any cell within hematopoiesis, both the size of the circulating clone and its lifetime depend on the location of the cell of origin in the hematopoietic hierarchy. Mutations in more primitive cells give rise to larger clones that survive for longer, taking also a longer time to appear in the circulation. On the contrary, the smaller clones caused by mutations of more differentiated precursors appear in the circulation much more rapidly after the causal mutation, but they are smaller and survive shorter. Three disease-causing mutations serve as illustrations: the breakpoint cluster region gene–V-abl Abelson murine leukemia viral oncogene homolog 1 gene (BCR–ABL) associated with chronic myeloid leukemia; mutations of the PIG-A gene associated with paroxysmal nocturnal hemoglobinuria; and the V617 F mutation in the JAK2 gene associated with myeloproliferative diseases. Among other, evidence is presented of existence of these mutations in asymptomatic individuals, speculatively, these are mutations in more differentiated precursors. Citing from Traulsen et al. [76]: "In general, we can expect that only a mutation in a hematopoietic stem cell will give long-term disease; the same mutation taking place in a cell located more downstream may produce just a ripple in the hematopoietic ocean."

Wilson et al. [77] argue based on a combination of flow cytometry with label-retaining assays (BrdU and histone H2B-GFP) that there exists a population of dormant mouse HSCs (d-HSCs) within the lin− Sca1+ cKit+ CD150+ CD48− CD34− population. Computational modeling suggests that d-HSCs divide about every 145 days, or five times per lifetime. d-HSCs harbor the vast majority of multilineage long-term self-renewal activity. They form a reservoir of the most potent HSCs during homeostasis, and are efficiently activated to self-renew in response to bone marrow injury or GCSF stimulation. After reestablishment of homeostasis, activated HSCs return to dormancy, suggesting that HSCs are not stochastically entering the cell cycle but they reversibly switch from dormancy to self-renewal under conditions of hematopoietic stress.

Becker et al. [78] show that Epo receptors have the ability to cope with steady-state and acute demand in the hematopoietic system. By mathematical modeling of quantitative data and experimental validation, these authors showed that rapid ligand depletion and replenishment of the cell surface receptor are characteristic features of the Epo receptor (EpoR). The amount of Epo–EpoR complexes and EpoR activation integrated over time corresponds linearly to ligand input.

Models of Mutations and Evolution of Disease

Carcinogenesis Models

Carcinogenesis modeling has had an established history of using stochastic models, beginning with the Knudson two-hit model. Successor models include the multi-hit model and eventually to the two-stage clonal expansion model of Moolgavkar [79]. With almost 1000 citations, this paper might be called one of the most influential ever mathematical models in cancer research. Concerning its application in leukemias, see, e.g., Radivoyevich et al. [80].

We will focus mostly on models of leukemogenesis, conceived in the genome-sequencing era. These models have the following features, which are a novelty due to both evolution of thinking in inflow of a large number of new variant data:

1. Mutations are identified as variants in studies in which whole exomes or even genomes are sequenced for each individual in the study.
2. Functionality of mutations is determined in two stages: first by bioinformatics algorithms (usually based on evolutionary comparisons) and then by wet-laboratory studies of pathways influenced by these mutations.
3. Progression of leukemogenesis is based on the concept of driver and passenger mutations. Driver mutations are selected for most advantageous phenotype of cancer cells, whereas the passenger mutations are neutral byproducts of carcinogenesis and serve as molecular clocks of the process.

A number of interesting models of mutations leading to cancer have recently been published (see references further on). They all explore models of proliferation, frequently using bp's, combining them with models of driver and passenger mutations. Driver mutations are those that, although they might have arisen spontaneously, provide selective advantage for the emerging cancer proliferation, particularly against the background of already existing inherited or acquired mutations. Passenger mutations are generally neutral and their accumulation may provide a molecular "clock" indicating how long it has been since the cancer cells deviated from normal cells.

Tumors are initiated by the first genetic alteration that provides a relative fitness advantage. In the case of leukemias, this might represent the first alteration of an oncogene, such as a translocation between BCR and ABL.

Recent paper by Ley et al. [81] addressed the issue of driver mutations contributing to the pathogenesis of AML, using analysis of the genomes of 200 adult cases of de novo AML, either whole-genome sequencing (50 cases) or whole-exome sequencing (150 cases), along with ribonucleic acid (RNA) and microRNA sequencing and DNA-methylation analysis. The conclusion was that AML genomes have fewer mutations than most other adult cancers, with an average of only 13 mutations found in genes. Of these, an average of five was in genes that are recurrently mutated in AML. A total of 23 genes were significantly mutated, and another 237 were mutated in two or more samples. Further analysis suggested strong biologic relationships among several of the genes and categories. Further studies of this kind are likely to lead to insights into the nature of these relationships, although with very few exceptions (such as Beekman et al. [64]) only a single time point in patient lifetime is usually available.

One of the important recent papers with this focus is the mathematical model of the relationship between accumulation of driver and passenger mutation in tumors published by Nowak and Vogelstein groups [82]. In most previous models of tumor evolution, mutations accumulate in cell populations of constant size or of variable size, but the models take into account only one or two mutations. Such models typically address certain aspects of cancer evolution, but not the whole process. In the model presented in Bozic et al. [82], it has been assumed that each new driver mutation leads to a slightly faster tumor growth rate. This model is as simple as possible, because the analytical results depend on only three parameters: the average driver mutation rate u, the average selective advantage associated with driver mutations s, and the average cell division time T. The model is based on G–W bp.

The hypotheses are as follows: At each time step, a cell can either divide or differentiate, senesce, or die. In the context of tumor expansion, there is no difference between differentiation, death, and senescence, because none of these processes will result in a greater number of tumor cells than present prior to that time step. It is assumed that driver mutations reduce the probability that the cell will become "stagnate," i.e., it will differentiate, die, or senesce, although the stagnant cells are not removed from the tumor. A cell with k driver mutations has a stagnation probability $d_k = (1-s)^k/2$. The division probability is $b_k = 1 - d_k$. The parameter s is the selective advantage provided by a driver mutation. When a cell divides, one of the daughter cells can acquire an additional driver mutation with probability u. The theory can accommodate any realistic mutation rate and the major numerical results are only weakly affected by varying the mutation rate.

We can calculate the average time between the appearances of successful cell lineages. Not all new mutants are successful, because stochastic fluctuations may lead to the extinction of a lineage. The lineage of a cell with k driver mutations survives only with a probability of approximately $1 - d_k/b_k \cong 2sk$. Assuming that $u \ll ks \ll 1$, the average time between the first successful cell with k and the first successful cell with $k+1$ driver mutations is given by

$$\tau_k = \frac{T}{ks}\log\frac{2ks}{u}.$$

This result is obtainable from the theory of the G–W process (by elementary means) and the derivation is found in the supplement to Bozic et al. [82]. The cumulative time to accumulate k mutations grows logarithmically with k. On the other hand, the average number of passenger mutations, $n(t)$, present in a tumor cell after t days is proportional to t, that is $n(t) = \nu t / T$, where ν is the rate of acquisition of neutral mutations. Combining the results for driver and passenger mutations, results in a formula for the number of passenger mutations that are expected in a tumor that has accumulated k driver mutations

$$n = \frac{\nu}{2s} \log \frac{4ks^2}{u^2} \log k.$$

Here, n is the number of passengers that were present in the last cell that clonally expanded. Bozic et al. [82] demonstrate that this dependence fits empirical data on several human cancers.

Distribution of Mutational Events in Various Phases of Tumor Growth

This question has been recently addressed by Tomasetti et al. [83]. The framework is not very different from that of Bozic et al. [82]. However, the paper describes mathematically the different phases in which somatic mutations occur in a tissue giving rise to a cancer. Starting from a single fertilized egg, all tissues are created via clonal expansion (development phase). The tissue is then subjected to periodic self-renewals. During development and tissue renewal, passenger mutations occur randomly, undergo clonal expansions and either become extinct or expand as successive passenger mutations accumulate. A driver gene mutation may initiate a tumor cell clone, which then can expand through subsequent driver mutations, eventually yielding a clinically detectable tumor mass (cancerous phase). Passenger mutations occur during this phase as well. The model makes the novel prediction, validated by empirical findings, that the number of somatic mutations in tumors of self-renewing tissues is positively correlated with the age of the patient at diagnosis. Importantly, the analysis indicates that half or more of the somatic (i.e., acquired, non-germline) mutations in tumors of self-renewing tissues occur prior to the onset of neoplasia. This is the case, among others for the chronic lymphocytic leukemia (CLL). The model also provides a novel way to estimate the in vivo tissue-specific somatic mutation rates in normal tissues directly from the sequencing data of tumors.

Stochastic models also allow modeling of eradication of leukemic stem cells while saving healthy stem cells. The paper by Sehl et al. [84] uses an impressive array of analytical and computational tools to analyze a pair of stochastic processes describing proliferation and death of healthy and cancer stem cells under chemotherapy. The question asked concerns the birth and death rates differential between these two cell types, required to eradicate the latter and preserve the former. Mutations, emergence of drug resistance, interactions of cancer and healthy cells, and other complicating factors are disregarded. Because the biological setup is simplified to the extreme, it allows effective mathematical analysis.

A review of gene copy number and loss of heterozygosity and gene mutation profiles demonstrated that relapsed AML invariably represented reemergence or evolution of a founder clone Parkin et al. [85]. Analysis of informative paired persistent AML disease samples uncovered cases with two coexisting dominant clones of which at least one was chemotherapy sensitive and one resistant, respectively. These data support the conclusion that incomplete eradication of AML founder clones rather than stochastic emergence of fully unrelated novel clones underlies AML relapse and persistence.

As a side note, it seems surprising that quite few papers offer estimates of absolute numbers of HSCs and committed cells. Against this background, the paper by Peixoto et al. [86] discusses the mathematics of hematopoiesis based on stochastic hypotheses. Mathematical model that describes normal hematopoiesis across mammals as a stable steady state of a hierarchical stochastic process is also used to understand the detailed dynamics of a range of blood disorders both in humans and in animal models. The paper includes comparative numerical estimates of the numbers of cells in different compartments.

Models of Spatial Effects and Structure

Niches

One of the most important realizations of recent years has been that there existed spatial organization in normal and malignant HSCs in bone marrow. This finding allowed correlating such states as proliferative activity, dormancy, and differentiation, not only with the locations in the bone marrow but also with proximity to some other cells type such as mesenchymal stem cells or osteoblasts. Such organization has been previously postulated in the papers by Roeder and coauthors; however, now it is based on biologically verifiable findings. The locations are referred to as *stem cell niches*. A recent review by Tieu et al. [28] states that the HSC niche is an important regulator of stem cell fate. Complex signaling pathways, such as those involving Notch, Wnt, and Hedgehog, regulate stem cell renewal, differentiation, and quiescence [87 – 89]. Mathematical models can be useful in studying the dynamics of stem cell maintenance and, specifically, spatial considerations related to the structural relationships between stem cells and their progeny with cells of the microenvironment. This paradigm will trigger development of a new generation of deterministic and stochastic models, in which the interaction of spatial and stochastic effects poses new mathematical problems.

Huang [90] discusses in qualitative terms three perspectives outside the realm of their familiar linear deterministic view: (i) state space, (ii) high dimensionality, and (iii) heterogeneity. These concepts jointly offer a new vista on stem cell regulation that naturally explains many novel, counterintuitive observations and their inherent inevitability, obviating the need for ad hoc explanations of their existence based on natural selection.

Bertolusso and Kimmel [91] considered the early carcinogenesis model originally proposed as a deterministic reaction-diffusion system of the following form:

$$\partial c/\partial t = (a(b,c) - d_c)c + \mu$$
$$\partial b/\partial t = \alpha(c)g - d_b b - db$$
$$\partial g/\partial t = \gamma^{-1}(\partial^2 g/\partial x^2) - \alpha(c)g - d_g g + \kappa(c) + db,$$

where c, g, and b are the spatial densities of cells, free-diffusing growth factor, and bound growth factor, respectively. The model explores the spatial effects stemming from growth regulation of precancerous cells by diffusing growth factor molecules. The model has been originally devised [92] for solid tumors spreading along linear or tubular structures such as in lung or breast cancer, but it might apply to systems of similar geometry, which are found in bone marrow. The original deterministic model exhibits Turing instability, producing transient spatial density spikes in cells which are model counterparts of emerging foci of malignant cells. However, the process of diffusion of growth factor molecules is by its nature a stochastic random walk. An interesting question emerges to what extent the dynamics of the deterministic diffusion model approximates the stochastic process generated by the model. Bertolusso and Kimmel [91] addressed this question using simulations with a new software tool called sbioPN (spatial biological Petri Nets). The conclusion is that whereas single-realization dynamics of the stochastic process is very different from the behavior of the reaction diffusion system, it is becoming more similar when averaged over a large number of realizations. The degree of similarity depends on model parameters. Interestingly, despite the differences, typical realizations of the stochastic process include spikes of cell density, which however are spread more uniformly and are less dependent of initial conditions than those produced by the reaction-diffusion system.

References

1. Whichard ZL, Sarkar CA, Kimmel M, Corey SJ. Hematopoiesis and its disorders: a systems biology approach. Blood. 2010;115(12):2339–47.
2. Snijder B, Pelkmans L. Origins of regulated cell-to-cell variability. Nat Rev Mol Cell Biol. 2011;12(2):119–25.
3. Pelkmans L. Using cell-to-cell variability-a new era in molecular biology. Science. 2012;336(6080):425–6.
4. Raue A, Becker V, Klingmüller U, Timmer J. Identifiability and observability analysis for experimental design in nonlinear dynamical models. Chaos. 2010;20(4):045105.
5. Laurent M, Deschatrette J, Wolfrom Claire M. Unmasking chaotic attributes in time series of living cell populations. PLoS ONE. 2010;5(2):e9346.
6. Kimmel M, Darzynkiewicz Z, Arino O, Traganos F. Analysis of a cell cycle model based on unequal division of metabolic constituents to daughter cells during cytokinesis. J Theor Biol. 1984;110(4):637–64.

7. Schroeder T. Long-term single-cell imaging of mammalian stem cells. Nat Methods. 2011;8(4):S30–5.
8. Takizawa H, Boettcher S, Manz MG. Demand-adapted regulation of early hematopoiesis in infection and inflammation. Blood. 2012;119(13):2991–3002.
9. Ogawa M. Stochastic model revisited. Int J Hematol. 1999;69(1):2.
10. Abkowitz JL, Catlin SN, Guttorp P. Evidence that hematopoiesis may be a stochastic process in vivo. Nat Med. 1996;2(2):190–7.
11. Brock A, Chang H, Huang S. Non-genetic heterogeneity—a mutation-independent driving force for the somatic evolution of tumours. Nat Rev Genet. 2009;10(5):336–42.
12. Cohen Ariel A, et al. Dynamic proteomics of individual cancer cells in response to a drug. Science. 2008;322(5907):1511–6.
13. Sigal A, et al. Variability and memory of protein levels in human cells. Nature. 2006;444(7119): 643–6.
14. Webb GF. Random transitions, size control, and inheritance in cell population dynamics. Math Biosci. 1987;85(1):71–91.
15. Arino O, Kimmel M. Asymptotic analysis of a cell cycle model based on unequal division. SIAM J Appl Math. 1987;47(1):128–45.
16. Tyson JJ, Hannsgen KB. Cell growth and division: a deterministic/probabilistic model of the cell cycle. J Math Biol. 1986;23(2):231–46.
17. Harnevo LE, Agur Z. Drug resistance as a dynamic process in a model for multistep gene amplification under various levels of selection stringency. Cancer Chemother Pharmacol. 1992;30(6):469–76.
18. Kimmel M, Axelrod DE. Mathematical models of gene amplification with applications to cellular drug resistance and tumorigenicity. Genetics. 1990;125(3):633–44.
19. Gupta PB, Fillmore CM, Jiang G, Shapira SD, Tao K, Kuperwasser C, Lander ES. Stochastic state transitions give rise to phenotypic equilibrium in populations of cancer cells. Cell. 2011;146(4):633–44.
20. Chang HH, Hemberg M, Barahona M, Ingber DE, Huang S. Transcriptome-wide noise controls lineage choice in mammalian progenitor cells. Nature. 2008;453(7194):544–7.
21. Laslo P, Spooner CJ, Warmflash A, Lancki DW, Lee H-J, Sciammas R, Gantner BN, Dinner AR, Singh H. Multilineage transcriptional priming and determination of alternate hematopoietic cell fates. Cell. 2006;126(4):755–66.
22. Muzzey D, van Oudenaarden A. When it comes to decisions, myeloid progenitors crave positive feedback. Cell. 2006;126(4):650–2.
23. Loose M, Swiers G, Patient R. Transcriptional networks regulating hematopoietic cell fate decisions. Curr Opin Hematol. 2007;14(4):307–14.
24. Raser JM, O'Shea EK. Noise in gene expression: origins, consequences, and control. Science. 2005;309(5743):2010–13.
25. Elowitz MB, Levine AJ, Siggia ED, Swain PS. Stochastic gene expression in a single cell. Sci Sig. 2002;297(5584):1183.
26. Lipniacki T, Paszek P, Marciniak-Czochra A, Brasier AR, Kimmel M. Transcriptional stochasticity in gene expression. J Theor Biol. 2006;238(2):348–67.
27. Samoilov MS, Price G, Arkin AP. From fluctuations to phnotypes: the physiology of noise. Sci Signal. 2006;2006(366):re17.
28. Tieu KS, Tieu RS, Martinez-Agosto JA, Sehl ME. Stem cell niche dynamics: from homeostasis to carcinogenesis. Stem Cells Int. 2012;2012:9.
29. Roeder I, Horn M, Glauche I, Hochhaus A, Mueller Martin C, Loeffler M. Dynamic modeling of imatinib-treated chronic myeloid leukemia: functional insights and clinical implications. Nat Med. 2006;12(10):1181–4.
30. Uguz A, Coskun M, Yuzbey S, Kizilors A, Karadogan I, Gura A, Yoldas B, Oygur N, Yegin O. Apoptosis of cord blood neutrophils and their response to colony-stimulating factors. Am J Perinatol. 2003;19(08):427–34.
31. Lajtha LG, Pozzi LV, Schofield R, Fox M. Kinetic properties of haemopoietic stem cells. Cell Prolif. 1969;2(1):39–49.

32. Haurie C, Dale DC, Mackey MC. Cyclical neutropenia and other periodic hematological disorders: a review of mechanisms and mathematical models. Blood. 1998;92(8):2629–40.
33. Wazewska-Czyzewska M, Lasota A. Mathematical models of the red cell system. Mat Stos. 1976;6:25–40.
34. Lo W-C, Chou C-S, Gokoffski KK, Wan FY-M, Lander AD, Calof AL, Nie Q. Feedback regulation in multistage cell lineages. Math Biosci Eng. 2009;6(1):59.
35. Arino, Ovide, and Marek Kimmel. "Stability analysis of models of cell production systems." *Mathematical Modelling* 1986;7(9):1269–1300.
36. Marciniak-Czochra A, Stiehl T, Ho AD, Jäger W, Wagner W. Modeling of asymmetric cell division in hematopoietic stem cells-regulation of self-renewal is essential for efficient repopulation. Stem Cells Dev. 2009;18(3):377–86.
37. Nakata Y, Getto P, Marciniak-Czochra A, Alarcón T. Stability analysis of multi-compartment models for cell production systems. J Biol Dyn. 2012;6(Supp. 1):2–18.
38. Getto P, Marciniak-Czochra A, Nakata Y, Vivanco M. dM. Global dynamics of two-compartment models for cell production systems with regulatory mechanisms. Math Biosci. 2013;245:258–68.
39. Stiehl T, Marciniak-Czochra A. Characterization of stem cells using mathematical models of multistage cell lineages. Math Comput Model. 2011;53(7):1505–17.
40. Marciniak-Czochra A, Stiehl T, Wagner W. Modeling of replicative senescence in hematopoietic development. Aging. 2009;1(8):723.
41. Lander AD, Gokoffski KK, Wan FYM, Nie Q, Calof AL. Cell lineages and the logic of proliferative control. PLoS Biol. 2009;7(1):e1000015.
42. Foley C, Mackey MC. Dynamic hematological disease: a review. J Math Biol. 2009;58(1-2): 285–322.
43. Bernard S, Bélair J, Mackey MC. Oscillations in cyclical neutropenia: new evidence based on mathematical modeling. J Theor Biol. 2003;223(3):283–98.
44. Stiehl T, Ho AD, Marciniak-Czochra A. The impact of CD34 + cell dose on engraftment after stem cell transplantations: personalized estimates based on mathematical modeling. Bone Marrow Transp. 2014;49:30–7.
45. Stiehl T, Marciniak-Czochra A. Mathematical modelling of leukemogenesis and cancer stem cell dynamics. Math Mod Natural Phenomena. 2012;7:166–202.
46. Stiehl T, Baran N, Ho AD, Marciniak-Czochra A. Clonal selection and therapy resistance in acute leukemias: mathematical modelling explains different proliferation patterns at diagnosis and relapse. J R Soc Interface. 2014;11:20140079. http://dx.doi.org/10.1098/rsif.2014.0079.
47. Walenda T, Stiehl T, Braun H, Froebel J, Ho AD, Schroeder T, Goecke T, Germing U, Marciniak-Czochra A, Wagner W. Feedback signals in myelodysplastic syndromes: increased self-renewal of the malignant clone suppresses normal hematopoiesis. PLoS Comput Biol. 2014;10(4):e1003599.
48. Kim PS, Lee PP. Levy Doron. A PDE model for imatinib-treated chronic myelogenous leukemia. Bull Math Biol. 2008;70(7):1994–2016.
49. Al-Hajj M, Clarke MF. Self-renewal and solid tumor stem cells. Oncogene. 2004;23(43): 7274–82.
50. Moore KA, Lemischka IR. Stem cells and their niches. Science. 2006;311(5769):1880–5.
51. Lord BI. Biology of the haemopoietic stem cell. In: Stem cells. 1997. pp. 401–22.
52. Uchida N, Fleming WH, Alpern EJ, Weissman IL. Heterogeneity of hematopoietic stem cells. Curr Opin Immunol. 1993;5(2):177–84.
53. Doumic M, Marciniak-Czochra A, Perthame B, Zubelli JP. A structured population model of cell differentiation. SIAM J Appl Math. 2011;71(6):1918–40.
54. Gwiazda P, Jamróz G, Marciniak-Czochra A. Models of discrete and continuous cell differentiation in the framework of transport equation. SIAM J Math Anal. 2012;44(2):1103–33.
55. Arino O, Kimmel M. Comparison of approaches to modeling of cell population dynamics. SIAM J Appl Math. 1993;53(5):1480–504.
56. Haccou P, Jagers P, Vatutin VA. Branching processes: variation, growth, and extinction of populations. Vol. 5. Cambridge: Cambridge University Press; 2005.

57. Kimmel M, Corey S. Stochastic hypothesis of transition from inborn neutropenia to AML: interactions of cell population dynamics and population genetics. Front Oncol. 2013;3:89.
58. Glaubach T, Corey SJ. From famine to feast: sending out the clones. Blood. 2012;119(22): 5063–4.
59. Dong F, Van Paassen M, Van Buitenen C, Hoefsloot LH, Lowenberg B, Touw IP. A point mutation in the granulocyte colony-stimulating factor receptor (G-CSF-R) gene in a case of acute myeloid leukemia results in the overexpression of a novel G-CSF-R isoform. Blood. 1995;85(4):902–11.
60. Donadieu J, Leblanc T, Bader Meunier B, Barkaoui M, Fenneteau O, Bertrand Y, Maier-Redelsperger M, et al. Analysis of risk factors for myelodysplasias, leukemias and death from infection among patients with congenital neutropenia. Experience of the French Severe Chronic Neutropenia Study Group. Haematologica. 2005;90(1):45–53.
61. Rosenberg PS, Alter BP, Bolyard AA, Bonilla MA, Boxer LA, Cham B, Fier C, et al. The incidence of leukemia and mortality from sepsis in patients with severe congenital neutropenia receiving long-term G-CSF therapy. Blood. 2006;107(12):4628–35.
62. Germeshausen M, Skokowa J, Ballmaier M, Zeidler C, Welte K. G-CSF receptor mutations in patients with congenital neutropenia. Curr Opin Hematol. 2008;15(4):332–7.
63. Beekman R, Touw IP. G-CSF and its receptor in myeloid malignancy. Blood. 2010;115(25): 5131–6.
64. Beekman R, Valkhof MG, Sanders MA, van Strien PMH, Haanstra JR, Broeders L, Geertsma-Kleinekoort WM, et al. Sequential gain of mutations in severe congenital neutropenia progressing to acute myeloid leukemia. Blood. 2012;119(22):5071–7.
65. Rosenberg PS, Zeidler C, Bolyard AA, Alter BP, Bonilla MA, Boxer LA, Dror Y, et al. Stable longâŁłterm risk of leukaemia in patients with severe congenital neutropenia maintained on G-CSF therapy. Br J Haematol. 2010;150(2):196–9.
66. Kimmel M, Axelrod DE. Branching processes in biology. Springer New York;2002.
67. Schroeder T. Asymmetric cell division in normal and malignant hematopoietic precursor cells. Cell Stem Cell. 2007;1(5):479–81.
68. Wu M, Kwon HY, Rattis F, Blum J, Zhao C, Ashkenazi R, Jackson TL, Gaiano N, Oliver T, Reya T. Imaging hematopoietic precursor division in real time. Cell Stem Cell. 2007;1(5):541–54.
69. Gardner TS, Cantor CR, Collins JJ. Construction of a genetic toggle switch in Escherichia coli. Nature. 2000;403(6767):339–42.
70. Vose JM, Armitage JO. Clinical applications of hematopoietic growth factors. J Clin Oncol. 1995;13(4):1023–35.
71. Awaya N, Uchida H, Miyakawa Y, Kinjo K, Matsushita H, Nakajima H, Ikeda Y, Kizaki M. Novel variant isoform of G-CSF receptor involved in induction of proliferation of FD-CPâŁł2 cells: relevance to the pathogenesis of myelodysplastic syndrome. J Cell Physiol. 2002;191(3):327–35.
72. Loinger A, Lipshtat A, Balaban NQ, Biham O. Stochastic simulations of genetic switch systems. Phys Rev E. 2007;75(2):021904.
73. Laslo P, Pongubala JMR, Lancki DW, Singh H. Gene regulatory networks directing myeloid and lymphoid cell fates within the immune system. Semin Immunol. 2008;20(4):228–35.
74. Michaels JL, Naudot V, Liebovitch LS. Dynamic stabilization in the PU1-GATA1 circuit using a model with time-dependent kinetic change. Bull Math Biol. 2011;73(9):2132–51.
75. Jaruszewicz J, Zuk PJ, Lipniacki T. Type of noise defines global attractors in bistable molecular regulatory systems. J Theor Biol. 2013;317:140–51.
76. Traulsen A, Pacheco JM, Luzzatto L, Dingli D. Somatic mutations and the hierarchy of hematopoiesis. Bioessays. 2010;32(11):1003–8.
77. Wilson A, Laurenti E, Oser G, van der Wath RC, Blanco-Bose W, Jaworski M, Offner S, et al. Hematopoietic stem cells reversibly switch from dormancy to self-renewal during homeostasis and repair. Cell. 2008;135(6):1118–29.
78. Becker V, Schilling M, Bachmann J, Baumann U, Raue A, Maiwald T, Timmer J, Klingmuller U. Covering a broad dynamic range: information processing at the erythropoietin receptor. Sci Signal. 2010;328(5984):1404.

79. Moolgavkar SH, Knudson AG. Mutation and cancer: a model for human carcinogenesis. J Natl Cancer Inst. 1981;66(6):1037–52.
80. Radivoyevitch T, et al. Quantitative modeling of chronic myeloid leukemia: insights from radiobiology. Blood. 2012;119(19):4363–71.
81. Ley TJ, et al. Genomic and epigenomic landscapes of adult de novo acute myeloid leukemia. N Engl J Med. 2013;368(22):2059–74.
82. Bozic I, Antal T, Ohtsuki H, Carter H, Kim D, Chen S, Karchin R, Kinzler KW, Vogelstein B, Nowak MA. Accumulation of driver and passenger mutations during tumor progression. Proc Natl Acad Sci U S A. 2010;107(43):18545–50.
83. Tomasetti C, Vogelstein B, Parmigiani G. Half or more of the somatic mutations in cancers of self-renewing tissues originate prior to tumor initiation. Proc Natl Acad Sci U S A. 2013;110(6):1999–2004.
84. Sehl M, Zhou H, Sinsheimer JS, Lange KL. Extinction models for cancer stem cell therapy. Math Biosci. 2011;234(2):132–46.
85. Parkin B, Ouillette P, Li Y, Keller J, Lam C, Roulston D, Li C, Shedden K, Malek SN. Clonal evolution and devolution after chemotherapy in adult acute myelogenous leukemia. Blood. 2013;121(2):369–77.
86. Peixoto D, Dingli D, Pacheco JM. Modelling hematopoiesis in health and disease. Math Comput Model. 2011;53(7):1546–57.
87. Morrison SJ, Kimble J. Asymmetric and symmetric stem-cell divisions in development and cancer. Nature. 2006;441(7097):1068–74.
88. Molofsky AV, Pardal R, Morrison SJ. Diverse mechanisms regulate stem cell self-renewal. Curr Opin Cell Biol. 2004;16(6):700–7.
89. Shenghui H, Nakada D, Morrison SJ. Mechanisms of stem cell self-renewal. Annu Rev Cell Dev. 2009;25:377–406.
90. Huang S. Systems biology of stem cells: three useful perspectives to help overcome the paradigm of linear pathways. Philos Trans R Soc Lond B Biol Sci. 2011;366(1575):2247–59.
91. Bertolusso R, Kimmel M. Modeling spatial effects in early carcinogenesis: stochastic versus deterministic reaction-diffusion systems. Math Model Nat Phenom. 2012;7(1):245–60.
92. Marciniak-Czochra A, Kimmel M. Reaction-difusion model of early carcinogenesis: the effects of influx of mutated cells. Math Model Nat Phenom. 2008;3(7):90–114.

Chapter 8
Systems Analysis of High-Throughput Data

Rosemary Braun

Abstract Modern high-throughput assays yield detailed characterizations of the genomic, transcriptomic, and proteomic states of biological samples, enabling us to probe the molecular mechanisms that regulate hematopoiesis or give rise to hematological disorders. At the same time, the high dimensionality of the data and the complex nature of biological interaction networks present significant analytical challenges in identifying causal variations and modeling the underlying systems biology. In addition to identifying significantly disregulated genes and proteins, integrative analysis approaches that allow the investigation of these single genes within a functional context are required. This chapter presents a survey of current computational approaches for the statistical analysis of high-dimensional data and the development of systems-level models of cellular signaling and regulation. Specifically, we focus on multi-gene analysis methods and the integration of expression data with domain knowledge (such as biological pathways) and other gene-wise information (e.g., sequence or methylation data) to identify novel functional modules in the complex cellular interaction network.

Keywords Statistical analysis · High-throughput data · Microarrays · Sequencing · NGS · Genomics · Machine learning · Network models

Introduction

The precise coordination of complex and adaptive living processes relies upon systems that regulate transcriptional, posttranscriptional, and epigenetic control of gene expression and protein production. In contrast to the simplified view of the "central dogma" of molecular biology, wherein transcription followed by translation leads linearly from DNA to RNA to protein, it is now understood that there exist feedback loops at each stage, forming a network of regulatory interactions (Fig. 8.1).

R. Braun (✉)
Biostatistics Division, Department of Preventive Medicine
and Northwestern Institute on Complex Systems, Northwestern University,
680 N. Lake Shore Dr., Suite 1400, Chicago, IL 60611, USA
e-mail: rbraun@northwestern.edu

S. J. Corey et al. (eds.), *A Systems Biology Approach to Blood*,
Advances in Experimental Medicine and Biology 844, DOI 10.1007/978-1-4939-2095-2_8

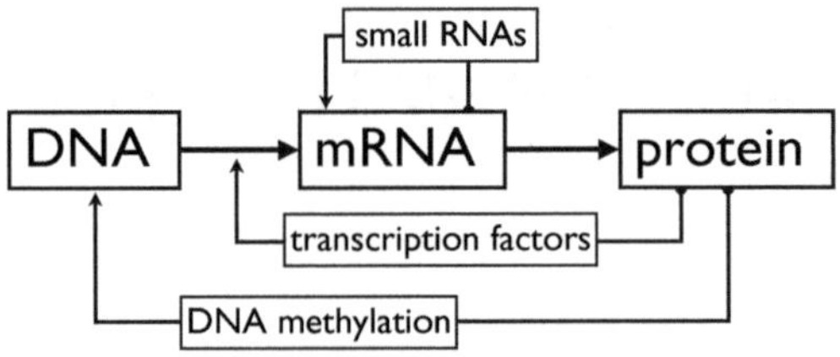

Fig. 8.1 Regulatory mechanisms in molecular biology. DNA is transcribed to mRNA and then translated into protein. The rate of transcription is controlled by a feedback loop in which the level of transcription factor proteins is regulated the activity of the transcriptional complex, and genes can be permanently silenced by methylation of cytosine in CpG promoter regions of the DNA sequence. More recently, it has been discovered that the expression of small noncoding RNA molecules (e.g., microRNAs) can downregulate entire sets of genes by binding to complementary sequences in the mRNA

Identifying functionally relevant genes and unraveling the systems governing their expression can elucidate the molecular mechanisms underlying development and disease, as well as facilitate the development of prognostic tests and therapeutic interventions [1, 2].

Although living organisms have long been thought of as complex systems comprising many strongly interdependent parts [3, 4], the study of biological processes at the systems level remained a theoretical practice until fairly recently. Prior to the completion of the Human Genome Project and the development of high-throughput technologies, limitations on the ability to exhaustively assay samples of interest required that each gene be probed one at a time, leading to a reductionist approach in which biological systems were investigated by examining their parts in isolation. In recent years, however, major technological advances have enabled assays that yield highly detailed genome-wide information for each sample (including sequence, expression, and epigenetic modifications). This unprecedented increase in our ability to probe how every gene is expressed in a particular tissue or responds to a particular environmental perturbation now makes systems biology possible. The wealth of data now being generated in high-throughput profiling studies not only allows gene-level analyses to be applied comprehensively across the entire genome, but provides an immense opportunity to augment reductionist one-gene-at-a-time techniques with systems-level analyses that treat the data in an integrative manner and elucidate the functional association between differentially expressed genes.

Complementing the advances in experimental technologies, advances in computing technology have ushered in an exciting era of computational systems biology. Broadly speaking, computational systems biology investigations may be classified into two groups, each with its own utility and set of challenges: the statistical analysis of high-dimensional data to infer differentially regulated network modules from experimental studies, and the dynamical simulation of these networks to model the occurrence of cellular events. Here, we focus on statistical and machine learning algorithms to draw inferences about regulatory networks from complex data sets. Combined with gene-level analyses, pathway-based methods provide comprehensive analyses of the functional modules that govern biological processes. The objective of this chapter is to provide theoretical and practical knowledge of how high-throughput data can be harnessed to yield mechanistic insights and build predictive models at the systems level.

Generating High-Throughput Data

The accuracy of any systems-level analysis will depend on the quality of the data being analyzed. This, in turn, depends upon the experimental design, the assay technology employed, and the preprocessing of the raw data. Although a full review of these considerations is beyond the scope of this chapter, a brief overview is presented for context.

Experimental Design

Experiments may be designed with several goals in mind:

Class comparison Identification of genes or gene sets behaving differently between predetermined "classes" of samples (e.g., cases and controls, different phenotypes, different stages of development, different treatments, etc.).

Time series Investigation of the dynamics of gene expression changes following an exposure (e.g., to examine how the expression profile changes over time and differs between growth phenotypes).

Class prediction (supervised machine learning) Identification of a minimal set of genes that can be used to categorize a new sample into one of several known types based on its molecular profile (e.g., with the goal of predicting treatment response). Also called supervised machine learning.

Class discovery (clustering/unsupervised ML) Identification of novel groups of samples on the basis of their molecular profiles (e.g., to identify disease subtypes among clinically similar cases that may correspond to differing prognoses).

Network Analysis Identification of differential relationships between molecules, either by analyzing the data in the context of putative interaction networks or by "reverse engineering" the underlying network based on experimental data.

Regardless of the question under consideration, several guiding principles should be observed. First, all high-throughput studies yield a measurements in a feature space (10^5–10^6 probes) that is of much higher dimensionality than the number of samples (often on the order of 10^2). From a mathematical modeling standpoint, these experiments are underdetermined, meaning there are many more variables (genes) than there are equations (samples), and different analysis methods may yield different results that are nevertheless equally valid/optimal fits. Second, despite improvements in quality control and experimental accuracy and precision, high-throughput technologies remain relatively noisy and are highly sensitive to batch effects (meaning that the same samples, assayed at two different labs or at two different times using identical protocols, may exhibit highly differentially expressed genes that are responding to extraneous biological variables). These two challenges underscore the need for biological replicates: both to increase the power of the many gene-wise statistical tests being performed, and to capture the natural level of variability between phenotypically identical samples.

Microarrays

There currently exist a number of different experimental modalities for genomic investigations, each with its own benefits and challenges. The oldest and best-established are microarrays, which measure the hybridization of fluorophore-labeled nucleic acid strands to complementary probe sequences on a chip. The intensity of fluorescence at a specific probe spot is proportional to the amount of bound nucleic acid strands. Microarray chips contain 10^5–10^6 different probes, permitting thousands of genes to be simultaneously assayed. These may be designed to measure mRNA abundance (gene expression profiling), microRNAs (miRNA profiling), or to detect single nucleotide polymorphisms (SNPs) in DNA. Chips functionalized with antibodies may be used in a similar fashion to assess protein abundance.

Before they can be analyzed, microarray data must be preprocessed and normalized. The preprocessing steps include the subtraction of background intensities, averaging across duplicated probes, thresholding or scaling to spiked-in controls or housekeeping genes, removal of probes that fail to meet QC criteria, and normalization to render each array comparable to the others. Normalization schemes rely upon the assumption that the vast majority of genes are not differentially modulated in the phenotype of interest, and attempt to remove chip-wide variations in gene expression that are likely due to technical factors alone. The choice of preprocessing and normalization algorithms can have a significant impact on the results of the statistical analysis, and the appropriate selection depends in part on the microarray technology; the reader is referred to the several comprehensive reviews [5–7] for additional guidance. Because the normalized abundances are approximately log-normally distributed, values expressed on a logarithmic scale are often tested using standard parametric statistics.

"Next Generation" Sequencing

The development of next generation sequencing (NGS) represents an important leap forward in identifying disease-specific genetic variants (DNAseq), epigenetic modifications (ChIPSeq of histone methylation), and transcriptional regulation and splicing (RNAseq). Combined, such genomic data provide a powerful means to identify the relationships between the genetic sequence, epigenetic marks, and expression of genes.

In contrast to microarrays, which probe regions of the genome with known sequences, NGS studies comprehensively assay the entire genome. The data produced are vast, and present different preprocessing challenges than those encountered in microarray studies. The experimental technique consists of fragmenting the DNA or RNA into short segments, which are then sequenced. These so-called "short reads" must then be aligned to a reference genome sequence in order to identify the genes to which they correspond. (Although NGS assays are highly comprehensive, the mapping of reads is a computationally challenging task, and the resulting data is often considerably noisier than that obtained by microarray.) The number of reads for a given genomic region is used as a measure of gene expression (in RNAseq) and to identify probable transcription-factor binding sites or epigenetic modifications (DNAseq, ChIPseq). For more details on sequencing, alignment, and variant calling in NGS studies, the reader is referred to two recent reviews [8, 9]. Once these steps are completed, the data may be analyzed to reveal disease-associated genetic variants, epigenetic modifications, and differential expression [10].

Gene-Level Statistical Analyses

While the focus of this chapter is to acquaint the reader with systems-level statistical analyses, it is useful to briefly review several common gene-level approaches.

Often, the first goal is to identify genes that behave differently in the sample groups of interest ("class comparison"). For mRNA and miRNA expression studies, where the gene level data are continuous, genes are tested for differential expression between groups using a t-test; where more covariates are involved (such as in studies investigating gene$\times$environment interactions), linear models may be used. Linear models may also be used in the context of time-series analysis to identify genes whose expression changes over time and detect those whose time-course profiles differ between sample classes. In SNP and sequence studies, where the covariates are categorical, χ^2 tests are used to identify SNP loci where minor allele frequencies differ significantly. These tests yield a statistic, one per gene/miRNA/locus, that quantifies the difference in expression or allele frequencies between the groups of interest. These statistics may then be compared against an appropriate distribution to yield a p-value and identify significant associations. (For expression microarrays, the `limma` package [11] in R [12] provides a user-friendly framework for gene level analyses. Other BioConductor utilities [13, 14] provide similar functionality for SNP arrays, NGS, and other experimental modalities.)

In all cases, the vast number of hypotheses being tested (at least one per gene, and often times more) necessitates a multiple testing correction [15] of the p-values. That is, at a significance threshold of $p \leq 0.05$, we expect that a gene will be falsely called significant 5 % of the time, leading to thousands of such false positives when the number of genes assayed is on the order of 10^5. While the simple Bonferroni adjustment may be used (in which the significance threshold is set to 0.05 divided by the number of genes assayed), it is considered to be excessively conservative.

Specifically, it is assumed in the Bonferroni adjustment that each gene is strictly independent of all others, an assumption well known to be false for genomic data (in the case of expression, co-regulated genes will exhibit correlated expression; in SNPs, patterns of recombination will lead to linkage disequilibrium, or a tendency for SNP alleles at one loci to be correlated with the alleles at another). Instead, the false positives should be controlled using the false discovery rate adjustment (FDR) [16], which has been proven to exert robust control over the error rate even when the hypotheses have dependencies [17]. Alternatively, assumption-free but computationally intensive permutation procedures [18] may be used.

Identifying Functional Modules

The lists of significant genes obtained by the analyses described above provide limited mechanistic insights without additional biological context. To gain an understanding of systems biology, it is necessary to assemble single-gene information to identify sets of genes and interactions that fulfill particular biological functions. Typically, this is done either by finding clusters of genes that behave in the same way in the experiment, or by incorporating expert knowledge from pathway databases to focus the analysis.

Clustering

It is well-accepted that genes interact with each other in transcriptional modules, and that these modules in turn interact with other modules [19, 20]. Because of these relationships, genes that function together often exhibit directly or inversely correlated expression. The simplest method for identifying those modules and connections is by clustering the genes to identify groups of genes whose expression is similar across the set of samples [21, 22].

The two most commonly used clustering algorithms are hierarchical clustering and k-means clustering. Their considerable popularity is due to their computational and conceptual simplicity. However, because both rely upon the user to specify the number of clusters, they are prone to artificially separating genes that should be in the same cluster (if the user specifies more clusters than are truly present) or speciously combining them (if the user specifies too few). They are limited in their ability to detect clusters with complex shapes. To address these limitations, refinements of both schemes have been proposed; we describe them below.

Hierarchical Clustering

The commonly used hierarchical clustering [23] technique agglomeratively sorts genes based on the similarity of their expression, producing a tree that can be cut into clusters. For each pair of genes i and j, hierarchical clustering computes a distance metric D_{ij}; then, starting with each gene as its own "cluster," iteratively merges clusters based on the smallest D_{ij} between them. Most frequently, a Euclidean distance metric (i.e., $D_{ij}=\sqrt{\sum_m (g_{i,m}^2+g_{j,m}^2)}$ where $g_{i,m}$ denotes the expression of gene i in sample m) is used, although non-Euclidean distances (e.g., Manhattan or Mahalanobis), correlation-based distances (e.g., $D_{ij}=1-\mathrm{Cor}(g_i, g_j)$), or information-theoretic metrics may also be used. The criteria for merging clusters is known as the linkage. Simply put, for any two clusters, we wish to consider merging, we examine the pairwise distances D_{ij} for the genes in the merged clusters; the linkage can be set to be the average, minimum ("single" linkage), or maximum ("complete" linkage) of the pairwise differences within the resulting cluster. The choice of which clusters to merge is then based on which cluster pairs yield the smallest linkage. At each iteration, those pairs of clusters are merged, forming a binary tree, and the number of clusters is determined by the height at which user cuts the tree.

However, while hierarchical clustering has a long history in microarray analysis, it is extremely sensitive to the choice of distance metric and the linkage method used to merge the clusters, since the "greedy" agglomeration causes slight inaccuracies to snowball. Hierarchical clustering should therefore be considered an exploratory tool rather than an analytical one.

k-Means Clustering

The well-established k-means clustering [24] technique provides a more stable partition of the genes. The algorithm iteratively finds points that define the centers of globular clusters: starting with a user-specified number of clusters k, it selects k genes at random as starting centroids, and clusters all the genes based on the centroid to which they are closest. For each of the resulting k clusters, new centers are computed based on the mean expression of the genes assigned to that cluster. The genes are reclustered with respect to the new centroids, and the process is repeated until the clustering assignments converge. In addition to being much less error prone than hierarchical clustering, k-means is also considerably faster. As with hierarchical clustering, however, the user must specify the number of clusters (which in the case of genes means guessing at the number of "modules"). In addition, k-means performs poorly when the genes do not form globular, linearly separable clusters.

Improved Approaches

To address these drawbacks, several refinements have been proposed. Graph-theoretic spectral techniques [25–30] are able to articulate clusters with nonlinear and nonconvex boundaries, allowing complex relationships between genes (such as those that oscillate differentially over the cell cycle) to be discerned. Variational clustering schemes [31, 32] achieve similar goals. Several schemes have also been proposed to estimate the number of clusters from the data itself rather than relying on user input [30, 33–36]. Combined, these methods produce robust partitions even in complex data sets.

One interesting and extensible approach, consensus clustering [36], is a method that may be wrapped around any clustering algorithm of choice (hierarchical, k-means, spectral clustering, etc.) to provide both an estimate of the number of clusters present in the data and a measure of the robustness of the clustering. In consensus clustering, the data is randomly subset so that only a portion of the genes and samples are used. The clustering algorithm of choice is then used to cluster the samples or genes into $k = 2, 3, 4, \ldots$ groups for multiple random subsets of the data. For each k, a consensus matrix is obtained where the i, jth entries are the fraction of times gene i and gene j were assigned to the same cluster across multiple subsets. For a truly robust partition, it is expected that the entries will all be close to 1 or 0, that is, either i and j are consistently placed in the same cluster, or they are consistently placed in different groups. (This reflects the intuition that if the algorithm only place objects in the same cluster together half the time, it is questionable whether a separate cluster truly exists.) The optimal k (number of clusters) is that for which the consensus matrix comes closest to the ideal of pure 1's and 0's. Wrapping consensus clustering around k-means or hierarchical clustering mitigates the limitations of those methods; moreover, because the consensus technique may be wrapped around any clustering engine, it can readily incorporate the advantages offered by the more sophisticated nonlinear clustering algorithms described above. Recently, consensus clustering was applied to identify molecular subtypes of diffused large B-cell lymphoma, leading to the identification of highly reproducible transcriptional signatures corresponding to differential signaling cascades [37].

Dimension Reduction

As the number of genes assayed is vast, it is often of interest to find a small number of representative patterns that describe most of the variation observed in the data and on which the gene expression may be modeled, rather than dealing with the whole data set. This problem is closely related to clustering: by identifying dominant patterns of gene expression (across samples or over time), one may then find clusters of genes that match particular patterns. Those pattern-based clusters may then be examined for common regulatory elements.

Principal Component Analysis

The simplest and best-known dimension reduction technique is principal component analysis (PCA) [38], which transforms a set of observations of possibly correlated variables (e.g., gene expression measurements) into a new set of variables, called the principal components (PCs), which are constructed such that the PCs are completely

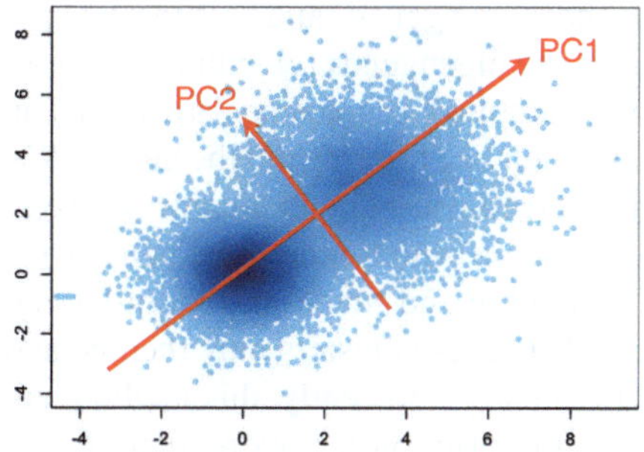

Fig. 8.2 In principal components analysis, the principal components are defined such that the first principal component (PC1) lies along the direction of greatest variation and each succeeding component (in two dimensions, only PC2) is defined to lie in an orthogonal direction with the highest variance. Geometrically, the PCA space is a rotation of the original axes

independent of each other. The transformation is defined such that the first principal component lies along the direction of greatest variation in the data, accounting for as much of the overall variation in the gene expression between samples as possible. Each succeeding component lies, in turn, along the direction of the highest variance under the constraint that it will be orthogonal to (i.e., uncorrelated with) the preceding components. Mathematically, the principal components are the eigenvectors of the covariance matrix; the associated eigenvalues indicate the amount of variance along each component. A graphical illustration in two dimensions is given in Fig. 8.2.

The transformation is linear, that is, the original coordinates (genes) are rotated in the PCA space, such that the bulk of the variation lies along the first PC, and so on, as shown in Fig. 8.2. Each gene may thus be described using a weighted combination of components (and vice versa). Because the bulk of the statistical variation in the data is contained in the first few components, it is possible to use just the first few PCs, rather than the full 10^5-dimensional feature space, when analyzing the data. The resulting clusters may then be examined for common regulatory elements. Recently, Chilarska and coworkers used this approach to identify combinatorial transcriptional control in a genome-wide study of blood stem/progenitor cells [39]. PCA has also been used to identify "fingerprints" of hematopoietic stem cell differentiation [40].

Eigengenes

The principal components transformation can be written in terms of another matrix factorization called the singular value decomposition (indeed, computation of the principal components is typically done from the SVD, rather than the mathematically equivalent but computationally costly eigendecomposition of the covariance matrix). While PCA yields a matrix containing the PCs (i.e., the eigenvectors) and a vector of loadings (the eigenvalues), SVD yields *two* matrices, each describing an orthogonal basis, and a vector of so-called singular values. When applied to gene expression data, the two matrices have the dimensions of the genes and samples, which are referred to as the "eigengenes" and "eigenarrays," respectively [41]. Like the principal components, the eigengenes (eigenarrays) are unique orthonormal superpositions of the genes (samples). Eigengenes/arrays that are inferred to represent noise may be filtered out, much like filtering out the higher PCs in PCA. Representing the data by the remaining eigengenes and eigenarrays gives a global picture of the dynamics of gene expression, in which individual genes (or samples) are clustered into groups of similar regulation and function (or similar cellular state and biological phenotype, respectively). These clusters may then be associated with observed genome-wide effects of regulators. Recently, this method has been used to uncover the combinatorial role of transcription factors regulating the yeast sulfur assimilation pathway [42] and combined with dynamical modeling; a similar approach could be used to link high-throughput data to dynamical models of blood stem cell fate (e.g., [43]).

Nonlinear Dimension Reduction

The patterns described by the principal component vectors or eigengenes are linear combinations of the gene expression measurements. However, if the biological patterns of interest have a nonlinear form, as is likely to arise from regulatory networks with feedback loops, neither classical PCA nor SVD can articulate those patterns. Instead, nonlinear dimension reduction (NLDR) techniques must be used. NLDR may be thought of as a nonlinear version of PCA where the coordinates are "threaded" along the direction of greatest variability. Optimally detecting those paths is a mathematically and computationally challenging task, and several methods have been proposed including kernel PCA, Laplacian eigenmaps, IsoMaps, and spectral embedding [44, 45]. Of these approaches, the neural-network-based self organizing map (SOM) [46] is the best represented in the genomics literature. Figure 8.3 provides an illustration of the first SOM coordinate versus the first PC for data lying on a curved manifold; while the first PC captures only 76.77 % of the variance, the first component of the SOM captures 93.14 % and provides a better description of the underlying pattern.

This property makes SOM (and NLDR generally) particularly well-suited for analyzing transcription dynamics, where the relationships between genes may not be strictly linear. SOM has been applied to detect and interpret gene expression patterns governing hematopoiesis [47]. For an in-depth mathematical treatment of various NLDR methods, the reader is referred to [48].

However, while NLDR provides a more accurate and possibly more biologically meaningful dimension reduction than PCA or SVD, it must also be noted that the transformation from the new, dimension-reduced space to that of the genes is not a straight-forward (or even necessarily possible) task. This is a direct consequence of their nonlinearity and places them in stark contrast to PCA and SVD, from which it is easy to recover the original coordinates. This, in turn, means that it is very difficult to say which genes are the ones driving the dominant pattern observed, which can pose problems when it comes time to identify specific genes for validation work. In short,

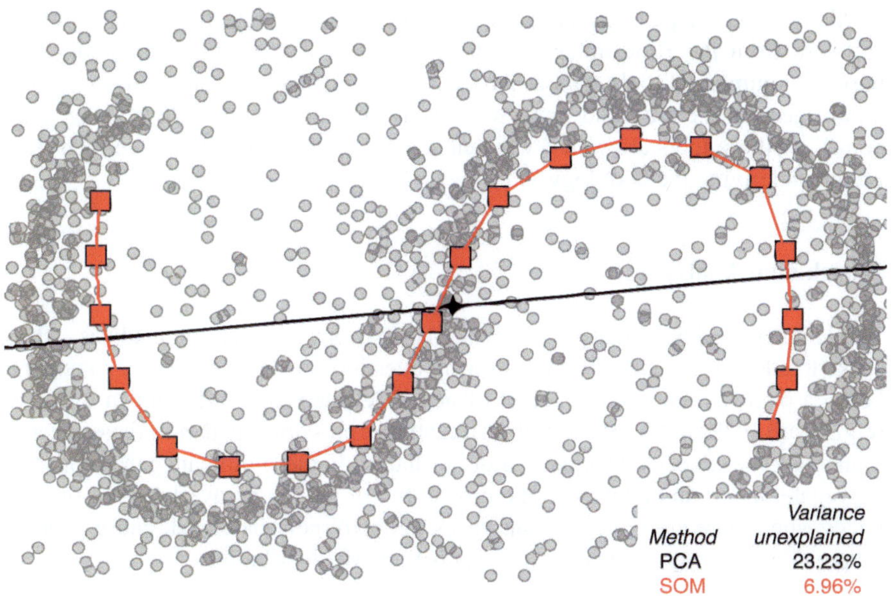

Fig. 8.3 Comparison of SOM versus PCA. While the first PC captures only 76.77 % of the variance, the first component of the SOM captures 93.14 % and provides a better description of the underlying pattern

what we gain in accuracy and representativeness in NLDR is lost in interpretability. The choice of dimension reduction should thus be undertaken with the end goal of the analysis in mind.

Pathway Analysis

Pathways, or networks of functionally related genes and molecules, provide a natural framework in which systems-level effects may be investigated in the context of existing "expert" knowledge. Pathway definitions may be extracted from a growing number of databases, including the Pathway Interaction Database [49], KEGG [50], Reactome [51], and InnateDB [52], among others. Many statistical computing packages, including R/BioConductor, have interfaces to these databases [53].

Analyzing high-throughput molecular measurements at the pathway level have two significant benefits. First, it permits the grouping of hundreds of thousands of genes (or other biomarkers) into several hundred pathways, reducing the complexity of the analysis. Second, identifying active pathways that differ between two conditions can provide more explanatory power and mechanistic insights than a simple list of genes. These benefits have given rise to a vast number of different pathway analysis approaches over the past decade [54].

Many tools for pathway analysis are available, including free, open-source R software from the BioConductor project [13] (http://www.bioconductor.org) and popular commercial tools such as Ariadne Genomics Pathway Studio (http://www.ariadnegenomics.com/) and Ingenuity Pathway Analysis (http://www.ingenuity.com/). As these tools are well-documented and constantly evolving, we focus here on the underlying methodology.

Enrichment Analyses

The simplest pathway analysis approach is an overrepresentation analysis, which seeks to address statistically the following question: given a set of genes known to be on a pathway, and given the list of genes detected to be different in the study (e.g., with FDR$\leq$0.05 in a test of differential expression), is there greater overlap than would be expected by chance alone? That is, do the significant genes appear to aggregate in certain pathways? The probability of having an overlap of m or more genes when there are M significant genes out of N genes assayed, and n genes in total on the pathway is given by the hypergeometric distribution,

$$\Pr(X \geq m|N,M,n) = \sum_{r=m}^{n} \frac{\binom{M}{r}\binom{N-M}{n-r}}{\binom{N}{n}} \tag{8.1}$$

which is easily computed for all gene sets of interest.

While simple, overrepresentation analysis has a significant limitation: because it uses only the most significant genes (e.g., those passing the arbitrary FDR$\leq$0.05 threshold), marginally less significant genes (e.g., FDR $= 0.051$) are discarded, resulting in information loss. In contrast, the popular Gene Set Enrichment Analysis (GSEA) algorithm [55, 56] uses the full list of all genes, ranked in order of significance, and uses a Kolmogorov–Smirnov running sum statistic to answer the question: what is the probability that the genes in this pathway lie as near the top of the ranked list as we observe them to be? Significance may be computed either by permuting the sample labels or permuting the genes included in the pathway [54, 57, 58]. These methods have been applied successfully to a variety of studies, including expression profiling of acute lymphocytic leukemia subtypes [59]; pathways involved in the activation of memory T cells, monocytes, and B cells [60]; and resistance to chemotherapy in acute myeloid leukemia cells [2].

Nevertheless, both simple overrepresentation analysis and GSEA have a common drawback: they rely upon the computation of gene-level statistics. It is well known that complex diseases exhibit considerable molecular heterogeneity, either due to causative mechanisms that can be deleteriously affected in a variety of ways (such that no particular alteration is dominant among the case samples) or to those that are only deleteriously affected through a specific combination of particular alterations (such that control samples may have some, but not all, the alterations necessary to produce the case phenotype). As a result, individual genes may fail to reach significance in univariate tests of significance, and pathway analyses that rely on single-gene association statistics may fail to detect significant pathways simply because the constituent genes do not exhibit *univariate* associations. Simply put, by relying on univariate, gene-level tests, overrepresentation and enrichment analyses still make the reductionist assumption that regulatory systems may be investigated by examining their parts in isolation.

Pathway Summary Statistics

An alternative is to compute a pathway-level "summary" statistic: a single value that summarizes the expression level (or other data) for all the genes in the pathway. For each pathway, a summary statistic is obtained for each sample based on its profile, and those summaries may then be tested for association with the condition of interest. A very crude example is to use the average expression value for the genes on the pathway, such that a sample in which many of the genes are upregulated will have a high pathway summary value, regardless of which genes happen to be upregulated. However, simply averaging the gene expression levels is a poor measure of pathway activity from a biological point of view, since these mechanisms involve both up- and downregulation for which coordinated gene expression (and hence correlations) are important. A more justifiable approach, therefore, is to use PCA for the genes in the pathway of interest, selecting the first PC as the "pathway summary." This has the effect of summarizing the expression (or other) values for all the genes on the pathway by a single number that describes the bulk of the variation and which mathematically accounts for the correlations in gene expression. This technique has been used identify differential pathways in leukemia [61] as well as other cancers.

Extending this idea, we proposed a method [62] in which the pathway summary values and genes not known to be on the pathway were tested for differential correlation. In the "Gene × Pathway Correlation (GPC) Score" method [62], we first computed pathway summary values based on the first PC for every pathway of interest, yielding for each pathway j a value $p_{j,m}$ summarizing sample m's expression across pathway j's genes. For each gene i with expression $g_{i,m}$ in sample m, we compute the GPC-score as the difference in the correlations of g and p in the case and control phenotype,

$$\text{GPC-score} = \text{Cor}_{m \in \text{Case}}(p_{j,m}, g_{i,m}) - \text{Cor}_{m \in \text{Control}}(p_{j,m}, g_{i,m}), \tag{8.2}$$

yielding a gene × pathway matrix of correlation differences for each gene-pathway pair. The significance of the correlation differences are assessed by randomly permuting the case and control labels. This method has the power to identify new regulatory interactions (by finding correlated gene-pathway pairs), as well as to detect those which are altered in disease (by identifying gene-pathway pairs with significant differences). An example gene-pathway pair identified in a prostate cancer study is given in Fig. 8.4.

Although this method was applied in [62] exclusively to mRNA expression data, the same technique may be applied as an integrative analysis using both mRNA

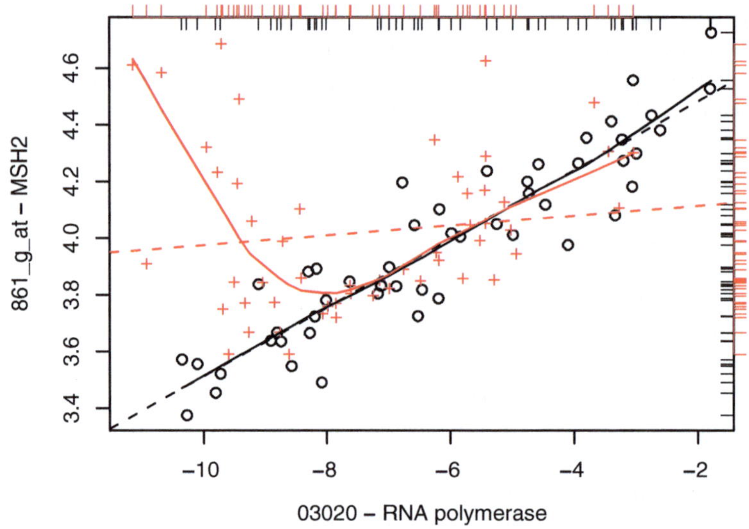

Fig. 8.4 GPC-Score identifies differential gene-pathway coexpression for the MSH2 (mismatch repair) gene and the RNA polymerase pathway for a subset of prostate tumor samples; these samples corresponded to worse clinical outcomes. (Image: [62])

and other genomic or environmental measurements. For instance, one can apply it to combined miRNA/mRNA data to search for differentially correlated miRNA-pathway pairs, thereby identifying miRNAs whose expression modulates the activity of regulatory networks.

These pathway summaries effectively amount to selecting the n genes on a given pathway and applying a dramatic dimension reduction to go from the n features down to a single one. As such, the same caveats about linearity described in Section "Dimension Reduction" apply, namely, linear methods cannot account for complex or oscillatory relationships between genes. Instead, NLDR such as kernel PCA or Laplacian eigenmaps may be used, providing a more accurate and biologically representative summary of the expression patterns across a pathway.

Sample Class Prediction and Class Discovery

In the previous section, our goal was to categorize genes into biologically relevant functional modules, either by grouping the genes into clusters of correlated expression or by pathway analysis. Here, we turn our attention to categorizing *samples* based on complex patterns in the experimental data, with the goal of predicting the status or outcome for a new sample or discovering sample subclasses that were previously unknown.

The gene-level tests described in Section "Generating High-Throughput Data" yield lists of differentially expressed genes and significantly associated genetic variants that are ubiquitously reported in genomic studies. However, while these genes are significantly associated with the phenotype of interest, they may not accurately classify or predict the outcome for a new sample. To develop predictors from high-throughput data, machine-learning algorithms are commonly used. These algorithms are first "trained" against a subset of the data for which the outcomes are known, and then evaluated for accuracy against an independent "testing" subset (for which the outcomes are also known). If the classifier performs well in the testing subset, it may then be validated in a distinct data set. The procedure of dividing the data into training/testing sets, known as cross-validation, ensures that the models are not overfit to technical nuances in the data. As the known sample labels (case/control, exposed/unexposed, etc.) are used to train the machine, these techniques are referred to as "supervised" classifiers.

The literature now contains many supervised machine learning algorithms; for a deep and comprehensive exposition, the reader is directed to Hastie and Tibshirani's *Elements of Statistical Learning* [48]. Here, we discuss three powerful techniques: one designed for continuous data (such as gene expression), one designed for categorical (SNP or sequence) data, and a third that can accommodate a mixture of covariates. From a systems-biology perspective, the predictive "signatures" obtained from these algorithms may be used to suggest functional modules, identify epistatic interactions between genetic variants, or provide an integrative analysis that combines genomic, epigenetic, and expression data. We also discuss unsupervised methods for class discovery (i.e., the identification of sample subtypes based solely on the high-throughput data). Such methods can articulate complex, systems-level similarities and differences that would be indetectable by association tests alone.

Nearest Shrunken Centroids

Given a set of samples comprising different categories (cases and controls, say), and a new sample whose categorization is unknown, a natural way to classify it is to ask which group, on average, the unknown sample is closer to. This approach is referred to as a "nearest centroid classifier"—the centroids represent the average gene expressions (or other data) in each sample class, and the new sample is classified based on the centroid to which it is nearest.

In the context of genome-wide expression profiling, the centroid for each sample class resides in a very high-dimensional space. If 10^5 genes have been assayed, the centroid for the cases is a 10^5-dimensional vector which gives the average expression for each gene across all the case samples; likewise the controls. As the vast majority of these genes are not biologically relevant, it is important to remove those which can be reasonably assumed to be noise. One powerful approach is to consider not only the average gene expression, but the variance as well, moving the centroid coordinates closer to each other by an amount proportional to the variance of the corresponding genes [63]. This procedure is referred to as "shrinking" the centroids.

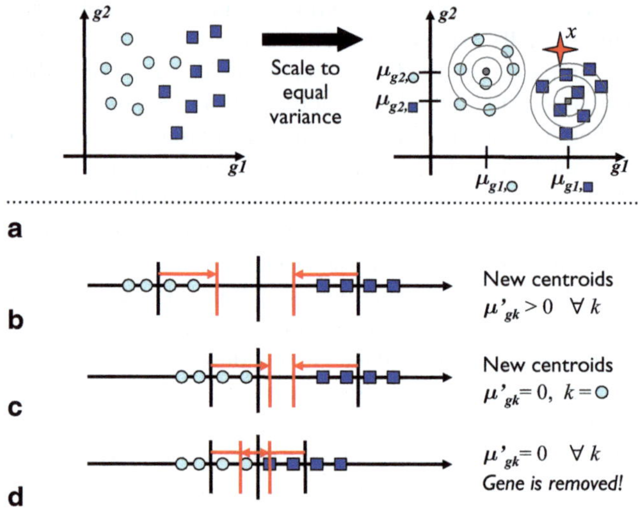

Fig. 8.5 Nearest shrunken centroids classifier. In **a**, the nearest centroid classifier in two dimensions is illustrated. There are two classes of samples k, shown as *light circles* and *dark squares*. After scaling each gene (here, $g1$ and $g2$) to unit variance within each group k, the unknown sample x is classified based upon the nearest centroid μ (in this case, the dark squares). (b)–(d) illustrate the shrinkage of the centroids for a gene g. Centroids μ_{gk}, shown as a black line, are moved in the direction of the center line to a new position μ'_{gk}. In **b**, neither cross the center line, and the new position is retained. In **c**, the centroid for the light circles crosses the center line and is thresholded to 0. In **d**, both centroids cross the center line and are thresholded to 0; because the new centroids are equal, the gene no longer contributes to the classification

A graphical illustration is given in Fig. 8.5. Mathematically, the nearest centroids procedure attempts to classify a new sample with gene expression $\mathbf{x}$ into group k such that $\delta_k(\mathbf{x})$ is minimized:

$$\delta_k(\boldsymbol{x}) = \boldsymbol{x}^\top \boldsymbol{\Sigma}^{-1} \boldsymbol{\mu}_k - \frac{1}{2} \boldsymbol{\mu}_k^\top \boldsymbol{\Sigma}^{-1} \boldsymbol{\mu}_k + \ln(\boldsymbol{\pi}_k), \tag{8.3}$$

where $\boldsymbol{\mu}_k$ are the centroids for each group k, $\boldsymbol{\Sigma}$ is the covariance matrix (across all groups), and $\boldsymbol{\pi}_k$ are the prior probabilities that $\mathbf{x}$ belongs to each group k. For instance, when building a classifier to detect a rare disease, $\boldsymbol{\pi}_k$ may be taken from the disease prevalence in the population, reflecting the low probability that the patient has the disease of interest.

In the "shrunken" approach [63], the gth entry of the vector $\boldsymbol{\mu}_k$ (i.e., the mean of gene g in group k) is moved from μ_{gk} to μ'_{gk} by an amount proportional to the pooled variance s_g (plus a slight offset s_0) for gene g:

$$\mu'_{gk} = \mu_{gk} - \Delta(s_g + s_0)\sqrt{1/n_k + 1/n}, \tag{8.4}$$

where n_k is the number of samples in group k, n is the total number of samples, and the degree of shrinkage is controlled by the parameter Δ. Genes that cross

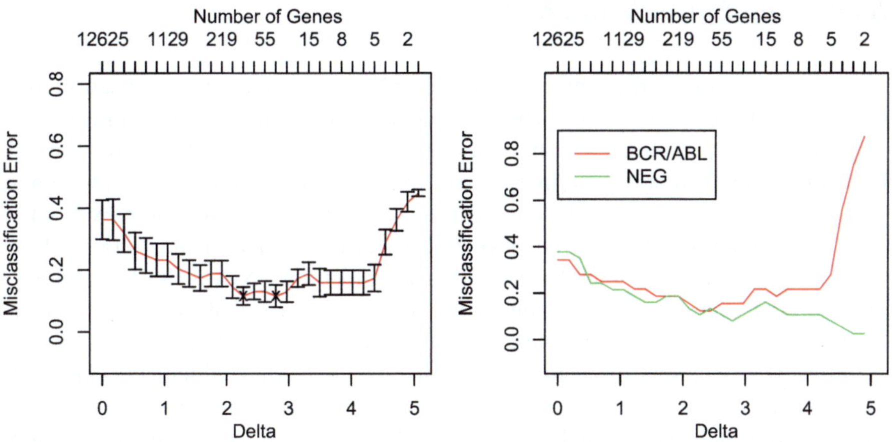

Fig. 8.6 Application of the nearest shrunken centroids classifier to distinguish cytogenetically normal cases ("NEG") from those with BCR/ABL fusion based on gene expression profiles of patients with acute lymphoblastic leukemia (ALL). The overall misclassification error is shown on the left, while the misclassification error for the known groups is shown on the right. As the shrinkage parameter Δ increases, fewer genes remain in the model. Initially, the removal of genes improves the accuracy as "noisy" genes are removed. Optimal values of Δ, corresponding to the smallest error observed in the cross-validation, are obtained at $\Delta = 2.272$ (115 genes) and $\Delta = 2.796$ (40 genes). Increasing Δ beyond 3 removes informative genes (only 20 remain at $\Delta = 3$), causing a dramatic increase in the error rate, particularly amongst BCR/ABL cases

the "overall" centroid across all groups (i.e., those for which $\mu'_{gk} - \mu_g \leq 0$) for all classes k do not contribute to the final model, resulting the removal of high-variance "noisy" genes from the classifier. The shrinkage parameter Δ controls the aggressiveness of the removal (higher Δ forces a greater degree of shrinkage and hence more genes are removed), and is optimized by cross-validation. The data set is split into multiple training and testing subsets, and samples in the testing subset are classified according to the shrunken centroids in the training subset. By varying the value of Δ over multiple training/testing splits, it is possible to choose Δ such that the error in the testing subset is minimized.

An example applied to gene expression data from well-known study [64] of acute myeloid leukemia (AML) and acute lymphoblastic leukemia (ALL) is given in Fig. 8.6. Here, the nearest shrunken centroids classifier, implemented in the R package `pamr` [65], has been applied to data from 12,625 genes in 95 ALL cases, of whom 42 are cytogenetically normal and 37 have BCR/ABL fusion. The classifier was trained using tenfold cross validation with increasing values of Δ ranging from $\Delta = 0$ (no shrinkage, all genes used) until no genes remained at $\Delta \approx 5$. As shown in Fig. 8.6, the misclassification error initially drops as Δ is increased and noisy genes are removed. The optimum $\Delta = 2.796$ yields an error rate of approximately 13 % using 40 genes. Further increasing Δ has the effect of removing informative genes, causing the error rate to rise again. This is particularly true for the BCR/ABL cases, which are frequently misclassified as cytogenetically normal as Δ is increased above the optimum.

Identifying Epistatic Interactions

As with expression profiling by microarray and NGS, genome-wide association studies (GWAS) have become a powerful and increasingly affordable tool to study genetic sequence variants associated with disease. Modern GWAS yield information on millions of single nucleotide polymorphism (SNPs) loci distributed across the human genome, and have already yielded insights into the genetic basis of complex diseases [66, 67]; a complete list of published GWAS can be found at the National Cancer Institute-National Human Genome Research Institute (NCI-NHGRI) catalog of published genome-wide association studies [68]. As described above, the data is typically analyzed by testing the alleles at each locus for association with case status; significant association is indicative of a nearby genetic variant which may play a role in the phenotype being studied. Genomic regions of interest may also be investigated by haplotype analysis, in which a handful of alleles transmitted together on the same chromosome are tested for association with disease; in this case, the loci which are jointly considered are located within a small genomic region, often confined to the neighborhood of a single gene.

Recently, however, there has been increasing interest in multilocus, systems-based analyses. This interest is motivated by a variety of factors. First, few loci identified in GWAS have large effect sizes (the problem of "missing heritability") and it is likely that the common-disease, common-variant hypothesis [69, 70] does not hold in the case of complex diseases. Second, single marker associations identified in GWAS often fail to replicate. This phenomenon has been attributed to underlying epistasis [71], and a similar problem in gene expression profiling has been mitigated through the use of gene-set statistics. Most importantly, it is now well understood that because biological systems are driven by complex biomolecular interactions, multi-gene effects will play an important role in mapping genotypes to phenotypes; recent reviews by Moore and coworkers describe this issue well [70, 72]. In addition, the finding that epistasis and pleiotropy appear to be inherent properties of biomolecular networks [73] rather than isolated occurrences motivates the need for systems-level understanding of human genetics.

Several multi-SNP GWAS analysis approaches have been described in the literature. Thorough reviews are provided in [74, 75], and we briefly describe several here. Building on the well-established Gene Set Enrichment Analysis [55] method initially developed for gene expression data, two articles have proposed an extension of GSEA for SNP data [76, 77] using the χ^2 SNP-level statistics. As in expression-based GSEA, the reliance on single-marker statistics means that systematic yet subtle changes in a gene set will be missed if the individual genes do not have a strong marginal association. In the case of a purely epistatic interaction between two SNPs in a set, the set may fail to reach significance altogether.

As an alternative, the notion that cases will more closely resemble other cases than they will controls has motivated a number of distance-based approaches for detecting epistasis. Multi-dimensionality reduction (MDR) has been proposed and applied to SNP data [78–80]. The technique is conceptually similar to the nearest Shrunken centroids classifier described above; here, sets of l SNPs are exhaustively searched for combinations that will best partition the samples by examining the 3^l cells in that space (corresponding to homozygous minor, heterozygous, and homozygous major alleles for each locus) for overrepresentation of cases. While this method finds epistatic interactions without requiring marginal effects and can be structured to incorporate expert knowledge, it is limited by the fact the the total number of loci to be combinatorially explored must be restricted to limit computational cost. To address this, an "interleaving" approach in which models are constructed hierarchically has been suggested [79] to reduce the combinatorial search space. A recent and powerful MDR implementation [81] taking advantage of the CUDA parallel computing architecture for graphics processors has made feasible a genome-wide analysis of pairwise SNP interactions. Still, MDR remains computationally challenging, such that expanding the search to other SNP set sizes (rather than restricting to pairwise interactions) can be impeded by combinatorial complexity if an exhaustive search is to be performed.

In order to narrow down the combinatorial complexity of discovering SNP sets using techniques such as MDR, feature selection may be employed. Of particular importance here is the distance-based approach of the Relief family of algorithms [82–85]. These are designed to identify features of interest by weighting each feature through a nearest-neighbor approach. The weights are constructed in the following way: for each SNP, one selects samples at random and asks whether the nearest neighbor (across *all* SNPs) from the same class and the nearest neighbor from the other class have the same or different values from the randomly chosen sample. Attributes for which in-class nearest neighbors tend to have the same value are weighted more strongly as being more representative of the underlying biology. Because the neighbor distances are computed across all attributes, Relief-type algorithms can identify SNPs that form part of an epistatic group and provide a means of filtering out unpredictive loci.

While these methods have so far been applied to finding small groups of interacting SNPs, one may instead be interested in whether cases and controls exhibit differential distance when considering a large number of genes. A multi-SNP statistic has been proposed in the literature [86–88] for determining whether a new sample of interest is on average (across a large number of SNPs) "closer" to one population (e.g., cases) than to another (e.g., controls). The method [86] is motivated by the idea that a subtle but systematic variation across a large number of SNPs can produce a discernible difference in the closeness of an individual to one population sample relative to another. Assuming an individual Y and two population groups F and G with minor allele frequencies y_i, f_i, and g_i ($y_i \in \{0, 0.5, 1\}$) for SNP i, respectively, we write the distance metric for SNP i as

$$D_{Y,i} = |y_i - f_i| - |y_i - g_i|, \tag{8.5}$$

and then consider the normalized mean across all SNPs of interest:

$$S_Y = \frac{\mathrm{E} D_{Y,i}}{\sqrt{\mathrm{Var} D_{Y,i}/l}}. \tag{8.6}$$

An illustration is given in Fig. 8.7.

$$D_{Y,i} = |y_i - f_i| - |y_i - g_i|;$$

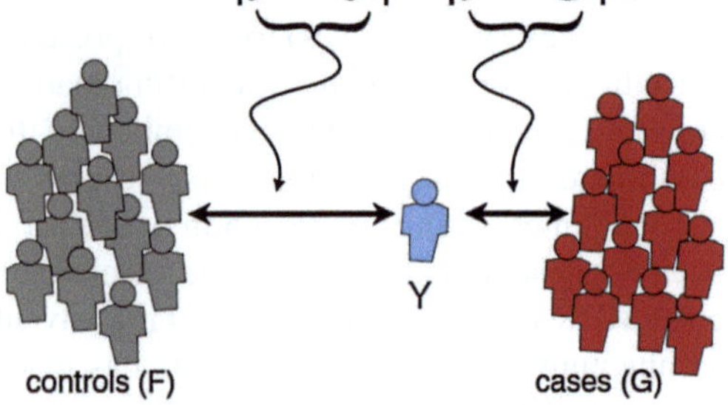

$$S_Y = \frac{\mathrm{E}(D_{Y,i})}{\sqrt{\mathrm{Var}(D_{Y,i})\,/\,l}}.$$

Fig. 8.7 Genetic distance metric (Eqs. 8.5–8.6). If Y is closer to G than to F for locus i, $D_{Y,i}$ is positive. If $D_{Y,i}$ is consistently positive across all l loci, S_Y will be so as well, indicating a tendency for Y to have more "G-like" patterns of genetic variation

It is clear from Eqs. 8.5 and 8.6 that individuals Y whose minor allele frequencies at locus i more closely resemble those of group G will have a positive $D_{Y,i}$ and vice versa. By chance alone, we would expect $D_{Y,i}$ to be as frequently positive as negative, yielding $S_Y \approx 0$. However, a slight but consistent tendency to be closer to one group than another across a set of SNPs will cause deviations in S_Y (Fig. 8.7). The significance of S_Y in Eq. 8.6 may be assessed either parametrically by assuming normality (only in the case of large l), or by resampling the F and G populations.

While this statistic was originally designed to identify group membership of individuals who were known to be in either F or G (and hence contributing to f_i and g_i), it was later shown in [87] that even out-of-pool breast cancer cases were in general "closer" to the population of other cases than to the controls, suggesting that the combination of multiple alleles has the potential to classify *new* samples. Building on these ideas, the PoDA [89] technique has been proposed to find pathway-based SNP-sets that distinguish cases from controls. The hypothesis is that if the SNPs come from a pathway that plays a role in disease, there will be greater in-class similarity than between-class similarity in the genotypes for those SNPs, i.e., a case will show greater genetic similarity to other cases than to controls for the SNPs on a disease-related pathway, but will be equidistant for the SNPs on a non-disease-related pathway. In order to identify the significant pathways, a leave-one-out cross-validation procedure is used: each sample in the study is treated as unknown, and the pathways with SNPs that most accurately classify the "unknown" samples are flagged. Because subtle but consistent $D_{Y,i}$'s will accumulate to give large values of S_Y in Eq. 8.6, PoDA can identify multi-SNP sets which differ systematically even when the single-SNP associations are not strong enough to be significant, making it useful for detecting epistatic interactions. By restricting the PoDA SNP sets to those defined based on known relationships (e.g., SNPs in genes sharing a common pathway), one may incorporate expert knowledge to reduce the search space and provide biological interpretability.

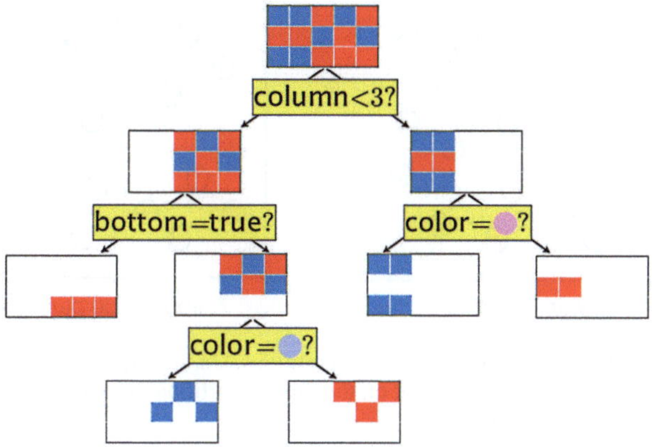

Fig. 8.8 Schematic of a decision tree. At each step, a variable and threshold is chosen to optimally partition the samples based on known labels. The decision rules may operate on continuous variables (like color here, with *blue* and *red* coming closest to the *mauve* and *periwinkle* ideals, respectively), categorical variables (like column, which can take on values 1–5), or booleans (like "bottom," which is either true or false). The partitioning stops when the nodes are pure. Variables may be used multiple places in the tree (such as color here), so long as they are not used along the same branch twice

Random Forests

The methods described above are designed to be applied to one type of data from a single experimental modality—continuous data, such as that obtained in expression profiling, or categorical data, such as that obtained by SNP or sequencing studies. Now that genome-wide experiments are growing increasingly affordable, it is becoming more common for a variety of assays to be run on the same set of samples, allowing the various measurements to be integrated in the analysis. The Random Forests algorithm [90] is a decision-tree based classifier that permits multiple types of data to be mixed a priori, enabling its use as an integrative predictor.

Decision trees are a conceptually simple supervised classifier. At each step, a variable and threshold is chosen to optimally partition the samples based on known labels. The decision rules may operate on continuous or categorical variables. The partitioning continues until either all nodes are pure or all variables have been used. An illustration is given in Fig. 8.8. Once the rules are established based on labeled samples (i.e., the tree is trained), the rules may be used to classify a new sample of unknown status.

Because at every level the decision tree partitions the samples completely, certain partitions are not possible to achieve. An example is given in Fig. 8.9; here, there is no way to place the cuts (corresponding to decision rules) to isolate the sample in the center and achieve pure partitions.

In order to address this issue, the "Random Forests" classifier was proposed [90]. As in consensus clustering (described above), in Random Forests, we also randomly

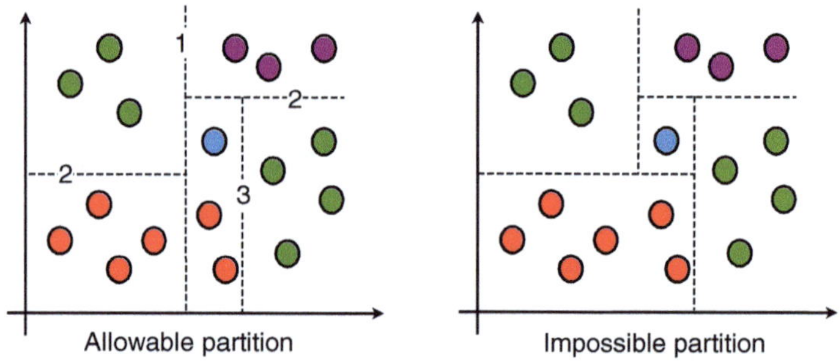

Fig. 8.9 Possible and impossible decision tree partitions. On the left, a possible partition (at levels 1, 2, and 3 in the decision tree) is shown; on the right, a partition that cannot be achieved with the classical decision tree algorithm

subset the data, selecting samples on which to grow the tree using a random sample of features (gene expression, SNP alleles, clinical covariates, or any other available information). The procedure is repeated for many different samplings, yielding a "forest" of decision trees, each trained on a random subset of the data (in much the same way that one obtains a multitude of clusterings of randomly subsetted data in consensus clustering). New samples are then classified according to a majority vote of the trees.

As the goal here is to *predict* rather than cluster, the measure of accuracy is not how well-clustered the selected samples are (as it is in consensus clustering), but how well the decision tree predicts the status of the samples *not* selected in the random sample. For each tree, one can compute the the prediction accuracy for the "out of bag" (OOB) samples—those not used in the training of that particular tree. The average OOB error rate is considered to be a good estimate of the testing error, since each OOB error rate computation is based on samples not used in that particular tree. The OOB error rate is also used to tune the size of the random subset of features. The more features are kept, the more similar the trees will be to one another (eventually converging to identical trees), leading to a forest that may be overfit. The smaller the size of the feature subset, the more diverse the trees are, but each tree will exhibit worse per-tree performance due to the lost information. By varying the size of the feature subset and examining the OOB error rate, these two competing forces may be optimally balanced to yield a forest of trees that are neither underdetermined nor overfit.

Random forests have a number of useful features as an integrative predictor: it can incorporate different data modalities, is invariant to transformations of the data, can handle missing data easily, has a tuning mechanism to prevent overfitting, and provides an estimate of its accuracy. In addition, by looking at the purity of the partition each time a particular feature is used across the entire forest of trees, one may obtain a measure of its importance, yielding a ranked list of discriminatory markers. The ranked list may then be used as an input to pathway enrichment analyses (see Section "Pathway Analysis"), providing further systems-level insight [91]. This approach both allows one to combine data types in the pathway analysis and indicates pathways that are not only "hit" by differential genes but by those that are *predictive* of the biological outcome. Alternatively, pathway summary statistics (as described in Section "Pathway Analysis") may be used as the features input to the Random Forests algorithm.

These features make Random Forests a powerful and highly accurate [92–94] algorithm for generating predictive models. Recently, it has been applied to public expression data as an in-silico screen to discover agents that eradicate leukemia stem cells [95]; applied to a SNP study to identify genomic variants that govern progression-free survival of myeloma patients [96]; and to elucidate transcription factor activity in hematopoietic stem cell differentiation [97].

Class Discovery

The prediction algorithms described above rely upon supervised training using a set of samples for which the true classification is known. However, as with clustering, our knowledge about the true structure of the data may be incomplete in the sense that there exist subtypes which are either unknown or not reflected by the training sample labels. In particular, if a set of samples comprise a single clinical phenotype but span several different molecular subtypes, classifying a new sample based on molecular data may be highly error-prone owing to the lack of a distinct pattern in the training set. As a result, it is often of interest to attempt to discover any existing molecular subtypes present in the data. To this end, the clustering and cluster-number determination algorithms described in Section "Identifying Functional Modules" may be applied to samples (as well as to genes) to discover the optimal number of sample clusters. As with genes, it is important to recognize that these clusters may not be linearly separable, and therefore nonlinear techniques are likely to be more accurate [98]. The application of these techniques may then be followed by training a supervised classifier on the detected molecular subtypes. (Note that if a nonlinear clustering method is used, it is necessary to ensure the appropriate nonlinear out-of-sample extension is used to project the test samples onto the nonlinear space defined by the training samples, as discussed in [44].)

The Partition Decoupling Method

One approach for identifying molecular subtypes at progressively finer scales without imposing linearity constraints is the partition decoupling method (PDM) [26, 30]. The PDM is able to reveal relationships between samples based on multigene expression profiles without requiring that the genes be differentially expressed (i.e., without requiring the samples to be linearly separable in the gene-expression space), as illustrated in Fig 8.10, and has the power to reveal relationships between samples at various scales, permitting the identification of phenotypic subtypes. The PDM consists of two iterated components: a spectral clustering step, in which the correlations

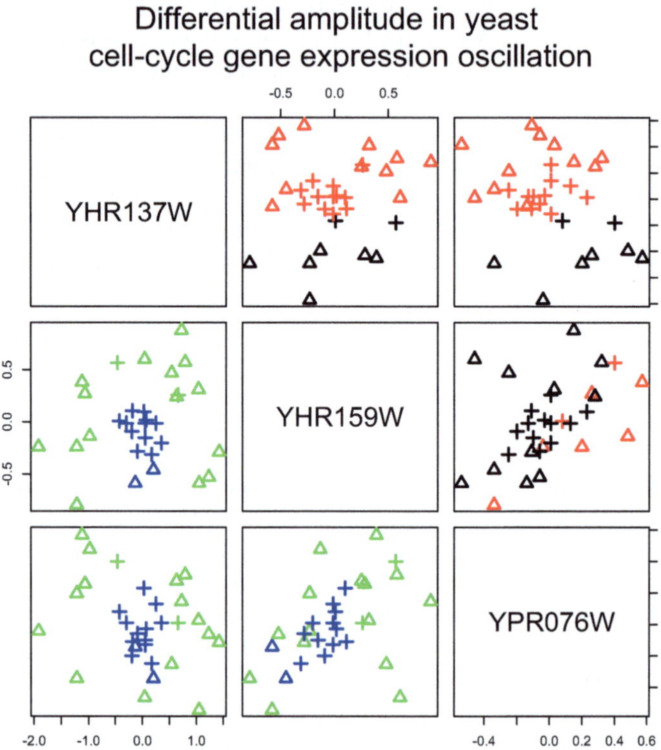

Fig. 8.10 Expression levels for three oscillatory yeast cell-cycle genes from two different treatments: +, elutriation-synchronized samples; Δ, CDC-28 synchronized samples. The samples have different amplitudes of expression oscillation, leading to a "bullseye" pattern (note that the means for each gene in the two groups is approximately the same). Cluster assignment for each sample is shown by color for linear *k* means clustering (*red/black*) above the diagonal, and nonlinear spectral clustering (*blue/green*) below the diagonal. Note the difference in accuracy. (Image: [30])

between samples in the high-dimensional feature space are used to partition samples into clusters, followed by a scrubbing step, in which the projection of the data onto the cluster centroids is subtracted so that the residuals may be clustered. As part of the spectral clustering procedure, a low-dimensional nonlinear embedding of the data is used, both reducing the effect of noisy features and permitting the partitioning of clusters with non-convex boundaries. The clustering and scrubbing steps are iterated until the residuals are indistinguishable from noise as determined by comparison to a resampled null model. This procedure yields "layers" of clusters that articulate relationships between samples at progressively finer scales.

The PDM has a number of satisfying features. The use of spectral clustering allows identification of clusters that are not necessarily separable by linear surfaces (such as the "bullseye" pattern in Fig. 8.10), permitting the identification of complex relationships between samples. This means that clusters of samples can be identified even in situations where the genes do not exhibit differential expression, a trait that makes it particularly well-suited to examining gene expression profiles of complex diseases. The PDM employs a low-dimensional embedding of the feature space, reducing the effect of noise in the data. As the data itself is used to determine both the optimal number of clusters and the optimal dimensionality in which the feature space is represented, the PDM provides an entirely unsupervised method for class discovery, without relying upon heuristics. Importantly, the use of a resampled null model to determine the optimal dimensionality and number of clusters prevents clustering when the geometric structure of the data is indistinguishable from chance. By scrubbing the data and repeating the clustering on the residuals, the PDM permits the resolution of relationships between samples at various scales; this is a particularly useful feature in the context of gene-expression analysis, as it permits the discovery of distinct sample types, subtypes, etc.

These features make the PDM a powerful tool for genomic data analysis. As we demonstrated in [30] and illustrated here in Fig. 8.11, PDM detects with near-perfect accuracy both the phenotype and exposure groups in a study of radiation response; application to a leukemia data set with "incomplete" sample labeling demonstrated the PDM's ability to detect ALL subtypes simply from the expression data alone, with higher accuracy than other algorithms.

As described in [30], the accuracy of the PDM can be applied to gene subsets defined by pathways to identify mechanisms that permit the partitioning of phenotypes. In Pathway-PDM, one subsets the genes by pathway, applies the PDM, and tests whether the unsupervised PDM cluster assignments reflect the known sample classes. Pathways that permit accurate partitioning contain genes with expression patterns that distinguish the classes, and may be inferred to play a role in the underlying biology. As the PDM does not require the pathway's constituent genes be differentially expressed, complex regulatory relationships within pathways may be detected (such as those giving rise to the pattern seen in Fig. 8.10). It was further demonstrated in [30] that this approach, due to its increase accuracy, is a useful meta-analytical tool that can improve cross-study concordance, allowing more robust findings to be culled from existing high-throughput datasets.

Network-Based Approaches

Further refinements to the analyses described here are achieved by examining the structure of interaction networks, rather than treating pathways as simple collections of genes. Network-based analyses fall into two broad categories: statistical analyses of high-throughput data in the context of putative interaction networks, and the inference of network topology from the data itself. A comprehensive introduction to network analysis in general may be found in [99]; case studies of its application to gene expression data are presented in [7], and a review of graph-theoretic concepts in the context of biology is provided in [100]. Here, we discuss a number of promising techniques.

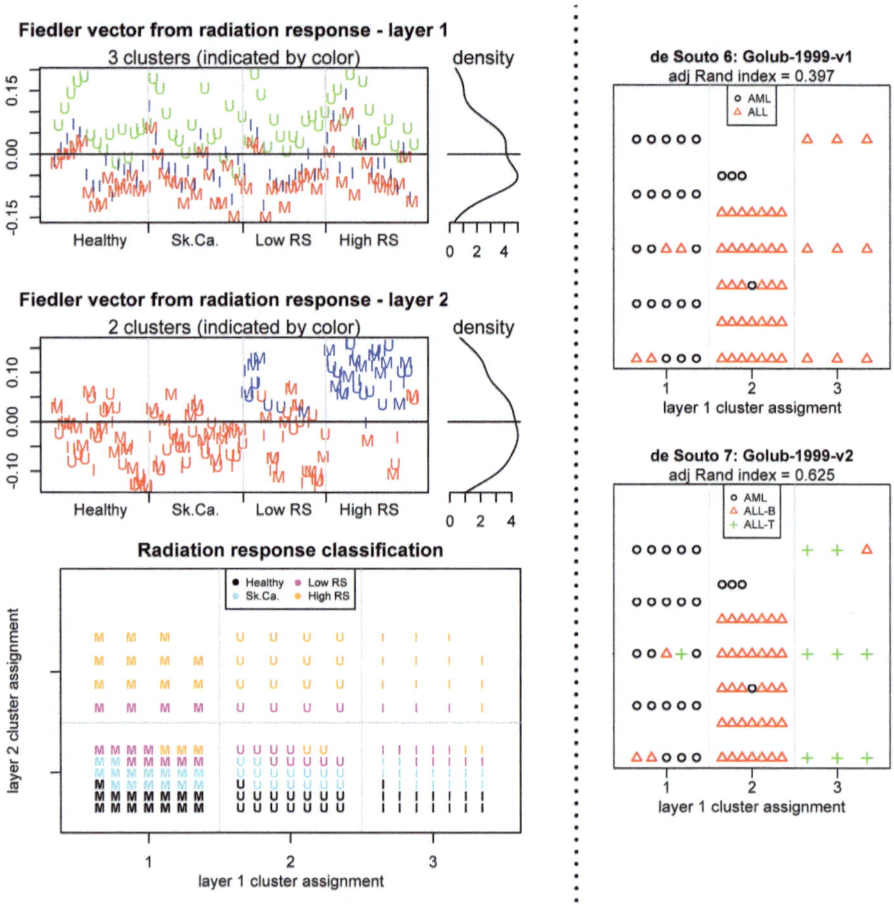

Fig. 8.11 Multilayered, highly accurate unsupervised class discovery using PDM. Left, two "layers" of clusters correspond to the radiation exposure (*U*V light, *I*onizing radiation, *M*ock) and the case (high-RS) group (versus three control groups) in a radiation sensitivity study. The number of clusters in each layer is determined by the PDM itself from the data yielding three clusters in the first layer (*top left panel*) and two in the second (*center left panel*); the resulting classification is near-perfect discrimination of both phenotype and exposure (*bottom left panel*). Right, we see the clustering for leukemia data from [64]. The PDM automatically detects three clusters; in the top panel, comparison against the provided labels (*AML/ALL*) shows that the ALL group has been split by PDM; in the lower panel, it is revealed that this corresponds to a subtype difference (*ALL-B, ALL-T*), demonstrating PDM's ability to identify sample subtypes even when they may be unknown or unannotated in the data. (Image: [30])

Network Statistics

To incorporate known interaction network topology with traditional pathway analyses (Section "Pathway Analysis"), several approaches have been proposed. These methods are based on gene-specific data (either the raw data itself or p-values derived from gene-level statistical tests) overlayed on biological networks obtained from databases such as Pathway Interaction Database [49], KEGG [50], Reactome [51], InnateDB [52], etc.

The relevance of network structure has long been appreciated. In [101], the authors presented systematic mathematical analysis of the topology of metabolic networks of 43 organisms representing all three domains of life, and found that despite significant variation in the pathway components, these networks share common mathematical properties which enhance error-tolerance. In [102], the authors compared the lethality of mutations in yeast with the positions of the affected protein in known pathways, and found that the biological necessity of the protein was well modeled by its connectivity in the network.

Based on such observations, Ideker et al. [103] proposed a method to identify subnetworks of pathways whose genes were enriched for highly significant genes. As the combinatorial problem of finding the maximum-scoring subnetwork is NP-hard (and hence computationally unfeasible for large networks), the authors introduced a simulated-annealing approximation. A related method, described in [104], searches for genes for which differential expression is present within the subnetwork of genes surrounding it. A more robust scoring approach improving upon [103] has recently been proposed [105], and is implemented in R/BioConductor as `BioNet` [106]. These techniques may be used to indicate subsets of interactions in a pathway that appear to be the most critical, and could be targeted in functional studies.

However, like non-network enrichment analyses (Section "Pathway Analysis"), these network-based enrichment analyses [102–104] rely upon the constituent genes displaying independent association with the phenotypes of interest and will fail to capture networks in which the individual gene expressions have similar distributions but altered coexpression characteristics. As an alternative, correlation- and co-expression-based approaches have been proposed in which the edges connecting two interacting genes are examined in the context of the surrounding network. In [107], the authors proposed an "activity" and "consistency" score for each interaction in a pathway. Beginning with a list of molecule input and outputs for every interaction in a biological pathway, Efroni et al. [107] defined the "activity" of the interaction as the joint probability of finding the interaction's genes in an overexpressed state and defined the "consistency" of the interaction as the probability of overexpression in the output conditioned on the activity of the inputs. Similarly, in ScorePAGE [108] the similarity between each pair of genes in a pathway is computed (e.g., correlation, covariance, etc.) and is averaged over the pathway weighted by the number of reactions needed to connect the two genes.

More recently, Signaling Pathway Impact Analysis (SPIA) [109, 110] was proposed. SPIA incorporates changes in gene expression with the types of interactions and the positions of genes in a pathway, defining a "perturbation factor" for each gene as the sum of its measured change in expression and a linear function of the perturbation factors of all the other genes in a pathway. Compared to GSEA [55], SPIA was found to have increased statistical power to detect altered pathways [110]. Similarly, the NetGSA method [111] also models each gene as a linear function of other genes in the network, but in addition accounts for a gene's baseline expression by representing it as a latent variable in the model. Both SPIA and NetGSA have been implemented as R BioConductor packages [13].

Network Inference

In the methods outlined above, pathway network descriptions are obtained from curated databases and used as a framework in which to analyze transcriptomic data at the systems level. While this enables existing biological knowledge to be incorporated into the analysis, it also has the drawback of presuming that the network of regulatory relationships is accurately represented by the pathway database. A complementary approach involves the inference of regulatory networks from the data without making assumptions about the underlying graph. Network inference methods are thus able to identify previously unknown relationships between genes, as well as incorporate elements (such as microRNAs) that are not represented in pathway databases.

Inference of the underlying network structure given a set of cell states [112–120] is a formidable task. Although some success in the reconstruction of large-scale gene regulatory networks (GRNs) has recently been achieved in some cases [119–123], the systematic reconstruction of large-scale networks describing regulatory function and direct interactions of genes from expression or other data remains a major challenge in systems biology. With the increasing feasibility of genome-wide assays, an increasing amount of systems biology research is concerned with attempting to infer GRNs from large scale data sets based upon correlations between expression levels under various experimental conditions. Methods developed for this task are faced with a fundamental difficulty: while direct regulatory relationships between genes typically yield a high degree of correlation in their expression, the reverse is not necessarily true. For instance, two non-interacting genes may share the same upstream regulator, causing their expression to be correlated despite not sharing a direct link. On a global scale, GRNs are known to be sparse, i.e., direct regulatory relationships are a small fraction of all possible connections, but correlations can be non-vanishing between any pair of genes.

Increasingly sophisticated techniques have been devised to tackle this difficulty and attempt to infer the topological properties of GRNs from correlations in gene expression [112, 119]. Prominent examples are simple thresholding techniques [118], the use of partial correlation [116], and mutual information [113]. It should, however, be noted that an essential drawback of these methods is the reliance on arbitrary thresholds or related external parameters that are not defined by the system, and the quality of inferences based on these techniques often depends sensitively on these parameters. For instance, choosing correlation thresholds too high or too low yields false negatives or false positives, respectively.

Reconstructed networks may also be compared across phenotypes to identify novel interactions. In [124], the authors describe a method in which pairs of genes connected by a common edge in the pathway network were examined for correlation in tumor and normal gene expression data in multiple cancers. Gene–gene edges with correlations that exceeded a threshold were kept, thus forming a correlation network in tumors and a separate correlation network in normal cells. Differences in the resulting correlation networks were then assessed through a permutation test, indicating pathways with significant differences in gene correlation. (This method could be regarded as an network based extension of [62, 125, 126].)

Recently, these network inference techniques have played a role in reverse engineering the regulatory networks of healthy human B cells [127] and chronic lymphocytic leukemia cancer cells [128], providing a richer description of the systems biology of blood. Because network inference approaches do not rely on assumptions about the pathway architecture, they are exceptionally well-suited to be applied to integrated data sets (e.g., combining both mRNA and microRNA expression data) to identify complex regulatory relationships. In a recent example, network inference techniques have revealed how the networks of miRNAs and target genes are reprogrammed in leukemia [129], further enriching our understanding of the systems biology underlying healthy and diseased hematopoietic processes.

Future Directions

Today, the feasibility of genome-wide assays, along with thousands of existing sequenced genomes [130] and hundreds of thousands of existing expression profiles [131] publicly available, provide exciting avenues for the investigation of developmental and disease processes in blood. To fully harness the power of this information, it is necessary not only to analyze the data at the gene-level, but also to examine it at the systems level. Driven by the abundance of experimental data, novel computational tools for systems-level investigations have been devised and implemented (including pathway enrichment analyses, methods for identifying functional gene-sets, and techniques for inferring regulatory networks), enabling a variety of complementary analytical techniques to be applied.

At the same time, a number of significant methodological challenges remain an area of active research, including improving the precision and accuracy of the knowledge contained in gene and pathways annotation databases, developing more efficient algorithms for combinatorially bound problems, and improving the robustness of network analysis and enrichment techniques. Just as the analytical methods will benefit experiment, so too will new experimental data inform methodological advances. We expect that these mutual advances will further improve the ability of computational and mathematical methods to model biological processes, predict clinical and experimental outcomes, and suggest therapeutic targets.

References

1. van den Akker-van Marle ME, Gurwitz D, Detmar SB, Enzing CM, Hopkins MM, de Mesa EG, Ibarreta D. Cost-effectiveness of pharmacogenomics in clinical practice: a case study of thiopurine methyltransferase genotyping in acute lymphoblastic leukemia in Europe. Pharmacogenomics. 2006;7(5):783–92.
2. Karajannis M, Vincent L, Direnzo R, Shmelkov S, Zhang F, Feldman E, Bohlen P, Zhu Z, Sun H, Kussie P, Rafii S. Activation of fgfr1beta signaling pathway promotes survival, migration and resistance to chemotherapy in acute myeloid leukemia cells. Leukemia. 2006.

3. Savageau MA, Rosen R. Biochemical systems analysis: a study of function and design in molecular biology, vol. 725. Reading: Addison-Wesley; 1976.
4. Von Bertalanffy L: Modern theories of development: an introduction to theoretical biology. In: Woodger JH, transl. Oxford University Press; 1933 (originally published 1928).
5. Irizarry RA, Hobbs B, Collin F, Beazer-Barclay YD, Antonellis KJ, Scherf U, Speed TP. Exploration, normalization, and summaries of high density oligonucleotide array probe level data. Biostatistics. 2003;4(2):249–64.
6. Bolstad BM, Irizarry RA, Åstrand M, Speed TP. A comparison of normalization methods for high density oligonucleotide array data based on variance and bias. Bioinformatics. 2003;19(2):185–93.
7. Parmigiani G. The analysis of gene expression data: methods and software. Springer; 2003.
8. Nielsen R, Paul JS, Albrechtsen A, Song YS. Genotype and snp calling from next-generation sequencing data. Nat Rev Genet. 2011;12(6):443–51.
9. Metzker ML. Sequencing technologies the next generation. Nat Rev Genet. 200911(1): 31–46.
10. Vazquez M, de la Torre V, Valencia A. Cancer genome analysis. PLoS Comput Biol. 2012 8(12):e1002824.
11. Smyth GK. Limma: linear models for microarray data. In: Bioinformatics and computational biology solutions using R and Bioconductor. Springer; 2005. pp. 397–420
12. R Development Core Team: R: A Language and Environment for Statistical Computing. R Foundation for Statistical Computing, Vienna, Austria. 2012. http://www.R-project.org/. ISBN 3-900051-07-0.
13. Gentleman RC, Carey VJ, Bates DM, Bolstad B, Dettling M, Dudoit S, Ellis B, Gautier L, Ge Y, Gentry J, et al. Bioconductor: open software development for computational biology and bioinformatics. Genome Biol. 2004;5(10):R80.
14. Gentleman, R., Carey, V., Huber, W., Irizarry, R., Dudoit, S.: Bioinformatics and computational biology solutions using R and Bioconductor, vol. 746718470. Springer; 2005.
15. Dupuy A, Simon RM. Critical review of published microarray studies for cancer outcome and guidelines on statistical analysis and reporting. J Natl Cancer Inst. 2007;99(2):147–57.
16. Benjamini Y, Hochberg Y. Controlling the false discovery rate: a practical and powerful approach to multiple testing. J R Stat Soc B. 1995; pp. 289–300.
17. Benjamini Y, Yekutieli, D. The control of the false discovery rate in multiple testing under dependency. Ann Stat. 2001; pp. 1165–88.
18. Han, B., Kang, H.M., Eskin, E.: Rapid and accurate multiple testing correction and power estimation for millions of correlated markers. PLoS Genet. 2009;5(4):e1000,456.
19. Csete ME, Doyle JC. Reverse engineering of biological complexity. Science 2002;295(5560), 1664–9.
20. Edelman GM, Gally JA. Degeneracy and complexity in biological systems. Proc Natl Acad Sci. 2001;98(24):13763–8.
21. D'haeseleer P. How does gene expression clustering work? Nat Biotechnol. 2005;23(12): 1499–501.
22. Datta S, Datta, S. Comparisons and validation of statistical clustering techniques for microarray gene expression data. Bioinformatics. 2003;19(4):459–66.
23. Eisen MB, Spellman PT, Brown PO, Botstein D. Cluster analysis and display of genome-wide expression patterns. Proc Natl Acad Sci. 1998;95(25):14863–8.
24. Hartigan, J, Wong M. Algorithm AS 136: A *k*-means clustering algorithm. J R Stat Soc C Appl Stat. 1979;28:100–8.
25. Ng A, Jordan M, Weiss Y. On spectral clustering: analysis and an algorithm. Adv Neur Inf Process Syst. 2002;2, 849–56.
26. Leibon G, Pauls S, Rockmore D, Savell R. Topological structures in the equities market network. Proc Natl Acad Sci. 2008;105(52):20589–594.
27. Chung F. Spectral graph theory. American Mathematical Society; 1997.
28. von Luxburg U. A tutorial on spectral clustering. Stat Comput. 2007;17(4):395–416.

29. Qiu P, Plevritis SK. Simultaneous class discovery and classification of microarray data using spectral analysis. J Comput Biol. 2009;16:935–44.
30. Braun R, Leibon G, Pauls S, Rockmore D. Partition decoupling for multi-gene analysis of gene expression profiling data. BMC Bioinformatics. 2011;12(497).
31. Kim D, Lee K, Lee D. Detecting clusters of different geometrical shapes in microarray gene expression data. Bioinformatics 2005;21(9):1927–34.
32. Baker S. Simple and flexible classification of gene expression microarrays via swirls and ripples. BMC Bioinformat. 2010;11(1):452
33. Fraley C, Raftery A. MCLUST: Software for model-based cluster analysis. J. Classification 1999;16(2):297–306.
34. Still S, Bialek W. How many clusters? An information-theoretic perspective. Neural Comput. 2004;16(12):2483–506.
35. Tibshirani R, Walther G, Hastie T. Estimating the number of clusters in a data set via the gap statistic. J R Stat Soc B. 2002;63(2):411–23.
36. Monti S, Tamayo P, Mesirov J, Golub T. Consensus clustering: a resampling-based method for class discovery and visualization of gene expression microarray data. Mach Learn. 2003;52 (1–2):91–118.
37. Monti S, Savage KJ, Kutok JL, Feuerhake F, Kurtin P, Mihm, M, Wu B, Pasqualucci L, Neuberg D, Aguiar RC, et al. Molecular profiling of diffuse large b-cell lymphoma identifies robust subtypes including one characterized by host inflammatory response. Blood. 2005;105(5):1851–61.
38. Jolliffe I. Principal component analysis. Wiley Online Library; 2005.
39. Wilson NK, Foster SD, Wang X, Knezevic K, Schütte J, Kaimakis P, Chilarska PM, Kinston S, Ouwehand WH, Dzierzak E, et al. Combinatorial transcriptional control in blood stem/progenitor cells: genome-wide analysis of ten major transcriptional regulators. Cell Stem Cell. 2010;7(4):532–44.
40. Chambers SM, Boles NC, Lin KYK, Tierney MP, Bowman TV, Bradfute SB, Chen AJ, Merchant AA, Sirin O, Weksberg DC, et al. Hematopoietic fingerprints: an expression database of stem cells and their progeny. Cell Stem Cell. 2007;1(5):578–91.
41. Alter O, Brown PO, Botstein D. Singular value decomposition for genome-wide expression data processing and modeling. Proc Natl Acad Sci. 2000;97(18):10101–6.
42. McIsaac RS, Petti AA, Bussemaker HJ, Botstein D. Perturbation-based analysis and modeling of combinatorial regulation in the yeast sulfur assimilation pathway. Mol Biol Cell 2012;23(15):2993–3007.
43. Narula J, Smith AM, Gottgens B, Igoshin OA. Modeling reveals bistability and low-pass filtering in the network module determining blood stem cell fate. PLoS Comput Biol. 2010;6(5):e1000771.
44. Bengio Y, Paiement J, Vincent P, Delalleau O, Le Roux N, Ouimet M. Out-of-sample extensions for LLE, IsoMap, MDS, eigenmaps, and spect ral clustering. Adv Neural Inf Process Syst. 2004;16:177–84.
45. Bengio Y, Delalleau O, Roux N, Paiement J, Vincent P, Ouimet M. Learning eigenfunctions links spectral embedding and kernel PCA. Neural Comput. 2004;16(10):2197–219.
46. Törönen P, Kolehmainen M, Wong G, Castrén E. Analysis of gene expression data using self-organizing maps. FEBS Lett. 1999;451(2):142–6.
47. Tamayo P, Slonim D, Mesirov J, Zhu Q, E Dmitrovsky SK, Lander ES, Golub TR. Interpreting patterns of gene expression with self-organizing maps: methods and application to hematopoietic differentiation. Proc Natl Acad Sci. 1999;96(6):2907–12.
48. Hastie T, Tibshirani R, Friedman J, Franklin J. The elements of statistical learning: data mining, inference and prediction. Springer; 2009.
49. Schaefer CF, Anthony K, Krupa S, Buchoff J, Day M, Hannay T, Buetow KH. PID: the Pathway Interaction Database. Nucleic Acids Res. 2009;37:D674–9.
50. Kanehisa M, Araki M, Goto S, Hattori M, Hirakawa M, Itoh M, Katayama T, Kawashima S, Okuda S, Tokimatsu T, Yamanishi Y. KEGG for linking genomes to life and the environment. Nucleic Acids Res. 2008;36(Database issue):D480–4.

51. Vastrik I, D'Eustachio P, Schmidt E, Gopinath G, Croft D, de Bono B, Gillespie M, Jassal B, Lewis S, Matthews L, Wu G, Birney E, Stein L. Reactome: a knowledge base of biologic pathways and processes. Genome Biol. 2007;8(3):R39.
52. Lynn DJ, Winsor GL, Chan C, Richard N, Laird MR, Barsky A, Gardy JL, Roche FM, Chan TH, Shah N, et al. Innatedb: facilitating systems-level analyses of the mammalian innate immune response. Mol Syst Biol. 2008;4(1).
53. Smedley D, Haider S, Ballester B, Holland R, London D, Thorisson G, Kasprzyk A. BioMart–biological queries made easy. BMC Genomics. 2009;10:22.
54. Khatri P, Sirota M, Butte AJ. Ten years of pathway analysis: current approaches and outstanding challenges. PLoS Comput Biol. 2012;8(2):e1002375.
55. Subramanian A, Tamayo P, Mootha VK, Mukherjee S, Ebert BL, Gillette MA, Paulovich A, Pomeroy SL, Golub TR, Lander ES, Mesirov JP. Gene set enrichment analysis: a knowledge-based approach for interpreting genome-wide expression profiles. Proc Natl Acad Sci. 2005;102(43):15545–50.
56. Jiang Z, Gentleman R. Extensions to gene set enrichment. Bioinformatics. 2007;23(3): 306–13.
57. Goeman JJ, Bühlmann P. Analyzing gene expression data in terms of gene sets: methodological issues. Bioinformatics. 2007;23(8):980–7.
58. Tian L, Greenberg SA, Kong SW, Altschuler J, Kohane IS, Park PJ. Discovering statistically significant pathways in expression profiling studies. Proc Natl Acad Sci U S A. 2005;102(38):13544–9.
59. Chiaretti S, Li X, Gentleman R, Vitale A, Vignetti M, Mandelli F, Ritz J, Foa R. Gene expression profile of adult t-cell acute lymphocytic leukemia identifies distinct subsets of patients with different response to therapy and survival. Blood. 2004;103(7):2771–8.
60. Grigoryev YA, Kurian SM, Avnur Z, Borie D, Deng J, Campbell D, Sung J, Nikolcheva T, Quinn A, Schulman H, et al. Deconvoluting post-transplant immunity: cell subset-specific mapping reveals pathways for activation and expansion of memory t, monocytes and b cells. PloS One. 2010;5(10):e13,358.
61. Ma S, Kosorok MR. Identification of differential gene pathways with principal component analysis. Bioinformatics. 2009;25(7):882–9.
62. Braun R, Cope L, Parmigiani G. Identifying differential correlation in gene/pathway combinations. BMC Bioinformatics. 2008;9:488.
63. Tibshirani R, Hastie T, Narasimhan B, Chu G. Class prediction by nearest shrunken centroids, with applications to dna microarrays. Stat Sci. 2003;104–17.
64. Golub TR, Slonim DK, Tamayo P, Huard C, Gaasenbeek M, Mesirov JP, Coller H, Loh ML, Downing JR, Caligiuri MA, et al. Molecular classification of cancer: class discovery and class prediction by gene expression monitoring. Science. 1999; 286(5439):531–7.
65. Hastie T, Tibshirani R, Narasimhan B, Chu G. pamr: Pam: prediction analysis for microarrays. 2011. http://CRAN.R-project.org/package=pamr. R package version 1.54.
66. Hirschhorn JN, Daly MJ. Genome-wide association studies for common diseases and complex traits. Nat Rev Genet. 2005;6(2):95–108.
67. McCarthy MI, Abecasis GR, Cardon LR, Goldstein DB, Little J, Ioannidis JPA, Hirschhorn JN. Genome-wide association studies for complex traits: consensus, uncertainty and challenges. Nat Rev Genet. 2008;9(5):356–69.
68. Hindorff LA, Sethupathy P, Junkins HA, Ramos EM, Mehta JP, Collins FS, Manolio TA. Potential etiologic and functional implications of genome-wide association loci for human diseases and traits. Proc Natl Acad Sci. 2009;106(23):9362–7.
69. Schork N, Murray S, Frazer K, Topol E. Common vs. rare allele hypotheses for complex diseases. Current Opin Genet Dev. 2009;19(3):212–9.
70. Moore J, Asselbergs F, Williams S. Bioinformatics challenges for genome-wide association studies. Bioinformatics. 2010;26(4):445.
71. Greene C, Penrod N, Williams S, Moore J. Failure to replicate a genetic association may provide important clues about genetic architecture. PLoS One. 2009;4(6):e5639.

72. Moore J. The ubiquitous nature of epistasis in determining susceptibility to common human diseases. Hum Hered. 2003;56(1–3):73–82.
73. Tyler A, Asselbergs F, Williams S, Moore J. Shadows of complexity: what biological networks reveal about epistasis and pleiotropy. BioEssays. 2009;31(2):220–7.
74. Holmans P. Statistical methods for pathway analysis of genome-wide data for association with complex genetic traits. Adv Genet. 2010;72:141.
75. Wang K, Li M, Hakonarson H. Analysing biological pathways in genome-wide association studies. Nat Rev Genet. 2010;11(12):843–54.
76. Wang K, Li M, Bucan M. Pathway-based approaches for analysis of genomewide association studies. Am J Hum Genet. 2007;81(6):1278.
77. Holden M, Deng S, Wojnowski L, Kulle B. GSEA-SNP: applying gene set enrichment analysis to SNP data from genome-wide association studies. Bioinformatics. 2008;24(23):2784–5.
78. Motsinger A, Ritchie M. Multifactor dimensionality reduction: an analysis strategy for modelling and detecting gene–gene interactions in human genetics and pharmacogenomics studies. Hum Genomics. 2006;2(5):318–28.
79. Moore J, Gilbert J, Tsai C, Chiang F, Holden T, Barney N, White B. A flexible computational framework for detecting, characterizing, and interpreting statistical patterns of epistasis in genetic studies of human disease susceptibility. J Theor Biol. 2006;241(2):252–61.
80. Cordell H. Detecting gene–gene interactions that underlie human diseases. Nat Rev Genet. 2009;10(6):392–404.
81. Greene C, Sinnott-Armstrong N, Himmelstein D, Park P, Moore J, Harris B. Multifactor dimensionality reduction for graphics processing units enables genome-wide testing of epistasis in sporadic als. Bioinformatics. 2010;26(5):694.
82. Kira K, Rendell L. A practical approach to feature selection. Proceedings of the Ninth International Workshop on Machine learning; 1992. pp. 249–56.
83. Robnik-Šikonja M, Kononenko I. An adaptation of relief for attribute estimation in regression. Proceedings of the International Conference on Machine Learning ICML-97; 1997. pp. 296–304.
84. Moore J. Genome-wide analysis of epistasis using multifactor dimensionality reduction: feature selection and construction in the domain of human genetics. Knowledge Discovery and Data Mining: Challenges and Realities with Real World Data; 2007. pp. 17–30.
85. Greene C, Penrod N, Kiralis J, Moore J. Spatially Uniform ReliefF (SURF) for computationally-efficient filtering of gene-gene interactions. BioData Mining. 2009;2:5.
86. Homer N, Szelinger S, Redman M, Duggan D, Tembe W, Muehling J, Pearson JV, Stephan DA, Nelson SF, Craig DW. Resolving individuals contributing trace amounts of DNA to highly complex mixtures using high-density SNP genotyping microarrays. PLoS Genet. 2008;4(8):e1000167.
87. Braun R, Rowe W, Schaefer C, Zhang J, Buetow K. Needles in the haystack: Identifying individuals present in pooled genomic data. PLoS Genet. 2009;5(10):e1000668.
88. Visscher PM, Hill WG. The limits of individual identification from sample allele frequencies: theory and statistical analysis. PLoS Genet. 2009;5(10):e1000628.
89. Braun R, Buetow K. Pathways of Distinction Analysis: a new technique for multi-SNP ana lysis of GWAS data. PLoS Genet. 2011;7(6):e1002101.
90. Breiman L. Random forests. Machine Learn. 2001;45(1):5–32.
91. Pang H, Lin A, Holford M, Enerson BE, Lu B, Lawton MP, Floyd E, Zhao H. Pathway analysis using random forests classification and regression. Bioinformatics. 2006;22(16):2028–36.
92. Díaz-Uriarte R, De Andres SA. Gene selection and classification of microarray data using random forest. BMC Bioinformatics. 2006;7(1):3.
93. Dettling M. Bagboosting for tumor classification with gene expression data. Bioinformatics. 2004;20(18):3583–93.
94. Lee JW, Lee JB, Park M, Song SH. An extensive comparison of recent classification tools applied to microarray data. Comput Stat Data Anal. 2005;48(4):869–85.
95. Hassane DC, Guzman ML, Corbett C, Li X, Abboud R, Young F, Liesveld JL, Carroll M, Jordan CT. Discovery of agents that eradicate leukemia stem cells using an in silico screen of public gene expression data. Blood. 2008;111(12):5654–62.

96. Van Ness B, Ramos C, Haznadar M, Hoering A, Haessler J, Crowley J, Jacobus S, Oken M, Rajkumar V, Greipp P, et al. Genomic variation in myeloma: design, content, and initial application of the bank on a cure snp panel to detect associations with progression-free survival. BMC Med. 2008;6(1):26.
97. Ackermann M, Sikora-Wohlfeld W, Beyer A. Elucidating the regulatory mechanisms of transcription factor activity in hematopoietic stem cell differentiation. In: Saxon Biotechnology Symposium; 2011. p. 79.
98. De Souto M, Costa I, De Araujo D, Ludermir T, Schliep A. Clustering cancer gene expression data: a comparative study. BMC Bioinformatics. 2008;9(1):497.
99. Kolaczyk ED. Statistical analysis of network data. Springer; 2009.
100. Barabasi AL, Oltvai ZN. Network biology: understanding the cell's functional organization. Nat Rev Genet. 2004;5(2):101–13.
101. Jeong H, Tombor B, Albert R, Oltvai ZN, Barabasi AL. The large-scale organization of metabolic networks. Nature. 2000;407(6804):651–4.
102. Jeong H, Mason SP, Barabasi AL, Oltvai ZN. Lethality and centrality in protein networks. Nature. 2001;411(6833):41–2.
103. Ideker T, Thorsson V, Ranish JA, Christmas R, Buhler J, Eng JK, Bumgarner R, Goodlett DR, Aebersold R, Hood L. Integrated genomic and proteomic analyses of a systematically perturbed metabolic network. Science. 2001;292(5518):929–34.
104. Nacu S, Critchley-Thorne R, Lee P, Holmes S. Gene expression network analysis and applications to immunology. Bioinformatics. 2007;23(7):850–8.
105. Dittrich MT, Klau GW, Rosenwald A, Dandekar T, Müller T. Identifying functional modules in protein–protein interaction networks: an integrated exact approach. Bioinformatics. 2008;24(13):i223–31.
106. Beisser D, Klau GW, Dandekar T, Müller T, Dittrich MT. Bionet: an r-package for the functional analysis of biological networks. Bioinformatics. 2010;26(8):1129–30.
107. Efroni S, Schaefer CF, Buetow KH. Identification of key processes underlying cancer phenotypes using biologic pathway analysis. PLoS One. 2007;2(5):e425.
108. Jörg R, Jochen M, Thomas L, et al. Calculating the statistical significance of changes in pathway activity from gene expression data. Stat Appl Genet Mol Biol. 2004;3(1):1–31.
109. Draghici S, Khatri P, Tarca AL, Amin K, Done A, Voichita C, Georgescu C, Romero R. A systems biology approach for pathway level analysis. Genome Res. 2007;17(10):1537–45.
110. Tarca AL, Draghici S, Khatri P, Hassan SS, Mittal P, Kim Js, Kim CJ, Kusanovic JP, Romero R. A novel signaling pathway impact analysis. Bioinformatics. 2009;25(1):75–82.
111. Shojaie A, Michailidis G. Penalized principal component regression on graphs for analysis of subnetworks. In: Advances in neural information processing systems; 2010. pp. 2155–63.
112. Bansal M, Belcastro V, Ambesi-Impiombato A, Di Bernardo D. How to infer gene networks from expression profiles. Mol Syst Biol. 2007;3(1).
113. Margolin A, Nemenman I, Basso K, Wiggins C, Stolovitzky G, Favera R, Califano A. ARACNE: an algorithm for the reconstruction of gene regulatory networks in a mammalian cellular context. BMC Bioinformatics. 2006;7(Suppl 1):S7.
114. Gardner T, Faith J. Reverse-engineering transcription control networks. Phys Life Rev. 2005;2(1):65–88.
115. Meyer P, Lafitte F, Bontempi G. minet: An R/Bioconductor package for inferring large transcriptional networks using mutual information. BMC Bioinformatics. 2008;9(1):461.
116. de la Fuente A, Brazhnik P, Mendes P. Linking the genes: inferring quantitative gene networks from microarray data. TRENDS Genet. 2002;18(8);395–8.
117. Gardner T, di Bernardo, D, Lorenz D, Collins J: Inferring genetic networks and identifying compound mode of action via expression profiling. Sci Signal. 2003;301(5629):102.
118. Rice J, Tu Y, Stolovitzky G. Reconstructing biological networks using conditional correlation analysis. Bioinformatics. 21(6):765–73.
119. Marbach D, Prill R, Schaffter T, Mattiussi C, Floreano D, Stolovitzky G: Revealing strengths and weaknesses of methods for gene network inference. Proc Natl Acad Sci. 2010;107(14):6286–91.

120. Altay G, Emmert-Streib F: Revealing differences in gene network inference algorithms on the network level by ensemble methods. Bioinformatics. 2010;26(14):1738–44.
121. Dodd IB, Micheelsen MA, Sneppen K, Thon G. Theoretical analysis of epigenetic cell memory by nucleosome modification. Cell. 2007;129(4):813–22.
122. Sedighi M, Sengupta AM. Epigenetic chromatin silencing: bistability and front propagation. Phys Biol. 2007;4(4):246–55.
123. Graf T, Enver T. Forcing cells to change lineages. Nature. 2009;462:(7273):587–94.
124. Choi JK, Yu U, Yoo OJ, Kim S. Differential coexpression analysis using microarray data and its application to human cancer. Bioinformatics. 2005;21(24):4348–55.
125. Ho YY, Cope L, Dettling M, Parmigiani G. Statistical methods for identifying differentially expressed gene combinations. In: Gene function analysis. Springer; 2007. pp. 171–91.
126. Dettling M, Gabrielson E, Parmigiani G. Searching for differentially expressed gene combinations. Genome Biol. 2005;6:R88.
127. Basso K, Margolin AA, Stolovitzky G, Klein U, Dalla-Favera R, Califano A. Reverse engineering of regulatory networks in human b cells. Nat Genet. 2005;37(4):382–90.
128. Vallat L, Kemper CA, Jung N, Maumy-Bertrand M, Bertrand F, Meyer N, Pocheville A, Fisher JW, Gribben JG, Bahram S. Reverse-engineering the genetic circuitry of a cancer cell with predicted intervention in chronic lymphocytic leukemia. Proc Natl Acad Sci. 2013;110(2):459–64.
129. Volinia S, Galasso M, Costinean S, Tagliavini L, Gamberoni G, Drusco A, Marchesini J, Mascellani N, Sana ME, Jarour RA, et al. Reprogramming of miRNA networks in cancer and leukemia. Genome Res. 2010;20(5):589–99.
130. Sayers EW, Barrett T, Benson DA, Bolton E, Bryant SH, Canese K, Chetvernin V, Church DM, DiCuccio M, Federhen S, et al. Database resources of the national center for biotechnology information. Nucleic Acids Res. 2011;39(Suppl 1):D38–51.
131. Barrett T, Troup DB, Wilhite SE, Ledoux P, Rudnev D, Evangelista C, Kim IF, Soboleva A, Tomashevsky M, Marshall KA, et al. NCBI GEO: archive for high-throughput functional genomic data. Nucleic Acids Res. 2009;37(Suppl 1):D885–90.

Chapter 9
Developing a Systems-Based Understanding of Hematopoietic Stem Cell Cycle Control

Ka Tat Siu and Alex C. Minella

Abstract To maintain hematologic homeostasis, hematopoietic stem cells (HSCs) undergo multiple rounds of cell division throughout their lives. Under steady-state conditions, adult HSCs are relatively quiescent and reside primarily in hypoxic bone marrow niches. In response to physiologic stimuli, normal HSCs either reenter the cell division cycle or remain in quiescence. A large body of work has focused on understanding the mechanistic underpinnings balancing differentiation against self-renewal programs in cycling HSCs. Numerous reports from genetically engineered mouse models harboring mutations in key pathways governing proliferation control, DNA damage responses, and metabolic regulation indicate the critical roles these processes play in determining HSC self-renewing versus blood-lineage-reconstituting divisions. In this chapter, we integrate these findings and highlight the cellular networks that control HSC function and fitness by regulating HSC cycling.

Keywords Hematopoietic stem cell · Hematopoietic progenitor cell · Cell cycle · Self-renewal · Differentiation · Quiescence · Metabolism · Aging · DNA damage

Introduction

The lifelong regeneration of blood cells is sustained by a small population of relatively quiescent hematopoietic stem cells (HSCs), which infrequently enter cell cycle to either self-renew or give rise to multi-potential progenitor cells. These progenitor cells undergo more rapid proliferation to produce cells with progressively greater lineage restriction, ultimately allowing repletion of terminally differentiated blood and bone marrow cells that are short-lived. HSCs are found predominantly in bone marrow niches, where most are maintained in a quiescent state [1]. Under conditions of hematopoietic homeostasis, the size of the adult HSC pool remains relatively constant, with striking evolutionary conservation of absolute stem cell

A. C. Minella (✉) · K. T. Siu
Northwestern University Feinberg School of Medicine,
303 East Superior Street, Lurie 5-115, Chicago, IL 60611, USA
Tel.: +312-503-2041
e-mail: alex.minella@bcw.edu

S. J. Corey et al. (eds.), *A Systems Biology Approach to Blood*,
Advances in Experimental Medicine and Biology 844, DOI 10.1007/978-1-4939-2095-2_9

numbers demonstrated in mammals from mouse to man [2, 3]. A major goal of molecular hematology has been to determine the critical genes that regulate the major outcomes of HSC mitotic divisions: towards the generation of two new HSCs (symmetric, self-renewing) or one HSC and one hematopoietic progenitor cell (HPC) with multi-lineage potential (asymmetric). The objectives of these studies are both to understand fundamental HSC biology and discover mechanistic underpinnings of bone marrow stem cell diseases, including the myelodysplastic syndromes (MDS) and aplastic anemias.

In experimental models, purported HSCs must fulfill the functional criteria that define adult tissue stem cells, namely long-term self-renewal and multi-lineage reconstitution capabilities. Although the functional characteristics of HSCs have long been well defined, the strategies for enrichment of bone marrow cells for highly purified HSCs have been refined considerably over the last decade. Phenotypic isolation of nearly pure populations of HSCs is now made possible with the use of signaling lymphocytic activation molecules (SLAM), including CD48 and CD244 (for negative selection) and CD150 (for positive selection) [4]. These SLAM markers have been demonstrated to remain stably expressed across a variety of conditions such as mobilization, development, aging, and hematopoietic reconstitution, making them optimal markers of HSCs [5, 6]. The "gold standard" for assessing HSC function in vivo involves the use of serial transplantation assays, which require the transfer of purified HSCs (or HSC-enriched donor bone marrow cells) into irradiated recipient animals and the subsequent reiterative transfer of engrafted, donor-derived bone marrow cells from primary to secondary recipients and beyond [7]. The rigorous nature of these experiments has necessitated a "reductionist" approach to studying the regulation of HSC self-renewal and reconstitution capacities, whereby the expression of genes in relative isolation is perturbed, followed by exhaustive characterization of functional defects in self-renewal and/or multi-lineage reconstitution associated with this manipulation. In the following sections, we highlight a number of studies representing the accumulated knowledge from over a decade's worth of work in mouse models now permitting the conceptualization of networks connecting the regulation of HSC cycling to cellular programs governing metabolism, differentiation, and self-renewal.

HSC Cycling Dynamics

Relative quiescence is a hallmark feature of adult HSCs. The Weissman group in 1999 reported that under steady-state conditions, approximately three-quarters of bone marrow HSCs with long-term reconstitution potential (LT-HSCs) are quiescent. Of those transiting the mitotic cell cycle, the majority are found in G_1-phase, with only about 5 % of all LT-HSCs identifiable within $S/G_2/M$-phases [8]. Calculations from in vivo labeling experiments using the thymidine analogue 5-bromo-2-deoxyuridine (BrdU) led to conclusions that mouse HSCs periodically enter and exit the cell cycle, with the entire HSC pool turning over every 57 days [8]. In later experiments employing limiting dilution competitive reconstitution assays in which donor

LT-HSCs were sorted into G_0 (quiescent), G_1, and $S/G_2/M$ subpopulations, Weissman and colleagues showed that the long-term repopulating capacity of HSCs resides in the quiescent fraction [9]. This result supports the concept that bone marrow HSCs are heterogeneous with respect to their functional capability, which intrinsically is coupled to their proliferative status. This conclusion is further supported by work performed by the Trumpp group, who applied mathematical modeling to data from long-term BrdU label-retaining assays. In this work, they concluded that two distinct subsets of HSCs are contained within the lineage-negative, Sca-1-positive, c-kit-positive, CD34-negative, CD48-negative, CD150-positive, BrdU-retaining cell population: one population of actively dividing HSCs that turns over every 36 days and another population of dormant HSCs, which divides every 145 days [10]. Their modeling further predicts that dormant HSCs only divide five times during the lifespan of a normal C57BL/6 mouse [10]. Serial transplantation assays revealed critical functional differences between the two HSC populations. Dormant HSCs retained long-term self-renewing capacity, whereas cycling HSCs had limited self-renewing potential [10]. The investigators further hypothesized that the dormant HSCs function chiefly to respond to bone marrow injury. Upon hematologic stress induced by 5-fluorouracil (5-FU) treatment, dormant HSCs reenter the cell cycle to generate proliferative progenitor cells, which give rise to mature cells to replenish the bone marrow space [10]. Finally, BrdU pulse-chase studies of 5-FU treated HSCs confirm that injury-activated HSCs are able to return to dormancy after reestablishment of hematologic homeostasis [10]. Thus, these findings have several important implications for HSC biology. First, they support the hypothesis that limiting rounds of mitoses is crucial to the maintenance of the HSC pool [11]. Second, they provide functional evidence for the existence of a dormant population of HSCs that acts as a reservoir to maintain lifelong hematopoiesis. Lastly, these data demonstrate that self-renewing HSCs are highly responsive to external stimuli and are able to reversibly switch from dormant to proliferative states during steady state and stress hematopoiesis [10].

Regulators of HSC Proliferation

A complex network controlling quiescence and self-renewal appears to be required for maintenance of a stable HSC pool, responsive to potential hematologic stress throughout an organism's lifespan. Some transcriptional regulators, such as Bmi-1 and MEF/ELF4, oppose quiescence programs to promote HSC self-renewal. The polycomb group protein Bmi-1 is preferentially expressed in HSCs compared to differentiated cells [12] and plays a central role in HSC self-renewal. Bmi-1 actively represses the *Ink4a* locus, which encodes the cyclin D-cdk4/6 inhibitor, $p16^{Ink4a}$, and the tumor suppressor, $p19^{ARF}$ that positively regulates p53 [13]. Loss of Bmi-1 results in upregulation of $p16^{Ink4a}$ and $p19^{ARF}$, and Bmi-1-deficient mice die of bone marrow failure due to exhaustion of HSCs within 2 months of birth [12, 14]. Deletion of both $p16^{Ink4a}$ and $p19^{ARF}$ rescues the hematopoietic defects in Bmi-1-deficient animals

[12, 15], suggesting the critical activities of Bmi-1 in maintaining HSC self-renewal are restraining p16^{Ink4a} and p19^{ARF} expression. Myeloid elf-1-like factor (MEF), also known as ELF4, is a member of the E26 transformation-specific (ETS) family of winged helix–turn–helix transcription factors [16]. Conditional deletion of *Elf4* in hematopoietic cells leads to accumulation of quiescent HSCs in the bone marrow [17]. MEF/ELF4-deficient bone marrows also demonstrate reduced sensitivity to myelotoxic stress induced by 5-FU, possibly consistent with an HSC-protective effect of enforced quiescence. Furthermore, MEF-deficient HSCs outnumber wild-type HSCs in competitive repopulation assays, consistent with increased overall fitness [17]. These results support the conclusion that quiescent HSCs are intrinsically more effective than cycling HSCs in long-term engraftment assays [9] and that MEF promotes HSC self-renewal by facilitating entry of quiescent HSCs into the cell cycle. Taken together, these data underscore the importance of limiting cell cycle reentry in maintaining lifelong hematopoiesis.

Disruption of HSC quiescence programs may promote HSC exhaustion. Depletion of hematopoietic stem and/or progenitor cells (HSPCs) resulting from ablating a single cell cycle regulator was first reported by the Scadden group using p21^{Cip1} knockout mice [18]. These knockout mice demonstrate enhanced proliferation of lineage-negative bone marrow cells, at the expense of their self-renewal capacity, manifested as progressive bone marrow failure and death starting with tertiary transplantation [18]. The role of p21 as a major regulator of HSC quiescence was revisited in a different model, in which HSCs from mice bearing a deletion of the Gfi-1 transcriptional repressor were studied. In *Gfi1*-null mice, Orkin and colleagues show HSPCs are increased in number compared to wild-type controls [19]. In addition, Gfi-1-deficient HSPCs are defective in repopulating bone marrows of lethally irradiated recipients in transplantation assays and show reduced long-term fitness in contributing to erythropoiesis in chimeric animals. Thus, Gfi-1 promotes HSPC quiescence and preserves HSPC function, findings the investigators suggest are associated with its positive regulation of p21 expression. Conditional inactivation of another member of the Cip/Kip family of cyclin-dependent kinases inhibitors (CKIs), p57^{Kip2}, in hematopoietic cells leads to a reduction in the quiescent HSC pool size and impaired repopulating capacity after transplantation [20]. Concomitant deletion of either p21^{Cip1} or p27^{Kip1} exacerbates the HSC abnormalities of p57^{Kip2}-deficient mice [20, 21], and knocking the p27^{Kip1} gene into the p57^{Kip2} locus reverses the defects in multi-lineage reconstitution and HSPC pool size associated with p57 loss [20]. These data indicate that p57^{Kip2} is an important regulator of HSC quiescence and also that functional redundancy exists among the Cip/Kip proteins in HSCs [11].

Similarly, Retinoblastoma (Rb) family proteins show functional redundancy in developing and adult tissues. Allelic disruption of all three Rb family members—pRb, p107, and p130 triple knockout (TKO)—in adult mice results in a cell-autonomous myeloproliferative phenotype [22]. This phenotype is accompanied by increased HSPC proliferation and impaired HSC self-renewal, as evidenced by the failure of TKO HSCs to support long-term engraftment. Thus, Rb family members cooperatively maintain HSC quiescence. Finally, the Morrison and Li groups showed conditional deletion of the *Pten* tumor suppressor gene in hematopoietic cells increases HSC cell cycling and results in depletion of HSCs. *Pten*-deficient HSCs are also defective in multi-lineage reconstitution [23, 24], and rapamycin, a pharmacological inhibitor of the mTOR pathway, rescues the multi-lineage reconstitution HSC defects associated with *Pten* loss [23]. Therefore, PTEN maintains HSC quiescence through modulating mTOR signaling.

Although the aforementioned studies may lead one to conclude that hyper-proliferation is intrinsically linked to stem cell exhaustion, other experimental models seem to contradict this notion. For one, loss of the p53 transcription factor and tumor suppressor protein drives HSCs into the cell cycle, actually resulting in enhanced HSC self-renewal, at least through two rounds of transplantation [25–27]. Thus, p53 normally maintains HSC in a quiescent state and restrains self-renewing cell divisions. Enhanced self-renewal of p53-null HSCs, however, may be associated with diminished multi-lineage reconstitution capacity [25, 27]. On the other hand, results from the Dick laboratory demonstrate human HSPCs with reduced p53 expression treated with moderately dosed ionizing radiation have enhanced self-renewal kinetics in primary transplants, compared to p53-intact control cells, though the former ultimately exhaust after secondary transplantation with increased DNA damage foci. Thus, p53 may not only restrict HSC self-renewal by promoting quiescence but also regulate long-term HSC function by enabling repair of and resolution of DNA damage foci [28]. Interestingly, data from the Nimer group suggest $p21^{Cip1}$ is not the major mediator of p53-dependent HSC quiescence [22]. Rather, they nominate Necdin and Gfi-1 as major mediators of the p53-dependent quiescence program [26].

Loss of $p18^{Ink4c}$ leads to an increased numbers of actively cycling HSPCs, while competitive repopulation assays show that $p18^{Ink4c}$-deficient HSPCs outnumber their wild-type counterparts after secondary transplantation [29, 30], suggesting that HSC self-renewal function is not impaired. Similarly, HoxB4 overexpression in HPCs enhances both in vivo and ex vivo expansion of murine and human HSCs [31–33]. Moreover, HoxB4-transduced bone marrow progenitor cells exhibit a competitive advantage over wild-type cells in competitive reconstitution assays, suggesting that HoxB4-associated HSC hyper-proliferation is not coupled to a functional deficit [31, 33]. Finally, the Passegué group reported a *Junb* knockout mouse model in which increased HSC proliferation was not found accompanied by evidence of stem cell exhaustion [34]. The authors attribute HSC hyper-proliferation in their model to disruption of Notch- and TGF-β-dependent signaling, possibly rendering HSCs insensitive to proliferation limits. Thus, under certain conditions HSC proliferation may be uncoupled from self-renewal. Interestingly, in contrast to its effect in HSCs, *Junb* loss induces aberrant expansion of myeloid progenitors and the development of myeloproliferative disease originating from the HSC compartment [34]. Since HSCs must transit through the cell cycle to undergo either self-renewal or differentiation, the identification of regulators unique to these divergent fates will further illuminate the mechanistic relationship between HSC proliferation and functionality.

Interplay Between HSC Proliferation and Metabolic Regulatory Controls

Adult HSCs are primarily localized in specialized microenvironments known as bone marrow niches. These niches are relative hypoxic [35]. One advantage for localizing quiescent HSCs to a hypoxic niche is to minimize accumulation of reactive oxygen species (ROS), which could promote premature differentiation and exhaustion [36]. Tolerance of HSCs to hypoxic microenvironments is attributable to induction of hypoxia-inducible factor-1α (HIF-1α) protein, a master transcriptional regulator of the hypoxic response [37] that promotes a cellular switch from mitochondrial oxidative metabolism to glycolysis [38]. The Suda laboratory identified a key role for HIF-1α in the regulation of HSC quiescence by demonstrating HIF-1α deficiency results in loss of HSC quiescence and exhaustion following various stress conditions including bone marrow transplantation, 5-FU-induced injury, and aging, in a $p16^{Ink4a}$ and $p19^{Arf}$-dependent manner [39]. Overexpression of Bmi-1, which suppresses the *Ink4a* gene products ($p16^{Ink4a}$ and $p19^{Arf}$), reverses the senescence-like HSC exhaustion phenotype associated with HIF-1α loss [39]. These results suggest that HIF-1α protects HSCs against cellular senescence.

Recently, the Bardeesy, DePinho, and Morrison laboratories independently demonstrated that liver kinase B1 (LKB1), a master regulator of cellular metabolism, promotes HSC quiescence by restricting cell cycle entry [40–42]. *Lkb1*, a known tumor suppressor gene, encodes a serine–threonine kinase that restricts cell growth under energy-deprived conditions [43]. *Lkb1* inactivation in adult mice results in loss of HSC quiescence followed by rapid depletion of HSCs and a marked reduction of their repopulating capacity [40–42]. Transcriptome analysis of *Lkb1*-null HSPCs reveals significant enrichment of genes involved in G_1/S-phase cell cycle checkpoint regulation and reduced expression of peroxisome-proliferator-activated receptor γ (PPARγ) co-activators PGC-1α and PGC-1β, master transcriptional regulators of mitochondrial biogenesis [40, 44]. These results are consistent with the observations that *Lkb1*-null HSPCs display enhanced cell cycling and decreased mitochondrial membrane potential and mitochondrial DNA content [40]. Another report from the Pandolfi group demonstrates a role for lipid catabolism in the maintenance of HSCs. Here, investigators show that promyelocytic leukemia (PML)-PPAR-δ-fatty acid oxidation (FAO) pathway maintains HSC function by sustaining ATP levels in HSCs and promoting asymmetric division of HSCs [45].

Elevated ROS levels can promote DNA damage and activation of senescence mechanisms in the HSC compartment [46]. Importantly, studies of ataxia-telangiectasia-mutated (ATM)-deficient mice demonstrated a role for ROS in the regulation of HSC function. The *Atm* gene encodes a serine–threonine kinase that maintains genomic stability by activating cell cycle checkpoints in response to DNA double-strand breaks (DSBs) [47, 48]. Ionizing radiation induces ATM-dependent cell cycle arrest in HSCs, followed by engagement of DNA repair mechanisms [49]. *Atm* deletion leads to ROS-mediated depletion of HSCs and progressive bone marrow failure in older mice [50]. Accumulation of ROS in ATM-null HSCs activates the p16^{Ink4a}–Rb pathway, resulting in suppression of their multi-lineage reconstitution activity [50]. Importantly, treatment of *Atm*-null mice with the antioxidant N-acetyl-cysteine (NAC) improves self-renewal and reconstitution of their HSCs [50, 51]. Like ATM, Forkhead box-O (FoxO) transcription factors are key modulators of oxidative stress in the hematopoietic system. FoxO3a deficiency results in elevation of ROS in HSPCs and reduced quiescence associated with an increase in cells in the G_2/M-phase [52, 53]. Additionally, mice engineered to express conditional alleles of *Foxo1, Foxo3,* and *Foxo4* show increased HSPC proliferation and reduced HSC numbers following inactivation of the three alleles, phenotypes reversed by NAC treatment [54]. Thus, the FoxO proteins cooperatively preserve HSC quiescence at least in part by mitigating ROS accumulation. Together, the reports summarized in this section strongly implicate a direct link between cellular metabolism and hematopoietic homeostasis.

Aging as a Paradigm for Systems-Based Understanding of HSC Cycle Regulation

Aging is characterized by reduced ability to regenerate tissues following damage or failure to maintain tissue homeostasis after exposure to stress [55]. Thus, studying stem cell exhaustion phenotypes should enhance our understanding of age-related HSC diseases, including the MDS and leukemias. For example, conditional inactivation of *Fbxw7*, which encodes the substrate-binding domain of the SCF^{Fbw7} ubiquitin ligase that promotes destruction of multiple regulators of cell proliferation, including c-Myc, Notch, cyclin E, and c-Jun [56], impairs HSC quiescence and repopulating capacity in transplantation experiments and causes premature loss of HSCs as a result of p53-dependent apoptosis [57, 58]. In the context of p53 deletion, Fbw7 loss in hematopoietic progenitors promotes development of T cell acute lymphoblastic leukemia (T-ALL) [57, 59]. Thus, Fbw7 appears to constitute a barrier against both HSC exhaustion and leukemogenesis. Moreover, p53 mutation may be subject to positive selective pressure in hematopoietic progenitors with impaired SCF^{Fbw7} activity.

Accumulated DNA damage has been proposed as one underlying mechanism causing age-associated stem cell dysfunction [60]. Although quiescent HSCs possess intrinsic barriers against genomic instability, including their utilization of glycolytic metabolism reducing the risk of ROS accumulation [54], the quiescent state may also pose unique risks to HSC genomes. For one, quiescence mandates that HSCs use the error-prone nonhomologous end joining (NHEJ) pathway to repair DNA damage, which could lead to chromosomal rearrangements [49, 61]. Studies of aged mice deficient in nucleotide excision repair (NER), NHEJ, and telomere maintenance pathways reveal that DNA damage accrual severely impairs the capacity of HSCs to self-renew and repopulate the hematopoietic system under conditions of stress [62]. These results are consistent with the observation that aged mice possess increased numbers of immunophenotypically identifiable HSCs, but with diminished

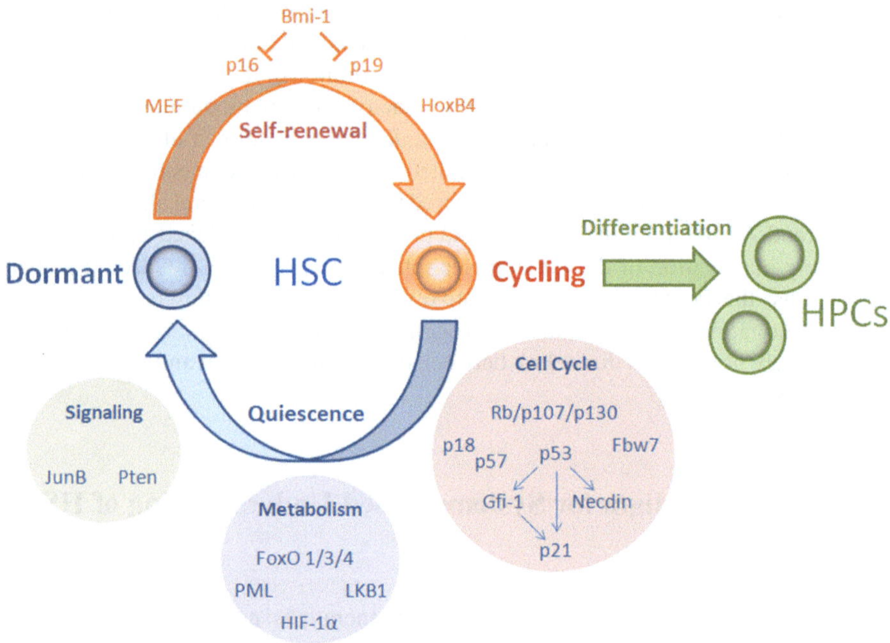

Fig. 9.1 Dynamic regulation of hematopoietic stem cell (*HSC*) cycling towards distinct fates. During mitoses, HSCs can generate two progeny with stem cell characteristics, two hematopoietic progenitor cells (*HPCs*) capable of multi-lineage differentiation, or one cell of each type. Gene products enforcing HSC quiescence or self-renewing cell cycling are highlighted and organized into concept-based clusters

functional capacity [55]. Older HSCs, indeed, exhibit increased incidences of DNA DSBs [62, 63] and undergo fewer cell divisions, due to increased $p16^{Ink4a}$ levels, compared to younger counterparts [64, 65]. Since increased expression of $p16^{Ink4a}$ is associated with cellular senescence [66], a reasonable conclusion is that cellular senescence accounts for decreased functional capacity of old HSCs. However, a recent report suggests no correlation between $p16^{Ink4a}$ levels and HSC aging [67]. Thus, progressive genome damage may impair HSC function without engaging classical DNA-damage-associated senescence pathways [68, 69].

Studies of late-generation telomerase knockout mice provide evidence that telomere shortening limits the repopulating capacity of HSCs under serial transplantation [70]. However, the mechanism that impairs regenerative capacity of aged HSCs remains largely unknown. Using in vivo RNAi screening, the Rudolph group identified a gene-encoding basic leucine zipper transcription factor, activating transcription factor (ATF)-like (*Batf*) as a major barrier against HSC self-renewal in response to telomere dysfunction [71]. *Batf* loss enhances HSC self-renewal but also results in DNA damage accumulation in HSCs in response to telomere shortening [71]. Additionally, BATF protein induction by DNA damage promotes lymphoid differentiation of HSPCs [71]. *BATF* expression levels in human HSPCs isolated from the MDS patients inversely correlate with telomere length and directly correlate with activation of DNA damage signals [71]. Together these results demonstrate that BATF controls a critical checkpoint that limits HSC proliferation in response to DNA damage [71].

Conclusion

In HSCs, proliferation controls, along with metabolic and signaling regulatory modules, are highly integrated into self-renewal and quiescence-enforcing mechanisms and have key roles in determining HSC fitness (summarized in Fig. 9.1). In order to maintain hematopoietic homeostasis, HSCs switch reversibly between quiescence and active cycling in response to physiologic stimuli. Aging phenotypes can be manifestations of effective engagement of checkpoint mechanisms that restrict propagation of HSCs in the setting of genome damage or oncogenic stress. Hematopoietic failure occurs when normal HSCs cannot replace the function of diseased counterparts, while the loss of critical checkpoints against inappropriate self-renewal in the setting of oncogenic or genotoxic insults permits the emergence of malignant clones. Understanding how to restore or replace physiologic barriers restricting malignant HSC self-renewal represents a key challenge in applying a systems-based understanding of HSC biology to the major clinical problem of HSC-based diseases.

Acknowledgements This work was supported by NIH grant R01HL098608.

References

1. Wilson A, Trumpp A. Bone-marrow haematopoietic-stem-cell niches. Nature reviews. Immunology. 2006;6:93–106.
2. Catlin SN, Busque L, Gale RE, Guttorp P, Abkowitz JL. The replication rate of human hematopoietic stem cells in vivo. Blood. 2011;117:4460–6.
3. Shepherd BE, et al. Hematopoietic stem-cell behavior in nonhuman primates. Blood. 2007;110:1806–13.
4. Kiel MJ, Yilmaz OH, Iwashita T, Terhorst C, Morrison SJ. SLAM family receptors distinguish hematopoietic stem and progenitor cells and reveal endothelial niches for stem cells. Cell. 2005;121:1109–21.
5. Kim I, He S, Yilmaz OH, Kiel MJ, Morrison SJ. Enhanced purification of fetal liver hematopoietic stem cells using SLAM family receptors. Blood. 2006;108:737–44.
6. Yilmaz OH, Kiel MJ, Morrison SJ. SLAM family markers are conserved among hematopoietic stem cells from old and reconstituted mice and markedly increase their purity. Blood. 2006;107:924–30.
7. Domen J, Weissman IL. Self-renewal, differentiation or death: regulation and manipulation of hematopoietic stem cell fate. Mol Med Today. 1999;5:201–8.
8. Cheshier SH, Morrison SJ, Liao X, Weissman IL. In vivo proliferation and cell cycle kinetics of long-term self-renewing hematopoietic stem cells. Proc Natl Acad Sci U S A. 1999;96:3120–5.

9. Passegue E, Wagers AJ, Giuriato S, Anderson WC, Weissman IL. Global analysis of proliferation and cell cycle gene expression in the regulation of hematopoietic stem and progenitor cell fates. J Exp Med. 2005;202:1599–611.
10. Wilson A, et al. Hematopoietic stem cells reversibly switch from dormancy to self-renewal during homeostasis and repair. Cell. 2008;135:1118–29.
11. Pietras EM, Warr MR, Passegue E. Cell cycle regulation in hematopoietic stem cells. J Cell Biol. 2011;195:709–20.
12. Park IK, et al. Bmi−1 is required for maintenance of adult self-renewing haematopoietic stem cells. Nature. 2003;423:302–5.
13. Sherr CJ. The INK4a/ARF network in tumour suppression. Nature reviews. Mol Cell Biol. 2001;2:731–7.
14. Lessard J, Sauvageau G. Bmi−1 determines the proliferative capacity of normal and leukaemic stem cells. Nature. 2003;423:255–60.
15. Oguro H, et al. Differential impact of Ink4a and Arf on hematopoietic stem cells and their bone marrow microenvironment in Bmi1-deficient mice. J Exp Med. 2006;203:2247–53.
16. Lacorazza HD, Nimer SD. The emerging role of the myeloid Elf−1 like transcription factor in hematopoiesis. Blood Cell Mol Dis. 2003;31:342–50.
17. Lacorazza HD, et al. The transcription factor MEF/ELF4 regulates the quiescence of primitive hematopoietic cells. Cancer Cell. 2006;9:175–87.
18. Cheng T, et al. Hematopoietic stem cell quiescence maintained by p21cip1/waf1. Science. 2000;287:1804–8.
19. Hock H, et al. Gfi−1 restricts proliferation and preserves functional integrity of haematopoietic stem cells. Nature. 2004;431:1002–7.
20. Matsumoto A, et al. p57 is required for quiescence and maintenance of adult hematopoietic stem cells. Cell Stem Cell. 2011;9:262–71.
21. Zou P, et al. p57(Kip2) and p27(Kip1) cooperate to maintain hematopoietic stem cell quiescence through interactions with Hsc70. Cell Stem Cell. 2011;9:247–61.
22. Viatour P, et al. Hematopoietic stem cell quiescence is maintained by compound contributions of the retinoblastoma gene family. Cell Stem Cell. 2008;3:416–28.
23. Yilmaz OH, et al. Pten dependence distinguishes haematopoietic stem cells from leukaemia-initiating cells. Nature. 2006;441:475–82.
24. Zhang J, et al. PTEN maintains haematopoietic stem cells and acts in lineage choice and leukaemia prevention. Nature. 2006;441:518–22.
25. Chen J, et al. Enrichment of hematopoietic stem cells with SLAM and LSK markers for the detection of hematopoietic stem cell function in normal and Trp53 null mice. Exp Hematol. 2008;36:1236–43.
26. Liu Y, et al. p53 regulates hematopoietic stem cell quiescence. Cell Stem Cell. 2009;4:37–48.
27. TeKippe M, Harrison DE, Chen J. Expansion of hematopoietic stem cell phenotype and activity in Trp53-null mice. Exp Hematol. 2003;31:521–7.
28. Milyavsky M, et al. A distinctive DNA damage response in human hematopoietic stem cells reveals an apoptosis-independent role for p53 in self-renewal. Cell Stem Cell. 2010;7:186–97.
29. Yu H, Yuan Y, Shen H, Cheng T. Hematopoietic stem cell exhaustion impacted by p18 INK4 C and p21 Cip1/Waf1 in opposite manners. Blood. 2006;107:1200–6.
30. Yuan Y, Shen H, Franklin DS, Scadden DT, Cheng T. In vivo self-renewing divisions of haematopoietic stem cells are increased in the absence of the early G1-phase inhibitor, p18INK4 C. Nat Cell Biol. 2004;6:436–42.
31. Thorsteinsdottir U, Sauvageau G, Humphries RK. Enhanced in vivo regenerative potential of HOXB4-transduced hematopoietic stem cells with regulation of their pool size. Blood. 1999;94:2605–12.
32. Antonchuk J, Sauvageau G, Humphries RK. HOXB4-induced expansion of adult hematopoietic stem cells ex vivo. Cell. 2002;109:39–45.
33. Antonchuk J, Sauvageau G, Humphries RK. HOXB4 overexpression mediates very rapid stem cell regeneration and competitive hematopoietic repopulation. Exp Hematol. 2001;29:1125–34.

34. Santaguida M, et al. JunB protects against myeloid malignancies by limiting hematopoietic stem cell proliferation and differentiation without affecting self-renewal. Cancer Cell. 2009;15:341–52.
35. Parmar K, Mauch P, Vergilio JA, Sackstein R, Down JD. Distribution of hematopoietic stem cells in the bone marrow according to regional hypoxia. Proc Natl Acad Sci U S A. 2007;104:5431–6.
36. Jang YY, Sharkis SJ. A low level of reactive oxygen species selects for primitive hematopoietic stem cells that may reside in the low-oxygenic niche. Blood. 2007;110:3056–63.
37. Danet GH, Pan Y, Luongo JL, Bonnet DA, Simon MC. Expansion of human SCID-repopulating cells under hypoxic conditions. J Clin Invest. 2003;112:126–35.
38. Simsek T, et al. The distinct metabolic profile of hematopoietic stem cells reflects their location in a hypoxic niche. Cell Stem Cell. 2010;7:380–90.
39. Takubo K, et al. Regulation of the HIF-1alpha level is essential for hematopoietic stem cells. Cell Stem Cell. 2010;7:391–402.
40. Gan B, et al. Lkb1 regulates quiescence and metabolic homeostasis of haematopoietic stem cells. Nature. 2010;468:701–4.
41. Gurumurthy S, et al. The Lkb1 metabolic sensor maintains haematopoietic stem cell survival. Nature. 2010;468:659–63.
42. Nakada D, Saunders TL, Morrison SJ. Lkb1 regulates cell cycle and energy metabolism in haematopoietic stem cells. Nature. 2010;468:653–8.
43. Shackelford DB, Shaw RJ. The LKB1-AMPK pathway: metabolism and growth control in tumour suppression. Nature reviews. Cancer. 2009;9:563–75.
44. Kelly DP, Scarpulla RC. Transcriptional regulatory circuits controlling mitochondrial biogenesis and function. Genes Dev. 2004;18:357–68.
45. Ito K, et al. A PML-PPAR-delta pathway for fatty acid oxidation regulates hematopoietic stem cell maintenance. Nat Med. 2012;18:1350–8.
46. Ghaffari S. Oxidative stress in the regulation of normal and neoplastic hematopoiesis. Antioxid Redox Signal. 2008;10:1923–40.
47. Meyn MS. Ataxia-telangiectasia and cellular responses to DNA damage. Cancer Res. 1995;55:5991–6001.
48. Shiloh Y. ATM and related protein kinases: safeguarding genome integrity. Nature reviews. Cancer. 2003;3:155–68.
49. Mohrin M, et al. Hematopoietic stem cell quiescence promotes error-prone DNA repair and mutagenesis. Cell Stem Cell. 2010;7:174–85.
50. Ito K, et al. Regulation of oxidative stress by ATM is required for self-renewal of haematopoietic stem cells. Nature. 2004;431:997–1002.
51. Ito K, et al. Reactive oxygen species act through p38 MAPK to limit the lifespan of hematopoietic stem cells. Nat Med. 2006;12:446–51.
52. Miyamoto K, et al. Foxo3a is essential for maintenance of the hematopoietic stem cell pool. Cell Stem Cell. 2007;1:101–12.
53. Yalcin S, et al. Foxo3 is essential for the regulation of ataxia telangiectasia mutated and oxidative stress-mediated homeostasis of hematopoietic stem cells. J Biol Chem. 2008;283:25692–705.
54. Tothova Z, et al. FoxOs are critical mediators of hematopoietic stem cell resistance to physiologic oxidative stress. Cell. 2007;128:325–39.
55. Rossi DJ, et al. Cell intrinsic alterations underlie hematopoietic stem cell aging. Proc Natl Acad Sci U S A. 2005;102:9194–9.
56. Minella AC, Clurman BE. Mechanisms of tumor suppression by the SCF(Fbw7). Cell Cycle. 2005;4:1356–9. (Georgetown, Texas)
57. Matsuoka S, et al. Fbxw7 acts as a critical fail-safe against premature loss of hematopoietic stem cells and development of T-ALL. Genes Dev. 2008;22:986–91.
58. Thompson BJ, et al. Control of hematopoietic stem cell quiescence by the E3 ubiquitin ligase Fbw7. J Exp Med. 2008;205:1395–408.
59. Onoyama I, et al. Conditional inactivation of Fbxw7 impairs cell-cycle exit during T cell differentiation and results in lymphomatogenesis. J Exp Med. 2007;204:2875–88.
60. Park Y, Gerson SL. DNA repair defects in stem cell function and aging. Annu Rev Med. 2005;56:495–508.

61. Sancar A, Lindsey-Boltz LA, Unsal-Kacmaz K, Linn S. Molecular mechanisms of mammalian DNA repair and the DNA damage checkpoints. Annu Rev Biochem. 2004;73:39–85.
62. Rossi DJ, et al. Deficiencies in DNA damage repair limit the function of haematopoietic stem cells with age. Nature. 2007;447:725–9.
63. Nijnik A, et al. DNA repair is limiting for haematopoietic stem cells during ageing. Nature. 2007;447:686–90.
64. Janzen V, et al. Stem-cell ageing modified by the cyclin-dependent kinase inhibitor p16INK4a. Nature. 2006;443:421–6.
65. Nygren JM, Bryder D. A novel assay to trace proliferation history in vivo reveals that enhanced divisional kinetics accompany loss of hematopoietic stem cell self-renewal. PloS ONE. 2008;3:e3710.
66. Sharpless NE, DePinho RA. How stem cells age and why this makes us grow old. Nature reviews. Mol Cell Biol. 2007;8:703–13.
67. Attema JL, Pronk CJ, Norddahl GL, Nygren JM, Bryder D. Hematopoietic stem cell ageing is uncoupled from p16 INK4 A-mediated senescence. Oncogene. 2009;28:2238–43.
68. Bartkova J, et al. Oncogene-induced senescence is part of the tumorigenesis barrier imposed by DNA damage checkpoints. Nature. 2006;444:633–7.
69. Mallette FA, Gaumont-Leclerc MF, Ferbeyre G. The DNA damage signaling pathway is a critical mediator of oncogene-induced senescence. Genes Dev. 2007;21:43–8.
70. Allsopp RC, Cheshier S, Weissman IL. Telomere shortening accompanies increased cell cycle activity during serial transplantation of hematopoietic stem cells. J Exp Med. 2001;193:917–24.
71. Wang J, et al. A differentiation checkpoint limits hematopoietic stem cell self-renewal in response to DNA damage. Cell. 2012;148:1001–14.

Chapter 10
A Systems Biology Approach to Iron Metabolism

Julia Chifman, Reinhard Laubenbacher and Suzy V. Torti

Abstract Iron is critical to the survival of almost all living organisms. However, inappropriately low or high levels of iron are detrimental and contribute to a wide range of diseases. Recent advances in the study of iron metabolism have revealed multiple intricate pathways that are essential to the maintenance of iron homeostasis. Further, iron regulation involves processes at several scales, ranging from the subcellular to the organismal. This complexity makes a systems biology approach crucial, with its enabling technology of computational models based on a mathematical description of regulatory systems. Systems biology may represent a new strategy for understanding imbalances in iron metabolism and their underlying causes.

Keywords Hydroxyl radical · Heme · Phagocytose · Constant decay rate · Petri nets · Homeostasis · Transferrin · Erythroid bone marrow · Erythrocytes · Phagocytosis · Plasma · Continuous versus discrete models

Introduction

Dysregulation of iron metabolism plays a role in a wide range of diseases [1], and understanding this role is crucial in the search for therapeutics. Fortunately, over the past decade, some key mechanisms involved in iron regulation have been uncovered, and a more complete picture of iron regulation is starting to emerge [2]. Complicating the matter, however, is the fact that iron regulation involves processes at scales

S. V. Torti (✉)
Department of Molecular Biology and Biophysics,
University of Connecticut Health Center, 263 Farmington Ave.,
Farmington, CT 06030-3305, USA
Tel.: 860-679-6503
e-mail: storti@uchc.edu

J. Chifman
Department of Cancer Biology, Wake Forest School of Medicine,
Winston-Salem, NC, USA

R. Laubenbacher
Center for Quantitative Medicine, University of Connecticut Health Center,
Farmington, CT, USA

S. J. Corey et al. (eds.), *A Systems Biology Approach to Blood*,
Advances in Experimental Medicine and Biology 844, DOI 10.1007/978-1-4939-2095-2_10

ranging from the organism to subcellular compartments, each of which interacts with the others. At each scale, the control system uses several intertwined feedback loops that also cross scales. Thus, it is crucial that, on the one hand, we understand iron metabolism as a multiscale control system, and on the other hand, we move beyond a purely descriptive static characterization of this control system.

Systems biology provides an approach and tool set to address both of these requirements. Systems biology integrates individual components of a system by tying them together through their interactions. This is done through the use of computational models that are capable of synthesizing all the different interactions between components into a dynamical system that captures global dynamic behavior. In particular, such a dynamical system representation allows the integration of regulatory events at different scales. System dynamics can then be probed using the computational model as a virtual laboratory, for the purpose of formulating hypotheses that can then be validated in the laboratory.

This chapter begins with a description of iron metabolism at the systemic and intracellular scales and discusses some of the most important diseases involving dysregulation of iron. We then take a systems biology view and describe some of the computational models of iron metabolism at both scales. Our aim is to present several different approaches to the construction of computational models, and the advantages and disadvantages of the different methods.

Iron Metabolism

The earliest accounts of iron being present in blood date back to as early as the eighteenth century, but it was not until the late 1930s that the first accounts of iron metabolism at the molecular level emerged and not until 1958 that the first comprehensive review of iron absorption was published [3, 4]. More recently, key findings have shaped our current view of iron metabolism. These include the discovery of the transferrin receptor (TfR), in the 1970s, the discovery of the iron-responsive element/iron-regulatory protein (IRE/IRP)regulatory axis in the 1980s, and the discovery of hemochromatosis gene (HFE), the gene mutated in hereditary hemochromatosis (HH), in 1996 [5–7]. Arguably, the most seminal finding in recent years was the discovery of the long-sought iron-regulatory hormone, *hepcidin,* and its target *ferroportin* (Fpn), in the early 2000s [8–13]. It is now apparent that many iron-associated disorders are attributable to genetic malfunctions that affect the *hepcidin–Fpn* axis. Nevertheless, our knowledge of iron biology remains incomplete.

The importance of iron to almost all living organisms is undeniable; iron is required for oxygen transport, energy production, DNA synthesis, and cellular respiration. For example, iron is a component of hemoglobin—an oxygen carrier that transports oxygen from the lungs to the peripheral tissue and then carries carbon dioxide back to the lungs. Likewise, iron is a constituent of myoglobin, an oxygen storage protein that provides oxygen to the muscle tissue. At the same time, excess iron can be toxic due to the ability of iron to exist in various oxidation states. The ability of iron to redox cycle can facilitate the formation of hydroxyl or lipid radicals, which in turn can damage proteins, DNA, and lipids. To maintain iron homeostasis at both the systemic and the cellular levels, vertebrates have developed an elaborate machinery to control iron intake, storage, utilization, and recycling. Our understanding of diseases associated with iron depends on our knowledge of iron homeostasis.

Systemic Iron Homeostasis

An adult well-nourished human contains approximately 3–5 g of iron. Nearly, 60 % of this iron is incorporated into hemoglobin, with 10 % in muscle myoglobin. The rest is stored in hepatocytes and reticuloendothelial macrophages. There is no known mechanism of iron excretion from the body. Roughly 1–2 mg of iron is lost daily through sweat, blood loss, sloughing of intestinal epithelial cells, and desquamation. To compensate for this loss, the body absorbs about 1–2 mg of dietary iron per day, but hemoglobin synthesis alone requires 20–25 mg of iron per day. To support hemoglobin synthesis and other metabolic processes, iron must be recycled and tightly regulated within the system. The circulating peptide hormone *hepcidin* together with its receptor Fpn primarily maintain systemic iron homeostasis, whereas IRPs play a primary role in the control of intracellular iron homeostasis. Recently, an intracellular iron network consisting of 151 chemical species, 107 reactions, and transport steps was identified [2]. Here, only key features are presented; for more details, comprehensive reviews, and current advances, the reader is encouraged to consult the articles [2, 14–19].

Iron Absorption

Inorganic, nonheme iron is available in many foods, e.g., eggs and vegetables, and is absorbed by duodenal enterocytes. Ferrireductase, *Cybrd1* (DcytB), reduces nonheme iron to Fe^{2+} before it is transported through the cellular membrane by the *divalent metal transporter 1* (DMT1; SLC11A2) [20–24]. The absorption of heme iron, found in red meats, is not fully understood. Once heme iron is absorbed, it is transported into the cytosol and released by *heme oxygenase 1* (HO1) [25]. Excess intracellular iron is stored in the storage protein *ferritin* (Ft). Ft oxidizes and sequesters excess ferrous iron into a ferrihydrite mineral core [26, 27]. Iron sequestered in the Ft of enterocytes is lost after a few days through the sloughing of intestinal epithelial cells. Dietary cytosolic iron is exported into the plasma by the basolateral iron exporter Fpn (SLC40A1) [8, 9, 11]. Export of iron from enterocytes into the circulation requires the ferroxidase *hephaestin* (HEPH), a multicopper oxidase, that oxidizes Fe^{2+} to Fe^{3+} [28]. In the plasma, Fe^{3+} circulates bound to Tf, a glycoprotein that has two binding sites for ferric iron and maintains iron in a soluble form. The discovery of Tf as a plasma iron transporter dates back to 1946 [29]. Tf has two important functions: It limits the formation of toxic radicals and delivers iron to cells. In

Table 10.1 Levels of transferrin saturation

Transferrin saturation (%)	Clinical interpretation
< 16	Iron deficiency
16–45	Normal levels
45–60	Signs of iron overload
> 60	Iron overload

healthy humans, about 1/3 of Tf is saturated with iron. Iron concentrations in healthy adults are approximately 14–32 μmol/L, with virtually all circulating iron bound to Tf. In conditions of iron overload, non-Tf-bound-iron (NTBI) accumulates. NTBI is thought to contribute substantially to the pathology associated with iron overload (Table 10.1) [17].

Iron Utilization, Recycling, and Storage

The principal consumer of iron is the *erythroid bone marrow,* and most of that iron comes from internal recycling by tissue *macrophages,* predominantly splenic macrophages. Erythroblasts acquire iron via a ubiquitous protein expressed on the cell surface, *transferrin receptor 1* (TfR1). Through receptor-mediated endocytosis, TfR1 transfers *iron-loaded Tf* (Holo-Tf) into acidified endosomes where iron dissociates from Tf with the assistance of *six transmembrane epithelial antigen of the prostate* (STEAP) proteins, and exits the endosome via DMT1 [30]. Tf and TfR are recycled back to the cell surface. Iron is imported into mitochondria from intracellular compartments by the inner membrane protein *mitoferrin 1* to form heme, the majority of which is then used for hemoglobin production [31]. Since excess heme is toxic and can lead to apoptosis, mechanisms must be in place to maintain heme at appropriate levels. It has been proposed that *feline leukemia virus subgroup C cellular receptor* (FLVCR) and ATP-binding cassette protein G2 (ABCG2) export excess heme, although this is not completely understood [32, 33].

Macrophages recapture iron from senescent and damaged erythrocytes by first phagocytosing erythrocytes and then catabolizing heme using heme oxygenase, to release iron. Ferrous iron is exported into the plasma via the iron exporter Fpn (SLC40A1) and unused iron is stored in macrophages, mainly in Ft [15, 17, 34]. Another major storage site for iron is the liver; the majority of iron entering the liver is stored in Ft and can be mobilized when required by the body. Hepatocytes acquire Holo-Tf through two receptors, TfR1 and TfR2, but TfR2 is believed to act mainly as a Tf saturation "sensor" and has much lower affinity for Holo-Tf than TfR1 [35–37]. Most importantly, when serum iron levels surpass the Tf-binding capacity, the liver becomes the major storage site for NTBI [15]. The mechanism by which hepatocytes acquire NTBI is not completely understood; one candidate for uptake of NTBI is *zinc transporter Zip14* (SLC39 A) [38]. Other tissues, such as heart and pancreas represent sites of iron accumulation in iron overload, and are also proposed to have mechanisms for NTBI uptake.

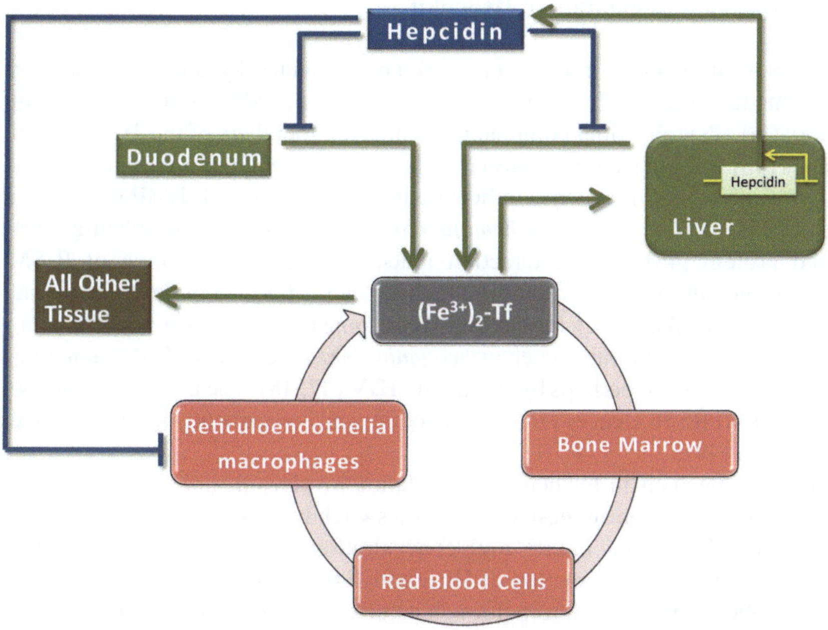

Fig. 10.1 Systemic iron homeostasis. Inorganic, nonheme iron is absorbed by duodenal enterocytes. In the plasma, iron circulates bound to transferrin (*Tf*). The principal consumer of iron is the *erythroid bone marrow,* and most of that iron comes from internal recycling by tissue *macrophages,* predominantly splenic macrophages. Liver is the major storage site of iron. Iron entering the liver is stored in ferritin (*Ft*) and can be mobilized when required by the body. Some iron is incorporated in other tissues. Hepcidin regulates systemic iron homeostasis by inhibiting iron release from duodenal enterocytes, macrophages, and hepatocytes

Regulation of Systemic Iron Homeostasis

To avoid iron overload or deficiency, an organism must maintain an internal equilibrium of iron and make iron available only when and where it is needed. The circulating peptide hormone *hepcidin* is a key molecule that regulates systemic iron homeostasis. It is predominantly produced by the liver, although studies indicate that other tissues also generate hepcidin [12, 13]. Hepcidin levels are modified in response to physiological stimuli that affect iron homeostasis, such as iron overload, hepatic iron stores, inflammation, iron deficiency, erythropoietic activity, and hypoxia. Higher levels of hepcidin reduce iron absorption and vice versa.

Hepcidin modulates serum iron levels and controls Tf saturation by inhibiting iron release from duodenal enterocytes, macrophages, and hepatocytes (Fig. 10.1). More precisely, hepcidin regulates iron efflux by binding to the iron exporter Fpn, triggering its internalization and degradation in lysosomes [39]. The mechanism was originally reported to be facilitated by *Janus kinase 2* (Jak2) [40]; although this has subsequently been challenged by work demonstrating a ubiquitin-mediated pathway of Fpn degradation [41,42].

Transcriptional Regulation of Hepcidin

Expression of hepcidin in the liver is primarily affected by transcriptional mechanisms mediated by the *bone morphogenetic protein* (BMP) family of transcription factors and other signaling components, which are members of the TGF-β family of ligands [43]. Recent studies suggest that the principal regulator of hepcidin is BMP6, which is increased in response to hepatic iron stores [44, 45]. BMP binds to its receptor (BMP-R) and coreceptor *hemojuvelin* (HJV), a glycosylphosphatidylinositol-linked protein [43]. This interaction induces the phosphorylation of R-SMAD proteins and subsequent formation of active transcription complexes involving the co-regulator SMAD4, which bind to BMP responsive elements in the hepcidin promoter [46]. The membrane receptor *neogenin* (NEO1) enhances BMP signaling and hepcidin expression, perhaps by stabilizing HJV [47, 48]. The transmembrane serine protease TMPRSS6 cleaves HJV, inactivates it, and consequently inhibits production of hepcidin [49].

Another mechanism for hepcidin regulation involves *hemochromatosis proteins* (HFEs). HFE has been suggested to act as a switch between two sensors of holo-Tf, TfR1, and TfR2. In this model, high concentrations of holo-Tf displace HFE from TfR1 and permit the interaction of HFE with TfR2. The HFE/TfR2 complex then promotes hepcidin transcription through an unknown mechanism [50–52].

Hepcidin expression is also induced by the inflammatory cytokine *interleukin-6* (IL-6) and other cytokines by activating *signal transducer and activator of transcription 3* (STAT3) [53–55]. STAT3 binds to specific sequences in the HAMP promoter. Cytokine-mediated induction of hepcidin is thought to contribute to the hypoferremia that frequently accompanies chronic infections, acute inflammation, and cancer [56].

Despite the substantial progress that has been made in defining key players in hepcidin regulation, the identification of critical components involved in hepcidin signaling and their functional relationships is far from complete. Mechanisms of hepcidin regulation mentioned above are depicted in Fig. 10.2.

Intracellular Iron Homeostasis

Free ferrous iron can be toxic, since it contributes to the formation of the hydroxyl radical through the Fenton reaction. Hence, intracellular iron must be maintained as meticulously as systemic iron. The regulatory mechanism that coordinates intracellular iron uptake, utilization, storage, and excretion is centered on the IRPs and utilizes IREs. What follows is a brief description of the mechanism for a "generic cell" that encompasses pathways that have been consistently observed in many cell types (Fig. 10.3). Further details can be found in [2].

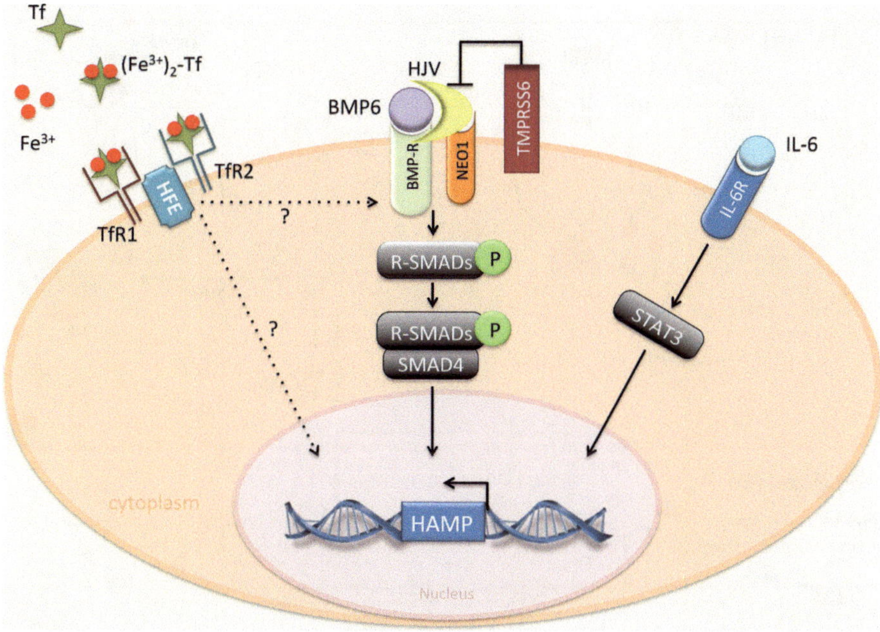

Fig. 10.2 Transcriptional regulation of hepcidin. Regulation of hepcidin by BMP/SMAD and IL-6/STAT3 pathways. Expression of hepcidin in the liver is mainly affected by transcriptional mechanisms mediated by the BMP family, primarily BMP6. BMP binds to its receptor (*BMPR*) in conjunction with the coreceptor HJV. This interaction induces the phosphorylation of R-SMAD proteins which interact with the common mediator SMAD4, bind specific sequences in the hepcidin promoter, and trigger hepcidin gene (*HAMP*) transcription. NEO1 may enhance BMP signaling by interacting with HJV. TMPRSS6 negatively regulates hepcidin by cleaving HJV. Hepcidin expression is also induced by IL-6 through activation of STAT3. STAT3 binds to specific sequences in the HAMP promoter. TfR2 and HFE are also involved in hepcidin activation through mechanisms that are incompletely defined. *BMP* bone morphogenetic protein; *BMPR* bone morphogenetic protein receptor; *HJV* hemojuvelin; *R-SMAD* receptor-regulated SMAD; *NEO1* neogenin; *STAT3* signal transducer and activator of transcription 3; *TMPRSS6* transmembrane serine protease; *TFR2* transferrin receptor 2

Iron Import

Mammalian cells acquire iron predominantly via TfR1. Following binding of Holo-Tf to TfR1, Tf-bound Fe is taken up by receptor-mediated endocytosis into acidified endosomes where ferric iron is reduced to Fe^{2+} by the transmembrane family of metalloreductases (STEAP) [30]. The DMT1 then facilitates the transport of ferrous iron from the endosomes into the cytoplasm. In some cells, e.g., enterocytes, DMT1 is also located on the cell surface and participates in the transport of extracellular iron. It is worth pointing out that the role of STEAP proteins has been studied and well defined in hepatocytes, macrophages, erythroid cells, and erythroblasts, while their role in peripheral tissues requires further investigation [14]. Following egress from the endosome, iron enters the so-called labile iron pool (LIP), a cytosolic pool of weakly bound iron available for a variety of interactions with other molecules.

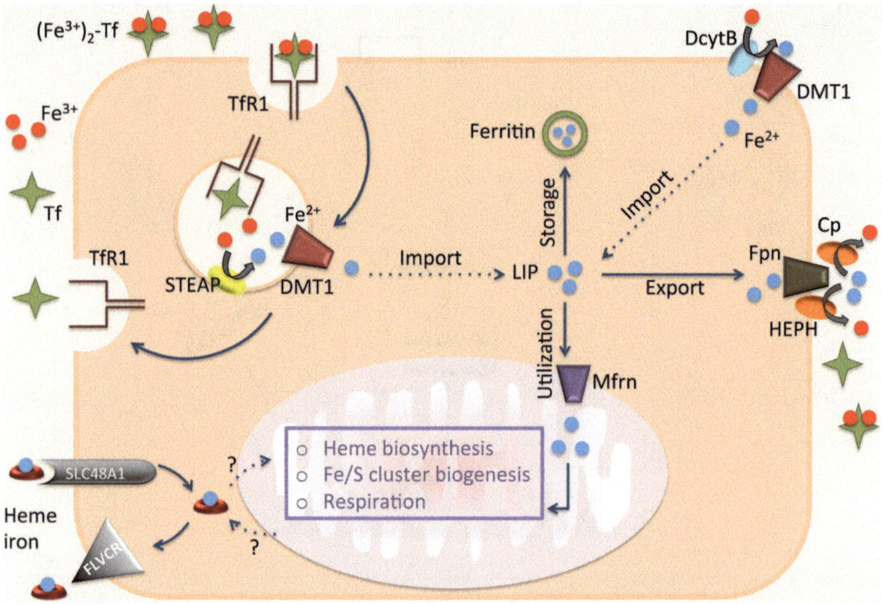

Fig. 10.3 Intracellular iron homeostasis of a generic cell. Cells acquire iron predominantly via TfR1. Ferric iron is reduced to Fe^{2+} by the transmembrane family of metalloreductases (*STEAP*). *DMT1* then facilitates the transport of ferrous iron from the endosomes into the cytoplasm. In some cells, e.g., enterocytes, *DMT1* participates in the transport of extracellular iron. *DcytB* reduces nonheme iron to Fe^{2+} before it is transported through the cellular membrane. Following egress from the endosome, iron enters the so-called labile iron pool (*LIP*). Ferroportin (*Fpn*) is believed to be the only ferrous iron exporter. It has been suggested that dietary heme iron can enter through *SLC48A1* and be exported via *FLVCR*. A mechanism by which heme is moved in and out of mitochondria (the major site of iron utilization) is poorly understood. Ferrous iron is imported into mitochondria for incorporation into bioactive heme by the SLC transporter mitoferrin (*Mfrn*). Iron that is not exported or utilized is stored in ferritin (*Ft*). *TfR1* transferrin receptor 1; *FLVCR* feline leukemia virus subgroup C cellular receptor; *DMT1* divalent metal transporter 1; *STEAP* six transmembrane epithelial antigen of the prostate

It has been suggested that dietary heme iron is transported by the *heme carrier protein 1* (SLC46A1) [57], but another study demonstrated that SLC46A1 is an important folate transporter [58, 59]. A year later SLC48A1 was identified as a possible candidate for heme import [60]. Some cells, like macrophages, acquire heme indirectly by phagocytosing erythrocytes and catabolizing heme to release iron. Hepatocytes have multiple mechanisms of iron entry, including TfR2 and a possible transporter for NTBI, *zinc transporter Zip14* (SLC39 A) [38].

Iron Export

While there is no known mechanism for iron excretion from the body, there is a well-organized and controlled regulation of iron excretion from cells. *Fpn,* located on the plasma membrane, is expressed in a wide variety of human tissue types and is believed to be the only ferrous iron exporter [8, 9, 11, 34]. It requires coordinated efforts of ferroxidases (ceruloplasmin; Cp, and/or HEPH) to assist iron oxidation and loading onto Tf. As mentioned above, cells also export iron in the form of heme through FLVCR and ABCG2 [32, 33].

Iron Utilization and Storage

The major site of iron utilization is the mitochondrion, where iron is used in synthesis of heme and iron–sulfur (Fe/S) cluster prosthetic groups, but the understanding of the mechanism by which iron is moved inside the cell is still incomplete. Iron is imported into the mitochondrion for incorporation into bioactive heme by the *SLC transporter mitoferrin* (SLC25A37) [31]. Intracellular heme regulates its own production through *delta aminolevulinate synthase* (ALAS) and its degradation by inducing HO1 [61], [62]. After heme is synthesized, it is exported via an unknown mechanism into the cytosol for integration into proteins.

Ferrous iron that is not exported or utilized is stored in *Ft,* a cytosolic protein whose main function is to oxidize and sequester excess ferrous iron into a ferrihydrite mineral core. Ft is a 24-subunit polymer comprised of heavy (Ft H) and light (Ft L) polypeptide chains in variable ratios. The subunit composition of Ft depends on cell type and physiological status [27]. Each Ft protein can accrue as many as 4500 iron atoms. Since free iron can promote formation of reactive oxygen species, Ft is crucial to preventing iron-mediated cell damage by keeping excess iron in a nonreactive form.

Intracellular Iron Regulation

Intracellular iron homeostasis is regulated posttranscriptionally by the *iron-regulatory proteins* IRP1 (ACO1) and IRP2 (IREB2) in response to changing iron levels. For a comprehensive review, the reader is encouraged to consult [7], [18].

IRP1 and IRP2 exert their effects by binding to IREs, *cis*-regulatory hairpin structures that are present in the *untranslated regions* (UTRs) of messenger RNA mRNAs involved in iron metabolism. The mRNAs encoding Ft, Fpn, ALAS2, *mitochondrial aconitase* (ACO2), and *hypoxia-inducible factor 2α* (HIF2α) contain a single IRE in their 5′ UTRs. The mRNA encoding TfR1 contains multiple IREs within the 3′ UTR, whereas the mRNAs encoding DMT1, *cell division cycle 14 homolog A* (Cdc14 A), *hydroxyacid oxidase 1* (HAO1), and MRCKα contain a single IRE in their 3′ UTRs.

When intracellular iron levels are low, IRPs bind to IREs with high affinity. Binding of IRPs to 5′-UTR IREs inhibits the translation of Ft and Fpn, while binding to the 3′-UTR IREs results in the stabilization of mRNA of the iron importer TfR1, thus increasing cytoplasmic iron levels. In iron-replete cells, the regulatory effect of IRPs stops: IRP2 is targeted for degradation, and IRP1 acquires a completed iron–sulfur cluster that impedes IRE binding (Fig. 10.4). The role of IRP regulation in the mechanisms and functions of other IRE-containing mRNAs has been less thoroughly studied.

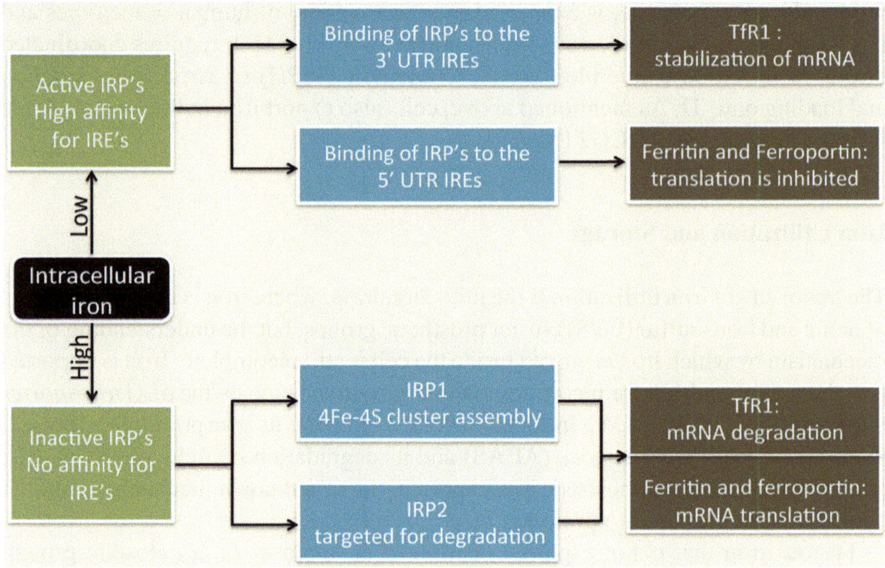

Fig. 10.4 Intracellular iron regulation

Diseases of Iron Metabolism

Iron is required for oxygen transport, energy production, DNA synthesis, and cellular respiration. Accordingly, inappropriately low or high levels of iron are detrimental and lead to a wide range of diseases.

Iron overload/deficiency is either hereditary or acquired. Levels of iron can be altered by the presence of mutated genes, diet that contains inappropriate amounts of iron (either insufficient or excessive), transfusion of red blood cells, iron injections, excessive blood loss, decreased intake or intestinal absorption of iron, and hemolysis.

Iron Overload

Excess iron leads to iron deposition in vital organs such as the liver, heart, pancreas, and endocrine glands. This propagates the formation of hydroxyl or lipid radicals, which damage proteins, DNA, cellular membranes, and can lead to cell death. Left untreated, chronic iron overload increases the risk of liver cirrhosis, cancer, hypogonadism, arthritis, cardiac arrhythmia, heart failure, retinal degeneration, diabetes mellitus, neurodegenerative diseases (Alzheimer's, Parkinson's, Huntington's), and premature death. Treatments for iron overload include phlebotomy and iron chelation therapy [63].

Table 10.2 Genes involved in iron-related disorders

Disorder	Genes	Protein	Protein function
Hemochromatosis	HFE	HFE	Involved in transcriptional regulation of hepcidin
Hemochromatosis	TfR2	Transferrin receptor 2	Holo-Tf sensor; at high Tf levels, HFE interaction with TfR2 is increased promoting hepcidin expression
Juvenile hemochromatosis	HJV	Hemojuvelin	Involved in transcriptional regulation of hepcidin; BMP coreceptor
Juvenile hemochromatosis	HAMP	Hepcidin	Modulates serum iron levels; regulates iron efflux by binding to the iron exporter ferroportin, triggering its internalization and degradation
Hemochromatosis (hepcidin resistance)	SLC40A1	Ferroportin	Iron exporter
Aceruloplasminemia	CP	Ceruloplasmin	Ferroxidase
Hypotransferrinemia	TF	Transferrin	Glycoprotein with two binding sites for ferric iron
IRIDA	TMPRSS6	Matriptase-2	Cleaves HJV, inactivates it, and, consequently, inhibits production of hepcidin

Tf Transferrin, *TfR2* transferrin receptor 2, *CP* ceruloplasmin, *TfR* transferring receptor, *BMP* bone morphogenetic protein, *TMPRSS6* transmembrane serine protease, *HJV* hemojuvelin, *IRIDA* Iron-refractory iron-deficiency anemia, *Holo-Tf sensor* iron-loaded Tf sensor

Hereditary Hemochromatosis

Hemochromatosis is the most common genetic iron overload disorder and results from mutations in several genes, all of which affect the Fpn/hepcidin regulatory axis (Table 10.2) [14, 16, 17]. The main characteristic of this disorder is increased absorption of dietary iron and its accumulation in the liver, heart, pancreas, endocrine glands, tissue and joints, where it causes injury and organ dysfunction, as described above. To date, researchers have identified five mutated genes associated with hemochromatosis, which can be grouped further into two cases: hepcidin deficiency and hepcidin resistance [16].

Hepcidin Deficiency Mutations in the genes encoding HFE, TfR2, hemojuvelin (HFE2, HJV), and hepcidin (HAMP) cause hemochromatosis by inactivating the pathway that upregulates hepcidin. The most common and mild form of HH is due to a missense mutation of the HFE gene, which is incompletely penetrant, and is influenced by environmental and other genetic factors [5, 64]. A severe and less common form of HH, with extremely low or absent hepcidin levels, is *juvenile hemochromatosis* (mutations in HJV or HAMP genes) that leads to hypogonadism, heart failure, and death [65–67]. Another rare form of HH but less severe is caused by a mutation in the TfR2 gene [68].

Hepcidin Resistance Missense mutations in the gene encoding Fpn obstruct hepcidin binding and result in insensitivity of Fpn to regulation by hepcidin, leading to hepatocyte iron accumulation and high plasma iron [69].

Aceruloplasminemia

Aceruloplasminemia is a disorder caused by mutations in the gene encoding Cp, a ferroxidase involved in the loading of Fe onto Tf following its release from cells [70, 71]. Low serum Cp levels and accumulation of iron in neural and glial cells of the brain, pancreatic islet cells, and hepatocytes characterize this disorder. The clinical outcome is retinal degeneration, diabetes mellitus, cerebellar ataxia, dementia, and neurologic diseases.

Other diseases related to iron overload and degenerative neurologic conditions are Hallervorden–Spatz disease and Friedreich's ataxia [63].

Hypotransferrinemia/Atransferrinemia

Practically undetectable plasma levels of Tf characterize hypotransferrinemia [72]. Deficiency in Tf allows NTBI to accumulate and deposit in the liver and other organs, leading to the accumulation of iron to toxic levels. On the other hand, reduction in Tf-bound iron impairs erythropoiesis in the bone marrow, which strictly depends on Tf-bound iron. Hypotransferrinemic patients also have a severe hepcidin deficiency, implying that Tf is somehow involved in the regulation of hepcidin [16].

Transfusional Siderosis

Repeated blood transfusions are a life-saving therapy in many conditions, but multiple transfusions can also lead to toxicity and chronic iron overload. Transfusions are used in patients with beta thalassemia, sickle cell anemia, bone marrow failure (aplastic anemia, myelodysplastic syndrome, red blood cell aplasia), and patients receiving aggressive cancer therapy. Each unit of transfused blood contains 200–250 mg of iron, which is more than a hundred times the amount absorbed daily from the diet (1–2 mg). At first, iron accumulates in reticuloendothelial macrophages and later in parenchymal tissue cells of the liver, pancreas, heart, and endocrine tissue, where it can lead to cardiomyopathy and other iron overload disorders [63].

Iron Deficiency

Iron deficiency is the major cause of anemia and a public health problem worldwide. Since roughly two thirds of total body iron is used in hemoglobin synthesis, deficiency in iron will affect the production of healthy red blood cells.

Iron-Deficiency Anemia

Approximately three billion people worldwide suffer from iron-deficiency anemia due to decreased dietary iron intake, poor absorption, and increased need for iron, which can result from blood loss; gastrointestinal bleeding; blood donations; pregnancy; and cancer of the esophagus, stomach, or colon. Children and women are at much greater risk. Iron deficiency can result in premature birth, poor growth development and cognitive skills, and also affects the nervous system. Patients may experience symptoms associated with anemia that include chronic fatigue, poor exercise tolerance, headaches, and problem concentrating [14, 63].

Left untreated, iron-deficiency anemia can lead to complications such as irregular heartbeat, angina and heart attack, low-birth weight, high risk of infection (childhood), and delayed growth (childhood) [63]. Changes in diet and iron supplements can treat minor iron deficiency, while severe cases may require transfusion of red blood cells, intravenous iron, or iron injections.

Iron-Refractory Iron-Deficiency Anemia

Iron-refractory iron-deficiency anemia (IRIDA) is triggered by a rare mutation in the gene TMPRSS6, which encodes matriptase-2 expressed in the liver. This mutation leads to reduced activity of TMPRSS6 and consequently high hepcidin levels. As a result, iron absorption from the intestine and iron release from macrophages is inhibited, causing severe iron deficiency [16, 17].

Anemia of Chronic Inflammation

Anemia of chronic inflammation, also called *anemia of chronic disease* (ACD), is a systemic iron disorder and occurs in association with malignancy, chronic infections, trauma, inflammatory disorders, and organ failure [56]. Iron stores in ACD are not exhausted, but iron is sequestered in macrophages. In addition, absorption of iron is reduced and hemoglobin synthesis is inhibited. The decrease in serum iron is a consequence of hepcidin increase in response to inflammation, which seems to be an attempt to restrict iron availability to invading microorganisms and tumor cells [73]. Hepcidin production is induced by the inflammatory cytokine *interleukin-6* (IL-6), bacterial pathogens, and lipopolysaccharide [74]. Anemia of chronic inflammation is considered to be mild-to-moderate anemia, and treatment is usually focused on the underlying disorder.

Iron Homeostasis and Cancer

Dysregulation of iron metabolism in cancer is well known, and it has been argued for years that excess iron and increased cancer incidence go hand in hand [75], although this is not always observed [76]. Links between excess iron and cancer are also suggested by the efficacy of dietary iron deprivation [77] and iron chelators [78] in cancer therapy. In the early 1980s, it was observed that levels of TfR1 are elevated in cancer, and use of TfR1 as a targeting ligand in the design of anticancer drugs was proposed [79, 80].

More recently, it was observed that hepcidin and Fpn are expressed in epithelial cells of peripheral tissue, such as the breast, where they exhibit the same regulatory interaction as in macrophages and liver cells [81]. Levels of hepcidin were increased and levels of Fpn were decreased in breast cancer cell lines when compared to non-malignant breast cells, and this was correlated with an increased LIP in malignant cells. Data collected from breast cancer patients also showed reduced levels of Fpn in malignant compared to nonmalignant breast tissue. Experimentally induced overexpression of Fpn reduced tumor growth of breast cancer xenografts, implying a direct relationship between intracellular iron and tumor growth. Importantly, low levels of Fpn in the tissues of breast cancer patients were associated with poor clinical outcome and reduced survival. Subsequent work determined that elevated TfR1 and reduced HFE (which would also be expected to elevate iron in tumor tissue) similarly predict poor survival in breast cancer patients [82].

Although cancer is of course more than an iron disorder, these findings indicate a clear and direct relationship between iron and cancer. Clarifying the precise nature of this relationship will require further study.

A Systems Biology Approach to Iron Metabolism

The complexity of iron regulation in mammals makes a systems biology approach crucial, with its enabling technology of computational models based on a mathematical description of the iron homeostasis control system. Before discussing specific mathematical models, we briefly summarize the role that models play in systems biology (Fig. 10.5) [83]. As discussed earlier, the primary role of mathematical models is to discover new biology. This is typically done in two steps. The first step, model building, consists of the description of the pertinent biology in mathematical terms. There are a variety of mathematical formalisms one can use for this purpose. Which one to choose for a given problem depends in large part on the type of information available and on the type of questions one would like to answer. A common approach is through a system of differential equations. In the case of a metabolic network, for instance, there will be one differential equation per molecular species in the network, viewing it as a biochemical reaction network. The equation for a given species describes the rate of change of the species in terms of the quantities of other species it depends on, together with a collection of kinetic parameters. Alternatively, one can view the network in terms of a collection of logical rules that govern the "decision making" of the node based on the states of the other molecular species. Rather than varying continuously, the species might take on categorical values, such as *low* or *high*.

The second step is model analysis, which, in many cases, relies on simulation. That is, the mathematical model is implemented as a computational algorithm in a computer. Simulation of the model consists of the choice of an initial concentration of all

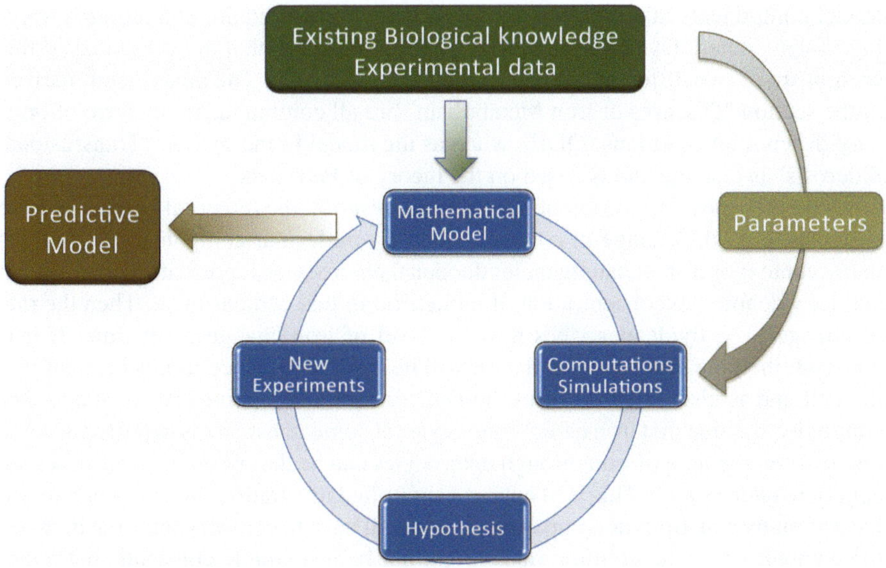

Fig. 10.5 The construction of a mathematical model. The starting point is a formulation of the problem and specific questions that the model will answer. Biological knowledge about components of the system, its structure, interactions, and any available experimental results must be gathered. Different types of experimental data are analyzed and integrated. This information is then used to construct a mathematical model. Since different models emphasize different features, the choice of mathematical model depends on the questions being asked. Its structure will also depend on the system description: organismal, cellular, or molecular. Some systems will have unknown biological parameters and will require detailed information about kinetic constants or time course data in order to estimate model parameters. Various computational techniques are used to assess if the model is in accordance with experimental results, and if not, hypotheses underlying the model need to be refined, and different types of experiments might be proposed. This iterative process is repeated and the model is refined until it accurately describes the relevant aspects of the system

the species in the network, from which the time evolution of the model is calculated. Observations might include which steady state is reached from this initialization or whether the simulation is robust with respect to small changes in the model parameters. The result of these simulations is an understanding of how the model behaves under certain perturbations of interest. If the model correctly captures the key features of the biology, then an understanding of the model can lead to a more targeted investigation of the actual network, leading to new biological insights. In the following section, we discuss some published computational models related to iron regulation.

Models of Iron Homeostasis

Biological systems can be described mathematically as dynamical systems, in terms of functional relationships between the variables, which govern the temporal evolution of the system. Some biological data are best modeled by systems in which the model components take on discrete states, while others require continuously varying system states. Dynamic models of iron homeostasis that are presented in this section are of two different types: *continuous* and *discrete*. The models summarized in the section "Diseases of Iron Metabolism" are all continuous in the form of ordinary differential equations (ODE), whereas the model in the section "Transfusional Siderosis" is discrete and is based on the theory of Petri nets.

Generally speaking, differential equations describe the temporal change of state variables, e.g., the change in concentration of a molecular component. To put this in the context of iron metabolism, let duodenal enterocytes represent a compartment and let y denote the concentration of labile iron in this compartment. Then the rate of change of y, dy/dt, describes how the level of iron changes over time. If it is constant, then the differential equation will be $dy/dt = 0$. Since iron is brought into the cell and is also exported from the cell, further assumptions can be made. For simplicity, assume that iron enters enterocytes at some rate a and is exported at some rate b. Then the flow of iron through enterocytes can be described by the differential equation $dy/dt = a\text{-}by$. This is of course a simplified illustration; in reality one would have as many equations as compartments and will have to consider shuttling between the compartments. In addition, rates might not be just simple constants, but rather complicated expressions, and would have to incorporate various regulations, for example, regulation of the iron exporter Fpn by hepcidin. From this simple example, one can see that the resulting ODE model will have many unknown parameters and will require detailed information about kinetic constants or time course data in order to estimate model parameters.

Compartmental Model of Iron Homeostasis

B. J. Lao and D. T. Kamei developed a simple compartmental model of iron homeostasis calibrated to mouse data [84]. The model consists of five compartments, each denoting the amount of iron at a specific location or in a particular state: *hepatocyte, diferric transferrin* (FeTf), *red blood cells* (RBC), *NTBI,* and *macrophage*. The system is then given by five ODEs that explicitly incorporate the roles of Fpn, hepcidin, TfR2, and HFE. We present only one of the equations, describing the RBC compartment.

$$\frac{d(RBC)}{dt} = k_{FeTf;RBC} - k_{RBC;macro} \times RBC$$

$k_{FeTf;RBC} = 36\frac{nmol}{h}$ (rate of iron transfer from FeTf to RBC compartment)

$k_{RBC;macro} = 1.6 \times 10^{-3}\text{h}^{-1}$ (rate of iron transfer from RBC compartment to macrophages)

These rates are based on findings published in [85], [86]. All parameters in the model were approximated using mouse data from various sources.

The resulting model was used to simulate anemia, iron overload, and erythropoiesis stimulation. The following conclusions were formed.

1. FeTf may be involved in determining the availability of iron for erythropoiesis.
2. To maintain proper iron absorption and intracellular iron levels, the IREs of Fpn are essential. Simulations performed without IREs resulted in normal FeTf and in iron accumulation in macrophages and hepatocytes.
3. Increased iron absorption by duodenal enterocytes replicated features of HFE hemochromatosis, and iron accumulation in hepatocytes was influenced by the uptake of NTBI. Moreover, increased levels of NTBI cause hepcidin decrease implying that removal of NTBI might revert hepcidin levels.

Systemic Model of Iron Homeostasis

In Ref. [87], a systemic model of iron metabolism was built based on data from normal mice (C57BL6) [88] under three diets: iron deficient, iron adequate, and iron loaded. The model consists of a plasma compartment and 15 peripheral organ compartments. Each compartment (pool) is represented by its iron content and is described by a balance equation.

$$\frac{dC_i}{dt} = \sum_j v_{ij} - \sum_j v_{ji} + v_{io} - v_{oi}$$

C_i iron content in ith compartment
v_{ij} the rate of iron influx from compartment *j* to *i*
v_{ji} the rate of iron outflux from compartment *i* to *j*
v_{io} the rate of iron flux from outside into compartment *i*
v_{oi} the rate of iron flux from compartment *i* out of the system

Time-course data were obtained by administering radioactive tracer (^{59}Fe) and then measuring at certain intervals over 28 days nonheme iron as well as hematocrit and hemoglobin content of blood. The experimental data from mice under three different diets were used to estimate parameters and calibrate the model. The authors argue that the resulting quantitative model reflects systemic properties of iron homeostasis. They conclude that this model could be used to study dietary iron perturbations and plan to use the model on genetically modified mice. Furthermore, the authors envision that this mathematical model of pools and fluxes will serve as a foundation for a whole-body model, which would ultimately include iron uptake, storage, secretion, heme synthesis, and regulatory structure.

Intracellular Model of Iron Homeostasis

A model of intracellular iron homeostasis was constructed in [89]. This model is specific to normal breast epithelial cells and represents the core control system of iron metabolism focused on iron import via TfR1, export (Fpn), sequestration (Ft), and regulation (IRPs). These proteins and the LIP are connected by several feedback loops that drive network dynamics. Each component in the model is defined by an ODE that describes changes in concentration with respect to time. The resulting ODE system has five equations and 15 parameters with two parameters being external: hepcidin and the iron saturation level of extracellular Tf. One of the assumptions made was that Ft is always bound to iron and undergoes natural degradation. As a result, Ft releases iron back into the LIP. It was represented by the following mechanism.

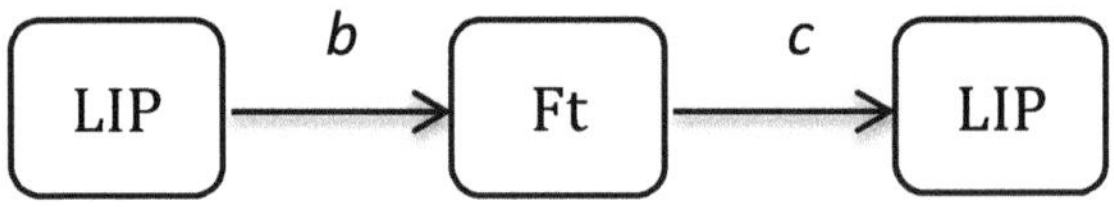

c constant decay rate
b hyperbolic rate that describes negative regulation of IRPs on Ft

$$b = \frac{a \times k}{k + \text{IRP}} .$$

Here, IRP represents an inhibiting state variable, k an activation threshold, and a the maximum production rate of the regulated protein (in this case Ft). Using this information, the rate of change of the LIP can be represented by the following equation.

$$\frac{d(LIP)}{dt} = a_1 Fe_{ex} \times (TfR1) + c \times (Ft) - a_2 \times (LIP) \times (Fpn) - b \times (LIP)$$

$a_1 Fe_{ex} \times (TfR1)$	iron import (via TfR1) with the rate a_1
$a_2 \times (LIP) \times (Fpn)$	iron export (via Fpn) with the rate a_2
Fe_{ex}	iron saturation level of extracellular Tf
b and c	as described above

The model was validated using data from overexpression of Fpn. Through a combination of analytical arguments and simulations, it was shown that the model has a unique stable steady state for any choice of parameters, agreeing with experimental evidence that cellular iron is tightly controlled [90].

Including additional relevant components in the model will be the next step, with the ultimate goal of identifying basic forces and key regulators that contribute to modifications in iron homeostasis as normal breast epithelial cells transition to malignancy.

Petri Net Model of Systemic Iron Homeostasis

Petri net theory uses a different approach to simulations and analysis, and has also been applied to iron metabolism. Carl A. Petri, a German mathematician and computer scientist, introduced and formally defined the concept in the 1960s. Today, Petri nets are used in computational biology to model gene-regulatory networks, metabolic pathways, signal transduction pathways, and biochemical networks.

Informally speaking, a Petri net is a directed graph that consists of two kinds of nodes, *places* and *transitions,* and *arcs* connecting them. Arcs only connect two nodes of different kinds; they do not join two places, or two transitions. Places, depicted by circles, represent passive elements such as proteins, protein complexes, or chemical compounds, while transitions, depicted by rectangles, are active elements and represent biological interactions or chemical reactions. In the context of iron metabolism, an example of a place might be Fe^{2+} and an example of a transition is oxidation of Fe^{2+} by HEPH. Some places are marked by black dots or natural numbers, called *tokens,* which are dynamic elements of the net and represent the concentration of a given species in terms of moles, molecules, or even abstract concentration levels such as high, medium, and low. Tokens are distributed over places to describe a systems state, e.g., the normal body iron physiological state. The distribution of tokens over places is called a *marking of the net*. For a certain biological reaction to happen, its places, e.g., proteins, must contain sufficient numbers of tokens. If all places are marked (contain tokens), then a transition may *fire* by removing one or more tokens from each place and moving it to another appropriate place (Fig. 10.6). This changes the marking of the net, i.e., the systems state. Enabled transitions do not have to fire, which makes Petri nets nondeterministic and the behavior of the system is established by all possible firing sequences. Of course, just following tokens does not represent the entire analysis. Comprehensive formal analysis must be performed to show behavioral and structural properties of the system.

In a series of papers, a Petri net model of systemic iron homeostasis was constructed [91–93]. The model consists of 47 places and 57 transitions and has been verified through extensive analysis. In the latest article in the series [92], the authors focused on some aspects of the anemia of chronic disorders and, based on their analysis of the net, they have made some observations, listed below. The predictions made by the model were validated using data from patients with chronic anemia that were treated with recombinant human erythropoietin (rHuEPO). The conclusions based on the model and laboratory tests are:

1. TfR levels are not influenced by inflammation.
2. There is a strong positive correlation between the dose of rHuEPO and soluble TfR.
3. There is a strong negative correlation between the dose of rHuEPO and hepcidin, suggesting a reverse relationship.
4. The TfR1 serum level was confirmed to be a suitable indicator of erythropoietic activity.

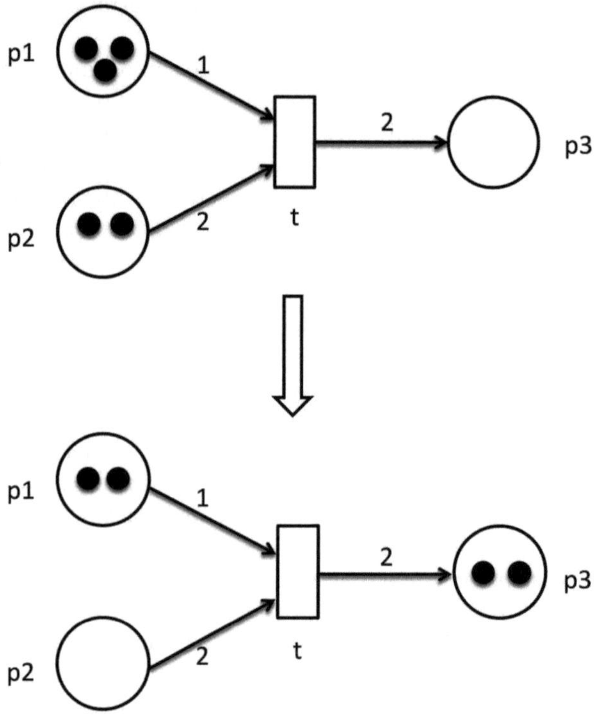

Fig. 10.6 A simple Petri net example. Initially, input places $p1$ and $p2$ contain three and two tokens, respectively. By firing transition t, one token will be removed from $p1$ and two tokens from $p2$. Transition t will consume tokens and place two of them into place $p3$. Transition t may fire since its preplaces $p1$ and $p2$ have sufficient number of tokens. After one firing step, the marking of the net is changed: $p1$ has only two tokens left, $p2$ has no tokens, and $p3$ has two tokens. Transition t cannot fire anymore

It is worth commenting on some differences among the mathematical models presented here. The majority of the models consist of a system of ODEs. These capture the continuous rate of change of the concentrations of the different molecular quantities over (continuous) time. The last model consists of a Petri net, that is, a graph structure of a certain type, together with rules that govern the state of the different nodes in this graph. In the model presented, the states are integer values, specifying how many tokens are placed at a particular node at a given time. Thus, the states are discrete, rather than continuous concentrations.

Furthermore, time progresses in discrete steps also. There are also other types of time- and/or state-discrete models in use in systems biology [83]. The different modeling methods each have their pros and cons. What particular modeling framework is best for a particular type of system depends on several factors, such as availability of experimental data, kinetic parameters, the type of question to be answered, and others.

Conclusion

Iron metabolism and its relationship to a variety of disorders and diseases is difficult or impossible to fully understand without a systems approach. Regulatory mechanisms in different parts of the organism, operating at different time and spatial scales, are connected to each other and interact through complex feedback loops. Without an understanding of how these interdependencies affect dynamic changes in iron homeostasis, systematic therapeutic approaches will remain elusive. Systems biology and mathematical modeling promise such a rigorous understanding. As detailed in this chapter, much has been discovered about the mechanistic foundations of iron regulation. However, key parts of the system remain poorly understood.

So-called reductionist biology has an important role to play in uncovering additional features of iron metabolism. These can then be integrated into system-level models, such as the case studies presented here. Systems biology brings another valuable approach to the problem through the generation and analysis of high-throughput "-omics" data, which have not yet been used extensively to study iron metabolism. Large-scale gene expression studies using DNA microarrays or high-throughput sequencing can help in discovering new genes involved in iron regulation and their connections to the known control network. Proteomics analyses using mass spectrometry-based methods can shed light on posttranscriptional regulation and further help identify important players in the network. The application of one or more of a variety of network-inference algorithms can be used to build up a more complete regulatory network structure that can be used to generate experimental hypotheses, to be validated in the laboratory.

Since iron regulation is a highly dynamic process, we require dynamic computational models for its study. This chapter describes some recent examples of such models. Given the complexity of the entire process, substantially more sophisticated models will be required. Their construction needs to be based on comprehensive time-resolved data at different scales and in different cell types. The confluence of new and improved mathematical and computational techniques, together with sophisticated new measurement techniques, brings such models into the realm of the possible. Thus, the promise of systems biology is yet to be fully realized in the study of iron metabolism and its relation to human health.

Acknowledgments This work was supported in part by Grants NIH R21 CA156133–01A1 (R.L., S.V.T.) and NIH Cancer Biology Training Grant T32-CA079448 (J.C.).

References

1. Kell DB. Iron behaving badly: inappropriate iron chelation as a major contributor to the aetiology of vascular and other progressive inflammatory and degenerative diseases. BMC Med Genomics. 2009;2:2.
2. Hower V, et al. A general map of iron metabolism and tissue-specific subnetworks. Mol Biosyst. 2009;5(5):422–43.
3. Josephs HW. Absorption of iron as a problem in human physiology; a critical review. Blood. 1958;13(1):1–54.
4. Laufberger V. Sur la cristallisation de la ferritine. Soc Chim Biol. 1937;19:1575–82.
5. Feder JN, et al. A novel MHC class I-like gene is mutated in patients with hereditary haemochromatosis. Nat Genet. 1996;13(4):399–408.

6. Trowbridge IS, Omary MB. Human cell surface glycoprotein related to cell proliferation is the receptor for transferrin. Proc Natl Acad Sci U S A. 1981;78(5):3039–43.
7. Hentze MW, Kuhn LC. Molecular control of vertebrate iron metabolism: mRNA-based regulatory circuits operated by iron, nitric oxide, and oxidative stress. Proc Natl Acad Sci U S A. 1996;93(16):8175–82.
8. Abboud S, Haile DJ. A novel mammalian iron-regulated protein involved in intracellular iron metabolism. J Biol Chem. 2000;275(26):19906–12.
9. Donovan A, et al. Positional cloning of zebrafish Fpn1 identifies a conserved vertebrate iron exporter. Nature. 2000;403(6771):776–81.
10. Krause A, et al. LEAP-1, a novel highly disulfide-bonded human peptide, exhibits antimicrobial activity. FEBS Lett. 2000;480(2–3):147–50.
11. McKie AT, et al. A novel duodenal iron-regulated transporter, IREG1, implicated in the basolateral transfer of iron to the circulation. Mol Cell. 2000;5(2):299–309.
12. Park CH, et al. Hepcidin, a urinary antimicrobial peptide synthesized in the liver. J Biol Chem. 2001;276(11):7806–10.
13. Pigeon C, et al. A new mouse liver-specific gene, encoding a protein homologous to human antimicrobial peptide hepcidin, is overexpressed during iron overload. J Biol Chem. 2001;276(11):7811–9.
14. Andrews NC. Forging a field: the golden age of iron biology. Blood. 2008;112(2):219–30.
15. Andrews NC, Schmidt PJ. Iron homeostasis. Annu Rev Physiol. 2007;69:69–85.
16. Ganz T. Hepcidin and iron regulation, 10 years later. Blood. 2011;117(17):4425–33.
17. Hentze MW, et al. Two to tango: regulation of mammalian iron metabolism. Cell. 2010;142 (1):24–38.
18. Muckenthaler MU, Galy B, Hentze MW. Systemic iron homeostasis and the iron-responsive element/iron-regulatory protein (IRE/IRP) regulatory network. Annu Rev Nutr. 2008;28: 197–213.
19. De Domenico I, Ward DM, Kaplan J. Hepcidin regulation: ironing out the details. J Clin Invest. 2007;117(7):1755–8.
20. Gunshin H, et al. Cloning and characterization of a mammalian proton-coupled metal-ion transporter. Nature. 1997;388(6641):482–8.
21. Fleming MD, et al. Microcytic anaemia mice have a mutation in Nramp2, a candidate iron transporter gene. Nat Genet. 1997;16(4):383–6.
22. Canonne-Hergaux F, et al. Cellular and subcellular localization of the Nramp2 iron transporter in the intestinal brush border and regulation by dietary iron. Blood. 1999;93(12):4406–17.
23. Oakhill JS, et al. Functional characterization of human duodenal cytochrome b (Cybrd1): redox properties in relation to iron and ascorbate metabolism. Biochim Biophys Acta. 2008;1777(3):260–8.
24. Turi JL, et al. Duodenal cytochrome b: a novel ferrireductase in airway epithelial cells. Am J Physiol Lung Cell Mol Physiol. 2006;291(2):L272–80.
25. Ferris CD, et al. Haem oxygenase-1 prevents cell death by regulating cellular iron. Nat Cell Biol. 1999;1(3):152–7.
26. Arosio P, Levi S. Cytosolic and mitochondrial ferritins in the regulation of cellular iron homeostasis and oxidative damage. Biochim Biophys Acta. 2010;1800(8):783–92.
27. Theil EC. Ferritin: at the crossroads of iron and oxygen metabolism. J Nutr. 2003; 133 (5 Suppl 1):1549S–53S.
28. Vulpe CD, et al. Hephaestin, a ceruloplasmin homologue implicated in intestinal iron transport, is defective in the sla mouse. Nat Genet. 1999;21(2):195–9.
29. Schade AL, Caroline L. An Iron-binding component in human blood plasma. Science. 1946;104(2702):340–1.
30. Ohgami RS, et al. The Steap proteins are metalloreductases. Blood. 2006;108(4):1388–94.
31. Shaw GC, et al. Mitoferrin is essential for erythroid iron assimilation. Nature. 2006;440(7080):96–100.

32. Krishnamurthy P, Xie T, Schuetz JD. The role of transporters in cellular heme and porphyrin homeostasis. Pharmacol Ther. 2007;114(3):345–58.
33. Keel SB, et al. A heme export protein is required for red blood cell differentiation and iron homeostasis. Science. 2008;319(5864):825–8.
34. Donovan A, et al. The iron exporter ferroportin/Slc40a1 is essential for iron homeostasis. Cell Metab. 2005;1(3):191–200.
35. Johnson MB, Enns CA. Diferric transferrin regulates transferrin receptor 2 protein stability. Blood. 2004;104(13):4287–93.
36. Kawabata H, et al. Molecular cloning of transferrin receptor 2. A new member of the transferrin receptor-like family. J Biol Chem. 1999;274(30):20826–32.
37. Robb A, Wessling-Resnick M. Regulation of transferrin receptor 2 protein levels by transferrin. Blood. 2004;104(13):4294–9.
38. Liuzzi JP, et al. Zip14 (Slc39a14) mediates non-transferrin-bound iron uptake into cells. Proc Natl Acad Sci U S A. 2006;103(37):13612–7.
39. Nemeth E, et al. Hepcidin regulates cellular iron efflux by binding to ferroportin and inducing its internalization. Science. 2004;306(5704):2090–3.
40. De Domenico I, et al. Hepcidin-induced internalization of ferroportin requires binding and cooperative interaction with Jak2. Proc Natl Acad Sci U S A. 2009;106(10):3800–5.
41. Qiao B, et al. Hepcidin-induced endocytosis of ferroportin is dependent on ferroportin ubiquitination. Cell Metab. 2012;15(6):918–24.
42. Ross SL, et al. Molecular mechanism of hepcidin-mediated ferroportin internalization requires ferroportin lysines, not tyrosines or JAK-STAT. Cell Metab. 2012;15(6):905–17.
43. Babitt JL, et al. Bone morphogenetic protein signaling by hemojuvelin regulates hepcidin expression. Nat Genet. 2006;38(5):531–9.
44. Andriopoulos B Jr, et al. BMP6 is a key endogenous regulator of hepcidin expression and iron metabolism. Nat Genet. 2009;41(4):482–7.
45. Meynard D, et al. Lack of the bone morphogenetic protein BMP6 induces massive iron overload. Nat Genet. 2009;41(4):478–81.
46. Wang RH, et al. A role of SMAD4 in iron metabolism through the positive regulation of hepcidin expression. Cell Metab. 2005;2(6):399–409.
47. Lee DH, et al. Neogenin inhibits HJV secretion and regulates BMP-induced hepcidin expression and iron homeostasis. Blood. 2010;115(15):3136–45.
48. Zhang AS, et al. Hemojuvelin-neogenin interaction is required for bone morphogenic protein-4-induced hepcidin expression. J Biol Chem. 2009;284(34):22580–9.
49. Silvestri L, et al. The serine protease matriptase-2 (TMPRSS6) inhibits hepcidin activation by cleaving membrane hemojuvelin. Cell Metab. 2008;8(6):502–11.
50. Gao J, et al. Interaction of the hereditary hemochromatosis protein HFE with transferrin receptor 2 is required for transferrin-induced hepcidin expression. Cell Metab. 2009;9(3):217–27.
51. Gao J, et al. Hepatocyte-targeted HFE and TFR2 control hepcidin expression in mice. Blood. 2010;115(16):3374–81.
52. Wallace DF, et al. Combined deletion of Hfe and transferrin receptor 2 in mice leads to marked dysregulation of hepcidin and iron overload. Hepatology. 2009;50(6):1992–2000.
53. Wrighting DM, Andrews NC. Interleukin-6 induces hepcidin expression through STAT3. Blood. 2006;108(9):3204–9.
54. Pietrangelo A, et al. STAT3 is required for IL-6-gp130-dependent activation of hepcidin in vivo. Gastroenterology. 2007;132(1):294–300.
55. Verga Falzacappa MV, et al. STAT3 mediates hepatic hepcidin expression and its inflammatory stimulation. Blood. 2007;109(1):353–8.
56. Andrews NC. Anemia of inflammation: the cytokine-hepcidin link. J Clin Invest. 2004; 113(9):1251–3.
57. Shayeghi M, et al. Identification of an intestinal heme transporter. Cell. 2005;122(5):789–801.
58. Qiu A, et al. Identification of an intestinal folate transporter and the molecular basis for hereditary folate malabsorption. Cell. 2006;127(5):917–28.

59. Andrews NC. When is a heme transporter not a heme transporter? When it's a folate transporter. Cell Metab. 2007;5(1):5–6.
60. Rajagopal A, et al. Haem homeostasis is regulated by the conserved and concerted functions of HRG-1 proteins. Nature. 2008;453(7198):1127–31.
61. Ferreira GC, Gong J. 5-Aminolevulinate synthase and the first step of heme biosynthesis. J Bioenerg Biomembr. 1995;27(2):151–9.
62. Yoshida T, et al. Human heme oxygenase cDNA and induction of its mRNA by hemin. Eur J Biochem. 1988;171(3):457–61.
63. Andrews NC. Disorders of iron metabolism. N Engl J Med. 1999;341(26):1986–95.
64. Allen KJ, et al. Iron-overload-related disease in HFE hereditary hemochromatosis. N Engl J Med. 2008;358(3):221–30.
65. Roetto A, et al. Mutant antimicrobial peptide hepcidin is associated with severe juvenile hemochromatosis. Nat Genet. 2003;33(1):21–2.
66. Papanikolaou G, et al. Mutations in HFE2 cause iron overload in chromosome 1q-linked juvenile hemochromatosis. Nat Genet. 2004;36(1):77–82.
67. Niederkofler V, Salie R, Arber S. Hemojuvelin is essential for dietary iron sensing, and its mutation leads to severe iron overload. J Clin Invest. 2005;115(8):2180–6.
68. Nemeth E, et al. Hepcidin is decreased in TFR2 hemochromatosis. Blood. 2005;105(4):1803–6.
69. Sham RL, et al. Hereditary hemochromatosis due to resistance to hepcidin: high hepcidin concentrations in a family with C326 S ferroportin mutation. Blood. 2009;114(2):493–4.
70. Harris ZL, et al. Aceruloplasminemia: molecular characterization of this disorder of iron metabolism. Proc Natl Acad Sci U S A. 1995;92(7):2539–43.
71. Yoshida K, et al. A mutation in the ceruloplasmin gene is associated with systemic hemosiderosis in humans. Nat Genet. 1995;9(3):267–72.
72. Beutler E, et al. Molecular characterization of a case of atransferrinemia. Blood. 2000;96(13):4071–4.
73. Nemeth E, et al. Hepcidin, a putative mediator of anemia of inflammation, is a type II acute-phase protein. Blood. 2003;101(7):2461–3.
74. Nemeth E, et al. IL-6 mediates hypoferremia of inflammation by inducing the synthesis of the iron regulatory hormone hepcidin. J Clin Invest. 2004;113(9):1271–6.
75. Torti SV, Torti FM. Ironing out cancer. Cancer Res. 2011;71(5):1511–4.
76. Kotsopoulos J, et al. Plasma micronutrients, trace elements, and breast cancer in BRCA1 mutation carriers: an exploratory study. Cancer Causes Control. 2012;23(7):1065–74.
77. Hann HW, Stahlhut MW, Blumberg BS. Iron nutrition and tumor growth: decreased tumor growth in iron-deficient mice. Cancer Res. 1988;48(15):4168–70.
78. Richardson DR. Iron chelators as therapeutic agents for the treatment of cancer. Crit Rev Oncol Hematol. 2002;42(3):267–81.
79. Faulk WP, Hsi BL, Stevens PJ. Transferrin and transferrin receptors in carcinoma of the breast. Lancet. 1980;2(8191):390–2.
80. Shan L, et al. Visualizing head and neck tumors in vivo using near-infrared fluorescent transferrin conjugate. Mol Imaging. 2008;7(1):42–9.
81. Pinnix ZK, et al. Ferroportin and iron regulation in breast cancer progression and prognosis. Sci Transl Med. 2010;2(43):43–56.
82. Miller LD, et al. An iron regulatory gene signature predicts outcome in breast cancer. Cancer Res. 2011;71(21):6728–37.
83. Laubenbacher R, et al. A systems biology view of cancer. Biochim Biophys Acta. 2009;1796(2):129–39.
84. Lao BJ, Kamei DT. A compartmental model of iron regulation in the mouse. J Theor Biol. 2006;243(4):542–54.
85. Rivera S, et al. Synthetic hepcidin causes rapid dose-dependent hypoferremia and is concentrated in ferroportin-containing organs. Blood. 2005;106(6):2196–9.
86. Hentze MW, Muckenthaler MU, Andrews NC. Balancing acts: molecular control of mammalian iron metabolism. Cell. 2004;117(3):285–97.

87. Lopes TJ, et al. Systems analysis of iron metabolism: the network of iron pools and fluxes. BMC Syst Biol. 2010;4:112.
88. Schumann K, et al. A method to assess 59Fe in residual tissue blood content in mice and its use to correct 59Fe-distribution kinetics accordingly. Toxicology. 2007;241(1–2):19–32.
89. Chifman J, et al. The core control system of intracellular iron homeostasis: a mathematical model. J Theor Biol. 2012;300:91–9.
90. Epsztejn S, et al. Fluorescence analysis of the labile iron pool of mammalian cells. Anal Biochem. 1997;248(1):31–40.
91. Formanowicz D, et al. Petri net based model of the body iron homeostasis. J Biomed Inform. 2007;40(5):476–85.
92. Formanowicz D, et al. Some aspects of the anemia of chronic disorders modeled and analyzed by petri net based approach. Bioprocess Biosyst Eng. 2011;34(5):581–95.
93. Sackmann A, et al. An analysis of the petri net based model of the human body iron homeostasis process. Comput Biol Chem. 2007;31(1):1–10.

Chapter 11
Innate Immunity in Disease: Insights from Mathematical Modeling and Analysis

Nabil Azhar and Yoram Vodovotz

Abstract The acute inflammatory response is a complex defense mechanism that has evolved to respond rapidly to injury, infection, and other disruptions in homeostasis. This robust responsiveness to biological stress likely endows the host with increased fitness, but over-robust or inadequate inflammation predisposes the host to various diseases. Importantly, well-compartmentalized inflammation is generally beneficial, but spillover of inflammation into the blood is a hallmark—and likely also a driver—of self-maintaining inflammation. The blood is also a key entry point and immunological interface for vectors of parasitic diseases, diseases that themselves incite systemic inflammation. The complex role of inflammation in health and disease has made this biological system difficult to understand comprehensively and modulate rationally for therapeutic purposes. Consequently, systems approaches have been applied in order to characterize dynamical properties and identify key control points in inflammation. This process begins with the collection of high-dimensional, experimental, and clinical data, followed by data reduction and data-driven modeling that finally informs mechanistic computational models for analysis, prediction, and rational modulation. These studies have suggested that the overall architecture of the inflammatory response includes a multiscale positive feedback from inflammation $\rightarrow$ tissue damage $\rightarrow$ inflammation, which is often inadequately controlled by negative feedback from anti-inflammatory mediators. Given the importance of the blood interface for the inflammatory response, and the accessibility of this compartment both as an immunological sampling reservoir for vectors as well as for diagnosis and therapy, we suggest that any rational efforts at modulating inflammation via the blood compartment must involve computational modeling.

Y. Vodovotz (✉)
Department of Surgery, University of Pittsburgh, W944 Biomedical Sciences Tower, 200 Lothrop St., Pittsburgh, PA 15213, USA
Tel.: 412-647-5609
e-mail: vodovotzy@upmc.edu

N. Azhar
University of Pittsburgh, 3064 Biomedical Sciences Tower 3, 3501 Fifth Ave., Pittsburgh, PA 15213, USA
Tel.: 412-383-7256

S. J. Corey et al. (eds.), *A Systems Biology Approach to Blood*,
Advances in Experimental Medicine and Biology 844, DOI 10.1007/978-1-4939-2095-2_11

Keywords Inflammation · Mathematical model · Systems biology · Sepsis · Trauma · Malaria

Abbreviations

ABM	Agent-based model
AsNOS	*Anopheles stephensi* nitric oxide synthase
DAMP	Damage-associated molecular pattern molecule
DBN	Dynamic Bayesian Networks
GMM	Genetically modified mosquito
MODS	Multiple organ dysfunction syndrome
ODE	Ordinary differential equations
PCA	Principal component analysis
RBM	Rule-based model
sTNFR	Soluble tumor necrosis factor-α receptor
TNF-α	Tumor necrosis factor-α

Introduction

Inflammation is an essential process in maintaining health and responding to disease. Acute inflammation is driven largely by the innate immune system, which not only serves as the first line of defense against invading pathogens but also functions to resolve tissue damage and restore homeostasis upon a variety of inflammatory conditions including sepsis, trauma, wound healing, and many more. A large aspect of the acute inflammatory response plays out in the blood, but usually only when inflammation is dysregulated. Dysregulated systemic inflammation also plays a significant role in the pathophysiology of other diseases that are not primarily attributed to innate immunity, such as cancer and diabetes. Although the list of diseases is broad and the processes important to each setting may differ in certain respects, the core architecture of the inflammatory response to biological stress is highly conserved [1]. An infection or a tissue injury/damage triggers an initially local cascade of events mediated by an array of cells (e.g., macrophages, neutrophils, dendritic cells, lymphocytes, etc.) and molecules (cytokines, free radicals, and damage-associated molecular pattern molecules (DAMPs)) that locate invading pathogens or stressed/damaged tissue, alert and recruit other cells and molecules, eliminate the offending agents, and finally restore the body to equilibrium [2]. When dysregulated or overexuberant, inflammation can be discerned in the systemic circulation in the form of altered levels of inflammatory cells and molecular mediators.

In sepsis and trauma, this response is concomitant with physiologic manifestations including changes in heart rate and body temperature, responses that act in a concerted fashion in order to help optimize host defense while minimizing tissue damage.

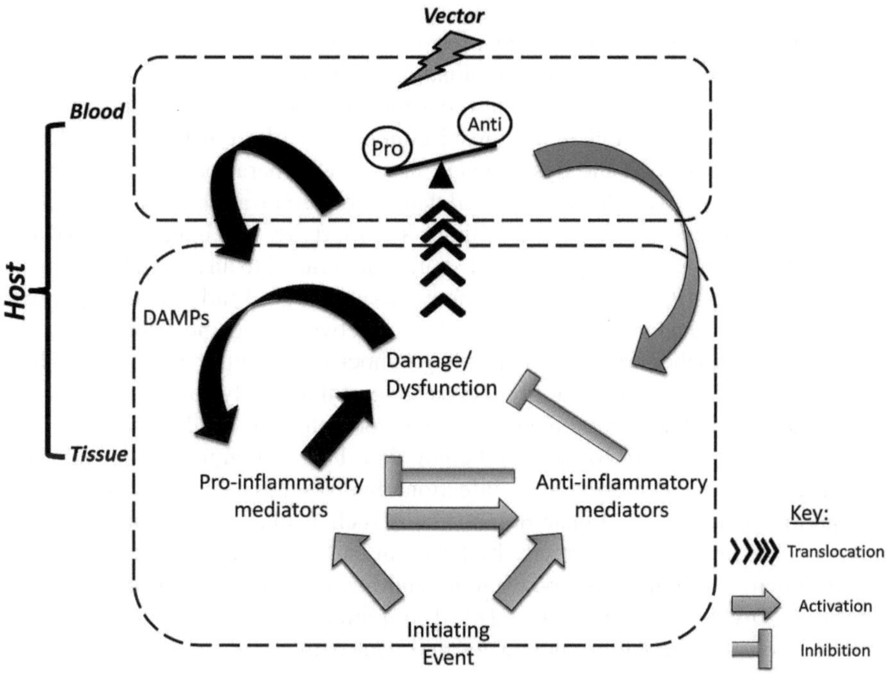

Fig. 11.1 Complex structure of the innate immune response to biological stress. Following an initiating event (e.g., trauma, hemorrhage, infection), both pro- and anti-inflammatory influences (e.g., chemokines, cytokines, lipid products, and free radicals) are elaborated, leading to tissue damage or dysfunction. These stressed tissues elaborate damage-associated molecular patterns (*DAMPs*), which further propagate innate immune mechanisms. When the pro-inflammatory mediators exceed defined thresholds, both pro- and anti-inflammatory mediators spillover into the blood and may cause inflammation to feedback and spread systemically to other organs as well; we refer to this process as an inflammatory tipping point. Inflammatory mediators are also transferred to blood-feeding vectors and can serve to communicate the infection status of the host, as well as modulating anti-parasite immunity in the vector. *Black curved arrows* represent the pro-inflammatory feedback and *gray curved arrows* represent feedback from anti-inflammatory mediators

Indeed, although a well-regulated inflammatory response is essential for proper healing and host defense, an overly exuberant response can become self-perpetuating and lead to organ dysfunction and death [3, 4]. These vastly different outcomes can be explained, at least in part, by the high-level architecture of the immune response, which includes a positive feedback loop from inflammation →damage/dysfunction → inflammation that can drive pathophysiology in inflammatory diseases (Fig. 11.1).

The detrimental effects of self-sustaining inflammation are likely responsible for the general perception of inflammation as an inherently harmful process [5, 6]. However, in addition to the aforementioned beneficial roles of inflammation in the resolution of tissue injury, recent studies suggest that morbidity and mortality are worse in animals and humans with low levels of early pro-inflammatory signals [7]. The emerging view of inflammation is indeed more nuanced, casting inflammation as a highly coordinated communication network that allows the body to sense and respond to challenges and subsequently restore homeostasis [8, 9]. One may consider the complexity resulting from this coordination to be an indicator of a well-regulated and properly orchestrated response, and consequently a less complex response would be indicative of a pathological dis- or misconnectivity of the network. Guided by insights from studies on the dysregulated physiology characteristic of sepsis and trauma/hemorrhage, which have reported that a decrease invariability/complexity of heart rate can presage increased morbidity and mortality, we have suggested that well-organized dynamic networks of mediators are crucial to an appropriate inflammatory response [10, 11]. Indeed, such networks are induced early in the response to experimental surgical trauma in mice, and these networks become disorganized and less complex with the addition of hemorrhagic shock to this minor trauma [10]. However, emerging studies from our group also suggest that overly-robust, and possibly self-sustaining, inflammation manifests as networks that are highly complex.

The current paradigm for acute inflammation, based in large part on studies in response to trauma, hemorrhage, or infection, involves a dynamic cascade of cellular and molecular events. Innate immune cells such as mast cells, neutrophils, and macrophages are activated directly by bacterial endotoxin or indirectly by various stimuli elicited systemically upon trauma and hemorrhage [12–15], including the release of DAMPs (Fig. 11.1) [16–18]. These stimuli enter the systemic circulation and activate circulating monocytes and neutrophils [19], which subsequently migrate to compromised tissue by following along a chemoattractant gradient induced at the site of injury/infection [20]. Activated macrophages and neutrophils produce and secrete effectors that activate a variety of immune cells (including further activating themselves) as well as nonimmune cells such as endothelial cells. Both DAMPs and pro-inflammatory cytokines—primary among them tumor necrosis factor-α (TNF-α) [21–27]—promote immune cell activation and affect important physiological functions that feedback positively to promote further production of inflammatory mediators. This behavior may lead to inflammatory tipping points—and concomitant spillover of inflammatory mediators into the blood—indicative of cascading system failure that occurs at multiple scales and across multiple compartments [28] (Fig. 11.1). In turn, dysregulated inflammation in the blood may itself become a driver of further inflammation in other tissues (Fig. 11.1).

Inflammation Is a Complex System

As evidenced by the preceding description, inflammation, like most biological systems, is a highly nonlinear system with multiple feedback loops that may be discerned even when viewed in a coarse-grained, relatively abstract fashion (Fig. 11.1). Positive feedback loops allow rapid ramping up of a response to biological stress, while the negative feedback works to suppress inflammation and restore homeostasis once the threat (infection, damaged tissue, etc.) has been eliminated. We suggest that, as has likely occurred in many other complex biological systems [29], inflammation has evolved to be robust to a broad range of perturbations but at a cost of fragility in key control nodes that may account for the tipping point behavior described above [16, 29] (Fig. 11.1). Failure at these points can lead to disease; therefore, characterizing these failure modes, and especially the tipping point phenotype, is paramount for the development of effective therapeutic interventions [28]. Another property of a complex nonlinear system is the ability to exhibit vastly different behaviors that depend on initial conditions and parameters (i.e., strengths and rates of interactions of components) [30, 31]. This heterogeneity, which recapitulates the clinical observation of patient-to-patient variability, complicates the prediction of individual patient outcomes using the current suite of statistically based tools [17, 28]. As described below, a systems approach to inflammation can be useful, indeed necessary, to explain the behavior of the innate immune response in an individual patient to various biological conditions and ultimately allow for the modulation of this response in pathological conditions.

Modeling Inflammation

Modeling Methods for A Systems Biology of Inflammation

Systems biology approaches span a broad range of techniques, and can be categorized roughly into correlative or causative approaches, with focus on either learning basic principles of system organization and function [32–34] or building predictive computational models [32, 35]. Although there is overlap between these areas, most efforts at elucidating biological mechanisms from high-dimensional data have traditionally focused on particular points along this spectrum of computational approaches. We suggest that gleaning translationally relevant insights into the inflammatory response and its interconnected (patho) physiology will require the successful navigation of this spectrum, in a logical progression from data to models to understanding and prediction [17, 28] (Fig. 11.2).

Correlative approaches, with which most biologists and clinicians are familiar, include regression techniques that build models predictive within the conditions of the data they were trained on [36]. Although these methods do not provide detailed mechanistic insight, these approaches can be used to understand abstract features of the response, such as the presence of nonlinearities and the order of the response. The main drawback of this class of models is that they are almost completely devoid of mechanistic insight, and can be very over-fit to the data on which they were trained. A less-utilized data-driven method is principal component analysis (PCA), which reduces a high-dimensional dataset into a few principal components that account for much of the observed variance in the data. When applied to time-series data, the variables (genes/proteins/etc.) that constitute these principal components may be interpreted as the principal drivers of the observed response and can give some mechanistic insights into the underlying process [10, 37]. In the setting of inflammation, correlative approaches such as PCA may facilitate the development of diagnostics by analyzing the cytokine milieu in the blood resulting from inflammatory spillover, in order to identify the health state of individuals and possibly inform patient-specific

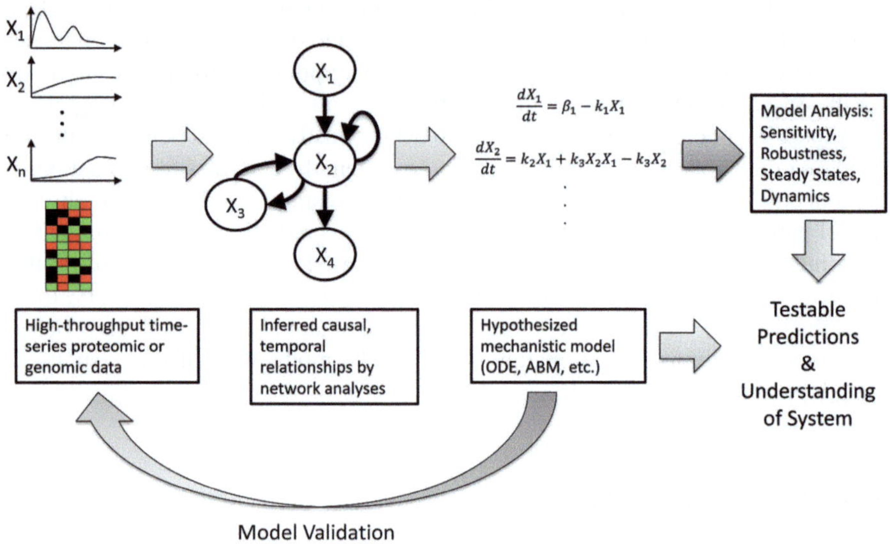

Fig. 11.2 Taming the data deluge: from high-dimensional data to data-driven and mechanistic models. Data-driven methods are used to reduce dimensionality and infer correlative and causal relationships among genes/proteins in the system. Quasi-mechanistic insights from these models, together expert knowledge and the network structure inferred by Dynamic Bayesian Networks inform mechanistic models that may be encoded as ODEs, ABMs, RBMs, etc. Predictions from simulation of these models are compared with experimental data under new conditions and the model is refined based on the discrepancies. Finally, the mechanistic model can be analyzed by a variety of methods to understand its dynamics and identify key control points. *ODE* ordinary differential equation, *ABM* agent-based model, *RBM* rule-based model

interventions [38]. While these methods correlate gene/protein levels to phenotype and can suggest relevant molecular players involved in a given inflammatory process, these methods do not provide much information about how the genes/proteins interact with each other [17, 33].

In order to better discern organizational aspects of interacting networks of mediators, such as co-regulation or auto-induction, a variety of methods have been developed. Hierarchical clustering and Bayesian methods use high-throughput genomic or proteomic data of several time points and/or conditions to correlate gene expression patterns with function and infer regulatory networks of correlated genes. Several developments in these methods over the past 15 years have yielded more informative networks that can be more easily translated into mechanistic models. Among these methods, Dynamic Bayesian Networks (DBNs) are particularly suited for inferring directed (causative) networks of interactions based on the probabilistic measure of how well the network can explain observed data. DBNs can be supplemented by additional experimental evidence and expert knowledge to hypothesize mechanistic models (Fig. 11.2).

Mechanistic models are derived from more detailed biological and physical descriptions of a system have a rich set of tools for both analysis and simulation. These models, based on causative interactions, can be constructed as ordinary differential equations (ODEs), rule-based models (RBMs), and agent-based models (ABMs) among other methods (including hybrid methods), and have the advantage of potentially being predictive outside the range of conditions/time points that they were calibrated on. Although it is often difficult to parameterize such models, they can unveil emergent phenomena that are not immediately obvious from the interactions that are encoded in the model. There are several analytic tools for ODE models especially that have been developed and used to decipher the organizational principles of networks (or subnetworks), the properties that explain the dynamics and robustness/sensitivity of a given complex system, and, perhaps most importantly, the critical points of control in the system [34] (Fig. 11.2). These tools are particularly important in order to help define the complex interplay between the inflammatory mediators in the blood and other compartments both within the host (organs/tissue) and without (e.g., in the case of interactions with blood-feeding vectors). Tools from dynamical systems theory allow identification of the possible steady state(s) of a system as well as the kinetics of the system's time evolution. These tools have been used extensively to explain (or predict, depending on the context) diverse behaviors such as bistability, hysteresis, and oscillations in a variety of biological systems [39]. Bifurcation diagrams, in particular, can be used to map out the effects of a particular parameter on the possible steady state behaviors of a system, and to indicate the transition from a healthy steady state to a pathological one [14, 40–42]. The relative importance of parameters can also be quantified by calculating the change in the model output in response to changes in the parameter values using sensitivity analysis [34, 43]. These methods work in a complementary fashion to identify the key points that can be modulated to change the behavior of a system (Fig. 11.2).

The analysis of ODE models of biological systems can be approached from a control theory perspective as well. Achieving robustness and efficiency are core principles of both evolution as well as engineering. Indeed, feedback, a pervasive biological phenomenon, is also a fundamental component of control strategies [29]. An ODE model is the equivalent of a state-space representation of a control system. Thus, it is possible to decompose the biological system into a control structure and analyze the role of each component using control theoretic tools that characterize their robustness and identify the key mediators that modulate the performance of such a control system [44]. These analyses are especially relevant given that the tipping point phenomenon in the inflammatory response is likely the result of a failure of the body's control structure to handle stress (Fig. 11.1).

Although ODE models are associated with a wide range of analytical tools, they are inappropriate descriptions for settings in which there are low numbers of molecules or in settings in which molecules are not well mixed and thus stochastic effects are at play. RBMs and ABMs (among others) are superior methods for such conditions, as they are able to handle stochastic simulations. Agent-based modeling software packages such as NetLogo [45] and SPARK [46] are especially useful because of their ability to encode and visualize spatially realistic effects as well. Hybrid models can be constructed to merge the advantages of both ODEs and ABMs, and are especially useful for describing phenomena that occur on different timescales. Moreover, the inflammatory response is a quintessential example of a system whose parts operate not only at various timescales but also across various compartments. Thus, a hybrid modeling approach that melds processes best modeled by ODE with processes best modeled by ABM/RBM is essential for a more complete description of inflammation. As a step in this direction, a hybrid model of pressure ulcer formation in spinal cord injury (SCI) patients was developed, abstracting the microscopic details of blood flow and oxygen availability as a series of resistors using differential equations, while encoding an abstracted cascade of inflammation and wound healing in response to simulated cycles of pressure on the peripheral tissue using ABM. Based on data on blood flow in noninjured human subjects versus SCI patients, the parameters gained from this hybrid model predicted the higher likelihood of pressure ulcer development in SCI patients (Solovyev et al., unpublished observations).

While we wish to navigate through process of data→data-driven model→ mechanistic model →prediction and understanding of the innate immune response, we seek to put it in the perspective of translational applications with a focus on clinical and preclinical settings. Much of the work in systems biology has understandably been in simpler, well-studied model organisms, but even among studies focused on preclinical science, there has been an overall lack of translation to the clinical arena. *Translational systems biology* is a framework with a focus on translational insights for novel diagnostic or therapeutic purposes and predictive mathematical models that inform in silico clinical trials [9, 47, 48]. Initially formulated to deal with the clinical challenge of integrating acute inflammation and organ dysfunction in critical illness, this work expanded to include healing of acute and chronic wounds and infections in various diseases, rational dynamic modulation of inflammation, and cross-species host–pathogen interactions.

Modeling and Rational Modulation of Inflammation in Sepsis and Trauma/Hemorrhage

Traumatic injury is often accompanied by hemorrhage and is a significant cause of morbidity and mortality in patients, especially among young people [49, 50]. These patients are particularly susceptible to multiple organ dysfunction syndrome (MODS), a poorly understood syndrome that may be partly attributed to excessive and dysregulated inflammation [4]. The complexity of the interactions between inflammation and organ physiology has likely stymied the development of therapies for MODS, and was the motivation for the development of both data-driven and mechanistic computational models [10, 12–14, 42, 51–54].

Several models of the acute inflammatory response to sepsis, trauma, and hemorrhage have been developed, models that provide insight into the mechanisms of inflammation at varying degrees of abstraction. Based on the typical progression of the inflammatory pathway described in the preceding section, an ODE model of acute inflammation consisting of pathogen, a single population of inflammatory cells, and a measure of global tissue damage/dysfunction interconnected the actions of pro-inflammatory cytokines and DAMPs for the first time, and described both recoverable infection and septic shock, as well as suggesting different therapeutic avenues for the diverse manifestations of sepsis [55]. With the addition of more interactions, including the positive feedback loop between inflammation and damage, a more complex model was used for simulating populations of patients in sepsis and anthrax [12, 52], and to test effects of probiotic treatment of necrotizing enterocolitis [56]. Another equation-based model was calibrated in various inflammatory scenarios in mice [12] and was calibrated on easily accessible circulating levels of inflammatory cytokines and nitric oxide (NO) reaction products, and explicitly included measurable physiological parameters such as blood pressure along with the more abstract global damage (a surrogate for both DAMPs and the health status of the individual; Fig. 11.1) in mice [12, 13, 51]. This calibrated model was capable of predicting, outside of its calibration set, dose ranges of endotoxin at which death is known to occur [12] in addition to predicting responses to combinations of insults on which it was not trained [12, 57, 58].

While investigating the role of initial trauma in the murine response to trauma/hemorrhagic shock, both correlative (transcriptomic analysis, PCA, regression) and causative (ODE) models were used in a complementary fashion, and suggested that the role of initial trauma is central in driving the inflammatory response, both systemically and in the liver [13]. Transcriptomic data indicated an overlap between the genes and pathways induced in trauma alone or trauma with hemorrhagic shock with differences in only the magnitude of expression. In agreement with this observation, a mechanistic mathematical model showed that using the same model with different initial conditions could differentiate the inflammatory responses to trauma versus hemorrhagic shock. Later, multivariate regression, PCA, and dynamic network analysis all suggested major mechanistic differences between sham cannulation and hemorrhagic shock and predicted that the majority of the inflammatory response to survivable trauma/hemorrhage was due mostly to the underlying tissue trauma induced by cannulation surgery [13]. The model was extended to include details of experimental trauma/hemorrhage in mice (e.g. bleeding rate and target blood pressure), and further validated using a unique, computerized platform for automated hemorrhage that was constructed specifically to test the behavior of this mathematical model [53].

The natural extension from understanding and predicting the inflammatory response is to modulate it in a rational fashion to reduce its detrimental effects. Whereas the modeling work described earlier can help identify targets for therapeutic intervention, and predictive models can be calibrated to account for individual variability while making therapeutic suggestions, synthetic biology can help drive further, clinically-useful developments. Indeed, recent advances have begun to lay the foundations for clinical applications of synthetic biology. These advances have focused on the engineering of synthetic biological circuits in bacterial cells that are introduced into the human host to sense and respond appropriately to transition the host from a diseased to a healthy state [59]. As noted in the review by Warren et al., advances need to be made in the use of mammalian synthetic biology in order to facilitate clinical translation [60–62] (see also Chap. 14).

In the setting of inflammatory diseases, key stumbling blocks to effective therapy involve the variability of individual responses to pro-inflammatory stimuli as well as the detrimental effects of an overly suppressed inflammatory response. Importantly, the blood compartment is both a component of the multiscale positive feedback loop of inflammation → damage → inflammation as well as being easily accessible for therapy [28]. Accordingly, we have envisioned a synthetic biological device using a human cell line to detect the circulating levels of pro-inflammatory mediators in the blood of an individual patient, and respond appropriately by producing an appropriate counter-stimulus—usually a neutralizing protein or receptor antagonist—for a given mediator. We have successfully created stably transfected human hepatocyte (HepG2-derived) cell lines expressing the mouse soluble TNF-α receptor (sTNFR) [63, 64], under control of the mouse variant of the central, TNF-α-responsive transcription factor NF-kB enhancer coupled to a reduced thymidine kinase promoter. These cells are housed in a bioreactor optimized for the growth and differentiation of hepatocytes, that directly connects the with host's circulatory system. Initial proof-of-concept studies using this bioreactor that produces sTNFR constitutively in a rat bacterial endotoxin infusion model (a quantitative paradigm of acute inflammation that mimics many of the features of sepsis) show promising results for the dynamic modulation of TNF and other pro-inflammatory mediators, as well as ameliorating organ pathophysiology [65]. We suggest that the combination of mechanistic mathematical modeling—of both a given inflammatory disease as well as the effect of this type of biohybrid device on the disease—could be combined to engineer the optimal use of this type of synthetic biohybrid device in order to modulate inflammation systemically.

Moving Beyond the Host: Cross-Species Immune Signaling

The aforementioned examples have focused on the host's inflammatory response to infection or injury. In the case of infectious diseases, however, the host is not an isolated system and instead part of an entire ecosystem involving the infectious agent/parasite as well as possible vectors. Infectious organisms have evolved alongside the host immune system and developed strategies for evasion and modulation of immunity in the host [66, 67]. In the case of diseases such as dengue and malaria, the addition of an invertebrate vector agent introduces a further layer of complexity in the disease process. Blood plays an expanded role in such diseases, serving not only as the site of immune system coordination within the host but also as an interface for communication and interaction among the parasite, vector, and host [66]. This complex ecology is being reassessed in light of the modern view of the vector as an organism that mounts an immune/inflammatory response in an attempt to control parasite growth, rather than as a willing partner in parasite transmission [66].

Studies show that in addition to the parasite *Plasmodium falciparum,* proteins and other biomolecules from the host are ingested and can persist in the mosquito vector *Anopheles stephensi* upon taking a blood meal [66]. Thus, the mosquito vector is likely to be sampling the current immune/inflammatory state of the vertebrate host, in essence getting a snapshot as well as early warning regarding the state of the host's inflammatory equilibrium (Fig. 11.1). Several such blood-derived factors have been identified, including insulin, insulin-like peptides, and the cytokine transforming growth factor (TGF)-β1[66]. Moreover, these host molecules induce signaling in the mosquito midgut cells and modulate protein expression [68, 69]. For example, mammalian TGF-β1 induces mosquito responses including mitogen-activated protein (MAP) kinase signaling [68]. More recently, we found using a Luminex™ assay for multiple cytokines and chemokines that interleukin (IL)-10 was selectively retained in the mosquito midgut for up to five hours post-blood meal (unpublished observations). *In vitro* studies also showed that administration of human IL-10 can alter MAPK signaling in mosquito cells (Luckhart et al. unpublished observations). The full gamut of interspecies signaling factors is likely to include DAMPs and other inflammation-related molecules as well (Fig. 11.1). Below, we discuss these findings in greater detail.

TGF-β1 has been identified as a central player in the immune response to parasite infection within the host [70]. However, much less is known about the converse, namely the possible role of TGF-β1 on mosquito immunity and physiology. Human TGF-β1 ingested by the *Anopheles stephensi* mosquito via a blood meal was shown to induce expression of the mosquito homolog of the inducible nitric oxide synthase, AsNOS [71]. Inducible NOS is often associated with mammalian host defense responses to malaria, and studies have shown that the mosquito also regulates parasite development through complex, multiphasic expression of AsNOS [72]. Several additional observations including evidence of feedback regulation by the MAPK MEK (MAP kinase kinase)/extracellular signal-regulated kinase (ERK), along with dichotomous dose-dependent effects of mammalian TGF-β1 on AsNOS induction and parasite growth suggested that computational modeling might be beneficial in clarifying the underlying mechanisms [71]. An initial Boolean model of the system predicted oscillations in AsNOS as well as MEK/ERK, which was one possible mechanistic model consistent with experimental data. An ODE model of the same system gave quantitative predictions that fit reasonably well with the data. This model also highlighted the necessity of a persistent presence of TGF-β1 to drive the multiphasic response. However, experimental data had previously suggested that the half-life of TGF-β1 may be much shorter than the observed multiphasic time course of AsNOS. This discrepancy was reconciled with the model-generated hypothesis of an endogenous mosquito TGF-β1-like molecule that is induced by exogenous mammalian TGF-β1 and can drive the long-term AsNOS response. Indeed, the mosquito homolog of TGF-β1, As60A [73] was shown to have the same multiphasic dynamics that the model predicted the hypothesized TGF-β1-like molecule must have in order to maintain the observed AsNOS response [74]. These studies begin to provide insight into some of the conserved, cross-species mechanisms of immune modulation between the mammalian host and mosquito vector. Notably, they highlight a new role for the blood as a medium for the interface and biological communication between species, with particular implications for vector-borne diseases.

In keeping with the goal of translational systems biology, we wish to use computational modeling and analysis for the development of new therapeutic strategies. In the decade since the publication of the genomes of the malaria parasite(s) and of *Anopheles gambiae,* another major malaria mosquito vector, there have been several functional and comparative genomic analyses that have helped uncover regulatory networks through correlative studies [75–80]. Some of these studies have focused on the interface between vector and parasite, identifying gene clusters/networks responsible for the mosquito's control of parasite growth [81]. However, most modeling work in malaria has been focused mainly on epidemiological aspects of the disease or very coarse-grained mechanistic modeling of host–pathogen–vector interactions [30], rather than models on the intra- and intercellular scale that build directly from the genomic studies or other quantitative experimental data. As outlined in Sect. 12.2.1, and partially illustrated in the preceding paragraph, a systematic approach starting with data-driven modeling and correlative studies that inform mechanistic models and analyses can help build a comprehensive understanding of the molecular and cellular mechanisms underlying the interspecies immune control of malaria parasite. This approach is essential for identifying master regulators in the mosquito vector that can point to therapeutic targets for disease control via genetic modification.

Just as we described the dynamic modulation of inflammation in the host via a bioreactor, we seek to modulate the interspecies immune response to infection via the use of transgenic mosquitoes, ideally at the blood-feeding interface. Genetically modified mosquito (GMM) vectors have become an attractive option for disease control in the past decade as efforts to eradicate mosquitoes or modulate human immunity to malaria infection have been met with reduced efficacy and other challenges. Key to the success of a strategy involving GMMs is ensuring that the modification remains dominant and spreads throughout the population while maintaining the fitness of the mosquito. Recent studies have generated mosquitoes with increased parasite killing but with detrimental effects on fitness [82, 83]. These studies are more descriptive of the phenotype than the underlying mechanism driving it, and much remains to be learned about the pathways driving the observed response. Thus, a systems-level understanding of blood factor-modulated immune response of the mosquito is needed to account for the trade-offs between parasite killing and mosquito fitness in potential interventions.

Conclusions and Future Prospects

The study of the inflammatory response dates back to Roman times, when it was first characterized by its physical manifestations of increased temperature, redness, pain, and swelling. Centuries of research have increased our understanding of inflammation beyond description of its symptoms, and unveiled an ever-increasing complexity underlying this primordial defense mechanism. The modern view of inflammation is that of a multifaceted communication process that manifests across multiple compartments of the body and multiple biological scales [28]. The blood is an important compartment among these, and serves at least three different functions in innate immunity. It is the medium through which inflammation progresses in its early stages as circulating monocytes and other inflammatory cells are recruited to sites of injury/infection. In settings of dysregulated or overexuberant inflammation, spillover of inflammatory mediators from the site of injury to the blood can contribute to a positive feedback, increasing systemic inflammation. Finally, inflammatory mediators in the host's blood can transfer to blood-feeding vectors and directly modulate the immune response of the vector in a complex host–vector–parasite interaction. The highly complex nature of the immune response to biological stress and the multifaceted role of the circulatory system in this response are perhaps to blame for the lack of efficient and/or successful therapies for diseases such as sepsis, trauma, and MODS. A systems approach to inflammation can be helpful, perhaps even necessary, for the identification of better therapeutic strategies by taking advantage of both data-driven and mechanistic modeling. The methods highlighted in this chapter can provide novel insights into the innate immune system, increasing our understanding to suggest targets for rational modulation of inflammation as well as providing predictive simulations on which to base further basic research, drug discovery, and clinical trials.

Future possibilities include design of synthetic biological circuits for dynamic, individualized modulation of inflammation in the host, as well as control of the host–pathogen–vector interface to eliminate parasite transmission in vector-borne disease. While much progress remains to be made in order to realize these far-reaching goals, recent advances offer a promising outlook for the future of translational systems biology of inflammation.

Acknowledgments This work was supported in part by the National Institutes of Health grants R01GM67240, P50GM53789, R33HL089082, R01HL080926, R01AI080799, R01HL76157, R01DC008290, and UO1 DK072146; National Institute on Disability and Rehabilitation Research grant H133E070024; a Shared University Research Award from IBM, Inc.; and grants from the Commonwealth of Pennsylvania, the Pittsburgh Life Sciences Greenhouse, and the Pittsburgh Tissue Engineering Initiative/Department of Defense.

References

1. Medzhitov R. Origin and physiological roles of inflammation. Nature. 2008;454(7203):428–35.
2. Brown KL, Cosseau C, Gardy JL, Hancock REW. Complexities of targeting innate immunity to treat infection. Trends Immunol. 2007;28(6):260–6.
3. Marshall JC. Inflammation, coagulopathy, and the pathogenesis of multiple organ dysfunction syndrome. Crit Care Med. 2001;29(7 Suppl):S99–S106.
4. Jarrar D, Chaudry IH, Wang P. Organ dysfunction following hemorrhage and sepsis: mechanisms and therapeutic approaches (Review). Int J Mol Med. 1999;4(6):575–83.
5. Waxman K. Shock: ischemia, reperfusion, and inflammation. New Horiz. 1996;4(2):153–60.
6. Peitzman AB, Billiar TR, Harbrecht BG, Kelly E, Udekwu AO, Simmons RL. Hemorrhagic shock. Curr Probl Surg. 1995;32(11):925–1002.

7. Namas R, Ghuma A, Torres A, Polanco P, Gomez H, Barclay D, et al. An adequately robust early TNF-a response is a hallmark of survival following trauma/hemorrhage. PLoS ONE. 2009;4(12):e8406.
8. Nathan C. Points of control in inflammation. Nature. 2002;420(6917):846–52.
9. Vodovotz Y, Csete M, Bartels J, Chang S, An G. Translational systems biology of inflammation. PLoS Comput Biol. 2008;4:1–6.
10. Mi Q, Constantine G, Ziraldo C, Solovyev A, Torres A, Namas R, et al. A dynamic view of trauma/hemorrhage-induced inflammation in mice: principal drivers and networks. PLoS ONE. 2011;6:e19424.
11. Namas R, Zamora R, Namas R, An G, Doyle J, Dick TE, et al. Sepsis: something old, something new, and a systems view. J Crit Care. 2012;27(3):314e1–11.
12. Chow CC, Clermont G, Kumar R, Lagoa C, Tawadrous Z, Gallo D, et al. The acute inflammatory response in diverse shock states. Shock. 2005;24:74–84.
13. Lagoa CE, Bartels J, Baratt A, Tseng G, Clermont G, Fink MP, et al. The role of initial trauma in the host's response to injury and hemorrhage: insights from a comparison of mathematical simulations and hepatic transcriptomic analysis. Shock. 2006;26:592–600.
14. Reynolds A, Rubin J, Clermont G, Day J, Vodovotz Y, Ermentrout BG. A reduced mathematical model of the acute inflammatory response: I. Derivation of model and analysis of anti-inflammation. J Theor Biol. 2006;242(1):220–36.
15. Torres A, Bentley T, Bartels J, Sarkar J, Barclay D, Namas R, et al. Mathematical modeling of post-hemorrhage inflammation in mice: studies using a novel, computer-controlled, closed-loop hemorrhage apparatus. Shock. 2008;32(2):172–8.
16. Vodovotz Y, An G. Systems biology and inflammation. In: Yan Q, editor. Systems biology in drug discovery and development: methods and protocols. Totowa:Springer; 2009. 181–201.
17. Mi Q, Li NYK, Ziraldo C, Ghuma A, Mikheev M, Squires R, et al. Translational systems biology of inflammation: potential applications to personalized medicine. Personal Med. 2010;7:549–59.
18. Chen GY, Nuez G. Sterile inflammation: sensing and reacting to damage. Nat Rev Immunol. 2010;10(12):826–37.
19. Parker SJ, Watkins PE. Experimental models of gram-negative sepsis. Br J Surg. 2001;88(1):22–30.
20. Bellingan G. Inflammatory cell activation in sepsis. Br Med Bull. 1999;55(1):12–29.
21. Jones AL, Selby P. Tumour necrosis factor: clinical relevance. Cancer Surv. 1989;8(4):817–36.
22. Cavaillon JM. Cytokines and macrophages. Biomed Pharmacother. 1994;48(10):445–53.
23. Kox WJ, Volk T, Kox SN, Volk HD. Immunomodulatory therapies in sepsis. Intensive Care Med. 2000;26 (Suppl 1):S124–8.
24. Dinarello CA. Proinflammatory cytokines. Chest. 2000;118(2):503–8.
25. Pinsky MR. Sepsis: a pro- and anti-inflammatory disequilibrium syndrome. Contrib Nephrol. 2001;(132):354–66.
26. Baugh JA, Bucala R. Mechanisms for modulating TNF alpha in immune and inflammatory disease. Curr Opin Drug Discov Dev. 2001;4(5):635–50.
27. Chen G, Goeddel DV. TNF-R1 signaling: a beautiful pathway. Science. 2002;296(5573):1634–5.
28. An G, Nieman G, Vodovotz Y. Computational and systems biology in trauma and sepsis: current state and future perspectives. Int J Burns Trauma. 2012;2:1–10.
29. Csete ME, Doyle JC. Reverse engineering of biological complexity. Science. 2002;295(5560):1664–9.
30. Mideo N, Day T, Read AF. Modelling malaria pathogenesis. Cell Microbiol. 2008;10(10):1947–55.
31. Vodovotz Y, Constantine G, Rubin J, Csete M, Voit EO, An G. Mechanistic simulations of inflammation: current state and future prospects. Math Biosci. 2009;217(1):1–10.
32. Mesarovic MD, Sreenath SN, Keene JD. Search for organising principles: understanding in systems biology. Syst Biol (Stevenage). 2004;1(1):19–27.

33. Janes KA, Yaffe MB. Data-driven modelling of signal-transduction networks. Nat Rev Mol Cell Biol. 2006;7(11):820–8.
34. Kitano H. Systems biology: a brief overview. Science. 2002;295(5560):1662–4.
35. Arkin A, Schaffer D. Network news: innovations in 21st century systems biology. Cell. 2011;144(6):844–9.
36. Mac Nally R. Regression and model-building in conservation biology, biogeography and ecology: the distinction between-and reconciliation of-"predictive" and "explanatory" models. Biodivers Conserv. 2000;9(5):655–71.
37. Janes KA, Gaudet S, Albeck JG, Nielsen UB, Lauffenburger DA, Sorger PK. The response of human epithelial cells to TNF involves an inducible autocrine cascade. Cell. 2006;124(6):1225–39.
38. Vodovotz Y, Constantine G, Faeder J, Mi Q, Rubin J, Bartels J, et al. Translational systems approaches to the biology of inflammation and healing. Immunopharmacol Immunotoxicol. 2010;32(2):181–95.
39. Angeli D, Ferrell JE, Sontag ED. Detection of multistability, bifurcations, and hysteresis in a large class of biological positive-feedback systems. Proc Natl Acad Sci U S A. 2004;101(7):1822–7.
40. Clermont G, Chow CC, Kumar R, Vodovotz Y. Mathematical simulation of the innate immune response. Crit Care Med. 2001;29(12Suppl):A111.
41. Bagci EZ, Vodovotz Y, Billiar TR, Ermentrout GB, Bahar I. Bistability in apoptosis: roles of Bax, Bcl−2, and mitochondrial permeability transition pores. Biophys J. 2006;90(5):1546–59.
42. Day J, Rubin J, Vodovotz Y, Chow CC, Reynolds A, Clermont G. A reduced mathematical model of the acute inflammatory response II. Capturing scenarios of repeated endotoxin administration. J Theor Biol. 2006;242(1):237–56.
43. Marino S, Hogue IB, Ray CJ, Kirschner DE. A methodology for performing global uncertainty and sensitivity analysis in systems biology. J Theor Biol. 2008;254(1):178–96.
44. Kurata H, El-Samad H, Iwasaki R, Ohtake H, Doyle JC, Grigorova I, et al. Module-based analysis of robustness tradeoffs in the heat shock response system. PLoS Comput Biol. 2006;2(7):e59.
45. Wilensky U. NetLogo. Center for connected learning and computer-based modeling, Northwestern University. Evanston, IL. 1999. http://ccl.northwestern.edu/netlogo/.
46. Solovyev A, Mikheev M, Zhou L, Dutta-Moscato J, Ziraldo C, An G, et al. SPARK: a framework for multi-scale agent-based biomedical modeling. Int J Agent Technol Syst. 2010;2:18–30.
47. An G, Hunt CA, Clermont G, Neugebauer E, Vodovotz Y. Challenges and rewards on the road to translational systems biology in acute illness: four case reports from interdisciplinary teams. J Crit Care. 2007;22:169–75.
48. An G, Faeder J, Vodovotz Y. Translational systems biology: introduction of an engineering approach to the pathophysiology of the burn patient. J Burn Care Res. 2008;29:277–85.
49. Kauvar DS, Wade CE. The epidemiology and modern management of traumatic hemorrhage: US and international perspectives. Crit Care. 2005;9 Suppl 5:S1–S9.
50. Kauvar DS, Lefering R, Wade CE. Impact of hemorrhage on trauma outcome: an overview of epidemiology, clinical presentations, and therapeutic considerations. J Trauma. 2006;60(6 Suppl):S3–11.
51. Prince JM, Levy RM, Bartels J, Baratt A, Kane JM III, Lagoa C, et al. In silico and in vivo approach to elucidate the inflammatory complexity of CD14-deficient mice. Mol Med. 2006;12(4–6):88–96.
52. Kumar R, Chow CC, Bartels JD, Clermont G, Vodovotz Y. A mathematical simulation of the inflammatory response to anthrax infection. Shock. 2008;29(1):104–11.
53. Torres A, Bentley T, Bartels J, Sarkar J, Barclay D, Namas R, et al. Mathematical modeling of posthemorrhage inflammation in mice: studies using a novel, computer-controlled, closed-loop hemorrhage apparatus. Shock. 2009;32(2):172–8.
54. Constantine G, Buliga M, Vodovotz Y, Bohnen N, Clermont G. Time varying patterns of organ failure. Int J Contemp Math Sci. 2010;5:2263–72.

55. Kumar R, Clermont G, Vodovotz Y, Chow CC. The dynamics of acute inflammation. J Theor Biol. 2004;230:145–55.
56. Arciero J, Rubin J, Upperman J, Vodovotz Y, Ermentrout GB. Using a mathematical model to analyze the role of probiotics and inflammation in necrotizing enterocolitis. PLoS ONE. 2010;5:e10066.
57. Vodovotz Y, Chow C, Bartels J, Lagoa C, Kumar R, Day J, Rubin J, Ermentrout B, Riviere B, Yotov I, Constantine G, Billiar T, Fink M, Clermont G. Mathematical simulations of sepsis and trauma. Proceedings of the 11th Congress of the European Shock Society. 2005.
58. Vodovotz Y, Chow CC, Bartels J, Lagoa C, Prince J, Levy R, et al. In silico models of acute inflammation in animals. Shock. 2006;26:235–44.
59. Ruder WC, Lu T, Collins JJ. Synthetic biology moving into the clinic. Science. 2011;333(6047):1248–52.
60. Arkin A, Fletcher D. Fast, cheap and somewhat in control. Genome Biol. 2006;7(8):114.
61. Karlsson M, Weber W, Fussenegger M. Design and construction of synthetic gene networks in mammalian cells. In: Weber W, Fussenegger M, editors. Synthetic gene networks, vol. 813. Humana Press (New York, NY).; 2012. pp. 359–76.
62. Weber W, Fussenegger M. Emerging biomedical applications of synthetic biology. Nat Rev Genet. 2012;13(1):21–35.
63. Fernandez-Botran R, Sun X, Crespo FA. Soluble cytokine receptors in biological therapy. Expert Opin Biol Ther. 2002;2(6):585–605.
64. Larrick JW, Wright SC. Native cytokine antagonists. Baillieres Clin Haematol. 1992;5(3):681–702.
65. Namas R, Mikheev M, Yin J, Over P, Young M, Constantine G, et al. Biohybrid device for the systemic control of acute inflammation. Disrupt Sci Technol. 2012;1(1).
66. Akman-Anderson L, Vodovotz Y, Zamora R, Luckhart S. Bloodfeeding as an Interface of mammalian and arthropod immunity. In: Beckage N, edtior. Insect Immunology. San Diego: Elsevier; 2007. pp. 149–177.
67. Gooding LR. Virus proteins that counteract host immune defenses. Cell. 1992;71(1):5–7.
68. Surachetpong W, Singh N, Cheung KW, Luckhart SMAPK. ERK signaling regulates the TGF-β1-dependent mosquito response to *Plasmodium falciparum*. PLoS Pathog. 2009;5(4):e1000366.
69. Surachetpong W, Pakpour N, Cheung KW, Luckhart S. Reactive oxygen species-dependent cell signaling regulates the mosquito immune response to *Plasmodium falciparum*. Antioxid Redox Signal. 2011;14(6):943–55.
70. Omer FM, Kurtzhals JAL, Riley EM. Maintaining the immunological balance in parasitic infections: a role for TGF-β. Parasitol Today. 2000;16(1):18–23.
71. Luckhart S, Lieber MJ, Singh N, Zamora R, Vodovotz Y. Low levels of mammalian TGF-β1 are protective against malaria parasite infection, a paradox clarified in the mosquito host. Exp Parasitol. 2008;118:290–6.
72. Luckhart S, Vodovotz Y, Cui L, Rosenberg R. The mosquito *Anopheles stephensi* limits malaria parasite development with inducible synthesis of nitric oxide. Proc Natl Acad Sci U S A. 1998;95:5700–5.
73. Crampton AL, Luckhart S. Isolation and characterization of As60 A, a transforming growth factor- β gene, from the malaria vector *Anopheles stephensi*. Cytokine. 2001;13(2):65–74.
74. Price I, Ermentrout B, Zamora R, Wang B, Azhar N, Mi Q, et al. In vivo, in vitro, and in silico studies suggest a conserved immune module that regulates malaria parasite transmission from mammals to mosquitoes. J Theor Biol. 2013;334:172–86.
75. Date SV, Stoeckert CJ. Computational modeling of the *Plasmodium falciparum* interactome reveals protein function on a genome-wide scale. Genome Res. 2006;16(4):542–9.
76. LaCount DJ, Vignali M, Chettier R, Phansalkar A, Bell R, Hesselberth JR, et al. A protein interaction network of the malaria parasite *Plasmodium falciparum*. Nature. 2005;438(7064):103–7.

77. Hu G, Cabrera A, Kono M, Mok S, Chaal BK, Haase S, et al. Transcriptional profiling of growth perturbations of the human malaria parasite *Plasmodium falciparum*. Nat Biotech. 2010;28(1):91–8.
78. Osta MA, Christophides GK, Vlachou D, Kafatos FC. Innate immunity in the malaria vector *Anopheles gambiae*: comparative and functional genomics. J Exp Biol. 2004;207(15):2551–63.
79. Zdobnov EM, von Mering C, Letunic I, Torrents D, Suyama M, Copley RR, et al. Comparative genome and proteome analysis of *Anopheles gambiae* and *Drosophila melanogaster*. Science. 2002;298(5591):149–59.
80. Winzeler EA. Applied systems biology and malaria. Nat Rev Micro. 2006;4(2):145–51.
81. Osta MA, Christophides GK, Kafatos FC. Effects of mosquito genes on Plasmodium development. Science. 2004;303(5666):2030–2.
82. Corby-Harris V, Drexler A, Watkins de Jong L, Antonova Y, Pakpour N, Ziegler R, et al. Activation of Akt signaling reduces the prevalence and intensity of malaria parasite infection and lifespan in *Anopheles stephensi* mosquitoes. PLoS Pathog. 2010;6(7):e1001003.
83. Dong Y, Das S, Cirimotich C, Souza-Neto JA, McLean KJ, Dimopoulos G. Engineered Anopheles immunity to Plasmodium infection. PLoS Pathog. 2011;7(12):e1002458.

Chapter 12
Modeling Biomolecular Site Dynamics in Immunoreceptor Signaling Systems

Lily A. Chylek, Bridget S. Wilson and William S. Hlavacek

Abstract The immune system plays a central role in human health. The activities of immune cells, whether defending an organism from disease or triggering a pathological condition such as autoimmunity, are driven by the molecular machinery of cellular signaling systems. Decades of experimentation have elucidated many of the biomolecules and interactions involved in immune signaling and regulation, and recently developed technologies have led to new types of quantitative, systems-level data. To integrate such information and develop nontrivial insights into the immune system, computational modeling is needed, and it is essential for modeling methods to keep pace with experimental advances. In this chapter, we focus on the dynamic, site-specific, and context-dependent nature of interactions in immunoreceptor signaling (i.e., the biomolecular site dynamics of immunoreceptor signaling), the challenges associated with capturing these details in computational models, and how these challenges have been met through use of rule-based modeling approaches.

Keywords Cell signaling · Immunoreceptor signaling · Mathematical/computational modeling · Rule-based modeling · TCR · BCR · IgE receptor · LAT · Multisite phosphorylation · Aggregation

W. S. Hlavacek (✉)
Los Alamos National Laboratory, Mail Stop K710, T-6, Los Alamos,
NM 87545, USA
Tel.: 505-665-1355
e-mail: wish@lanl.gov

L. A. Chylek
Department of Chemistry and Chemical Biology, Cornell University, Ithaca,
NY 14853, USA
Tel.: 607-255-4848
e-mail: lac269@cornell.edu

B. S. Wilson
Department of Pathology, University of New Mexico School of Medicine, Albuquerque,
NM 87131, USA
Tel.: 505-272-8852
e-mail: bwilson@salud.unm.edu

S. J. Corey et al. (eds.), *A Systems Biology Approach to Blood*,
Advances in Experimental Medicine and Biology 844, DOI 10.1007/978-1-4939-2095-2_12

Introduction

Immune cells must process information about their changing environment to respond to signs of damage and infection. These cells possess surface receptors that bind extracellular ligands and initiate intracellular signaling, with information propagating through complex networks of molecular interactions. Dysregulation of these networks, and resulting dysregulation of immune responses, can lead to pathological conditions such as allergies, asthma, and autoimmunity. Thus, an understanding of signaling in immune cells is needed for improved understanding and treatment of disease. The complexity of cell signaling challenges intuition, but computational modeling offers the possibility of expanding our reasoning capabilities to obtain a predictive understanding of how immune cells respond to stimuli [1, 2].

Intracellular signals are propagated through enzyme-catalyzed reactions and noncovalent interactions, which are mediated by specific sites within biomolecules. These biomolecules, especially proteins, tend to each contain multiple functional components or sites. Examples of biomolecular sites involved in cell signaling include tyrosine residues that undergo phosphorylation and bind SH2 domains, SH3 domains and proline-rich motifs that interact with one another, and PH domains that bind phospholipids. The outcomes of biomolecular interactions are changes in population levels of chemical species (e.g., multimolecular complexes). The biomolecular site dynamics of cell signaling are governed by the same laws of physics and chemistry that govern chemical reaction kinetics [3]. Chemical reaction kinetics have long been modeled through the formalism of ordinary differential equations (ODEs). Use of ODE models is widespread in systems biology [4–6] and has yielded useful insights [7]. However, formulation of an ODE model depends on the availability of a reaction network, for which one must enumerate all species that can be populated, and make definite statements about how these species are connected and influence each other. In the case of cell signaling systems, this requirement can become a significant obstacle to model specification. This difficulty arises from the multisite structures of biomolecules.

As an example, let us consider the T cell receptor (TCR)/CD3 complex. This receptor contains ten immunoreceptor tyrosine-based activation motifs (ITAMs) [8], each of which contains two tyrosine residues. Each of the 20 ITAM tyrosine residues has two possible states: phosphorylated or unphosphorylated. As a result, the TCR/CD3 complex has accessible to it $2^{20} \approx 1$ million possible phosphorylation states. In general, without quantitative characterization of phosphorylation kinetics, the exact phosphoform of a given receptor at a given time cannot be narrowed through logical reasoning alone to less than 1 million possible phosphoforms. A reaction network capturing the full spectrum of TCR phosphoforms would contain 1 million nodes (representing chemical species), corresponding to 1 million ODEs, which would be impractical to specify. Thus, despite a wealth of information available about this receptor and proteins associated with it, specifying a reaction network that fully captures the possible consequences of known protein interactions and modifications is a challenge. Even if the populated chemical species and active chemical reactions could be identified within an experimental system to narrow the scope of modeling, any changes of the system, such as protein copy number variations, could alter the populated chemical species and active chemical reactions. Thus, traditional modeling approaches are problematic when one is interested in the site-specific dynamics of a biomolecular interaction network, i.e., biomolecular site dynamics.

The problem is not that biomolecular site dynamics are impossible to capture in a model but rather that available mechanistic knowledge is difficult to translate into a traditional model form. Allowing more natural representation of available mechanistic knowledge of cell signaling is a formalism that leverages the modular, multisite nature of proteins and other biomolecules. This formalism is based on use of local rules to represent interactions between sites of binding partners, which are taken to be modular, meaning the interactions are assumed to be somewhat independent of molecular context. An interaction is modular if a rule can be specified to represent the interaction (i.e., to define when the interaction occurs and with what rate) and the rule does not completely define the reactants. For example, if ligand–receptor binding is independent of receptor phosphorylation, then the interaction can be said to be modular, and a rule can be specified for ligand–receptor binding that applies regardless of receptor phosphorylation state. Rule-based modeling allows the translation of mechanistic knowledge into computational models consistent with chemical reaction kinetics. Because an interaction can be represented without complete knowledge of the states of the participating reactants, there is no need to specify a reaction network, which eliminates a major barrier to modeling of biomolecular site dynamics.

A rule concisely describes the necessary and sufficient conditions required of reactants for a reaction to occur. A rule-based model captures the same chemical kinetics as an ODE-based model (up to assumptions of modularity, which may be relaxed as needed to accommodate empirical observations), while permitting simulation of chemical kinetics without pregeneration of a reaction network. The methodology of rule-based modeling has been reviewed in detail elsewhere [9–11] and comprehensive guides to rule-based modeling software tools, such as BioNetGen [12], are available. Here, rather than reviewing methodology or software tools, we will review how rule-based modeling has been used to study biomolecular site dynamics of immunoreceptor signaling systems.

These systems are characterized by at least two mechanisms that, as we will discuss, the rule-based approach is well suited to capture: aggregation (or multivalent binding) and multisite phosphorylation. Aggregation of receptors is induced by interactions with cells bearing multiple ligands and/or multivalent ligands, and serves to initiate signaling [13–15]. Following aggregation of the IgE receptor (FcεRI), for example, the first biochemically detectable event in intracellular signaling is multisite phosphorylation of receptor ITAMs. Each of these motifs contains two canonical tyrosine residues that, when phosphorylated, serve as docking sites for SH2 domains of Src- and Syk-family kinases, which trigger subsequent signaling events. Aggregation reemerges in the cytoplasm, as signaling complexes assemble through multivalent interactions of scaffold/adaptor/linker proteins [16]. Aspects of these complex processes have been formalized, simulated, and analyzed in the models that will be discussed below.

Comparison of Modeling Assumptions

An ODE model is specified by making statements about how concentrations of chemical species change with time. Thus, a modeler is steered towards making assumptions about which species can be populated. These assumptions must sometimes be *ad hoc*, and may be at odds with available data. For example, the number of phosphorylation states of a TCR/CD3 complex could be reduced through a "virtual phosphorylation site" assumption, as it has been called in a study of ErbB receptor signaling by Birtwistle et al. [17]. Under this assumption, multiple ITAM tyrosine residues would be treated as a single site. This assumption would be serviceable for some purposes, such as a model that aims to elucidate how overall features of TCR signaling are affected by ligand–receptor binding kinetics [18]. However, if a modeler aimed to investigate the dependence of signaling events on the number or identity of specific ITAMs, a question that has been investigated experimentally [19], a virtual phosphorylation site assumption would be limiting. A virtual phosphorylation site could also be problematic in cases where different phosphotyrosines interact with different binding partners, as is the case for the linker for activation of T cells (LAT), because if multiple sites are treated as one, a false competition between binding partners may arise. These issues are discussed further by Chylek [20].

In contrast, a rule-based model is specified by making statements about site-specific requirements that must be met by reactants for a reaction to occur. A modeler is then steered towards making assumptions about the modularity or cooperativity of interactions. Which set of assumptions is preferable depends on what type of information is available. Given that signaling proteins are generally composed of modular domains [21], specification of a model in the form of rules is often more straightforward and efficient than specification in the form of ODEs. However, the two approaches are complementary, mirroring alternate modeling formalisms that are found in other fields (e.g., use of Lagrangian and Eulerian coordinates in fluid dynamics).

Summary of Recent Modeling Work

Rule-based modeling has been used to investigate a number of biological systems. In Table 12.1 we summarize recent applications aimed at understanding immune signaling, and in Table 12.2, we summarize applications aimed at understanding other types of cell signaling systems, as well as general mechanisms of cell signaling. We consider only applications from 2007 to present; earlier applications have been reviewed elsewhere [9]. It is worth noting that rule-based approaches, although developed for and most commonly used for modeling of cell signaling systems, have been used to model other processes, including metabolism [22–24], dynamics of various complex chemical reaction mechanisms [25–27], viral capsid assembly [28, 29], and synthetic gene circuitry [30]. A number of algorithms and software tools

Table 12.1 Recent studies of immune signaling that have employed rule-based modeling

Reference	Topic of study	Software used
Lipniacki et al. [18]	Feedbacks in T cell receptor signaling	BioNetGen
Nag et al. [53]	LAT aggregation	Problem-specific code
An and Faeder [54]	Toll-like receptor signaling	BioNetGen
Nag et al. [55]	Serial engagement of FcεRI	BioNetGen
Monine et al. [56]	Steric effects on aggregation of FcεRI	Problem-specific code
Nag et al. [57]	Syk activation in mast cells	BioNetGen
Artymov et al. [58]	Coreceptors in T cell receptor signaling	SSC
Nag et al. [59]	LAT aggregation with varying valency	Problem-specific code
Barua et al. [60]	Interlocked feedbacks in BCR signaling	BioNetGen
Mukherjee et al. [61]	Effect of ligand valency on BCR signaling	SSC
Chylek [20]	Extension of model of Mukherjee et al. [61]	BioNetGen
Barua and Goldstein [62]	Role of lipid rafts in FcεRI signaling	BioNetGen
Mukhopadhyay et al. [63]	Ultrasensitivity in T cell receptor signaling	BioNetGen
Liu et al. [64]	FcεRI signaling in response to multivalent antigen	BioNetGen

LAT linker for activation of T cells, BCR B cell receptor

for rule-based modeling are available [12, 31–51]. In addition, a number of rule-based models have been developed as demonstrations in methodological work. For example, to demonstrate use of a model visualization tool, a rule-based model of the mitogen-activated protein (MAP) kinase signaling network in yeast was developed, with all parameter values set to 1 [52].

Modeling of LAT Aggregation

In Tables 12.1 and 12.2, we list several examples of applications of rule-based modeling. We now discuss one topic in detail: interactions of the linker protein LAT. This protein is subject to multisite phosphorylation and forms aggregates with other signaling proteins through multivalent interactions, exemplifying two processes commonly found in immunoreceptor signaling. We also discuss a series of recently developed models for investigation of LAT aggregation. These models were obtained through traditional modeling methods and rule-based modeling, which provides an opportunity to highlight differences in model specification and the type of information that can be gained from the different approaches.

LAT is a transmembrane protein that undergoes multisite phosphorylation following stimulation of immunoreceptor signaling. Its four distal tyrosine residues have well-characterized roles as binding sites for SH2 domains of other signaling proteins.

Table 12.2 Other recent studies of cell signaling that have employed rule-based modeling

Reference	Topic of study	Software used
Barua et al. [65]	Interactions of Shp-2	BioNetGen
Barua et al. [66]	Interactions of tandem SH2 domains	BioNetGen
Barua et al. [67]	Jak kinase activation	BioNetGen
Gong et al. [68]	HMGB1 signaling in cancer	BioNetGen
Ray and Igoshin [69]	Transcriptional feedback in bacteria	BioNetGen, Mathematica
Malleshaiah et al. [70] (see supplemental material of this paper)	Scaffold proteins and switch-like behavior	Facile, ANC, MATLAB
Dushek et al. [71]	Multisite phosphorylation	Smoldyn
Selivanov et al. [72]	Mitochondrial respiration	Problem-specific code
Sorokina et al. [73]	The postsynaptic proteome of the neuronal synapse	RuleStudio, jsim, R
Thomson et al. [74]	Yeast pheromone signaling	BioNetGen, MATLAB
Geier et al. [75]	Integrin activation	BioNetGen
Ghosh et al. [76]	Iron homeostasis in tuberculosis	KaSim, Cytoscape
Abel et al. [77]	Influence of the membrane environment on bistability	SSC
Deeds et al. [78]	Combinatorial complexity in protein interaction networks	KaSim
Kocieniewski et al. [79]	Dual phosphorylation in the MAP kinase cascade	BioNetGen
Michalski and Loew [80]	Activation of CaMKII	BioNetGen, VCell
Tsernyschkow et al. [81]	Kinetochore architecture	SRSim
Kesseler et al. [82]	DNA damage G2 checkpoint	BioNetGen
Kozer et al. [83]	EGFR oligomerization	BioNetGen
Falkenberg and Loew [84]	Rho GTPase cycling	BioNetGen
Kiselyov et al. [85]	Ligand binding of insulin receptor and IGF1 receptor	Problem-specific code

HMGB1 high-mobility group protein B1, EGFR epidermal growth factor receptor, IGF1 insulin-like growth factor 1, MAP mitogen-activated protein

One of these proteins is growth factor receptor-bound 2 (GRB2), whose SH2 domain can bind phosphorylated tyrosines 171, 191, and 226 in LAT. GRB2 also contains a pair of SH3 domains, which interact with proline-rich sequences in son of sevenless homolog 1 (SOS1). The presence of at least four proline-rich sequences in SOS1 allows it to cross-link two GRB2 molecules [86]. In this way, aggregates of LAT–GRB2–SOS1 can form. LAT aggregation has been observed following stimulation of T cells [87] and mast cells [88], and the GRB2 binding sites in LAT are required for this process [16]. Expression of a SOS1 proline-rich region that can bind only one GRB2 molecule inhibits LAT aggregation and attenuates downstream signaling events, including calcium mobilization [16]. These results indicate that LAT's capacity to aggregate is relevant for its physiological function.

LAT is subject to valence switching, i.e., the number of LAT sites that can bind GRB2 can vary from zero to three, and depends on how many LAT tyrosine residues are phosphorylated. Thus, aggregation of LAT is influenced by its phosphorylation state. This dependence has been explored quantitatively by Goldstein and coworkers, using a combination of equilibrium theory borrowed from polymer chemistry, and rule-based modeling.

Nag et al. [53] formulated an equilibrium continuum model to compare aggregation of bivalent LAT versus trivalent LAT. The molecules and interactions considered in the model are illustrated in Fig. 12.1. The equilibrium model predicted that an increase in valency from two to three leads to a dramatic increase in average LAT aggregate size. For a homogenous population of trivalent LAT, a sol–gel coexistence region is predicted if concentrations of LAT, GRB2, and SOS1 are within certain ranges. The following equation was derived for the fraction of LAT molecules in the gel, a super-aggregate containing a significant fraction of all LAT, that is in equilibrium with unclustered LAT and small LAT aggregates:

$$f_g = 1 - \frac{2(1+\beta)^2}{\alpha\sigma\chi\mu\theta g^2 s}, \tag{12.1}$$

where s is the fractional concentration of free SOS1 and σ is a negative cooperativity factor. The parameter β is given by the following equation:

$$\beta = K_{GL}G + 2K_{GL}K_{GS}GS + 2\sigma K_{GL}K_{GS}^2G^2S, \tag{12.2}$$

where K_{GL} is the solution equilibrium constant for GRB2 binding to LAT, K_{GS} is the solution equilibrium constant for GRB2 binding to free SOS1, G is the cytosolic concentration of GRB2 free of SOS1, and S is the cytosolic concentration of SOS1 free of GRB2. The nondimensional parameters α, χ, μ, and θ are defined as follows:

$$\alpha = 3\overline{K}_{GL}L_T \tag{12.3}$$

$$\chi = K_{GL}G_T \tag{12.4}$$

$$\mu = 2K_{GS}S_T \tag{12.5}$$

$$\theta = K_{GS}G_T. \tag{12.6}$$

Here $\overline{K}_{GL}$ is the surface equilibrium cross-linking constant for membrane-associated GRB2 binding to LAT at the end of a chain, and G_T, L_T, and S_T are the total concentrations of GRB2, LAT, and SOS1, respectively.

The equilibrium relations given above were found by enumerating all possible complexes of LAT, GRB2, and SOS1, with the exception of cyclic aggregates.

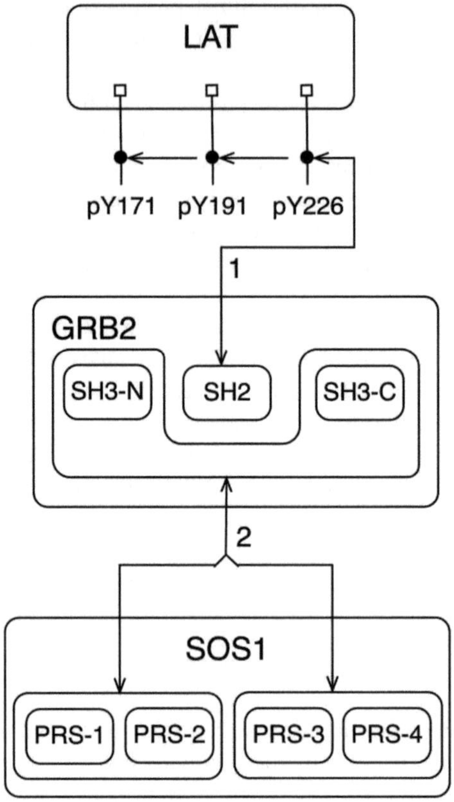

Fig. 12.1 A model of interactions among LAT, GRB2, and SOS1 [53, 59] represented as an extended contact map [111]. *Boxes* represent proteins, domains, and motifs. Double-headed arrows represent noncovalent interactions, and arrows are numbered to correspond to rules in Fig. 12.2. Posttranslational modifications are designated by small squares connected to flags that indicate the site and type of modification (e.g., "pY171" refers to phosphorylation of tyrosine 171). LAT contains three tyrosine residues that can be phosphorylated and serve as binding sites for the SH2 domain of GRB2. GRB2 also contains a pair of SH3 domains, which are taken to be a single site in the model. Four proline-rich sequences in SOS1 are taken to be a pair of sites that can bind *GRB2*. Thus, *GRB2* can bind *SOS1* to form a 1:1 complex, which can be bound by a second *GRB2* molecule to form *GRB2–SOS1–GRB2* complex. This ternary complex can cross-link two *LAT* molecules. An example of a molecule type definition is LAT(PY, PY, PY), which represents *LAT* containing three phosphorylated tyrosine residues, which by being assigned the same name are taken to be indistinguishable by convention. An example of a rule is LAT(PY) + GRB2(SH2,SH3) < − > LAT(PY!1)·GRB2(SH2!1,SH3). This rule indicates that free GRB2 can reversibly bind a free phosphosite in LAT via its SH2 domain when its *SH3* domain is free. It is assumed that the other two phosphotyrosines in LAT do not influence the interaction, as these sites are omitted from the rule. For a more complete model specification, see Fig 12.2. *LAT* linker for activation of T cells, GRB2 growth factor receptor-bound 2, SOS1 son of sevenless homolog 1

Enumeration of complexes requires insights from the field of branching processes, because binding of trivalent LAT to three GRB2 molecules can form a vast number of distinct linear chains and treelike networks. A statistical weight (relative concentration) is assigned to each complex. These weights enter into a partition function, which is a convergent infinite sum that can be reduced to an algebraic expression. The partition function can be used to obtain conservations laws for each molecule in the system, from which one can calculate the free concentrations of these molecules as well as concentrations of other species. The terms of the partition function are obtained by assuming detailed balance, which holds under equilibrium conditions. Thus, the model is silent about reaction kinetics. An ODE model for the chemical kinetics of this system cannot be easily specified because of the large number of distinct chemical species that can be populated [89]. The equilibrium model is analogous to an earlier model for binding of a trivalent ligand to a bivalent cell-surface receptor [90], and can be thought of as a model for binding of a soluble cytosolic bivalent ligand (GRB2–SOS1–GRB2) to a trivalent membrane receptor (LAT).

The equilibrium model of Nag et al. [53] is exact in the continuum limit, i.e., for a system of infinite size. However, a cell is of finite size. To evaluate the effect of finite system size, a rule-based model was developed and simulated (to steady state), and its predictions were compared to those of the equilibrium model. The rules of the model are presented in Fig 12.2; the model was simulated using a network-free simulation method [89]. At the time at which the model was developed, a problem-specific code was required to perform network-free simulation. General-purpose network-free simulators have since become available [36–38]. Steady-state results obtained from the rule-based model were found to be consistent with the equilibrium model. Information equivalent to Eq. 12.1 can be obtained by simulating the rule-based model to steady state and calculating the fraction of LAT molecules in the largest aggregate. The rule-based model is analogous to other rule-based models for ligand-induced receptor aggregation that have recently been studied [56, 89].

Although Nag et al. [53] focused on equilibrium behavior, their rule-based model enables study of the kinetics of aggregation, which are not captured in their equilibrium model. Kinetics of aggregation of multivalent binding partners have been modeled using traditional modeling methods (e.g., ODEs) by restricting the scope of the model to consideration of, for example, only ligand states [91]; consideration of the full spectrum of possible complexes requires a rule-based approach. In addition, a rule-based approach allows cyclic aggregates to be considered, as demonstrated by Monine et al. [56]. The approach of Perelson and DeLisi [91] becomes unwieldy when cyclic aggregates, or rings, can form [92].

A second set of models related to LAT aggregation has recently been used by Nag et al. [59] to evaluate the robustness of their earlier predictions. Rather than assuming a homogenous population of trivalent or bivalent LAT, a mixed population of trivalent, bivalent, and monovalent LAT was assumed. Monovalent LAT blocks aggregate growth, and bivalent LAT prevents branching. It was found that the presence of monovalent and bivalent LAT reduced the size of the sol–gel coexistence region in parameter space. Consideration of varying valency (or valence switching) is important, because a distribution of LAT phosphoforms is likely to be found in cells.

```
begin molecule types
LAT(PY,PY,PY)
GRB2(SH2,SH3)
SOS1(PRS,PRS)
end molecule types

begin reaction rules

# 1a: Free GRB2 binds free SOS1
GRB2(SH2,SH3)+SOS1(PRS,PRS) <->
GRB2(SH3!1,SH2).SOS1(PRS!1,PRS) kp1,km1

# 1b: Free GRB2 binds SOS1 bound to GRB2
GRB2(SH2,SH3)+SOS1(PRS,PRS!1).GRB2(SH3!1,SH2) <->
GRB2(SH3!2,SH2).SOS1(PRS!1,PRS!2).GRB2(SH3!1,SH2) s*kp1,km1

# 1c: Membrane-associated GRB2 binds free SOS1
GRB2(SH3,SH2!+)+SOS1(PRS,PRS) <->
GRB2(SH3!1,SH2!+).SOS1(PRS!1,PRS) kp1,km1

# 1d: Membrane-associated GRB2 binds SOS1 bounds to GRB2
GRB2(SH3,SH2!+)+SOS1(PRS,PRS!1).GRB2(SH3!1,SH2) <->
GRB2(SH3!2,SH2!+).SOS1(PRS!2,PRS!1).GRB2(SH3!1,SH2)
s*kp1,km1

# 1e: Free GRB2 binds membrane-associated SOS1
GRB2(SH3,SH2)+SOS1(PRS,PRS!1).GRB2(SH3!1,SH2!+) <->
GRB2(SH3!2,SH2).SOS1(PRS!2,PRS!1).GRB2(SH3!1,SH2!+)s*kp1,km1

# 1f: Membrane-associated GRB2 binds membrane-associated SOS1
GRB2(SH3,SH2!+)+SOS1(PRS,PRS!1).GRB2(SH3!1,SH2!+) <->
GRB2(SH3!2,SH2!+).SOS1(PRS!2,PRS!1).GRB2(SH3!1,SH2!+) kps,km1

# 2a: LAT binds free GRB2
LAT(PY)+GRB2(SH2,SH3) <->
LAT(PY!1).GRB2(SH2!1,SH3) kp2,km2

# 2b: LAT binds GRB2 bound to SOS1
LAT(PY)+GRB2(SH2,SH3!1).SOS1(PRS!1,PRS) <->
LAT(PY!2).GRB2(SH2!2,SH3!1).SOS1(PRS!1,PRS) kp2,km2

# 2c: LAT binds membrane-associated GRB2
LAT(PY)+GRB2(SH2,SH3!1).
SOS1(PRS!1,PRS!2).GRB2(SH2,SH3!2) <->
LAT(PY!3).GRB2(SH2!3,SH3!1).
SOS1(PRS!1,PRS!2).GRB2(SH2,SH3!2) kp2,km2

# 2d: LAT binds membrane-associated GRB2
LAT(PY)+GRB2(SH2,SH3!1).SOS1(PRS!1,PRS!2).
GRB2(SH2!3,SH3!2).LAT(PY!3) <->
LAT(PY!4).GRB2(SH2!4,SH3!1).SOS1(PRS!1,PRS!2).
GRB2(SH2!3,SH3!2).LAT(PY!3) kp2,km2

end reaction rules

# Call network-free simulator
simulate_nf({t_end=>1000,n_steps=>100});
```

Fig. 12.2 Abbreviated BioNetGen input file for a model of LAT, GRB2, and SOS1 interactions. The molecule types block specifies the molecules included in the model, and the components that each molecule contains. The reaction rules block contains rules that represent interactions that can

This distribution could shift as signaling progresses because of different kinetics of phosphorylation at different sites [93], which we discuss in more detail below.

Modeling Early Events in BCR Signaling

The rule-based approach has also been applied to study early events in B cell antigen receptor (BCR) signaling [60, 61]. In the study of Barua et al. [60], a rule-based model was used to study the roles of two related but distinct Src family kinases: LYN and FYN. These kinases are involved in interlocked positive and negative feedback loops. Positive feedback arises as receptor ITAMs are phosphorylated, generating binding sites for LYN, FYN, and SYK. Negative feedback arises as the adaptor protein PAG1 is phosphorylated. Phosphorylated PAG1 recruits CSK, a kinase that phosphorylates LYN and FYN at negative regulatory sites. By incorporating available mechanistic knowledge into a rule-based model capturing the interactions of the receptor (BCR), LYN, FYN, SYK, CSK, and PAG1, the site dynamics of this system were explored, and the effects of perturbations (e.g., protein knockdown and overexpression) were evaluated. It was found that oscillations in SYK activity could arise for certain ranges of the stimulatory signal, and for certain expression levels of LYN and FYN. It was also found that bistability could arise in cells lacking LYN or CSK. These results represent model predictions that are experimentally testable. Another recent study of BCR signaling [61], combining rule-based modeling with experimentation, focused on how spatial reorganization of the receptor and associated kinases contributes to engagement of positive feedback loops. For further discussion of this model, see Chylek [20].

Integration with Experimentation

Development of detailed computational models is justified by availability of detailed experimental data. Emerging technologies have enabled examination of site-specific aspects of cell signaling, in some cases at the single-molecule level. Insights and questions derived from these studies motivate modeling efforts to capture a comparable level of detail.

Fig. 12.2 (continued) occur in the model. Rules are numbered to correspond to *arrows* in Fig 12.1. Note that multiple rules correspond to each arrow. An arrow represents an interaction; the rules corresponding to a given arrow each represents the kinetics of the interaction in a unique molecular context. The simulation command at the bottom calls a network-free simulator. Not shown are the parameters block, in which parameters are assigned values; the seed species block, in which initial conditions are set; and the observables block, in which model outputs are specified. For more information about the contents and format of a BioNetGen input file, see Faeder et al. [12]. LAT linker for activation of T cells, GRB2 growth factor receptor-bound 2, SOS1 son of sevenless homolog 1

In the case of LAT, site-specific antibodies have been used to monitor phosphorylation kinetics of tyrosine residues 132 and 191 following TCR stimulation [93]. It was found that tyrosine 191 becomes phosphorylated significantly faster than tyrosine 132. This difference could have noteworthy consequences for regulation of downstream signaling, because these sites have distinct binding partners: phosphorylated tyrosine 132 binds phospholipase Cγ1 (PLCG1), whereas phosphorylated tyrosine 191 binds GRB2-related adapter protein 2 (GRAP2) and related adaptor proteins. These types of site-specific interactions are captured naturally in a rule-based model.

Another area in which site-specific phosphorylation has proven significant is partial phosphorylation of ITAMs. The consensus view is that, when doubly phosphorylated, ITAMs recruit the tandem SH2 domains of Syk family kinases. However, it is now clear that phosphorylation of a single ITAM tyrosine residue may lead to recruitment of a different set of signaling proteins, with distinct consequences for downstream signaling. By using mono-SH2 and dual-SH2 domain recombinant proteins as probes for singly and doubly phosphorylated ITAMs, it has been found that ITAM monophosphorylation is associated with anergy and activation of the negative regulators DOK1 and INPP5D in BCR signaling [94]. Activation of this inhibitory circuit may be linked to recruitment of the kinase LYN to singly phosphorylated ITAMs of the CD79A and CD79B subunits of the BCR.

Development of useful models will depend on suitable data sets for model parameterization. A potentially rich source of data is likely to come from quantitative high-resolution mass spectrometry (MS). MS can be used to detect phosphorylation of specific residues in an unbiased manner [95], enabling discovery of previously uncharacterized phosphorylation sites in receptor subunits and their downstream targets [96, 97]. MS can also be used to quantitatively track changes in phosphorylation levels of known phosphorylation sites as signaling progresses, including changes that occur on short time scales [98]. In the near future, novel methods for single-molecule MS may even offer the possibility of characterizing the phosphoforms of individual proteins [99]. We propose that rule-based modeling is well suited for mechanistic interpretation of large-scale MS data sets, because the rule-based formalism entails the same site-specific resolution [100].

Other advances in quantitative measurements will also provide modelers with systematic measurements of binding constants for protein domains involved in immunoreceptor signaling. High-throughput platforms have been developed and used to measure the affinities of SH2 domains of human proteins for specific phosphotyrosine peptides [101–104]. Such tools can potentially be used to determine the parameters needed for mechanistic modeling.

Lastly, super-resolution imaging techniques have enabled observation of signaling complexes on the nanoscale [105]. These studies have elucidated the spatial reorganization of proteins that occurs during immunoreceptor signaling, including aggregation of LAT following TCR stimulation [106]. The ability to image aggregation at high resolution, such that individual molecules can be distinguished, means that predictions from rule-based models of aggregation, such as a predicted aggregate size distribution, could potentially be tested in a direct manner.

Conclusions and Future Directions

A great deal of knowledge about immunoreceptor signaling has been accumulated. In addition, modern experimental methodologies allow us to obtain data that pertains to the site-specific details of molecular interactions. We believe that these data can be integrated to form a more complete, and more predictive, picture of how immune cells sense and respond to their environment. Our approach to piecing together this picture is to translate biological knowledge into models for chemical kinetics, using a formalism that naturally captures available biological knowledge. This formalism, in which rules are used to represent interactions and their contextual dependencies, allows us to capture biomolecular site dynamics (e.g., site-specific details of protein–protein interactions), more comprehensively simulate the reaction networks that mediate cell signaling, and manipulate specific features of cell signaling systems in silico. Until recently, only a subset of rule-based models could be simulated. With the development of network-free simulation methods [36–38, 89, 107, 108], simulation of a much wider array of models is now possible.

Mechanistic models have value as hypothesis generators and as vehicles of understanding [109]. A number of interesting biological questions can now be addressed via rule-based modeling, in part because this approach facilitates consideration of the full spectrum of possible phosphoforms for a protein of interest, which could be especially valuable for the study of ITAMs and related motifs that can have opposing regulatory functions that depend on phosphoform [110]. Rule-based modeling approaches are needed to address these and other questions that will emerge as the intricate machinery of immune signaling is explored further.

Acknowledgments We thank Byron Goldstein for helpful discussions. We acknowledge support from NIH grant P50 GM085273. L.A.C. acknowledges support from the Center for Nonlinear Studies that made visits to Los Alamos possible.

References

1. Germain RN, Meier-Schellersheim M, Nita-Lazar A, Fraser ID. Systems biology in immunology: a computational modeling perspective. Annu Rev Immunol. 2011;29:527–85.
2. Chakraborty AK, Das J. Pairing computation with experimentation: a powerful coupling for understanding T cell signalling. Nat Rev Immunol. 2010;10:59–71.
3. Kholodenko BN. Cell-signalling dynamics in time and space. Nat Rev Mol Cell Biol. 2006;7:165–76.
4. Hucka M, Finney A, Sauro HM, Bolouri H, Doyle JC, Kitano H, and the rest of the SBML Forum. The systems biology markup language (SBML): a medium for representation and exchange of biochemical network models. Bioinformatics. 2003;19:524–31.
5. Le Novère N, Finney A, Hucka M, Bhalla US, Campagne F, Collado-Vides J, Crampin EJ, Halstead M, Klipp E, Mendes P, Nielsen P, Sauro H, Shapiro B, Snoep JL, Spence HD, Wanner BL. Minimum information requested in the annotation of biochemical models (MIRIAM). Nat Biotechnol. 2005;23:1509–15.
6. Le Novère N, Bornstein B, Broicher A, Courtot M, Donizelli M, Dharuri H, Li L, Sauro H, Schilstra M, Shapiro B, Snoep JL, Hucka M. BioModels Database: a free, centralized database

of curated, published, quantitative kinetic models of biochemical and cellular systems. Nucleic Acids Res. 2006;34:D689–91.
7. Kreeger PK, Lauffenburger DA. Cancer systems biology: a network modeling perspective. Carcinogenesis 2010;31:2–8.
8. Cambier JC. Antigen and Fc receptor signaling. The awesome power of the immunoreceptor tyrosine-based activation motif (ITAM). J Immunol. 1995;155: 3281–85.
9. Hlavacek WS, Faeder JR, Blinov ML, Posner RG, Hucka M, Fontana W. Rules for modeling signal-transduction systems. Sci STKE. 2006;344: re6.
10. Hlavacek WS. Two challenges of systems biology. In: Stumpf MPH, Balding DJ, Girolami M, editors. Handbook of statistical systems biology. NJ: Wiley; 2011. pp. 3–14.
11. Chylek LA, Stites EC, Posner RG, Hlavacek WS. Innovations of the rule-based modeling approach. In: Prokop A, Csukás B, editors. Systems Biology: integrative biology and simulation tools. Vol 1. Dordrecht: Springer; 2013. pp. 273–300.
12. Faeder JR, Blinov ML, Hlavacek WS. Rule-based modeling of biochemical systems with BioNetGen. Methods Mol Biol. 2009;500:113–67.
13. Metzger H. Transmembrane signaling: the joy of aggregation. J Immunol. 1992;149:1477–87.
14. Dintzis HM, Dintzis RZ, Vogelstein B. Molecular determinants of immunogenicity: the immunon model of immune response. Proc Natl Acad Sci U S A. 1976;73:3671–75.
15. Dintzis RZ, Middleton MH, Dintzis HM. Studies on the immunogenicity and tolerogenicity of T-independent antigens. J Immunol. 1983;131:2196–203.
16. Houtman JC, Yamaguchi H, Barda-Saad M, Braiman A, Bowden B, Appella E, Schuck P, Samelson LE. Oligomerization of signaling complexes by the multipoint binding of GRB2 to both LAT and SOS1. Nat Struct Mol Biol. 2006;13:798–805.
17. Birtwistle MR, Hatakeyama M, Yumoto N, Ogunnaike BA, Hoek JB, Kholodenko BN. Ligand-dependent responses of the ErbB signaling network: experimental and modeling analyses. Mol Syst Biol. 2007;3:144.
18. Lipniacki T, Hat B, Faeder JR, Hlavacek WS. Stochastic effects and bistability in T cell receptor signaling. J Theor Biol. 2008;254:110–22.
19. Holst J, Wang H, Eder KD, Workman CJ, Boyd KL, Baquet Z, Singh H, Forbes K, Chruscinski A, Smeyne R, van Oers NS, Utz PJ, Vignali DA. Scalable signaling mediated by T cell antigen receptor-CD3 ITAMs ensures effective negative selection and prevents autoimmunity. Nat Immunol. 2008;9:658–66.
20. Chylek LA. Decoding the language of phosphorylation site dynamics. Sci Signal. 2013;6:jc2.
21. Pawson T, Nash P. Assembly of cell regulatory systems through protein interaction domains. Science. 2003;300:445–452.
22. Hatzimanikatis V, Li C, Ionita JA, Henry CS, Jankowski MD, Broadbelt LJ. Exploring the diversity of complex metabolic networks. Bioinformatics. 2005;21:1603–09.
23. Mu F, Williams RF, Unkefer CJ, Unkefer PJ, Faeder JR, Hlavacek WS. Carbon-fate maps for metabolic reactions. Bioinformatics. 2007;23:3193–99.
24. Asztalos A, Daniels M, Sethi A, Shen T, Langan P, Redondo A, Gnanakaran S. A coarse-grained model for synergistic action of multiple enzymes on cellulose. Biotechnol Biofuels. 2012;5:55.
25. Faulon JL, Carbonell P. Reaction network generation. In: Faulon JL, Bender A, editors. Handbook of chemoinformatics algorithms. Boca Raton: Chapman & Hall/CRC Press; 2010. pp. 317–41.
26. Rangarajan S, Bhan A, Daoutidis P. Language-oriented rule-based reaction network generation and analysis: description of RING. Comput Chem Eng. 2012a;45:114–23.
27. Rangarajan S, Bhan A, Daoutidis P. Language-oriented rule-based reaction network generation and analysis: applications of RING. Comput Chem Eng. 2012b;46:141–52.
28. Jamalyaria F, Rohlfs R, Schwartz R. Queue-based method for efficient simulation of biological self-assembly systems. J Comput Phys. 2005;204:100–20.
29. Zhang T, Rohlfs R, Schwartz R. Implementation of a discrete event simulator for biological self-assembly systems. In: Kuhl ME, Steiger NM, Armstrong FB, Joines JA, editors. Proc 2005 Winter Simulat Conf; 2005. pp. 2223–31.

30. Marchisio MA, Colaiacovo M, Whitehead E, Stelling J. Modular, rule-based modeling for the design of eukaryotic synthetic gene circuits. BMC Syst Biol. 2013;7:42.
31. Bugenhagen SM, Beard DA. Specification, construction, and exact reduction of state transition system models of biochemical processes. J Chem Phys. 2012;137:154108.
32. Moraru II, Schaff JC, Slepchenko BM, Blinov ML, Morgan F, Lakshminarayana A, Gao F, Li Y, Loew LM. Virtual Cell modelling and simulation software environment. IET Syst Biol. 2008;2:352–62.
33. Mallavarapu A, Thomson M, Ullian B, Gunawardena J. Programming with models: modularity and abstraction provide powerful capabilities for systems biology. J R Soc Interface. 2009;6:257–70.
34. Lok L, Brent R. Automatic generation of cellular reaction networks with Moleculizer 1.0. Nat Biotechnol. 2005;23:131–6.
35. Lis M, Artyomov MN, Devadas S, Chakraborty AK. Efficient stochastic simulation of reaction-diffusion processes via direct compilation. Bioinformatics. 2009;25:2289–91.
36. Colvin J, Monine MI, Faeder JR, Hlavacek WS, Von Hoff DD, Posner RG. Simulation of large-scale rule- based models. Bioinformatics. 2009;25:910–7.
37. Colvin J, Monine MI, Gutenkunst R, Hlavacek WS, Von Hoff DD, Posner RG. RuleMonkey: software for stochastic simulation of rule-based models. BMC Bioinformatics. 2010;11:404.
38. Sneddon MW, Faeder JR, Emonet T. Efficient modeling, simulation, and coarse-graining of biological complexity with NFsim. Nat Methods. 2011;8:177–83.
39. Xu W, Smith AM, Faeder JR, Marai GE. RuleBender: a visual interface for rule-based modeling. Bioinformatics. 2011;27:1721–22.
40. Clarke EM, Faeder JR, Harris LA, Langmead CJ, Legay A, Jha SK. Statistical model checking in BioLab: applications to the automated analysis of T-cell receptor signaling pathway. Lect Notes Comput Sci. 2008;5307:231–50.
41. Ollivier JF, Shahrezaei V, Swain P. Scalable rule-based modeling of allosteric proteins and biochemical networks. PLoS Comput Biol. 2010;6:e1000975.
42. Gruenert G, Ibrahim B, Lenser T, Lohel M, Hinze T, Dittrich P. Rule-based spatial modeling with diffusing, geometrically constrained molecules. BMC Bioinformatics. 2010;11:307.
43. Smith AM, Xu W, Sun Y, Faeder JR, Marai GE. RuleBender: integrated modeling, simulation and visualization for rule-based intracellular biochemistry. BMC Bioinformatics. 2012;13:S3.
44. Meier-Schellersheim M, Xu X, Angermann B, Kunkel E, Jin T, Germain RN. Key role of local regulation chemosensing revealed by a new molecular interaction-based modeling method. PLoS Comput Biol. 2006;2:e82.
45. Angermann BR, Klauschen F, Garcia AD, Prustel T, Zhang F, Germain RN, Meier-Schellersheim M. Computational modeling of cellular signaling processes embedded into dynamic spatial contexts. Nat Methods. 2012;9:283–9.
46. Boutillier P, Feret J, Krivine J. KaSim3 reference manual. https://github.com/Kappa-Dev/KaSim. Accessed 6 Oct 2014.
47. Andrews SS, Addy NJ, Brent R, Arkin AP. Detailed simulations of cell biology with Smoldyn 2.1. PLoS Comput Biol. 2010;6:e1000705.
48. Andrews SS. Spatial and stochastic cellular modeling with the Smoldyn simulator. Methods Mol Biol. 2012;804:519–42.
49. Zhang F, Angermann BR, Meier-Schellersheim M. The Simmune Modeler visual interface for creating signaling networks based on bi-molecular interactions. Bioinformatics. 2013;29:1229–30.
50. Klinke DJ II, Finley SD. Timescale analysis of rule-based biochemical reaction networks. Biotechnol Prog. 2012;28:33–44.
51. Lopez CF, Muhlich JL, Bachman JA, Sorger PK. Programming biological models in Python using PySB. Mol Syst Biol. 2013;9:646.
52. Tiger CF, Krause F, Cedersund G, Palmér R, Klipp E, Hohmann S, Kitano H, Krantz M. A framework for mapping, visualisation and automatic model creation of signal-transduction networks. Mol Syst Biol. 2012;8:578.

53. Nag A, Monine MI, Faeder JR, Goldstein B. Aggregation of membrane proteins by cytosolic cross-linkers: theory and simulation of the LAT-Grb2-SOS1 system. Biophys J. 2009;96:2604–23.
54. An GC, Faeder JR. Detailed qualitative dynamic knowledge representation using a BioNetGen model of TLR−4 signaling and preconditioning. Math Biosci. 2009;217:53–63.
55. Nag A, Monine MI, Blinov ML, Goldstein B. A detailed mathematical model predicts that serial engagement of IgE-FcεRI complexes can enhance Syk activation in mast cells. J Immunol. 2010;185:3268–76.
56. Monine MI, Posner RG, Savage PB, Faeder JR, Hlavacek WS. Modeling multivalent ligand-receptor interactions with steric constraints on configurations of cell-surface receptor aggregates. Biophys J. 2010;98:48–56.
57. Nag A, Faeder JR, Goldstein B. Shaping the response: the role of FcεRI and Syk expression levels in mast cell signaling. IET Syst Biol. 2010;4:334–47.
58. Artyomov MN, Lis M, Devadas S, Davis MM, Chakraborty AK. CD4 and CD8 binding to MHC molecules primarily acts to enhance Lck delivery. Proc Natl Acad Sci U S A. 2010;107:16916–21.
59. Nag A, Monine M, Perelson AS, Goldstein B. Modeling and simulation of aggregation of membrane protein LAT with molecular variability in the number of binding sites for cytosolic Grb2-SOS1-Grb2. PLoS ONE. 2012;7:e28758.
60. Barua D, Hlavacek WS, Lipniacki T. A computational model for early events in B cell antigen receptor signaling: analysis of the roles of Lyn and Fyn. J Immunol. 2012;189:646–58.
61. Mukherjee S, Zhu J, Zikherman J, Parameswaran R, Kadlecek TA, Wang Q, Au-Yeung B, Ploegh H, Kuriyan J, Das J, Weiss A. Monovalent and multivalent ligation of the B cell receptor exhibit differential dependence upon Syk and Src family kinases. Sci Signal. 2013;6:ra1.
62. Barua D, Goldstein B. A mechanistic model of early FcεRI signaling: lipid rafts and the question of protection from dephosphorylation. PLoS ONE. 2012;7:e51669.
63. Mukhopadhyay H, Cordoba SP, Maini PK, van der Merwe PA, Dushek O. Systems model of T cell receptor proximal signaling reveals emergent ultrasensitivity. PLoS Comput Biol. 2013;9:e1003004.
64. Liu Y, Barua D, Liu P, Wilson BS, Oliver JM, Hlavacek WS, Singh AK. Single-cell measurements of IgE-mediated FcεRI signaling using an integrated microfluidic platform. PLoS ONE. 2013;8:e60159.
65. Barua D, Faeder JR, Haugh JM. Structure-based kinetic models of modular signaling protein function: focus on Shp2. Biophys J. 2007;92:2290–300.
66. Barua D, Faeder JR, Haugh JM. Computational models of tandem SRC homology 2 domain interactions and application to phosphoinositide 3-kinase. J Biol Chem. 2008;283:7338–45.
67. Barua D, Faeder JR, Haugh JM. A bipolar clamp mechanism for activation of Jak-family protein tyrosine kinases. PLoS Comput Biol. 2009;5:e1000364.
68. Gong H, Zuliani P, Komuravelli A, Faeder JR, Clarke EM. Analysis and verification of the HMGB1 signaling pathway. BMC Bioinformatics. 2010;11(Suppl 7):S10.
69. Ray JC, Igoshin OA. Adaptable functionality of transcriptional feedback in bacterial two-component systems. PLoS Comput Biol. 2010;6:e1000676.
70. Malleshaiah MK, Shahrezaei V, Swain PS, Michnick SW. The scaffold protein Ste5 directly controls a switch-like mating decision in yeast. Nature. 2010;465:101–5.
71. Dushek O, van der Merwe PA, Shahrezaei V. Ultrasensitivity in multisite phosphorylation of membrane-anchored proteins. Biophys J. 2011;100:1189–97.
72. Selivanov VA, Votyakova TV, Pivtoraiko VN, Zeak J, Sukhomlin T, Trucco M, Roca J, Cascante M. Reactive oxygen species production by forward and reverse electron fluxes in the mitochondrial respiratory chain. PLoS Comput Biol. 2011;7:e1001115.
73. Sorokina O, Sorokin A, Armstrong JD. Towards a quantitative model of the post-synaptic proteome. Mol BioSyst. 2011;7:2813–23.
74. Thomson TM, Benjamin KR, Bush A, Love T, Pincus D, Resnekov O, Yu RC, Gordon A, Colman-Lerner A, Endy D, Brent R. Scaffold number in yeast signaling system sets tradeoff between system output and dynamic range. Proc Natl Acad Sci U S A. 2011;108:20265–70.

75. Geier F, Fengos G, Iber D. A computational analysis of the dynamic roles of talin, Dok1, and PIPKI for integrin activation. PLoS ONE. 2011;6:e24808.
76. Ghosh S, Prasad KV, Vishveshwara S, Chandra N. Rule-based modelling of iron homeostasis in tuberculosis. Mol BioSyst. 2011;7:2750–68.
77. Abel SM, Roose JP, Groves JT, Weiss A, Chakraborty AK. The membrane environment can promote or suppress bistability in cell signaling networks. J Phys Chem B. 2012;116:3630–40.
78. Deeds EJ, Krivine J, Feret J, Danos V, Fontana W. Combinatorial complexity and compositional drift in protein interaction networks. PLoS ONE. 2012;7:e32032.
79. Kocieniewski P, Faeder JR, Lipniakci T. The interplay of double phosphorylation and scaffolding in MAPK pathways. J Theor Biol. 2012;295:116–24.
80. Michalski PJ, Loew LM. CaMKII activation and dynamics are independent of the holoenzyme structure: an infinite subunit holoenzyme approximation. Phys Biol. 2012;9:036010.
81. Tschernyschkow S, Herda S, Gruenert G, Döring V, Görlich D, Hofmeister A, Hoischen C, Dittrich P, Diekmann S, Ibrahim B. Rule-based modeling and simulations of the inner kinetochore structure. Prog Biophys Mol Biol. 2013;113:33–45.
82. Kesseler KJ, Blinov ML, Elston TC, Kaufmann WK, Simpson DA. A predictive mathematical model of the DNA damage G2 checkpoint. J Theor Biol. 2013;320:159–69.
83. Kozer N, Barua D, Orchard S, Nice EC, Burgess AW, Hlavacek WS, Clayton AH. Exploring higher-order EGFR oligomerisation and phosphorylation-a combined experimental and theoretical approach. Mol Biosyst. 2013;9:1849–63.
84. Falkenberg CV, Loew LM. Computational analysis of Rho GTPase cycling. PLoS Comput Biol. 2013;9:e1002831.
85. Kiselyov VV, Versteyhe S, Gauguin L, De Meyts P. Harmonic oscillator model of the insulin and IGF1 receptors' allosteric binding and activation. Mol Syst Biol. 2009;5:243.
86. Sethi A, Goldstein B, Gnanakaran S. Quantifying intramolecular binding in multivalent interactions: A structure-based synergistic study on Grb2-Sos1 complex. PloS Comp Biol. 2011;7:e1002192.
87. Bunnell SC, Hong DI, Kardon JR, Yamazaki T, McGlade CJ, Barr VA, Samelson LE. T cell receptor ligation induces the formation of dynamically regulated signaling assemblies. J Cell Biol. 2002;158:1263–75.
88. Wilson BS, Pfeiffer JR, Surviladze Z, Gaudet EA, Oliver JM. High resolution mapping of mast cell membranes reveals primary and secondary domains of FcεRI and LAT. J Cell Biol. 2001;154:645–58.
89. Yang J, Monine MI, Faeder JR, Hlavacek WS. Kinetic Monte Carlo method for rule-based modeling of biochemical networks. Phys Rev E. 2008;78:031910.
90. Goldstein B, Perelson AS. Equilibrium theory for the clustering of bivalent cell surface receptors by trivalent ligands: Application to histamine release from basophils. Biophys J. 1984;45:1109–23.
91. Perelson AS, DeLisi C. Receptor clustering on a cell surface. I. Theory of receptor cross-linking by ligands bearing two chemically identical functional groups. Math Biosci. 1980;48:71–110.
92. Posner RG, Wofsy C, Goldstein B. The kinetics of bivalent ligand-bivalent receptor aggregation: ring formation and the breakdown of the equivalent site approximation. Math Biosci. 1995;126:171–90.
93. Houtman JCD, Houghtling RA, Barda-Saad M, Toda Y, Samelson LE. Early phosphorylation kinetics of proteins involved in proximal TCR-mediated signaling pathways. J Immunol. 2005;175:2449–58.
94. O'Neill SK, Getahun A, Gauld SB, Merrell KT, Tamir I, Smith MJ, Dal Porto JM, Li QZ, Cambier JC. Monophosphorylation of CD79a and CD79b ITAM motifs initiates a SHIP-1 phosphatase-mediated inhibitory signaling cascade required for B cell anergy. Immunity. 2011;35:746–56.
95. Cox J, Mann M. Quantitative, high-resolution proteomics for data-driven systems biology. Annu Rev Biochem. 2011;80:273–99.

96. Nguyen V, Cao L, Lin JT, Hung N, Ritz A, Yu K, Jianu R, Ulin SP, Raphael BJ, Laidlaw DH, Brossay L, Salomon AR. A new approach for quantitative phosphoproteomic dissection of signaling pathways applied to T cell receptor activation. Mol Cell Proteomics. 2009;8:2418–31.
97. Brockmeyer C, Paster W, Pepper D, Tan CP, Trudgian DC, McGowan S, Fu G, Gascoigne NR, Acuto O, Salek M. T cell receptor (TCR)-induced tyrosine phosphorylation dynamics identifies THEMIS as a new TCR signalosome component. J Biol Chem. 2011;286:7535–47.
98. Dengjel J, Akimov V, Olsen JV, Bunkenborg J, Mann M, Blagoev B, Andersen JS. Quantitative proteomic assessment of very early cellular signaling events. Nat Biotechnol. 2007;25:566–8.
99. Naik AK, Hanay MS, Hiebert WK, Feng XL, Roukes ML. Towards single-molecule nanomechanical mass spectrometry. Nat Nanotechnol. 2009;4:445–50.
100. Creamer MS, Stites EC, Aziz M, Cahill JA, Tan CW, Berens ME, Von Hoff DD, Hlavacek WS, Posner RG. Visualization, annotation and simulation of a large rule-based model for ErbB receptor signaling. BMC Syst Biol. 2012;6:107.
101. Jones RB, Gordus A, Krall JA, MacBeath G. A quantitative protein interaction network for the ErbB receptors using protein microarrays. Nature. 2006;439:168–74.
102. Kaushansky A, Gordus A, Chang B, Rush J, MacBeath G. A quantitative study of the recruitment potential of all intracellular tyrosine residues on EGFR, FGFR1 and IGF1R. Mol BioSyst. 2008;4:643–53.
103. Hause RJ Jr, Leung KK, Barkinge JL, Ciaccio MF, Chuu CP, Jones RB. Comprehensive binary interaction mapping of SH2 domains via fluorescence polarization reveals novel functional diversification of ErbB receptors. PLoS ONE. 2012;7:e44471.
104. Koytiger G, Kaushansky A, Gordus A, Rush J, Sorger PK, MacBeath G. Phosphotyrosine signaling proteins that driver oncogenesis tend to be highly interconnected. Mol Cell Proteomics. 2013;12:1204–13.
105. Huang B, Babcock H, Zhuang X. Breaking the diffraction barrier: super-resolution imaging of cells. Cell. 2010;143:1047–58.
106. Sherman E, Barr V, Manley S, Patterson G, Balagopalan L, Akpan I, Regan CK, Merrill RK, Sommers CL, Lippincott-Schwartz J, Samelson LE. Functional nanoscale organization of signaling molecules downstream of the T cell antigen receptor. Immunity. 2011;35:705–20.
107. Danos V, Feret J, Fontana, W, Harmer R, Krivine J. Rule-based modelling of cellular signalling. Lect Notes Comput Sci. 2007;4703:17–41.
108. Yang J, Hlavacek WS. The efficiency of reactant site sampling in network-free simulation of rule-based models for biochemical systems. Phys Biol. 2011;8:055009.
109. Lander AD. The edges of understanding. BMC Biol. 2010;8:40.
110. Blank U, Launay P, Benhamou M, Monteiro RC. Inhibitory ITAMs as novel regulators of immunity. Immunol Rev. 2009;232:59–71.
111. Chylek LA, Hu B, Blinov ML, Emonet T, Faeder JR, Goldstein B, Gutenkunst RN, Haugh JM, Lipniacki T, Posner RG, Yang J, Hlavacek WS. Guidelines for visualizing and annotating rule-based models. Mol BioSyst. 2011;7:2779–95.

Chapter 13
Structure and Function of Platelet Receptors Initiating Blood Clotting

Elizabeth E. Gardiner and Robert K. Andrews

Abstract At the clinical level, recent studies reveal the link between coagulation and other pathophysiological processes, including platelet activation, inflammation, cancer, the immune response, and/or infectious diseases. These links are likely to underpin the coagulopathy associated with risk factors for venous thromboembolic (VTE) and deep vein thrombosis (DVT). At the molecular level, the interactions between platelet-specific receptors and coagulation factors could help explain coagulopathy associated with aberrant platelet function, as well as revealing new approaches targeting platelet receptors in diagnosis or treatment of VTE or DVT. Glycoprotein (GP)Ibα, the major ligand-binding subunit of the platelet GPIb-IX-V complex, that binds the adhesive ligand, von Willebrand factor (VWF), is co-associated with the platelet-specific collagen receptor, GPVI. The GPIb-IX-V/GPVI adheso-signaling complex not only initiates platelet activation and aggregation (thrombus formation) in response to vascular injury or disease but GPIbα also regulates coagulation through a specific interaction with thrombin and other coagulation factors. Here, we discuss the structure and function of key platelet receptors involved in thrombus formation and coagulation in health and disease, with a particular focus on platelet GPIbα.

Keywords Platelets · Coagulation · GPIb-IX-V · GPVI

Introduction: Coagulation and Platelets

Coagulation of human plasma is initiated by activation of the intrinsic (FXII-dependent) or extrinsic (tissue factor-dependent) pathways (Fig. 13.1a) [1]. In vivo, release of activated tissue factor at sites of damaged vasculature provides a triggering

---

R. K. Andrews (✉)
Australian Centre for Blood Diseases, Monash University, Alfred Medical Research & Education Precinct, Level 6, Burnet Building, 89 Commercial Road, Melbourne, VIC 3004, Australia
Tel.: 61-3-9903-0136
e-mail: rob.andrews@monash.edu

E. E. Gardiner
Australian Centre for Blood Diseases, Monash University, Melbourne, VIC, Australia

S. J. Corey et al. (eds.), *A Systems Biology Approach to Blood*,
Advances in Experimental Medicine and Biology 844, DOI 10.1007/978-1-4939-2095-2_13

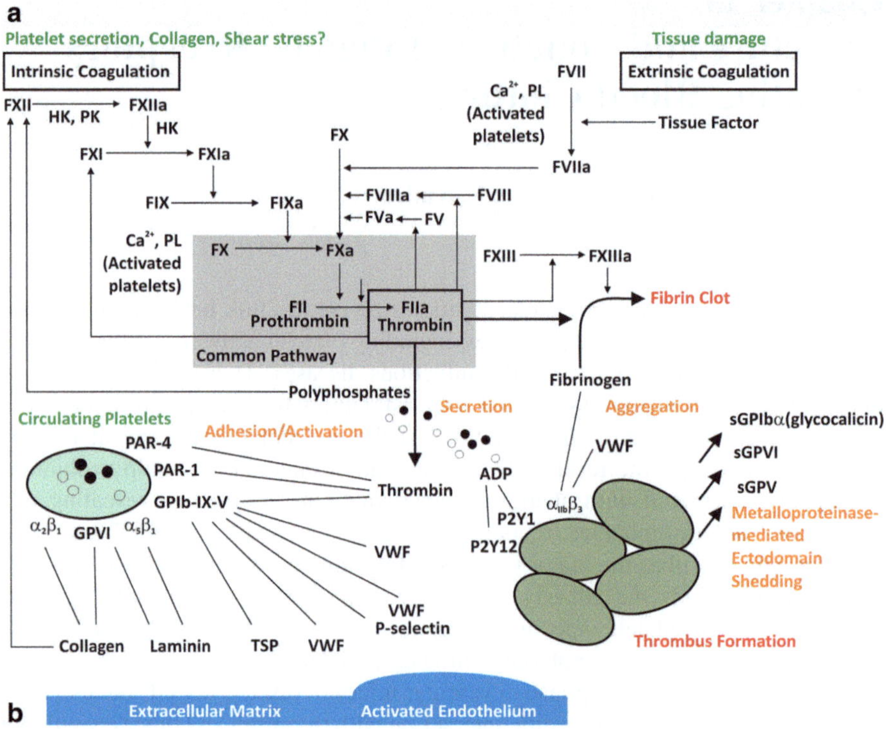

Fig. 13.1 Coagulation and platelet function. **a** Intrinsic (FXII-dependent) and extrinsic (tissue factor-dependent) coagulation pathways leading to the common pathway of thrombin (FIIa) generation. Thrombin induces clotting via conversion of fibrinogen to fibrin, a process accelerated by activated platelets, and **b** can bind to platelet GPIbα (of the GPIb-IX-V complex) and activate platelets via GPIbα signaling when GPV (a thrombin substrate) is removed, and G-protein-coupled protease-activated thrombin receptors, PAR-1 and PAR-4. Thrombin can thereby enhance thrombus formation following adhesion of circulating platelets to extracellular matrix or activated endothelium, or under shear stress, leading to secretion of ADP and procoagulant factors such as polyphosphates, increased expression of platelet surface phospholipids, and activation of integrin $\alpha_{IIb}\beta_3$ that binds fibrinogen or VWF and mediates platelet aggregation. Thrombin can be activated by vascular damage (releasing tissue factor) or by activation of FXII (intrinsic pathway) by collagen exposure, platelet secretion of procoagulant factors such as polyphosphates, collagen (that binds FXII under some conditions or activates prekallikrein under hyperglycemic conditions), or possibly by pathological shear stress **a**. *F* factor, *GP* glycoprotein, *HK* high molecular weight kininogen, *PK* prekallikrein, *PL* phospholipids, *TSP* thrombospondin, *VWF* von Willebrand factor

mechanism for initiating coagulation [2]. For the intrinsic pathway, however, while contact activation by negative surfaces is known to trigger coagulation in vitro, recent evidence suggests that collagen exposure, release of procoagulant polyphosphates from activated platelets [3], and possibly even pathological shear stress could provide a mechanism for activating factor XII (FXII) in vivo [4]. Other mechanisms could involve activation of prekallikrein that acts on FXII. Another intriguing possibility is that antibacterial leukocyte DNA-containing neutrophil extracellular traps (NETs) or associated proteins could activate intrinsic coagulation. Both intrinsic and extrinsic pathways activate the common coagulation pathway, involving activation of FX to FXa which converts prothrombin (FII) to active thrombin (FIIa; Fig. 13.1a).

However coagulation is initiated, there is a clear role for platelets in spatial and temporal regulation of coagulation at prothrombotic sites [5]. Circulating platelets in the bloodstream are rapidly activated following vascular damage by exposure of collagen, von Willebrand factor (VWF), or other adhesive ligands in the subendothelial matrix or ruptured atherosclerotic plaque [6], VWF/P-selectin on activated endothelium, or VWF in stenotic vessels (Fig. 13.1b) [7]. Platelet glycoprotein (GP)Ibα of the GPIb-IX-V complex binds VWF [8] or thrombospondin (TSP), GPVI binds collagen or laminin (facilitated by activated platelet integrins, $\alpha_2\beta_1$, or $\alpha_5\beta_1$, respectively, facilitating adhesion or platelet activation via GPVI) [9]. Engagement of GPIb-IX-V/GPVI leads to activation of the integrin $\alpha_{IIb}\beta_3$, which binds fibrinogen or VWF and mediates platelet aggregation and fibrin formation [10]. Activated platelets secrete agonists such as ADP which acts on purinergic G-protein-coupled receptors [11], and secrete procoagulant factors such as polyphosphates [3] which promote coagulation and generation of active thrombin [12]; expression of phosphatidylserine or other procoagulant phospholipids on the surface of activated platelets also accelerates coagulation by localization and assembly of coagulation complexes (Fig. 13.1b). Thrombin activates platelets by using GPIbα as a cofactor in the activation of platelets via G-protein-coupled protease-activated receptors, PAR-1 or PAR-4, which in turn promote platelet activation and degranulation. Platelet activation is also associated with time-dependent metalloproteinase-mediated ectodomain shedding of platelet receptors, GPIbα ("glycocalicin"), GPV, and GPVI [13] (Fig. 13.1b). In this regard, elevated levels or plasma soluble GPVI (sGPVI) associated with disseminated intravascular coagulation (DIC) correlate with increased levels of coagulation markers [14]. Interestingly, GPV may be shed via cleavage at separate sites either by platelet sheddases or by thrombin [13, 15], with loss of surface GPV being associated with increased platelet activation by the interaction of thrombin with GPIbα.

Ligand Binding to Platelet GPIbα

Coagulation factors including thrombin, FXII, FXI, and high molecular weight kininogen (HK) bind to the same ligand-binding domain of GPIbα involved in binding VWF, TSP, and other ligands [16, 17]. GPIbα is a multifunctional receptor which binds prothrombotic and procoagulant ligands within a versatile "shear-activated" ligand-binding region (Fig. 13.2). The absence or deficiency of GPIbα causes the inherited bleeding disorder, Bernard–Soulier Syndrome (BSS) [18] and along with the loss of high-shear- and VWF-dependent platelet-to-platelet interactions [19]; platelets from individuals with BSS are generally thought to have ablated procoagulant function [20]. GPIbα ($\sim$ 135 kDa) consists of an N-terminal globular domain ($\sim$ 40 kDa), a sialomucin core, an extracellular membrane-proximal tandem Cys sequence which forms disulfide bonds to 2 GPIbβ subunits (forming GPIb) [21],

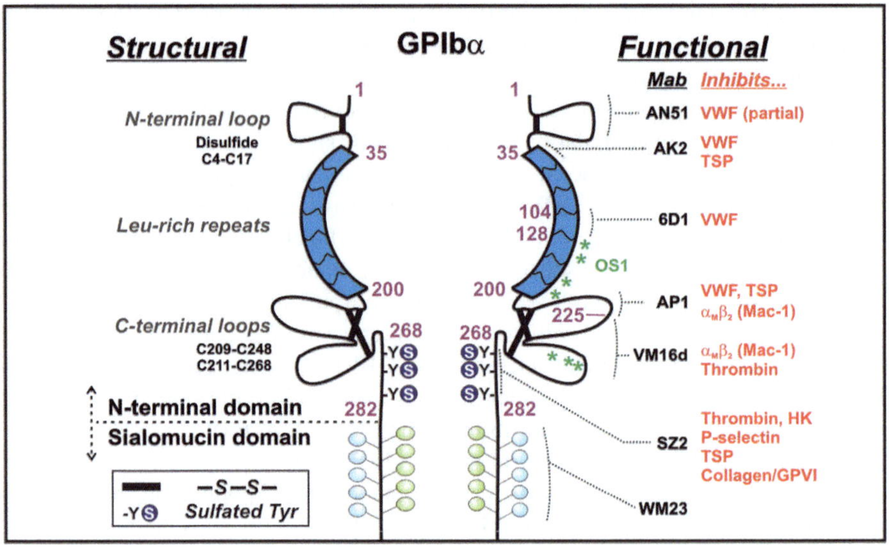

Fig. 13.2 N-terminal ligand-binding region of platelet GPIbα Structural regions based on primary sequence and crystal structure, and functional mapping of anti-GPIbα monoclonal antibodies which inhibit binding of one or more ligands. Antibodies against different epitopes, and the small-molecule allosteric inhibitor, the cyclic peptide OS1, interact with different regions within the 282-residue N-terminal domain illustrating the globular conformationally dependent binding sites for VWF and possibly other ligands. Coagulation factors of the intrinsic pathway (FXII, FXI, high molecular weight kininogen, with FXI being a substrate for both FXII and thrombin) also interact with the N-terminal domain of GPIbα

transmembrane domain, and cytoplasmic tail containing binding sites for intracellular signaling/cytoskeletal proteins. GPIb is noncovalently associated with GPIX and GPV, all members of the leucine-rich repeat family. The N-terminal globular domain of GPIbα (His1–Glu282) contains four important structural domains: the tandem leucine-rich repeats ($\sim$ 24 residues, each spanning the sequence 36–200), the N-terminal (residues 1–35) and C-terminal (201–268) disulfide-looped sequences, and an anionic sulfated tyrosine-rich sequence (269–282; Fig. 13.2) [22]. Using enzymes which specifically cleave at 282/283 (mocarhagin) [23] or inhibitory anti-GPIbα monoclonal antibodies mapped to specific structural regions [22, 24] as well as other approaches, it has been shown that GPIbα 1–282 contains discontiguous but overlapping binding sites for VWF [25], TSP [26], thrombin [27–29], FXII [30], FXI [31], HK [32], and counter receptors on activated endothelial cells (P-selectin) [33] or leukocytes ($\alpha_M\beta_2$; Mac-1) [34]. The sulfated tyrosine sequence also associates with the ectodomain of the immunoreceptor family protein, GPVI, on human platelets [35]. In addition, the procoagulant protein, recombinant FVIIa has been reported to bind to the sialomucin domain of GPIbα, downstream of Glu282 which could also localize thrombin generation to activated platelets.

Extensive biochemical, crystallographic, and molecular simulation studies have analyzed binding of VWF to human GPIbα, revealing that the leucine-rich repeat sequence 60–128 (repeats "2–4") is critical for interacting with the VWF A1 domain [28, 36–39]. Compared to the structure of the ligand-binding domain under resting conditions, as the shear stress increases from low to high physiological or pathological levels, the C-terminal disulfide-looped domain alters its conformation when complexed with VWF A1. A small molecular weight inhibitor, OS1, allosterically inhibits VWF binding to GPIbα by preventing the formation of the active conformation [40], while gain-of-function mutations within the C-terminal disulfide loop also increase binding to VWF-A1.

Specialized electrostatic "catch-slip" bonding facilitates high-affinity adhesion of VWF to receptor as shear rate increases, thereby enabling platelets to roll, skip, or firmly adhere to immobilized VWF in a shear-dependent manner [4, 38]. At high physiological or pathological shear rates such as encountered in a sclerotic or blocked artery, platelet adhesion becomes entirely GPIbα dependent [41]. However, examination of arterial thrombus formation in experimental models in vivo shows a significantly greater dependence on platelet GPIbα than VWF [42, 43], suggesting other ligands are also important. The extent to which conformational activation of GPIbα regulates interaction of ligands other than VWF is unknown, and precise binding sites for other ligands, including coagulation factors, are yet to be fully resolved. It is clear, however, that ligands such as TSP, P-selectin, and $\alpha_M\beta_2$ not only bind to the N-terminal domain of GPIbα under static conditions but also support GPIbα-dependent adhesion under flow conditions. The interaction of GPIbα with $\alpha_M\beta_2$ involves a domain of α_M ("I-domain") homologous to the GPIbα-binding A1 domain of VWF, although the binding sites for the two ligands are not identical. VWF A1 competes for binding of TSP, and some anti-GPIbα antibodies differentially block VWF or other ligands. It has been determined that the sulfated sequence (269–282) is critically involved in thrombin binding, with this interaction facilitating thrombin-dependent activation of platelet PAR-1. The extent to which co-localization of FXII, thrombin, and the common substrate FXI on a single receptor or adjacent copies of GPIbα within the GPIb-IX-V/GPVI complex is yet to be definitively established, although binding to GPIbα promotes activation of FXI by thrombin. Interestingly, regions of GPIbα beyond the 45-kDa N-terminal portion may be involved in platelet procoagulant function, as specific enzymatic removal of this region in murine washed platelets did not interfere with thrombin generation [44]. Targeted disruption of the cytoplasmic portion of GPIb-IX-V, for example, by site-directed mutagenesis, may help elucidate how the receptor complex modulates platelet procoagulant activity.

Functional Role of Interactions Involving GPIbα

Considering together the network of potential interactions of platelet GPIbα illustrates the potential for this receptor to co-localize, sequester, or otherwise regulate different components of platelet thrombus formation, coagulation, and platelet–leukocyte and platelet–endothelial cell interactions (Fig. 13.3). These interactions,

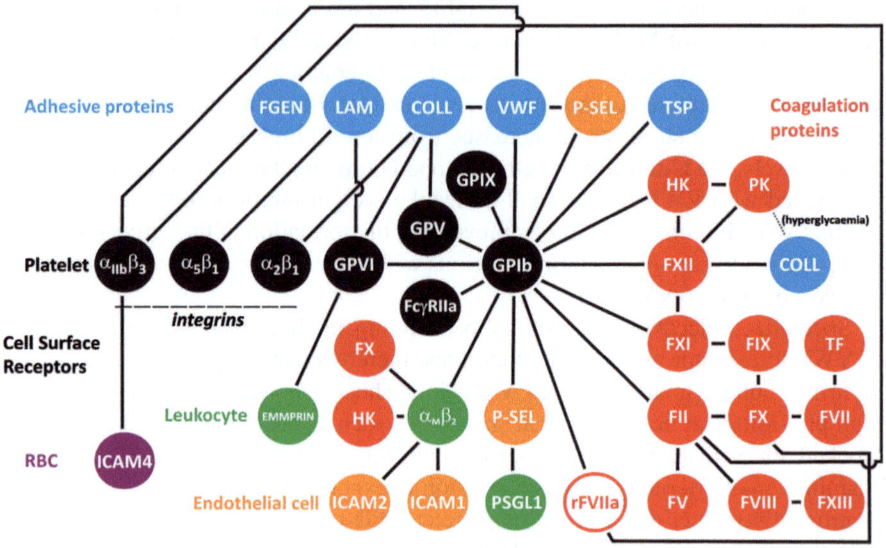

Fig. 13.3 Interactions involving GPIbα In addition to other platelet surface receptors (*black*) with which it is co-associated (GPVI, GPV, GPIX, FcγRIIa), GPIbα interacts with at least ten other purported binding partners, including adhesive proteins (*blue*), coagulation factors (*red*), and receptors on either leukocytes (*green*) or activated endothelial cells (*orange*). Indirectly, the GPIbα-related network includes over 30 proteins. What this interaction map does not show is the spatial-temporal nature of these interactions or how these interactions are regulated under static or high-shear conditions (for example, when VWF binding is enhanced). These features are likely to control a coordinated, localized thrombotic, inflammatory, and coagulation response to injury, atheroma, immune disease, or infectious diseases

rarely studied in combination, suggest how coagulation associated with platelet thrombus formation and inflammatory responses involving activated platelets and leukocytes in immune or infectious diseases could be coordinated by interactions involving GPIbα and adhesive or procoagulant ligands or counter-receptors under resting or activated conditions. GPIbα could provide a common regulatory receptor controlling time-dependent transition from initial platelet adhesion, activation and aggregation, to coagulation and inflammation in response to vascular injury or disease, for example, by progressively binding to VWF/TSP, thrombin/FXII/FXI/HK, or P-selectin/$\alpha_M\beta_2$, respectively. The capacity of platelets to rapidly adhere, become activated and degranulate in flowing blood mediated by GPIbα and other receptors would be a key property enabling the coordination of these pathophysiological processes. It is only recently that the role of platelets and platelet receptors has been investigated in detail in inflammation [45], coagulation [46], cancer [47–49], and infectious diseases [50, 51].

One potential link between coagulation factors and platelet receptors involves findings by Renne and colleagues [52], showing that while deficiency of FXII/FXI has minimal impact on bleeding times, there is marked inhibition of occlusive platelet thrombus formation at high shear in the arterial circulation in experimental models. It is unclear how FXII is activated under these conditions, but interaction with platelet GPIbα or GPIbα-dependent adhesion localizing platelet activation and secretion, and phosphatidylserine exposure could provide the means for stable occlusive thrombus formation in the presence of FXII. The role of HK in GPIbα binding and activation of FXII is also interesting in terms of possible GPIbα-mediated activation of FXII in vivo, and the link between coagulation and platelet–leukocyte adhesion as HK also engages the GPIbα counter-receptor on leukocytes, $\alpha_M\beta_2$ [53]. The HK binding site on $\alpha_M\beta_2$ overlaps a fibrinogen-binding site, and increased the capacity for binding GPIbα, providing a potential mechanism for HK-dependent enhancement of platelet–leukocyte adhesion [53]. The FXII activator, plasma kallikrein (PK), also interacts with collagen under hyperglycemic conditions [54], such that collagen exposure could lead to activation of PK/FXII as well as localizing platelet GPIb-IX-V/GPVI via interactions with collagen/VWF. Together, this would provide a mechanism for bridging platelets, leukocytes, and the subendothelial matrix leading to the activation of coagulation [55].

More recently, clear roles for leukocytes in the direct upregulation of platelet procoagulant function have emerged. While the majority of circulating microparticles in healthy individuals are platelet or megakaryocyte derived [46], leukocyte-derived microparticles originating from neutrophils, monocytes/macrophages, or lymphocytes as well as endothelial-derived microparticles are significantly upregulated in all stages of atherosclerosis and circulate at a high level in the bloodstream of patients with high atherothrombotic risk [56, 57]. Microparticles have been demonstrated to associate with resting platelets via CD36, lowering the required threshold concentration of agonist to activate platelets [58] and also via engagement of platelet GPIbα by active $\alpha_M\beta_2$ on microparticles derived from activated neutrophils [59]. In the second study, engagement of GPIbα by $\alpha_M\beta_2$–bearing microparticles triggered signaling pathways that led to surface expression of P-selectin and activation of $\alpha_{IIb}\beta_3$-mediating platelet aggregation. Both interactions provide a clear and distinct mechanistic link between platelet prothrombotic and leukocyte inflammatory states where microparticles from unstimulated versus activated neutrophils differentially facilitate interaction with either activated platelets (via PSGL-1/P-selectin) or resting platelets (via active $\alpha_M\beta_2$/GPIbα), respectively.

Platelet GPIbα also interacts with bacterial proteins, such as the *Staphylococcal* superantigen-like protein 5 (SSL5) via the sulfated-tyrosine sequence and carbohydrate moieties of GPIbα [60, 61]. SSL5 also interacts with extracellular immunoglobulin domains of GPVI [61]. These types of interactions could be involved in platelet activation associated not only with bacterial infection and increased thrombotic risk but also with the coagulopathy commonly associated with sepsis and other infections. Bacterial-induced activation of leukocytes also releases DNA-containing NETs, which may limit dispersal of bacterial, but are also associated with release of nuclear proteins such as histones [62]. NETs have been linked to the development of venous "red" (platelet-deficient) thrombus in experimental models of deep vein thrombosis (DVT) [63]; however, NETs and associated proteins such as VWF A1 domain-binding histones [64] could also promote platelet activation, secretion, and leukocyte recruitment in arterial "white" (platelet-rich) thrombus.

On the platelet surface, GPVI interacts with the sulfated region of GPIbα that binds thrombin [35], and could also influence coagulation in other ways. GPVI contains two extracellular immunoglobulin domains, and is co-associated with the accessory signaling receptor, FcRγ, required for GPVI surface expression [65–67]. The anti-GPIbα monoclonal antibody SZ2, inhibits collagen-dependent platelet activation via GPVI [68]. Masking GPVI also attenuates collagen-induced or tissue factor-dependent thrombin generation, thrombus formation [69], or pulmonary thromboembolism [70]. GPVI engagement could promote phosphatidylserine exposure on activated platelets [71–73], induce procoagulant platelet-derived microparticles [73], or activate platelets leading to secretion of procoagulant factors such as polyphosphates. However, GPVI blockade can also inhibit tissue factor-mediated coagulation in the absence of collagen or other known GPVI ligands [69], while GPVI ligands also induce dose-dependent increases in FXa and thrombin generation, regulated by a subpopulation of platelets with increased coagulation factor binding that is not related to increased phosphatidylserine exposure [71]. These mechanisms require further analysis in combination with interactions involving coagulation factors and GPIbα, which is co-associated with GPVI [35]. GPVI also binds to extracellular matrix metalloproteinase (MMP) inducer (EMMPRIN; CD147), like GPVI, a member of the immunoreceptor family expressed on activated platelets, monocytes, and tumor cells [74]. The GPVI–EMMPRIN interaction could contribute to platelet-mediated coagulation at sites of monocyte recruitment, for example, at atherosclerotic sites of the vasculature [75], or in the context of tumor growth.

Plasma GPIbα and GPVI

Although the extracellular domain of platelet GPIbα is important for ligand binding, constitutive ectodomain shedding of GPIbα results in high levels of soluble GPIbα ectodomain (glycocalicin) in normal plasma (approximately two thirds of total GPIbα in blood) [76]. The functional consequences of glycocalicin-binding ligands are not addressed by existing studies, and it possibly has a regulatory role in some circumstances or is less efficacious than surface-expressed GPIbα within the GPIb-IX-V/GPVI adheso-signaling complex. Similarly, shedding of platelet GPVI liberating plasma soluble GPVI [77, 78] could downgrade the capacity for GPVI-dependent platelet activation or microparticle generation. Unlike GPIbα, levels of sGPVI in healthy plasma are relatively low (approximately one sixth of total blood GPVI) [79], but are elevated under prothrombotic [80–82] or procoagulant [14] conditions and may serve as a platelet-specific biomarker as an indicator of risk, for example, in the case of infectious diseases or immune disease [78]. In this regard, the platelet Fc receptor, FcγRIIa, utilises intracellular signaling pathways equivalent to GPVI/FcRγ, and engagement of FcγRIIa induces GPVI shedding [83]. Through this mechanism, antiplatelet autoantibodies, for example, targeting platelet factor 4/heparin complexes as seen in heparin-induced thrombocytopenia can activate platelets via FcγRIIa [84, 85], to release platelet-derived microparticles [86] and increase platelet thrombin generation [87].

Conclusions: Targeting Platelets in Human Disease

Whether selectively targeting platelet GPIb-IX-V/GPVI ligand binding, platelet activation or secretion to inhibit the impact of platelets on coagulation to aid in the therapeutic control of thrombosis [88, 89], or coagulopathy where platelets and platelet/leukocyte or platelet/endothelium interactions are implicated in procoagulant activity, warrants further investigation [90]. Further analysis, particularly centered on studies in human vascular systems of interactive sites, changes in binding under shear conditions, and the influence of other ligands under different conditions is required [91] to exploit these possibilities. It is also worth noting how multifunctional interactions of thrombin, FXa, and other factors beyond coagulation, broaden the range of interactions of platelet GPIb-IX-V/GPVI in human pathophysiology.

References

1. Levi M, van der Poll T. Inflammation and coagulation. Crit Care Med. 2010;38:S26–34.
2. del Conde I Shrimpton CN Thiagarajan P López JA. Tissue-factor-bearing microvesicles arise from lipid rafts and fuse with activated platelets to initiate coagulation. Blood. 2005;106(5):1604–11.
3. Müller F, Mutch NJ, Schenk WA, et al. Platelet polyphosphates are proinflammatory and procoagulant mediators in vivo. Cell. 2009;139(6):1143–56.
4. Colace TV, Tormoen GW, McCarty OJ, Diamond SL. Microfluidics and coagulation biology. Ann Rev Biomed Eng. 2013;15:283–303.
5. Jackson SP, Nesbitt WS, Westein E. Dynamics of platelet thrombus formation. J Thromb Haemost. 2009;7(s1):17–20.
6. Winckers K, ten Cate H, Hackeng TM. The role of tissue factor pathway inhibitor in atherosclerosis and arterial thrombosis. Blood Rev. 2013;27(3):119–32.
7. Versteeg HH, Heemskerk JW, Levi M, Reitsma PH. New fundamentals in hemostasis. Physiol Rev. 2013;93(1):327–58.
8. Chow TW, Hellums JD, Moake JL, Kroll MH. Shear stress-induced von Willebrand factor binding to platelet glycoprotein Ib initiates calcium influx associated with aggregation. Blood. 1992;80(1):113–20.
9. Andrews RK, Shen Y, Gardiner EE, Berndt MC. Platelet adhesion receptors and (patho)physiological thrombus formation. Histol Histopathol. 2001;16(3):969–80.
10. Cosemans JMEM, Schols SEM, Stefanini L, et al. Key role of glycoprotein Ib/V/IX and von Willebrand factor in platelet activation-dependent fibrin formation at low shear flow. Blood. 2011;117(2):651–60.
11. van der Meijden PEJ, Schoenwaelder SM, Feijge MAH, et al. Dual $P2Y_{12}$ receptor signaling in thrombin-stimulated platelets-involvement of phosphoinositide 3-kinase β but not γ isoform in Ca^{2+} mobilization and procoagulant activity. FEBS J. 2008;275(2):371–85.

12. Choi SH, Smith SA, Morrissey JH. Polyphosphate is a cofactor for the activation of factor XI by thrombin. Blood. 2011;118(26):6963–70.
13. Gardiner EE, Karunakaran D, Shen Y, Arthur JF, Andrews RK, Berndt MC. Controlled shedding of platelet glycoprotein (GP)VI and GPIb-IX-V by ADAM family metalloproteinases. J Thromb Haemost. 2007;5(7):1530–7.
14. Al-Tamimi M, Grigoriadis G, Tran H, et al. Coagulation-induced shedding of platelet glycoprotein VI mediated by Factor Xa. Blood. 2011;117(14):3912–20.
15. Rabie T, Strehl A, Ludwig A, Nieswandt B. Evidence for a role of ADAM17 (TACE) in the regulation of platelet glycoprotein V. J Biol Chem. 2005;280(15):14462–68.
16. Andrews RK, Gardiner EE, Shen Y, Whisstock JC, Berndt MC. Glycoprotein Ib-IX-V. Int J Biochem Cell Biol. 2003;35(8):1170–4.
17. Canobbio I, Balduini C, Torti M. Signalling through the platelet glycoprotein Ib-V-IX complex. Cell Signal. 2004;16(12):1329–44.
18. López JA, Andrews RK, Afshar-Kharghan V, Berndt MC. Bernard-Soulier syndrome. Blood. 1998;91(12):4397–418.
19. Cranmer SL, Ashworth KJ, Yao Y, et al. High shear-dependent loss of membrane integrity and defective platelet adhesion following disruption of the GPIbα-filamin interaction. Blood. 2011;117(9):2718–27.
20. Dicker IB, Pedicord DL, Seiffert DA, Jamieson GA, Greco NJ. Both the high affinity thrombin receptor (GPIb-IX-V) and GPIIb/IIIa are implicated in expression of thrombin-induced platelet procoagulant activity. Thromb Haemost. 2001;86(4):1065–9.
21. Luo S-Z, Mo X, Afshar-Kharghan V, Srinivasan S, Lopez JA, Li R. Glycoprotein Ibα forms disulfide bonds with 2 glycoprotein Ibβ subunits in the resting platelet. Blood. 2007;109(2):603–9.
22. Andrews RK, Berndt MC, López JA. The glycoprotein Ib-IX-V complex. In Michelson AD, editor. Platelets. San Diego: Academic; 2006:145–63.
23. Ward CM, Andrews RK, Smith AI, Berndt MC. Mocarhagin, a novel cobra venom metalloproteinase, cleaves the platelet von Willebrand factor receptor glycoprotein Ibα. Identification of the sulfated tyrosine/anionic sequence Tyr-276-Glu-282 of glycoprotein Ibα as a binding site for von Willebrand factor and α-thrombin. Biochemistry. 1996;35(15):4929–38.
24. Cauwenberghs N, Vanhoorelbeke K, Vauterin S, et al. Epitope mapping of inhibitory antibodies against platelet glycoprotein Ibα reveals interaction between the leucine-rich repeat N-terminal and C-terminal flanking domains of glycoprotein Ibα. Blood. 2001;98(3):652–60.
25. Shen Y, Romo GM, Dong JF, et al. Requirement of leucine-rich repeats of glycoprotein (GP) Ibα for shear-dependent and static binding of von Willebrand factor to the platelet membrane GP Ib-IX-V complex. Blood. 2000;95(3):903–10.
26. Jurk K, Clemetson KJ, de Groot PG, et al. Thrombospondin-1 mediates platelet adhesion at high shear via glycoprotein Ib (GPIb): an alternative/backup mechanism to von Willebrand factor. FASEB J. 2003;17(11):1490–2.
27. Li CQ, Vindigni A, Sadler JE, Wardell MR. Platelet glycoprotein Ibα binds to thrombin anion-binding exosite II inducing allosteric changes in the activity of thrombin. J Biol Chem. 2001;276(9):6161–8.
28. Huizinga EG, Tsuji S, Romijn RAP, et al. Structures of glycoprotein Ibα and its complex with von Willebrand Factor A1 domain. Science. 2002;297(5584):1176–9.
29. Dumas JJ, Kumar R, Seehra J, Somers WS, Mosyak L. Crystal structure of the GPIbα-thrombin complex essential for platelet aggregation. Science.2003;301(5630): 222–6.
30. Bradford HN, Pixley RA, Colman RW. Human factor XII binding to the glycoprotein Ib-IX-V complex inhibits thrombin-induced platelet aggregation. J Biol Chem. 2000;275(30):22756–42.
31. Yun TH, Baglia FA, Myles T, et al. Thrombin activation of factor XI on activated platelets requires the interaction of factor XI and platelet glycoprotein Ibα with thrombin anion-binding exosites I and II, respectively. J Biol Chem. 2003;278(48):48112–9.
32. Joseph K, Nakazawa Y, Bahou WF, Ghebrehiwet B, Kaplan AP. Platelet glycoprotein Ib: a zinc-dependent binding protein for the heavy chain of high-molecular-weight kininogen. Mol Med. 1999;5(8):555–63.

33. Romo GM, Dong JF, Schade AJ, et al. The glycoprotein Ib-IX-V complex is a platelet counterreceptor for P-selectin. J Exp Med. 1999;190(6):803–14.
34. Simon DI, Chen Z, Xu H, et al. Platelet glycoprotein Ibα is a counterreceptor for the leukocyte integrin Mac-1 (CD11b/CD18). J Exp Med. 2000;192(2):193–204.
35. Arthur JF, Gardiner EE, Matzaris M, et al. Glycoprotein VI is associated with GPIb-IX-V on the membrane of resting and activated platelets. Thromb Haemost. 2005;93(4):716–23.
36. Cruz MA, Diacovo TG, Emsley J, Liddington R, Handin RI. Mapping the glycoprotein Ib-binding site in the von Willebrand factor A1 domain. J Biol Chem. 2000;275(25):19098–105.
37. Shen Y, Dong Jf JF, Romo GM, et al. Functional analysis of the C-terminal flanking sequence of platelet glycoprotein Ibα using canine-human chimeras. Blood. 2002;99(1):145–50.
38. Shen Y, Cranmer SL, Aprico A, et al. Leucine-rich repeats 2-4 (Leu60-Glu128) of platelet glycoprotein Ibα regulate shear-dependent cell adhesion to von Willebrand factor. J Biol Chem. 2006;281(36):26419–26423.
39. Dumas JJ, Kumar R, McDonagh T, et al. Crystal structure of the wild-type von Willebrand Factor A1-glycoprotein Iba complex reveals conformation differences with a complex bearing von Willebrand Disease mutations. J Biol Chem. 2004;279(22):23327–34.
40. McEwan PA, Andrews RK, Emsley J. Glycoprotein Ibα inhibitor complex structure reveals a combined steric and allosteric mechanism of von Willebrand factor antagonism. Blood. 2009;114(23):4883–5.
41. Yago T, Lou J, Wu T, et al. Platelet glycoprotein Ibα forms catch bonds with human WT vWF but not with type 2B von Willebrand disease vWF. J Clin Invest. 2008;118(9):3195–207.
42. Bergmeier W, Piffath CL, Goerge T, et al. The role of platelet adhesion receptor GPIbα far exceeds that of its main ligand, von Willebrand factor, in arterial thrombosis. Proc Natl Acad Sci U S A. 2006;103(45):16900–5.
43. Bergmeier W, Chauhan AK, Wagner DD. Glycoprotein Ibα and von Willebrand factor in primary platelet adhesion and thrombus formation: lessons from mutant mice. Thromb Haemost. 2008;99(2):264–70.
44. Ravanat C, Strassel C, Hechler B, et al. A central role of GPIb-IX in the procoagulant function of platelets that is independent of the 45-kDa GPIbα N-terminal extracellular domain. Blood. 2010;116(7):1157–64.
45. Boulaftali Y, Hess PR, Getz TM, et al. Platelet ITAM signaling is critical for vascular integrity in inflammation. J Clin Invest. 2013;123(2):908–16.
46. Owens AP, Mackman N. Microparticles in hemostasis and thrombosis. Circ Res. 2011;108(10):1284–97.
47. Gay LJ, Felding-Habermann B. Contribution of platelets to tumour metastasis. Nat Rev Cancer. 2011;11(2):123–34.
48. Jain S, Harris J, Ware J. Platelets: linking hemostasis and cancer. Arterioscler Thromb Vasc Biol. 2010;30(12):2362–7.
49. Goubran HA, Burnouf T, Radosevic M, El-Ekiaby M. The platelet-cancer loop. Eur J Internal Med. 2013 24(5):393–400.
50. Semple JW, Italiano JE, Freedman J. Platelets and the immune continuum. Nat Rev Immunol. 2011;11(4):264–74.
51. Clark SR, Ma AC, Tavener SA, et al. Platelet TLR4 activates neutrophil extracellular traps to ensnare bacteria in septic blood. Nat Med. 2007;13(4):463–9.
52. Renne T, Pozgajova M, Gruner S, et al. Defective thrombus formation in mice lacking coagulation factor XII. J Exp Med. 2005;202(2):271–81.
53. Chavakis T, Santoso S, Clemetson KJ, et al. High molecular weight kininogen regulates platelet-leukocyte interactions by bridging Mac-1 and glycoprotein Ib. J Biol Chem. 2003;278(46):45375–81.
54. Liu J, Gao B-B, Clermont AC, et al. Hyperglycemia-induced cerebral hematoma expansion is mediated by plasma kallikrein. Nat Med. 2011;17(2):206–10.
55. Morrissey JH, Choi SH, Smith SA. Polyphosphate: an ancient molecule that links platelets, coagulation, and inflammation. Blood. 2012;119(25):5972–9.

56. Shantsila E, Kamphuisen PW, Lip GYH. Circulating microparticles in cardiovascular disease: implications for atherogenesis and atherothrombosis. J Thromb Haemost. 2010;8(11):2358–68.
57. Morel O, Toti F, Hugel B, et al. Procoagulant microparticles. Arterioscler Thromb Vasc Biol. 2006;26(12):2594–604.
58. Ghosh A, Li W, Febbraio M, et al. Platelet CD36 mediates interactions with endothelial cell-derived microparticles and contributes to thrombosis in mice. J Clin Invest. 2008;118(5):1934–43.
59. Pluskota E, Woody NM, Szpak D, et al. Expression, activation, and function of integrin αMβ2 (Mac-1) on neutrophil-derived microparticles. Blood. 2008;112(6):2327–35.
60. De Haas CJC Weeterings C Vughs MM De Groot PG Van Strijp JA Lisman T. Staphylococcal superantigen-like 5 activates platelets and supports platelet adhesion under flow conditions, which involves glycoprotein Ibα and αIIbβ3. J Thromb Haemost. 2009;7(11):1867–74.
61. Hu H, Armstrong PCJ, Khalil E, et al. GPVI and GPIbα mediate Staphylococcal Superantigen-like protein 5 (SSL5) induced platelet activation and direct toward glycans as potential inhibitors. PLoS ONE. 2011;6(4):e19190.
62. Brinkmann V, Reichard U, Goosmann C, et al. Neutrophil extracellular traps kill bacteria. Science. 2004;303(5663):1532–5.
63. Brill A, Fuchs TA, Savchenko AS, et al. Neutrophil extracellular traps promote deep vein thrombosis in mice. J Thromb Haemost. 2012;10(1):136–44.
64. Ward CM, Tetaz TJ, Andrews RK, Berndt MC. Binding of the von Willebrand factor A1 domain to histone. Thromb Res. 1997;86(6):469–77.
65. Ezumi Y, Shindoh K, Tsuji M, Takayama H. Physical and functional association of the Src family kinases Fyn and Lyn with the collagen receptor glycoprotein VI-Fc receptor γ chain complex on human platelets. J Exp Med. 1998;188(2):267–76.
66. Jandrot-Perrus M, Busfield S, Lagrue A-H, et al. Cloning, characterization, and functional studies of human and mouse glycoprotein VI: a platelet-specific collagen receptor from the immunoglobulin superfamily. Blood. 2000;96(5):1798–807.
67. Suzuki-Inoue K, Tulasne D, Shen Y, et al. Association of Fyn and Lyn with the proline-rich domain of glycoprotein VI regulates intracellular signaling. J Biol Chem. 2002;277(24):21561–6.
68. Ruan CG, Du XP, Xi XD, Castaldi PA, Berndt MC. A murine antiglycoprotein Ib complex monoclonal antibody, SZ2, inhibits platelet aggregation induced by both ristocetin and collagen. Blood. 1987;69(2):570–7.
69. Ohlmann P, Hechler B, Ravanat C, et al. Ex vivo inhibition of thrombus formation by an anti-glycoprotein VI Fab fragment in non-human primates without modification of glycoprotein VI expression. J Thromb Haemost. 2008;6(6):1003–11.
70. Schulte V, Reusch HP, Pozgajova M, Varga-Szabo D, Gachet C, Nieswandt B. Two-phase antithrombotic protection after anti-glycoprotein VI treatment in mice. Arterioscler Thromb Vasc Biol. 2006; 26 (7): 1640–1647.
71. Gilio K, van Kruchten R, Braun A, et al. Roles of platelet STIM1 and Orai1 in glycoprotein VI- and thrombin-dependent procoagulant activity and thrombus formation. J Biol Chem. 2010;285(31):23629–38.
72. Heemskerk JW, Siljander P, Vuist WM, et al. Function of glycoprotein VI and integrin α2β1 in the procoagulant response of single, collagen-adherent platelets. Thromb Haemost. 1999;81(5):782–92.
73. Siljander P, Farndale RW, Feijge MAH, et al. Platelet adhesion enhances the glycoprotein VI-dependent procoagulant response: Involvement of p38 MAP kinase and calpain. Arterioscler Thromb Vasc Biol. 2001;21(4):618–27.
74. Seizer P, Borst O, Langer HF, et al. EMMPRIN (CD147) is a novel receptor for platelet GPVI and mediates platelet rolling via GPVI-EMMPRIN interaction. Thromb Haemost. 2009;101(4):682–6.
75. Schulz C, von Bruhl ML, Barocke V, et al. EMMPRIN (CD147/basigin) mediates platelet-monocyte interactions in vivo and augments monocyte recruitment to the vascular wall. J Thromb Haemost. 2011;9(5):1007–19.

76. Coller BS, Kalomiris E, Steinberg M, Scudder LE. Evidence that glycocalicin circulates in normal plasma. J Clin Invest. 1984;73(3):794–9.
77. Gardiner EE, Arthur JF, Kahn ML, Berndt MC, Andrews RK. Regulation of platelet membrane levels of glycoprotein VI by a platelet-derived metalloproteinase. Blood. 2004;104(12):3611–7.
78. Al-Tamimi M, Arthur JF, Gardiner EE, Andrews RK. Focusing on plasma glycoprotein VI. Thromb Haemost. 2012;107(4):648–55.
79. Al-Tamimi M, Mu FT, Moroi M, Gardiner EE, Berndt MC, Andrews RK. Measuring soluble platelet glycoprotein VI in human plasma by ELISA. Platelets. 2009;20(3):143–9.
80. Al-Tamimi M, Gardiner EE, Thom JY, et al. Soluble glycoprotein VI is raised in the plasma of patients with acute ischemic stroke. Stroke. 2011;42(2):498–500.
81. Bigalke B, Potz O, Kremmer E, et al. Sandwich immunoassay for soluble glycoprotein VI in patients with symptomatic coronary artery disease. Clin Chem. 2011;57(6):898–904.
82. Al-Tamimi M, Tan C, Qiao J, et al. Pathological shear triggers shedding of vascular receptors: a novel mechanism for downregulation of platelet glycoprotein (GP)VI in stenosed coronary vessels. Blood. 2012;119(18):4311–20.
83. Gardiner EE, Karunakaran D, Arthur JF, et al. Dual ITAM-mediated proteolytic pathways for irreversible inactivation of platelet receptors: De-ITAM-izing FcγRIIa. Blood. 2008;111(1):165–74.
84. Gardiner EE, Al-Tamimi M, Mu FT, et al. Compromised ITAM-based platelet receptor function in a patient with immune thrombocytopenic purpura. J Thromb Haemost. 2008;6(7):1175–82.
85. Reilly MP, Taylor SM, Hartman NK, et al. Heparin-induced thrombocytopenia/thrombosis in a transgenic mouse model requires human platelet factor 4 and platelet activation through FcγRIIA. Blood. 2001;98(8):2442–7.
86. Warkentin TE, Hayward CP, Boshkov LK, et al. Sera from patients with heparin-induced thrombocytopenia generate platelet-derived microparticles with procoagulant activity: an explanation for the thrombotic complications of heparin-induced thrombocytopenia. Blood. 1994;84(11):3691–9.
87. Tardy-Poncet B, Piot M, Chapelle C, et al. Thrombin generation and heparin-induced thrombocytopenia. J Thromb Haemostas. 2009;7(9):1474–81.
88. Andrews RK, Du X, Berndt MC. The 14-3-3ζ-GPIb-IX-V complex as an antiplatelet target. Drug News Perspect. 2007;20(5):285–92.
89. Clemetson KJ, Clemetson JM. Platelet GPIb complex as a target for anti-thrombotic drug development. Thromb Haemost. 2008;99(3):473–9.
90. Jackson SP. Arterial thrombosis-insidious, unpredictable and deadly. Nat Med. 2011;17(11):1423–36.
91. Wei AH, Schoenwaelder SM, Andrews RK, Jackson SP. New insights into the haemostatic function of platelets. Br J Haematol. 2009;147(4):415–30.

Part III
Clinical Applications

Translating our understanding of blood cell systems into novel strategies for improving clinical treatment of disease will require the development and integration of novel experimental, computational, and conceptual tools. This section introduces the reader to contemporary approaches and pressing challenges at this frontier through both general overviews and in-depth applications of such methodologies to specific clinical objectives.

In Chap. 14, Lei and Mackey apply the tools of systems biology to the clinical challenge of treating cytopenia. The authors first use computational modeling to build understanding of patient responses to treatment, and these prediction are then extended to guide the design of novel therapeutic agents and strategies for their administration. In Chap. 15, Tomasetti provide an overview of the use of mathematical modeling to understand drug resistance and then explore a specific case study describing the dynamics of drug resistance in leukemia. In Chap. 16, Radivoyevitch, Li and Sachs use stochastic models to explore several processes in leukemia. They first explore the potential role of stem-like cancer cells in disease progression and consider potential consequences for therapeutic dosing and then consider a model by which chromosomal lesions may lead to disease progression through a series of coordinated stochastic steps occurring over a wide range of time scales. In Chap. 17, Stiehl, Ho, and Marciniak-Czochra develop a dynamic model describing engraftment and recovery of normal hematopoiesis following stem cell transplantation. They use this model to identify mechanisms and therapeutic strategies predicted to reduce risks of graft failure and minimize time required to generate normal blood cell populations. Finally, in Chap. 18, Dudek, Chuang, and Leonard discuss the prospect of engineering human cells to carry out customized therapeutic objectives as an leading front in the field of personalized, design-driven medicine. They illustrate this paradigm by considering the historical development, recent progress, and current challenges facing the use of engineered T cells for immunotherapy of cancer. Furthermore, they consider how the emerging field of synthetic biology may dovetail with advances in systems biology to accelerate the generation and development of novel therapeutic strategies and technologies.

Seth Joel Corey, MD, MPH
Chicago, IL
Mark Kimmel, PhD
Houston, TX
Joshua N. Leonard, PhD
Evanston, IL

Chapter 14
Understanding and Treating Cytopenia Through Mathematical Modeling

Jinzhi Lei and Michael C. Mackey

Abstract Here, we briefly review how the study of dynamic hematological diseases with mathematical modeling tools has led to a better understanding of the origin of some types of neutropenia and thrombocytopenia and to improved treatment strategies. In addition, we have briefly discussed how these models suggest improved ways to minimize and/or treat cytopenia induced by chemotherapy.

Keywords Anemia · Chemotherapy · Granulocyte colony-stimulating factor · Neutropenia · Thrombocytopenia · Thrombopoietin

Introduction

All blood cells arise from a common origin in the bone marrow, the hematopoietic stem cells (HSC). HSC are morphologically undifferentiated cells which can either proliferate or differentiate to produce all types of blood cells (erythrocytes, neutrophils, and platelets). The proliferation of the stem cells and progenitor cells is controlled by a negative feedback system mediated by hematopoietic cytokines. Erythropoietin (EPO) is the hormone that mediates the red blood cell (RBC) production, granulocyte colony-stimulating factor (G-CSF) controls the regulation of neutrophils, and thrombopoietin (TPO) known as c-mpl ligand or megakaryocyte growth and development factor is the primary regulator of thrombopoicsis.

Hematopoiesis is a homeostatic system and, consequently, most disorders of its regulation lead to transient or chronic failures in the production of either all or only one blood cell type. Among the wide range of diseases affecting the blood

M. C. Mackey (✉)
Department of Physiology and CAMBAM, Room 1124, 3655 Promenade Sir William Osler, Monteral, QC H3G 1Y6, Canada
Tel: 514-398-4336
e-mail: michael.mackey@mcgill.ca

J. Lei
Zhou Pei-Yuan Center for Applied Mathematics, Tsinghua University, Beijing 100084, China
Tel: 86-10-62795156
e-mail: jzlei@tsinghua.edu.cn

S. J. Corey et al. (eds.), *A Systems Biology Approach to Blood*,
Advances in Experimental Medicine and Biology 844, DOI 10.1007/978-1-4939-2095-2_14

cells, there are some which are characterized by predictable oscillations in one or more cellular elements of the blood. They are called periodic or dynamical diseases [1]. The investigation of their dynamic character offers an opportunity to enrich our knowledge about the regulatory processes controlling blood cell production and may suggest better therapeutic strategies [2]. Cyclical neutropenia (CN) [2–3], periodic chronic myelogenous leukemia (CML) [3, 7], periodic autoimmune hemolytic anemia (AIHA) [8], and cyclical thrombocytopenia (CT) [9, 10] are some classical examples of dynamical hematological diseases. Diseases like periodic chronic myelogenous leukemia (PCML) and CN, which involve fluctuations in all major blood cell lines with the same period on a given subject, are believed to arise in the stem cell compartment in the bone marrow. Since in CT or periodic AIHA, besides oscillations in one type of cell count, the patient hematological profile has been consistently normal, a destabilization of a peripheral control mechanism might play an important role in the genesis of these disorders.

Dynamic Hematological Disease

Cyclical Neutropenia

Neutrophils are critical for the immune response, and low absolute neutrophil counts (ANC) in the blood can lead to infections. Neutropenia is a term that designates a low number of neutrophils, thus indicating that the individual is less effective at fighting infections. The severity of neutropenia in patients can be classified based on the ANC [11]: mild neutropenia ($1.0 \leq \text{ANC} < 1.5 \times 10^9$ cells/L) with minimal risk of infection, moderate neutropenia ($0.5 \leq \text{ANC} < 1.0 \times 10^9$ cells/L) with moderate risk of infection, and severe neutropenia ($\text{ANC} < 0.5 \times 10^9$ cells/L) with severe risk of infection. Patients with severe neutropenia are often seen with symptoms such as mouth ulcers, fever, pharyngitis, sinusitis, otitis, and other infections, some of which can sometimes be life threatening.

CN is characterized by oscillations in the number of neutrophils from normal to very low levels (less than 0.5×10^9 cells/L, also called severe neutropenia). The period of these oscillations is usually around 3 weeks for humans, although periods up to 45 days have been observed [6]. One major characteristic of CN is that the oscillations are present not only in neutrophils but also in platelets, monocytes, and reticulocytes [4]. For CN patients, the period of severe neutropenia usually lasts for about 3–5 days within every 3-week period [5, 12].

CN also occurs in gray collies with the same characteristics as in humans and with oscillation periods on the order of 11–16 days [4, 6, 13]. This animal model has provided extensive experimental data that would be difficult, if not impossible, to obtain in humans.

CN was first reported as an inherited disease by Reimann [14], and later confirmed in studies of Australian families by Morley et al. [15]. In families, the severity of symptoms and the severity of neutropenia may vary. Furthermore, the disease is more severe in children and is ameliorated by unknown factors as they grow older

[16]. Through family studies and linkage analysis, mutations in the gene encoding neutrophil elastase (ELA2) are recognized as causing cyclic neutropenia [17].

Cyclical Thrombocytopenia

Platelets are blood cells whose function is to take part in the clotting process, and the term thrombocytopenia denotes a reduced platelet (thrombocyte) count. In CT, platelet counts oscillate generally from very low values (1×10^9 cells/L) to normal ($150-450 \times 10^9$ platelets/L blood) or above normal levels (2000×10^9 cells/L) [9]. These oscillations have been observed with periods varying between 20 and 40 days [18]. In addition, patients may exhibit a variety of clinical symptoms indicative of impaired coagulation such as purpura, petechiae, epistaxis, gingival bleeding, menorrhagia, easy bruising, possibly premenstrually, and gastrointestinal bleeding [9]. There are two proposed origins of CT. One is of autoimmune origin and most prevalent in females. The other is of amegakaryocytic origin, more common in males.

Autoimmune CT is characterized by a shortened platelet lifespan at the time of decreasing platelet counts [19]. This is consistent with normal to high levels of bone marrow megakaryocytes and with an increased destruction rate of circulating platelets [9]. Autoimmune CT has also been postulated to be a rare form of idiopathic (immune) thrombocytopenic purpura (ITP) [19].

The amegakaryocytic form of CT is characterized by oscillations in bone marrow megakaryocytes preceding the platelet oscillations [20–23]. In this second type of CT, platelet oscillations are thought to be due to a cyclical failure in platelet production [18, 21–25]. The platelet lifespan is usually normal [25] and antibodies against platelets are not detected [24]. Although it has been suggested that the failure of platelet production could arise at the stem cell level [26], it is generally thought that the cycling originates at the megakaryocyte level [22, 24]. For a more detailed review of CT, see [9, 27].

It has been hypothesized that autoimmune and amegakaryocytic CT have different dynamic origins [27]. This is supported by Swinburne and Mackey [9], who noted that the patients diagnosed as having the autoimmune CT generally have shorter periods (13–27 days) than those classified as amegakaryocytic (27–65 days). Moreover, they reported that autoimmune patients typically show platelet oscillations from low to normal levels, whereas amegakaryocytic subjects generally show oscillations from above normal to below normal levels of platelets.

Periodic Anemia

Examples of periodic anemia are relatively rare in the clinical literature, though there are a few well-documented examples [28, 29]. Although periodic fluctuations of erythroid precursors in the bone marrow are well documented in CN and some cases of periodic leukemia (see below), the rarity of reports of actual periodic anemia is presumably due to the extremely long lifespan of circulating erythrocytes in humans.

There are, however, well-documented examples of cyclical anemia in mice following either the administration of a single dose of 89Sr [30–32] or after whole-body irradiation [33, 34].

AIHA results from an abnormality of the immune system that produces autoantibodies, which attack red blood cells (RBC) as if they were substances foreign to the body. It leads to an abnormally high destruction rate of the red blood cells. Periodic AIHA is a rare form of hemolytic anemia in humans [28] characterized by oscillatory erythrocyte numbers about a depressed level. The origin of the disease is unclear. Periodic AIHA, with a period of 16–17 days in hemoglobin and reticulocyte counts, has been induced in rabbits by using RBC autoantibodies [35]. Mackey [36] showed that the laboratory characteristics of this induced disorder were consistent with model predictions using a mathematical formulation like those explored in the section "Mathematical Model Development".

Periodic Leukemia

Leukemia is a cancer of the blood or bone marrow characterized by an abnormal proliferation of blood cells, usually leucocytes. CML is distinguished from other leukemias by the presence of a genetic abnormality in blood cells, called the Philadelphia chromosome, which is a translocation between chromosomes 9 and 22 that leads to the formation of the BcrAbl fusion protein [37]. This protein is thought to be responsible for the dysfunctional regulation of myelocyte growth and other features of CML [38]. (For more details about CML, see [39]).

A dynamical disease of particular interest is PCML, characterized by oscillations in circulating cell numbers that occur primarily in leucocytes, but may also occur in the platelets and reticulocytes [7]. The leucocyte count varies periodically, typically between values of 30 and 200×10^9 cells/L, with a periods ranging from 40 to 80 days. In addition, oscillation of platelets and reticulocytes may occur with the same period as the leucocytes, around normal or elevated numbers [7, 40]. As in cyclical neutropenia, the hypothesis that the disease originates from the stem cell compartment is supported by the presence of oscillations in more than one cell lineage.

Mathematical Model Development

As is clear from the preceding section describing the periodic cytopenias, the hematopoietic system is capable of displaying interesting dynamical properties in pathophysiological situations. These dynamics have been instrumental in guiding the development of a multiplicity of mathematical models of hematopoietic dynamics. Many of these have been reviewed in [41] and [42].

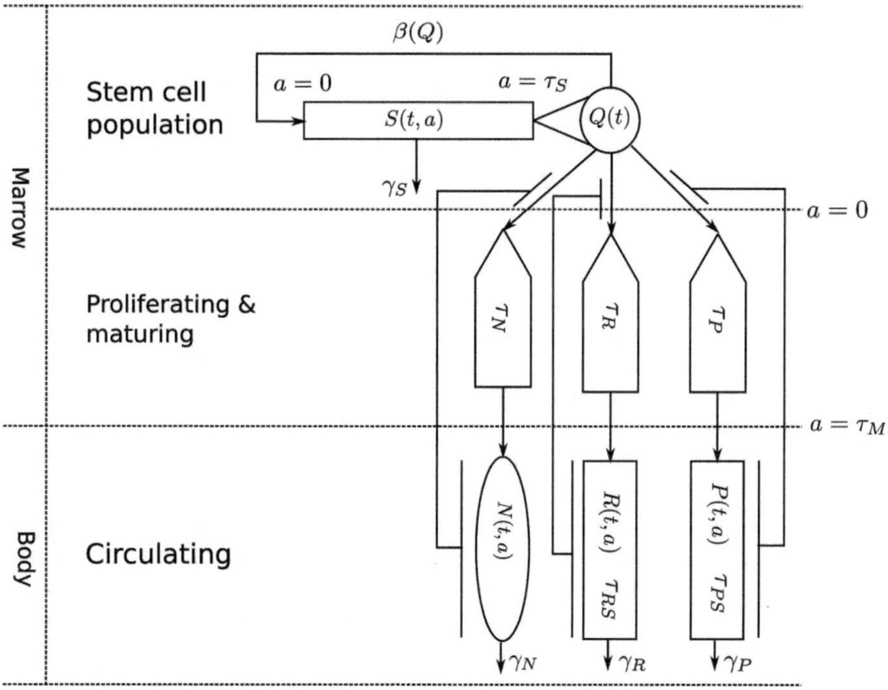

Fig. 14.1 A cartoon representation of the age-structured model of hematopoiesis. See text for details and notations. (Adapted from [44])

Model Description

Although the regulation of blood cell production is complicated [4, 43], and its understanding constantly evolving, the broad outlines are clear. Fig. 14.1 contains a cartoon representation of hematopoiesis.

There are four linages, including the hemopoietic stem cells and three differentiated cell lines, leukocytes, erythrocytes, and platelets.

Hemopoietic stem cells are classified as either proliferating or resting phase [45]. The proliferating stem cells undergo mitosis at a fixed time after entry into that state, and are lost randomly during the proliferating phase [46]. Each normal cell generates two resting-phase cells at the end of mitosis. The resting-phase cells can either reenter the proliferative phase at a rate that involves a negative feedback, or develop to mixed myeloid progenitor cell, which further differentiate into precursors of any of the three cell lines, leukocytes (white blood cells (WBC)), erythrocytes (RBC), or thrombocytes (platelets). The rates of differentiation into these three lines depend on the numbers of circulating cells of the relevant type that encapsulate the feedback between the circulating cell numbers and the production. The feedback is always negative so that when the number of circulating mature cells of a given line falls, the relevant differentiation rate has a corresponding compensatory increase.

There are two stages for each of the circulating cell lines after the differentiation, first the amplification/maturation of precursor cells in the bone marrow, and next circulation of mature cells throughout the whole body. In the stage of amplification/maturation, the precursor cells undergo many stages of cell division and randomly die so that the number of precursors increases rapidly in a period: one stem cell is capable of producing about 10^6 mature blood cells after 20 cell divisions [47]. In the circulation stage, mature blood cells are removed randomly at a fixed rate. In addition, the circulating erythrocytes and platelets are actively destroyed at a fixed time from the entry into the circulating compartment [48, 49].

The proliferation and differentiation of hematopoietic cells and the function of mature blood cells are regulated by a variety of cytokines, including EPO [50], which mediates the regulation of erythrocyte production, G-CSF [51], which regulates neutrophil number, and TPO [52, 53], which regulates production of platelets and other cell lineages.

For the red blood cells, EPO mediates a negative feedback loop that helps to regulate erythrocyte production [50, 54]. A decrease in the numbers of circulating erythrocytes leads to a decrease in tissue pO_2 levels, which in turn triggers the production and release of EPO by renal macula cells. This elevation of EPO increases the net production of primitive erythroid precursors (colony-forming units-erythroid, CFU-E) and, ultimately, an increase in the number of circulating erythrocytes and hence the tissue pO_2 levels.

The regulation of platelet production (thrombopoiesis) involves similar feedback mechanisms mediated by TPO [55]. A decrease in circulating platelet counts results in an increased level of TPO, which then stimulates maturation of the platelet progenitor cells (CFU-megakaryocyte (CFU-Meg)), and eventually leads to an increase in platelet production.

There are three types of leucocytes, namely the lymphocytes, the granulocytes, and the monocytes. Here, we focus on granulopoiesis (production of granulocytes) and more specifically on neutrophils. The mechanisms of regulating granulopoiesis involve G-CSF, which is the main regulator of neutrophil production [56]. It stimulates the formation of neutrophils from HSC, accelerates the formation of neutrophils in bone marrow, and stimulates their release from the bone marrow into the blood. Although the exact mechanisms by which G-CSF acts is still unclear, it has been shown that the neutrophils regulate their own production through a negative feedback [51]: An increase (decrease) in the number of circulating neutrophils would induce a decrease (increase) in the production of neutrophils throughout the adjustment of the G-CSF levels. Several studies have shown an inverse relationship between the serum levels of G-CSF and the number of circulation neutrophils [57–60].

Formulation

In the past several decades, different mathematical tools have been used in hematological modeling, including differential equations (partial, ordinarily, or delay), stochastic processes, Boolean networks, Bayesian theory, multivariate statistics, decision trees, etc. For reviews, see [41, 61, 62]. The choice of the mathematical tools often depends on the desired level of description of the model. Here, we focus on models that originate from age-structured models, which offer a natural way to model hematopoietic dynamics, and are widely used in the study of dynamical blood diseases [3, 10, 44, 63–65].

Age-Structured Model

We refer the model illustrated in Fig. 14.1. Variables used in the following equations and typical values for hematologically normal individuals are summarized in Table 14.1.

Let $Q(t)$ (cells/kg) denote the population of resting-phase stem cells and $s(t,a)$ (cells/kg) the population of stem cells in the proliferating phase, with age $a = 0$ for their time of entry into the proliferative state. For the other three cell lines, let $n(t,a), r(t,a)$, and $p(t,a)$ (cells/kg) represent the populations of leukocytes, erythrocytes, and platelets, respectively, with age $a = 0$ for the time point of differentiating from stem cells. Let

$$N(t) = \int_{\tau_N}^{+\infty} n(t,a)\mathrm{d}a, \quad R(t) = \int_{\tau_R}^{\tau_{R_{\mathrm{sum}}}} r(t,a)\mathrm{d}a, \quad P(t) = \int_{\tau_P}^{\tau_{P_{\mathrm{sum}}}} p(t,a)\mathrm{d}a \quad (14.1)$$

which are the populations of circulating cells. Hereinafter, we set

$$\tau_{R_{\mathrm{sum}}} = \tau_R + \tau_{RS}, \quad \tau_{P_{\mathrm{sum}}} = \tau_P + \tau_{PS}$$

Moreover, the differentiation rates κ_N, κ_R, and κ_P are assumed to depend on circulating cell populations through negative feedback functions denoted by $\kappa_N(N), \kappa_R(R)$, and $\kappa_P(P)$, respectively.

Using the above notation, the age-structured model of hematopoiesis is then described by the following partial differential equations [44]:

$$\begin{aligned}
&\nabla s(t,a) = -\gamma_S(t)s(t,a) \qquad (t > 0,\ 0 \le a \le \tau_S)\\
&\frac{\mathrm{d}Q}{\mathrm{d}t} = 2s(t,\tau_S) - (\beta(Q) + \kappa_N(N) + \kappa_R(R) + \kappa_P(P))Q \qquad (t > 0)\\
&\nabla n(t,a) = \begin{cases} \eta_N(t,a)n(t,a) \ (t > 0,\ 0 \le a \le \tau_N) \\ -\gamma_N(t,a)n(t,a) \quad (t > 0,\ \tau_N \le a) \end{cases}\\
&\nabla r(t,a) = \begin{cases} \eta_R(t,a)r(t,a) \quad (t > 0,\ 0 \le a \le \tau_R) \\ -\gamma_R(t,a)r(t,a) \ (t > 0,\ \tau_R \le a \le \tau_{R_{\mathrm{sum}}}) \end{cases}\\
&\nabla p(t,a) = \begin{cases} \eta_P(t,a)p(t,a) \quad (t > 0,\ 0 \le a \le \tau_P) \\ -\gamma_P(t,a)p(t,a) \ (t > 0,\ \tau_P \le a \le \tau_{P_{\mathrm{sum}}}) \end{cases}
\end{aligned} \qquad (14.2)$$

Table 14.1 Variables used in the model equations and typical value for hematologically normal individuals [44]

Variable	Definition	Value	Unit
Stem cell compartment			
$Q(t)$	Population of resting-phase stem cells	1.12	10^6 cells/kg
$s(t,a)$	Population of proliferating-phase stem cells	–	cells/kg
β	Rate of reentering the proliferative phase	0.0433	day^{-1}
τ_S	Duration of mitosis	2.83	days
γ_S	Apoptosis rate of proliferating stem cells	0.1013	day^{-1}
Neutrophil compartment			
$n(t,a)$	Population of neutrophils	–	cells/kg
$N(t)$	Population of circulating neutrophils	5.59	10^8 cells/kg
κ_N	Differentiation rate from stem cells to neutrophils	0.0077369	day^{-1}
η_N	Amplification rate of neutrophil precursor cells	2.2887	day^{-1}
τ_N	Duration of neutrophil precursor amplification/maturation	12.6	days
γ_N	Apoptosis rate of circulating neutrophils	2.4	day^{-1}
Erythrocyte compartment			
$r(t,a)$	Population of erythrocytes	–	cells/kg
$R(t)$	Population of circulating erythrocytes	3.5	10^{11} cells/kg
κ_R	Differentiation rate from stem cells to erythrocytes	0.005271	day^{-1}
η_R	Amplification rate of erythrocyte precursor cells	1.8114	day^{-1}
τ_R	Duration of erythrocyte precursor amplification/maturation	6	days
γ_R	Apoptosis rate of circulating erythrocytes	0.001	day^{-1}
τ_{RS}	Lifetime of circulating erythrocytes	120	days
$\tau_{R_{\mathrm{sum}}}$	$\tau_R + \tau_{RS}$	126	days
Platelet compartment			
$p(t,a)$	Population of platelets	–	cells/kg
$P(t)$	Population of circulating platelets	1.3924	10^{10} cells/kg
κ_P	Differentiation rate from stem cells to platelets	0.0087074	day^{-1}
η_P	Amplification rate of platelet precursor cells	1.7928	day^{-1}
τ_P	Duration of platelet precursor amplification/maturation	7	days
γ_P	Apoptosis rate of circulating platelets	0.15	day^{-1}
τ_{PS}	Lifetime of circulating platelets	9.5	days
$\tau_{P_{\mathrm{sum}}}$	$\tau_P + \tau_{PS}$	16.5	days

Table 14.2 Parameters for the Hill functions (3) [44]

Parameter name	Value	Unit
Function $\beta(Q)$		
k_0	8.0	day^{-1}
θ_2	0.0826	10^6 cells/kg
s_2	2	(none)
Function $\kappa_N(N)$		
f_0	0.154744	day^{-1}
θ_1	0.2942	10^8 cells/kg
s_1	1	(none)
Function $\kappa_R(R)$		
$\bar{\kappa}_r$	1.23744	day^{-1}
K_r	0.0382	$(10^{-11}\ \text{cells / kg})^{-S_3}$
s_3	6.96	day^{-1}
Function $\kappa_P(P)$		
$\bar{\kappa}_p$	0.2802	day^{-1}
K_p	20.343	$(10^{10}\ \text{cells / kg})^{-S_4}$
s_4	1.29	day^{-1}

where

$$\nabla = \frac{\partial}{\partial t} + \frac{\partial}{\partial a}$$

The negative feedback functions are represented by Hill functions [64]:

$$\kappa_N(N) = f_0 \frac{\theta_1^{s_1}}{\theta_1^{s_1} + N^{s_1}},\ \beta(Q) = k_0 \frac{\theta_2^{s_2}}{\theta_2^{s_2} + Q^{s_2}},\ \kappa_R(R) = \frac{\bar{\kappa}_r}{1 + K_r R^{s_3}},$$
$$\kappa_P(P) = \frac{\bar{\kappa}_p}{1 + K_p P^{p_4}} \tag{14.3}$$

Typical parameter values are given in Table 14.2.

The boundary conditions at $a = 0$ are given by

$$\begin{aligned} s(t,0) &= \beta(Q(t))Q(t), \\ n(t,0) &= \kappa_N(N(t))Q(t), \qquad (t \geq 0) \\ r(t,0) &= \kappa_R(R(t))Q(t), \\ p(t,0) &= \kappa_P(P(t))Q(t), (t \geq 0) \end{aligned} \tag{14.4}$$

according to the negative feedback loops. Moreover, we have

$$\lim_{a\to\infty} n(t,a) = 0. \tag{14.5}$$

The initial conditions are

$$\begin{aligned} s(0,a) &= g_S(a), && (0 \le a \le \tau_S) \\ Q(0) &= Q_0 \\ n(0,a) &= g_N(a), && (0 \le a \le +\infty) \\ r(0,a) &= g_R(a), && (0 \le a \le \tau_{R_{\text{sum}}}) \\ p(0,a) &= g_P(a), && (0 \le a \le \tau_{P_{\text{sum}}}) \end{aligned} \tag{14.6}$$

where $g_S(a), g_N(a), g_R(a)$, and $g_P(a)$ give the initial population distributions of proliferating-phase stem cells, and the precursors of neutrophils, erythrocytes, and platelets, respectively.

Equations (14.1)–(14.6) define the initial boundary value problem for the age-structured model of hematopoietic regulation, and is the basis of the following simplified model and analysis.

Delay Differential Equation Model

In hematological modeling, we are mainly interested at the dynamics of circulating blood cell populations $N(t)$, $R(t)$, and $P(t)$. This can be modeled by delay differential equations obtained from the above age-structured model. We assume the apoptosis rates $\gamma_N, \gamma_R, \gamma_P$ are constants. Using the method of characteristics with Eq. (14.2), and using the boundary conditions Eq. (14.4) and Eq. (14.5), we obtain the following equations when $t > \tau_{\max} = \max\{\tau_S, \tau_N, \tau_{R_{\text{sum}}}, \tau_{P_{\text{sum}}}\}$:

$$\begin{aligned} \frac{dQ}{dt} &= 2e^{-\tau_S \hat{\gamma}_S(t-\tau_S)} \beta(Q_{\tau_S}) Q_{\tau_S} - (\beta(Q) + \kappa_N(N) + \kappa_R(R) + \kappa_P(P))Q, \\ \frac{dN}{dt} &= -\gamma_N N + e^{\tau_N \hat{\eta}_N(t-\tau_N)} \kappa_N(N_{\tau_N}) Q_{\tau_N}, \\ \frac{dR}{dt} &= -\gamma_R R + e^{\tau_R \hat{\eta}_R(t-\tau_R)} \kappa_R(R_{\tau_R}) Q_{\tau_R} - e^{-\gamma_R \tau_{RS}} e^{\tau_R \hat{\eta}_R(t-\tau_{R\text{sum}})} \kappa_R(R_{\tau_{R_{\text{sum}}}}) Q_{\tau_{R_{\text{sum}}}}, \\ \frac{dP}{dt} &= -\gamma_P P + e^{\tau_P \hat{\eta}_P(t-\tau_P)} \kappa_P(P_{\tau_P}) Q_{\tau_P} - e^{-\gamma_P \tau_{PS}} e^{\tau_P \hat{\eta}_P(t-\tau_{P\text{sum}})} \kappa_P(P_{\tau_{P_{\text{sum}}}}) Q_{\tau_{P_{\text{sum}}}}, \end{aligned} \tag{14.7}$$

where

$$\gamma_S = \frac{1}{\tau_S} \int_0^{\tau_S} \gamma_S(t+s)ds, \quad \hat{\eta}_k(t) = \frac{1}{\tau_k} \int_0^{\tau_k} \eta_k(t+s,s)ds, \qquad (k = N, R, P) \tag{14.8}$$

Here, the subscripts on the dependent variables indicate delayed arguments, i.e., $Q_{\tau_S} = Q(t - \tau_S)$.

The delay differential Eq. (14.7) determines the dynamic behavior for the circulating blood cell populations. Here, we note that when $t < \tau_{\max}$, Eq. (14.7) is not equivalent to the original age-structured model Eq. (14.2). In this case, the initial conditions of Eq. (14.6) have to be involved in the dynamical equation. Refer to [44] for a detailed discussion of this point.

For hematologically normal individuals, we assumed the apoptosis rate γ_S and amplification rates $\eta_k, (k = N, R, P)$ are constants, and hence $\hat{\gamma}_S = \gamma_S, \hat{\eta}_k = \eta_k$, $(k = N, R, P)$. Thus, we obtain the following delay differential equations:

$$\begin{aligned}
\frac{\mathrm{d}Q}{\mathrm{d}t} &= 2\mathrm{e}^{-\tau_S\gamma_S}\beta(Q_{\tau_S})Q_{\tau_S} - (\beta(Q) + \kappa_N(N) + \kappa_R(R) + \kappa_P(P))Q, \\
\frac{\mathrm{d}N}{\mathrm{d}t} &= -\gamma_N N + \mathrm{e}^{\tau_N\eta_N}\kappa_N(N_{\tau_N})Q_{\tau_N}, \\
\frac{\mathrm{d}R}{\mathrm{d}t} &= -\gamma_R R + \mathrm{e}^{\tau_R\eta_R}\left(\kappa_R(R_{\tau_R})Q_{\tau_R} - \mathrm{e}^{-\gamma_R\tau_{RS}}\kappa_R\left(R_{\tau_{R\mathrm{sum}}}\right)_{\tau_{R\mathrm{sum}}}\right), \\
\frac{\mathrm{d}P}{\mathrm{d}t} &= -\gamma_P P + \mathrm{e}^{\tau_P\eta_P}\left(\kappa_P(P_{\tau_P})Q_{\tau_P} - \mathrm{e}^{-\gamma_P\tau_{PS}}\kappa_P\left(P_{\tau_{P\mathrm{sum}}}\right)Q_{\tau_{P\mathrm{sum}}}\right).
\end{aligned} \tag{14.9}$$

Eq. (14.9) was first presented in [64], and has been used to study different types of dynamical blood diseases [3, 10, 44, 64].

To study the effect of clinical treatments, such as chemotherapy and G-CSF administration, which are known to affect hematopoiesis in the bone marrow, we further divide the amplification/maturation compartment of each cell line into two sub-compartments, corresponding to amplification and maturation, respectively. Let

$$\tau_k = \tau_{kP} + \tau_{kM}, \qquad (k = N, R, P) \tag{14.10}$$

where τ_{kP} is the duration of the amplification stages, and τ_{kM} is the time for maturation. The amplification rates $\eta_k(k = N, R, P)$ are defined separately in the two stages:

$$\eta_k(t, a) = \begin{cases} \eta_{kP}(t) & 0 \le a \le \tau_{kP} \\ -\gamma_{kM}(t) & \tau_{kP} \le a \le \tau_k \end{cases} \tag{14.11}$$

where η_{kP} is the amplification rate in the amplification stage and γ_{kM} is the apoptosis rate in the maturation stage, and are assumed to be independent of the age a. Thus, $\hat{\eta}_k$ defined by (8) can be written as

$$\hat{\eta}_k = \frac{1}{\tau_k}\left[\int_0^{\tau_{kP}} \eta_{kP}(t + s)\mathrm{d}s - \int_{\tau_{kP}}^{\tau_k} \gamma_{kM}(t + s)\mathrm{d}s\right]. \tag{14.12}$$

For hematologically normal individuals whose rates η_{kP} and γ_{kM} are constant, we have

$$\eta_k = (\eta_{kP}\tau_{kP} - \gamma_{kM}\tau_{kM})/\tau_k, \quad (k = N, R, P). \tag{14.13}$$

Parameters for the neutrophil compartment can be referred to [66], and parameters for the erythrocyte compartment and platelet compartment are not known yet.

Cyclical Neutropenia

Modeling of Cyclical Neutropenia

Although it is a rare disorder, CN is probably the most extensively studied periodic hematological disorder due to its interesting dynamics and its clinical and laboratory manifestations. A number of mathematical models have been put forward in an attempt to understand this disorder, and these fall into two major categories according to the origin of CN (see Fig. 14.1 to place them in perspective). For other reviews, see [4, 41, 67–69].

The first group of models identifies the origin of CN with a loss of stability in the peripheral control loop. Typical examples are [70–85], all of which have postulated an alteration in the feedback on immature precursor production from the mature cell population number. However, the work of [86] casts doubt on this explanation, by showing that any alternations of parameters in the peripheral control system consistent with the extant laboratory and clinical data on CN are unable to reproduce either the characteristics of clinical CN or its laboratory counterpart in the gray collie [43, 86].

The second group of models builds upon the existence of oscillations in many of the peripheral cellular elements (neutrophils, platelets, and erythroid precursors, see Fig. 14.1) and postulates that the origin of CN is in the common HSC population. Mackey [45] has suggested that the oscillations originate in a loss of stability in the HSC. This hypothesis allowed the quantitative calculation of the period of oscillation when the stability was lost due to an abnormally large-cell apoptosis rate within the proliferating compartment. Some mathematical models coupled a stem cell compartment with the peripheral loop for granulocytes [4, 6, 87], whereas others present a more complex model showing the stem cells coupled to all major cell lines [3, 44, 88]. For recent reviews, see [67] and [65].

Here, we introduce several models, from simple to sophisticated, that have given significant insight into the origin and dynamical features of CN. Then, we show how these models have been used to understand and improve the effects of CN treatments.

Origin of CN

Mackey [45] presented the following delay differential equation

$$\frac{\mathrm{d}Q}{\mathrm{d}t} = -(\beta(Q) + \kappa)Q + 2\mathrm{e}^{-\gamma_S \tau_S} \beta(Q_{\tau_S}) Q_{\tau_S} \tag{14.14}$$

for the resting-phase HSC populations, which can be obtained from Eq. (14.9) by omitting the cell lines of neutrophil, erythrocyte, and platelet, and writing κ the total HSC differentiation rate.

The model (14) was sufficiently simple that it was possible to perform a complete bifurcation analysis [45]. This model has two possible steady states. There is a steady state corresponding to no cells ($Q_0 = 0$), which is stable if it is the only steady state. The second positive steady state Q_* exists for small HSC apoptosis rate γ_S. The stability of the positive steady state depends on the value of γ_S. When $\gamma_S = 0$, this steady state cannot be destabilized to produce oscillatory dynamics of CN. For $\gamma_S > 0$, increases in γ_S lead to a decrease in the HSC numbers, and destabilize the steady state when a critical value of γ_S is reached, $\gamma_S = \gamma_{\text{crit},1}$, at which a supercritical Hopf bifurcation occurs. When γ_S is further increased, a reverse bifurcation occurs at a critical value $\gamma_S = \gamma_{\text{crit},2}$, where the positive steady state becomes stable, and approaches the zero steady state as γ_S increases. For all values of γ_S satisfying $\gamma_{\text{crit},1} < \gamma_S < \gamma_{\text{crit},2}$, there is a periodic solution of Eq. (14.14) whose period is in good agreement with that seen in CN [45]. These results suggest that CN might be related to defects, possibly genetic, within the HSC population that lead to an abnormal apoptotic loss of cells from the proliferative phase of the cell cycle.

Bernard et al. [87] presented a two-variable delay differential equation model that couples the above HSC population model with the neutrophil compartment:

$$\begin{aligned} \frac{\mathrm{d}Q}{\mathrm{d}t} &= 2\mathrm{e}^{-\tau_S\gamma_S}\beta(Q_{\tau_S})Q_{\tau_S} - (\beta(Q) + \kappa_N(N))Q, \\ \frac{\mathrm{d}N}{\mathrm{d}t} &= -\gamma_N N + \mathrm{e}^{\tau_N\eta_N}\kappa_N(N_{\tau_N})Q_{\tau_N}, \end{aligned} \tag{14.15}$$

where

$$\kappa_N(N) = f_0\frac{\theta_1^{s_1}}{\theta_1^{s_1} + N^{s_1}}, \qquad \beta(Q) = k_0\frac{\theta_2^{s_2}}{\theta_2^{s_2} + Q^{s_2}}.$$

This model is derived from Eq. (14.9) by simply neglecting the compartments for the erythrocytes and platelets. First, we note that this model has a unique positive steady state for Q and N if

$$f_0 < k_0(2\mathrm{e}^{-\gamma_S\tau_S} - 1). \tag{14.16}$$

This condition states that the rate of HSC differentiation must be smaller than the net increase due to one cell division times the proliferation rate k_0 [87]. Using a combination of mathematical analysis and computational tools, [87] showed that the origin of CN is probably due to an increased apoptosis rate in the stem cell compartment (γ_S) and in the neutrophil precursors (which leads to a decrease in η_N), leading to a destabilization of the HSC compartment through a supercritical Hopf bifurcation. This has the effect of generating oscillations in the HSC population. This result was in accordance with previous modeling studies [6] and agrees with experimental data on gray collies. Moreover, numerical analysis showed that

Eq. (14.15) has bistability, i.e., coexistence of a stable steady state and a stable oscillatory solution when γ_S and η_N take values from a certain range. This bistability is essential for understanding the diverse effect of G-CSF treatment on CN as we will see in the next section.

A more sophisticated model that includes not only the neutrophils and HSC but also the platelets and RBC was developed by [3, 64] (Fig. 14.1). This model combines a number of compartmental models: the stem cell and neutrophil dynamics [87], and the erythrocyte and platelet compartment models [48, 89]. The circulating cells are coupled to each other via their common origin in the stem cell compartment. The model consists of a set of four coupled delay differential equations as given by Eq. (14.9).

In [3], the authors used a simulated annealing approach to fit clinical and laboratory data (from humans and dogs) to estimate the model parameters that would reproduce the characteristics of CN. The results supported the hypothesis on the origin of CN in [87] that realistic CN oscillations in neutrophils and platelets can result from an increased apoptosis rate in the neutrophil precursors. Furthermore, in order to mimic the data, it was also necessary to decrease the rate of differentiate into the neutrophil line, and the changes of the apoptosis rate of stem cells in the proliferative phase.

In [44], the authors further investigate the model numerically for the possible solutions of the model equations Eq. (14.9) with respect to changes in the parameters as well as the initial conditions. The results confirmed the findings in [3] that decreasing the proliferation rate of neutrophil precursors or increasing the stem cell death rate are two possible mechanisms to induce CN, and the periods of the resulting oscillations are independent of the changed parameters. In particular, the results suggested that by either decreasing the neutrophil precursor proliferation rate to 3–15 % less than the normal value or increasing the HSC apoptosis to 40–100 % larger than the normal value, it is possible to induce oscillations reminiscent of those in CN patients. Furthermore, simulations with changing initial conditions showed that the hematopoietic system possesses multistability over a wide range of parameters values, including typical parameter values for a healthy state. In this multistable region, the hematopoietic system may display the coexistence of a stable steady state along with an oscillatory state. This result is crucial for understanding the effects of CN patient treatment. Because of the multistability, CN that is caused by changes in system parameters may not recover to the healthy state even if the parameters are taken back to their normal values by therapy, for example through G-CSF treatment.

Modeling of Chemotherapy Induced CN

Before introducing models for studying different G-CSF treatment strategies for CN in the next section, we show how a simple model can be used to explore neutrophil dynamics in response to chemotherapy.

Chemotherapy is frequently accompanied by hematopoietic side effects due to the myelosuppressive character of the drugs used. These side effects commonly include neutropenia and, to a lesser extent, thrombocytopenia and/or anemia. In [90] and [66], the authors presented a two-compartment mathematical model of the combined dynamics of the HSC and the differentiated neutrophil progeny, modified from the model of [87] for the stem cell and neutrophil dynamics.

This model contains the HSC compartment, as well as a neutrophil compartment. The neutrophil compartment is further divided into three sub-compartments corresponding to proliferating, maturating, and circulating neutrophils, respectively. The erythrocytes and platelets are not included in the model other than to assume that the total differentiation rate of HSCs into these two cell lines is a constant κ_δ (days^{-1}). An illustration of the model is showed in Fig. 14.2. The equations describing the dynamics of this model can be obtained from Eq. (14.7) and Eq. (14.12) and are given below:

$$\begin{aligned} \frac{\mathrm{d}Q}{\mathrm{d}t} &= -(\beta(Q) + \kappa_N(N) + \kappa_\delta)Q + A_Q(t)\beta(Q_{\tau_S})Q_{\tau_S}, \\ \frac{\mathrm{d}N}{\mathrm{d}t} &= -\gamma_N N + A_N(t)\kappa_N(N_{\tau_N})Q_{\tau_N} \end{aligned} \tag{14.17}$$

and

$$\begin{aligned} \kappa_N(N) &= f_0 \frac{\theta_1^{s_1}}{\theta_1^{s_1} + N^{s_1}}, \\ \beta(Q) &= k_0 \frac{\theta_2^{s_2}}{\theta_2^{s_2} + Q^{s_2}}, \\ A_Q(t) &= 2\exp\left[-\int_0^{\tau_S} \gamma_S(t - \tau_S + s)\mathrm{d}s \right], \\ A_N(t) &= \exp\left[\int_0^{\tau_{NP}} \eta_{NP}(t - \tau_N + s)\mathrm{d}s - \int_{\tau_{NP}}^{\tau_N} \gamma_{NM}(t - \tau_N + s)\mathrm{d}s \right], \\ \tau_N &= \tau_{NP} + \tau_{NM}. \end{aligned}$$

Chemotherapy increases apoptosis in both proliferative HSCs and proliferative neutrophil precursors [91] leading to an increase in γ_S and a decrease in η_{NP}. Chemotherapy is often administered with a fixed period T(days) so that the rates $\gamma_S(t)$ and $\eta_{NP}(t)$ are periodic functions and dependent on the chemotherapy administration. There are many different chemotherapeutic drugs currently in use, and therefore different methods for modeling pharmacokinetics. Here, we present the simple model as in [90] in which the effect of chemotherapy is maintained for 1 day, and assumes square wave temporal functions for the apoptosis rate γ_S and the neutrophil precursor proliferative rate η_{NP} of the following form:

$$\gamma_S(t) = \begin{cases} \gamma_S^{\max} & \text{if } 0 \le t - kT < 1, \\ \gamma_S & \text{otherwise} \end{cases} \tag{14.18}$$

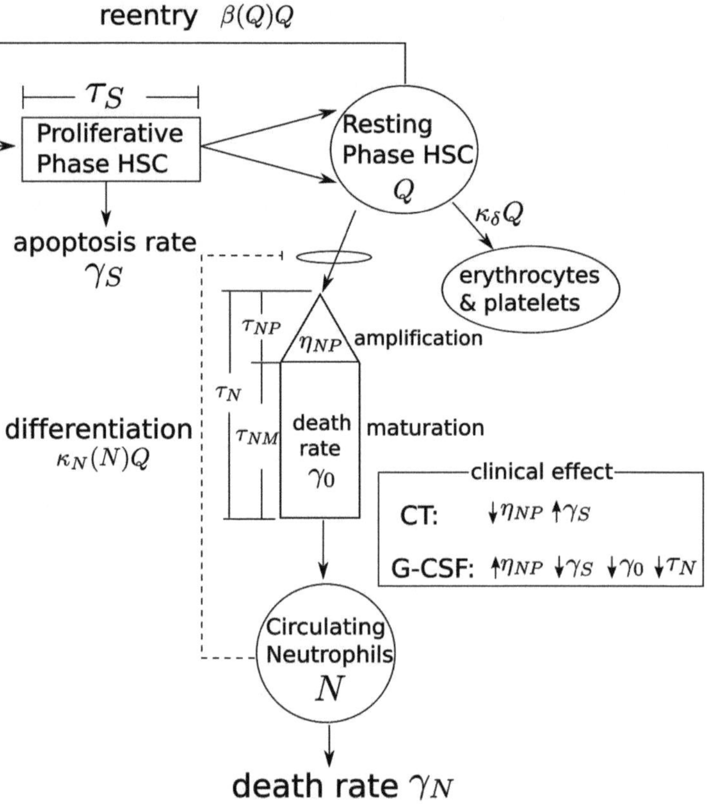

Fig. 14.2 A cartoon representation of the model of neutrophil production is investigated here. The model dynamics include those of the hematopoietic stem cells (HSC) as well as differentiated cells committed to the neutrophil line. Quiescent (resting phase) HSCs can either remain in Q, exit into the proliferative HSC phase at a rate β, or differentiate into the committed neutrophil compartment at a rate κ_N, or into the combined megakaryocyte/erhthrocyte lines at a rate κ_δ. Cells in the HSC proliferative phase are assumed to undergo apoptosis at a rate γ_S and the duration of the proliferative phase is taken to be τ_S. Cells in the neutrophil pathway are amplified by successive divisions for a time τ_{NP}, and then enter a purely maturation (no proliferation) compartment for a period of time τ_{NM} before they enter the circulation. The circulating neutrophils (N) die at a random rate γ_N so their average lifespan is γ_N^{-1}. The differentiation rate of HSC to neutrophils is controlled by the circulating neutrophil population through the differentiation rate κ_N, while the HSC proliferation is controlled by the resting HSC population with proliferation rate β. (Adapted from [90])

and

$$\eta_{NP} = \begin{cases} \eta_{NP}^{\min} & \text{if } T_1 \le t - kT < 1, \\ \eta_{NP} & \text{otherwise.} \end{cases} \tag{14.19}$$

Here, k is an integer, and $t = 0$ for the starting time of the first chemotherapy period. An expanded model with more realistic chemotherapy dynamics is presented in [66].

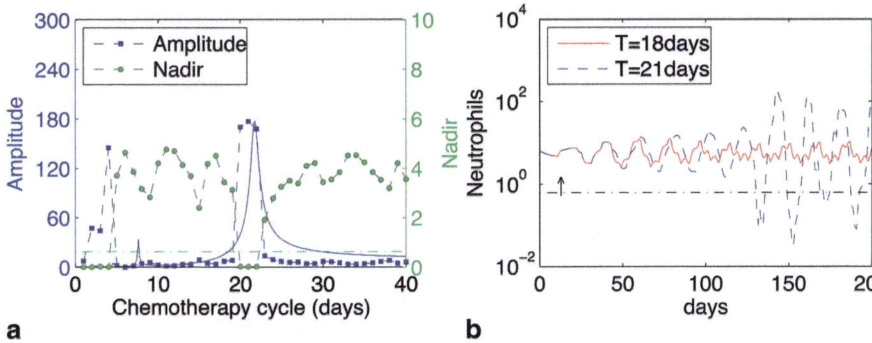

Fig. 14.3 Numerical simulation results for the neutrophil compartment model with chemotherapy alone. **a** The amplitude (*left hand ordinate*) in neutrophil response (*blue squares connected with a dashed blue line*) as well as the nadir (*right hand ordinate and green circles connected with a dashed green line*) as a function of the period T of chemotherapy. The *horizontal green dash-dot line* indicates the level for severe neutropenia (0.63×10^8 cells/kg). Note that the major peaks in the amplitude coincide with the minima in the nadir. The *solid blue line* is the computed linear frequency response function (refer [90]; rescaled to compare with the amplitude). **b** Simulated neutrophil levels from Eq. (14.17) in response to chemotherapy with a period of either $T = 18$ days or 21 days. Neutrophil levels are in units of 10^8 cells/kg, the *dashed-dot horizontal* line again indicates the level for severe neutropenia, and the *arrow* shows the first neutrophil nadir. (Adapted from [90])

Using a combination of simulation and mathematical analysis, Zhuge et al. [90] studied the neutrophil response to chemotherapy as a function of the period T. Simulations showed that the neutrophil amplitude varies with the period T of chemotherapy, with a peak at $T = 21$ days, and the neutrophil nadir has a minimum at the same period (Fig. 14.3). Figure 14.3b shows a computed time series for the neutrophils at two different periods of chemotherapy administration. The model predicts substantial differences in the dynamic response of the system as severe neutropenia was produced in the model at $T = 21$ days but not at $T = 18$ days.

According to [90], a possible reason for the occurrence of a significant peak in the amplitude and minimum in the nadir at a specific T is resonance between the periodic perturbation to the system and the intrinsic characteristic frequency in the neutrophil production dynamics. This hypothesis is confirmed by comparing the amplitude response with the frequency response function obtained analytically from the linearizing the model Eq. (14.17) around the steady state (shown by a solid line in Fig. 14.3a). The frequency response function has a maximum at $T = 21.8$ days in agreement with the simulation results. Furthermore, an analysis of the linear response function predicts that the resonant period for the model is given by twice the average neutrophil lifetime (defined as the average time τ_N spent in marrow proliferation and maturation following commitment from the HSC plus the average lifetime γ_N^{-1} in the circulation). If this simple relationship is found to hold clinically, then it offers a way to tailor chemotherapy for individuals. Namely, using the techniques employed by [51], determine τ_N and γ_N for a specific patient, and then compute the resonant period T to be avoided in any delivery of myelosuppressive agents.

We note that in Fig. 14.3a, there is a peak in the amplitude response and minimum in the nadir at $T = 4$ days that cannot be explained by resonance. The mechanism for the occurrence of this peak remains unknown.

Modeling of G-CSF Administration

CN in humans is often treated using G-CSF [92], which is known to interfere with apoptosis [93–96], and it has the overall effect of decreasing the period of severe neutropenia by increasing the nadir and the amplitude of the oscillations as well as decreasing their period [4]. However, G-CSF is expensive (about US \$ 40,000 per year for a 70-kg adult treated daily) and may cause undesirable side effects [97, 98]. In this section, we show how mathematical modeling can illuminate the effects of different G-CSF treatment schemes. For another review, see [65].

In [87], five parameters in the model Eq. (14.15) are modified to mimic the effects of G-CSF in CN: decreased apoptosis in both the HSC (decrease γ_S) and in the neutrophil precursor compartment (decrease η_N), decrease in the duration of both the proliferative and differentiating phases (τ_N and τ_S) as well as increasing the parameter θ_1 in the feedback function. Interesting dynamical features of the model were found. The bifurcation analysis agreed with the clinical aspects of G-CSF administration that results in an increase of amplitude and a decrease in the period of the oscillations [6, 92]. In clinical observations, some cases have been reported in which G-CSF treatment abolished significant oscillations [4, 5, 92]. This is also seen in the model of [87], in which a stable steady state (corresponding to annihilation of oscillations) coexists with a stable large amplitude oscillation. This bistability is interesting since it suggests that it is possible to stabilize the neutrophil count by properly designing the treatment administration scheme and could potentially reduce the amount of G-CSF required in treatment.

In [2], the model of [87] was explored with different G-CSF treatment protocols. The authors showed that, depending on the starting time of G-CSF treatment, the neutrophil count could either be stabilized or show large amplitude oscillations. This is also seen in the comprehensive model given by Eq. (14.9) that includes erythrocyte and platelet dynamics [44, 88]. Using computer simulations, the authors also showed that other G-CSF treatment schemes (such as administering G-CSF every other day) could be effective while using less G-CSF, hence reducing the cost of treatment and side effects for patients.

In [88], the authors used the comprehensive model as in [3] coupled with a two-compartment model for G-CSF pharmacokinetics. They fitted their model with clinical data for neutrophils and platelets and explored the effects of different treatment schedules. They found that different initial conditions or temporary interventions may lead to dramatically different long-term behaviors.

G-CSF is frequently used to deal with neutropenia induced by chemotherapy [65, 97]. However, the clinical administration schedule of G-CSF after chemotherapy is typically determined by trial and error and it is not clear if there is an optimal way of giving G-CSF [99, 100]. In [65], the authors present a delay differential equation model for the regulation of neutrophil production that accounts for the effect of G-CSF. Using a combination of analysis and numerical simulations, they used this model to study the effects of delaying G-CSF treatment following chemotherapy for two recombinant forms of G-CSF (filgrastim and pegfilgrastim). They found that varying the starting day or the duration of G-CSF treatment can lead to different qualitative responses in the neutrophil count.

In [90], the authors presented a simple model based on Eq. (14.17) that coupled changes in $\gamma_S, \gamma_{NM}, \eta_{NP}$, and τ_{NM} due to 1-day G-CSF administration. They found that the neutrophil dynamics response to G-CSF is highly variable, depending on the time of G-CSF delivery after chemotherapy at each cycle. In particular, there are specific times in the chemotherapy cycle when G-CSF can have positive effects in terms of ameliorating or even eliminating severe neutropenia. However, there are also broad ranges of administration times that will lead to a worsening by G-CSF of the neutropenia induced by the chemotherapy. These results are in general agreement with results presented in [65], but await confirmation until more realistic G-CSF kinetic are included in the modeling (for example, refer to [66]).

In summary, these studies have showed complicated dynamical properties of hematopoiesis after G-CSF treatment. Understanding the effects of G-CSF is difficult since G-CSF is known to affect the neutrophil maturation time in the bone marrow, whose detailed dependence is unknown, and further clinical investigations are needed to characterize this important facet of neutrophil regulation.

Discussion

Here, we have given a brief survey of how the study (using mathematical models) of dynamic hematological diseases in which there is a period cytopenia has not only given an insight into the physiological origin of these diseases but also afforded investigators an opportunity to see how to better treat these diseases. As an unexpected by-product of these investigations which have extended over four decades, mathematical biologists in collaboration with hematologists and oncologists are now starting to address the all-important question of "How can the severe side effects of myelosuppressive therapy on the hematopoietic system be either mitigated or avoided altogether?" This later question is, in our minds, one of the more important by-products of the modeling venture and offers a potentially exciting opportunity to use insight from mathematics to better the delivery of medical care for those needing it.

Acknowledgments This work was supported by the Natural Sciences and Engineering Research Council (NSERC, Canada) and the Mathematics of Information Technology and Complex Systems (MITACS, Canada), and carried out in Beijing and Montreal.

References

1. Glass L, Mackey MC. From clock to chaos. Princeton: Princeton University Press; 1988.
2. Foley C, Bernard S, Mackey MC. Cost-effective G-CSF therapy strategies for cyclical neutropenia: mathematical modeling based hypotheses. J Theor Biol. 2006;238:754–63. doi:10.1016/j.jtbi.2005.06.021.
3. Colijn C, Mackey MC. A mathematical model of hematopoiesis: II. cyclical neutropenia. J Theor Biol. 2005;237:133–46.
4. Haurie C, Dale DC, Mackey MC. Cyclical neutropenia and other periodic hematological diseases: a review of mechanisms and mathematical models. Blood. 1998;92:2629–40.
5. Haurie C, Dale DC, Mackey MC. Occurrence of periodic oscillations in the differential blood counts of congenital, idiopathic and cyclical neutropenic patients before and during treatment with G-CSF. Exp Hematol. 1999;27:401–9.
6. Haurie C, Dale DC, Rudnicki R, Mackey MC. Modeling complex neutrophil dynamics in the grey collie. J Theor Biol. 2000;204:505–19.
7. Fortin P, Mackey MC. Periodic chronic myelogenous leukaemia: spectral analysis of blood cell counts and aetiological implications. Br J Haematol. 1999;104:336–45.
8. Mackey MC, Glass L. Oscillation and chaos in physiological control systems. Science. 1977;197:287–9.
9. Swinburne J, Mackey MC. Cyclical thrombocytopenia: characterization by spectral analysis and a review. J Theor Med. 2000;2:81–91.
10. Apostu R, Mackey MC. Understanding cyclical thrombocytopenia: a mathematical modeling approach. J Theor Biol. 2008;251:297–316.
11. Hsieh MM, Everhart JE, Byrd-Holt DD, Tisdale JF, Rodgers GP. Prevalence of neutropenia in the U.S. population: age, sex, smoking status, and ethnic differences. Ann Intern Med. 2007;146:486–92.
12. Gill M, Ockelford P, Morris A, Bierre T, Kyle C. Diagnostic handbook-the interpretation of laboratory tests. Auckland: Diagnostic Medlab; 2000.
13. Haurie C, Person R, Dale DC, Mackey MC. Hematopoietic dynamics in grey collies. Exp Hematol. 1999;27:1139–48.
14. Reimann HA, de Beradinis CT. Periodic(cyclic) neutropenia. an entity. A collection of sixteen cases. Blood. 1949;4:1109–16.
15. Morley AA, Carew JP, Baikie AG. Familial cyclical neutropenia. Br J Haematol. 1967;13:719–38.
16. Palmer SE, Stephens K, Dale DC. Genetics, phenotype, and natural history of autosomal dominant cyclic hematopoiesis. Am J Med Genet. 1996;88:335–40.
17. Norwitz M, Benson K, Person R, Aprikyan A, Dale D. Mutations in ELA2, encoding neutrophil elastase, define a 21-day biological clock in cyclic haematopoiesis. Nat Genet. 1999;23:433–6.
18. Cohen T, Cooney DP. Cyclical thrombocytopenia: case report and review of literature. Scand J Haemat. 1974;16:133–8.
19. Beutler E, Lichtman MA, Coller BS, Kipps T. Williams hematology. New York: McGraw-Hill; 1995.
20. Balduini C, Stella C, Rosti V, Bertolino G, Noris P, Ascari E. Acquired cyclic thrombocytopenia thrombocytosis with periodic defect of platelet function. Brit J Haematol. 1993;85:718–22.
21. Bernard J, Caen J. Purpura thrombopénique et megacaryocytopénie cycliques mensuels. Nouv Rev franc Hémat. 1962;2:378–86.
22. Dan K, Inokuchi K, An E, Nomura T. Cell mediated cyclic thrombocytopenia treated with azathioprine. Brit J Haematol. 1991;77:365–79.
23. Engstrom K, Lundquist A, Soderstrom N. Periodic thrombocytopenia or tidal platelet dysgenesis in a man. Scand J Haemat. 1966;3:290–2.
24. Hoffman R, Bridell RA, van Besien K, Srour EF, Guscar T, Hudson NW, et al. Acquired cyclic amegakaryocytic thrombocytopenia associated with an immunoglobulin blocking the action of granulocyte-macrophage colony-stimulating factor. N Engl J Med. 1989;321:97–102.

25. Lewis ML. Cyclical Thrombocytopenia: a thrombopoietin deficiency? J Clin Path. 1974;27:242–6.
26. Kimura F, Nakamura Y, Sato K, Wakimoto N, Kato T, Tahara T, et al. Cyclic change of cytokines in a patient with cyclic thrombocytopenia. Br J Haemat. 1996;94:171–4.
27. Santillán M, Bélair J, Mahaffy JM, Mackey MC. Regulation of platelet production: the normal response to perturbation and cyclical platelet disease. J Theor Biol. 2000;206:585–603. doi:10.1006/jtbi.2000.2149.
28. Ranlov P, Videbaek A. Cyclic haemolytic anaemia synchronous with Pel-Ebstein fever in a case of Hodgkin's disease. Acta Medica Scandinavica. 1963;100.
29. Gordon RR, Varadi S. Congenital hypoplastic anemia (pure red-cell anemia) with periodic erythroblastopenia. Lancet. 1962;1:296–9.
30. Gurney CW, Simmons EL, Gaston EO. Cyclic erythropoiesis in W/Wv mice following a single small dose of 89Sr. Exp Hemat. 1981;9:118–22.
31. Gibson CM, Gurney CW, Gaston EO, Simmons EL. Cyclic erythropoiesis in the S1/S1d mouse. Exp Hemat. 1984;12:343–8.
32. Gibson CM, Gurney CW, Simmons EL, Gaston EO. Further studies on cyclic erythropoiesis in mice. Exp Hemat. 1985;13:855–60.
33. Vácha J, Znojil V. Application of a mathematical model of erythropoiesis to the process of recovery after acute X-irradiation of mice. Biofizika. 1975;20:872–9.
34. Vácha J. Postirradiational oscillations of erythropoiesis in mice. Acta Sc Nat Brno. 1982;16(2):1–52.
35. Orr JS, Kirk J, Gray KG, Anderson JR. A study of the interdependence of red cell and bone marrow stem cell populations. Br J Haemat. 1968;15:23–4.
36. Mackey MC. Periodic auto-immune hemolytic anemia: an induced dynamical disease. Bull Math Biol. 1979;41:829–34.
37. O'Dwyer M, Druker BJ, Mauro M, Talpaz M, Resta D, Peng B, et al. STI571: a tyrosine kinase inhibitor for the treatment of CML. Ann Oncol. 2000;11:155.
38. Melo J. The diversity of BCR-ABL fusion proteins and their relationship to leukemia phenotype. Blood. 1996;88:2375.
39. Grignani F. Chronic myelogenous leukemia. Crit Rev Oncol Hematol. 1985;4:31–66.
40. Henderson ES, Lister TA, Greaves MF, editors. Leukemia. Philadelphia: Saunders; 1996.
41. Foley C, Mackey MC. Dynamic hematological disease: a review. J Math Biol. 2009;58:285–322.
42. Wichard ZL, Sarkar CA, Kimmel M, Corey SJ. Hematopoiesis and its disorders: a systems biology approach. Blood. 2011;115:2339–47.
43. Mackey MC, Haurie C, Bélair J. Cell replication and control. In: Beuter A, Glass L, Mackey MC, Titcombe MS, editors. Nonlinear dynamics in physiology and medicine. New York: Springer; 2003, pp. 233–69.
44. Lei J, Mackey MC. Multistability in an age-structured model of hematopoiesis: cyclical neutropenia. J Theor Biol. 2011;270:143–53.
45. Mackey MC. Unified hypothesis for the origin of aplastic anemia and periodic haematopiesis. Blood. 1978;51:941–56.
46. Mackey MC. Mathematical models of hematopoietic cell replication and control. In: Othmer H, Adler F, Lewis M, Dallon J, editors. Case studies in mathematical modeling-ecology, physiology, and cell biology. New Jersey: Prentice-Hall; 1996. pp. 149–78.
47. Hoffbrand AV, Pettit JE, Moss PAH. Essential haematology. 4th ed. Milan: Blackwell Science; 2011.
48. Mahaffy JM, Bélair J, Mackey MC. Hematopoietic model with moving boundary condition and state dependent delay: application in erythropoiesis. J Theor Biol. 1998;190:135–46.
49. Bélair J. Mackey MC. A model for the regulation of mammalian platelet. Ann N Y Acad Sci. 1987;504:280–2.
50. Adamson JW. The relationship of erythropoietin and iron metabolism to red blood cell production in humans. Semin Oncol. 1994;2:9–15.

51. Price TH, Chatta GS, Dale DC. Effect of recombinant granulocyte colony-stimulating factor on neutrophil kinetics in normal young and elderly humans. Blood. 1996;88:335–40.
52. Ratajczak MZ, Ratajczak J, Marlicz W, Jr Pletcher WC, Machalinshi B, Moore J, et al. Recombinant human thrombopoietin(TPO) stimulates erythropoiesis by inhibiting erythroid progenitor cell apoptosis. Br J Haematol. 1997;98:8–17.
53. Tanimukai S, Kimura T, Stakabe H, Ohmizono Y, Kato T, Miyazaki H, et al. Recomninant human c-Mpl ligand (thrombopoietin) not only acts on megakaryocyte progenitors, but also on erythroid and multipotential progenitors in vitro. Exp Hematol. 1997;25:1025–33.
54. Silva M, Grillot D, Benito A, Richard C, Nunez G, Fernandez-Luna J. Erythropoietin can promote erythroid progenitor survival by repressing apoptosis through bcl-1 and bcl-2. Blood. 1996;88:1576–82.
55. Ritchie A, Gotoh A, Gaddy J, Braun S, Broxmeyer H. Thrombopoietin upregulates the promoter conformation of p53 in a proliferation-independent manner coincident with a decreased expression of bax: potential mechanisms for survival enhancing effects. Blood. 1997;90:4394–402.
56. Kaushansky K, Lin N, Grossmann A, Humes J, Sprugel K, Broudy V. Thrombopoietin expands erythoid, granulocyte-macrophage, and megakaryocyte progenitor cells in normal and myelosuppressed mice. Exp Hematol. 1996;24:256–69.
57. Kearns CM, Wang WC, Stute N, Ihle J, Evans W. Disposition of recombinant human granulocyte colony-stimulating factor in children with severe chronic neutropenia. J Pediatr. 1993;123(3):471–9.
58. Mempel K, Pietsch T, Menzel T, Zeidler C, Welte K. Increased serum levels of granulocyte colony stimulating factor in patients with severe congenital neutropenia. Blood. 1991;77:1919–22.
59. Takatani H, Soda H, Fukuda M, Watanabe M, Kinoshita A, Nakamura T, et al. Levels of recombinant human granulocyte colony stimulating factor in serum are inversely correlated with circulating neutrophil counts. Antimicrob Agents Chemother. 1996;40:988–91.
60. Watari K, Asano S, Shirafuji N, Kodo H, Ozawa K, Takaku F, et al. Serum granulocyte colony stimulating factor levels in healthy volunteers and patients with various disorders as estimated by enzyme immunoassay. Blood. 1989;73:117–22.
61. Roeder I. Quantitative stem cell biology: computational studies in the hematopoietic system. Curr Opin Hematol. 2006;13:222–8.
62. Viswanathan S, Zandstra PW. Towards predicitive models of stem cell fate. CytoTechnol. 2003;41:75–92.
63. Bélair J, Mackey MC, Mahaffy JM. Age-structuredand two-delay models for erythropoiesis. Math Biosci. 1995;128:317–46.
64. Colijn C, Mackey MC. A mathematical model of hematopoiesis: I. periodic chronic myelogenous leukemia. J Theor Biol. 2005;237:117–32.
65. Foley C, Mackey MC. Mathematical model for G-CSF administration after chemotherapy. J Theor Biol. 2009;257:27–44.
66. Brooks G, Provencher-Langlois G, Lei J, Mackey MC. Neutrophil dynamics after chemotherapy and G-CSF: the role of pharmacokinetics in shaping the response. J Theor Biol. 2012;315:97–109.
67. Colijn C, Dale DC, Foley C, Mackey MC. Observations on the pathophysiology and mechanisms for cyclic neutropenia. Math Model Nat Phenomena. 2006;1:45–69.
68. Dunn CDR. Cyclical hematopoiesis: the biomathematics. Exp Hematol. 1983;11:779–91.
69. Fisher G. An introduction to chaos theory and some haematological applications. Comp Haematol Int. 1993;3:43–51.
70. Kazarinoff ND, van den Driessche P. Control of oscillations in hematopoiesis. Science. 1979;203:1348–50.
71. King-Smith EA, Morley A. Computer simulation of granulopoiesis: normal and impaired granulopoiesis. Blood. 1970;36:254–62.
72. MacDonald N. Cyclical neutropenia: models with two cell types and two time lags. In: Valleron AJ, Macdonald PDM, editors. Biomathematics and cell kinetics. Amsterdam: Elsevier; 1978. pp. 287–95.

73. Morley A. A platelet cycle in normal individuals. Aust Ann Med. 1969;18:127–9.
74. Morley A. Blood-cell cycles in polycythaemia vera. Aust Ann Med. 1969;18:124.
75. Morley A, Stohlman F. Cyclophosphamide induced cyclical neutropenia. N Engl J Med. 1970;282:643–6.
76. Morley A. Cyclic hemopoiesis and feedback control. Blood Cells.1979;5:283–96.
77. Reeve J. An analogue model of granulopoiesis for the analysis of isotopic and other data obtained in the non-steady state. Br J Haematol. 1973;25:15–32.
78. Schmitz S. Ein mathematisches Modell der zyklischen Haemopoese. Ph.D thesis, Universitat Koln; 1988.
79. Schmitz S, Franke H, Brusis J, Wichmann HE. Quantification of the cell kinetic effects of G-CSF using a model of human granulopoiesis. Exp Hematol. 1993;21:755–60.
80. Schmitz S, Franke H, Loeffler M, Wichmann HE, Diehl V. Reduced variance of bone-marrow transit time of granulopoiesis: a possible pathomechanism of human cyclic neutropenia. Cell Prolif. 1994;27:655–67.
81. Schmitz S, Loeffler M, Jones JB, Lange RD, Wichmann HE. Synchrony of bone marrow proliferation and maturation as the origin of cyclic haemopoiesis. Cell Tissue Kinet. 1990;23:425–41.
82. Schmitz S, Franke H, Wichmann HE, Diehl V. The effect of continuous G-CSF application in human cyclic neutropenia: a model analysis. Br J Haematol. 1995;90:41–7.
83. Shvitra D, Laugalys R, Kolesov YS. Mathematical modeling of the production of white blood cells. In: Marchuk G, Belykh LN, editors. Mathematical modeling in immunology and medicine. Amsterdam: North-Holland; 1983. pp. 211–23.
84. von Schulthess GK, Mazer NA. Cyclic neutropenia(CN): a clue to the control of granulopoiesis. Blood. 1982;59:27–37.
85. Wichmann HE, Loeffler M, Schmitz S. A concept of hemopoietic regulation and its biomathematical realization. Blood Cells.1988; 14:411–29.
86. Hearn T, Haurie C, Mackey MC. Cyclical neutropenia and the peripherial control of white blood cell production. J Theor Biol. 1998;192:167–81.
87. Bernard S, Bélair J, Mackey MC. Oscillations in cyclical neutropenia: new evidence based on mathematical modeling. J Theor Biol. 2003;223:283–98. doi:10.1016/S0022-5193(03)00090-0.
88. Colijn C, Foley C, Mackey MC. G-CSF treatment of canine cyclical neutropeina: a comprehensive mathematical model. Exp Hematol. 2007;37:898–907.
89. Santillan M, Mahaffy JM, Bélair J, Mackey MC. Regulation of platelet productin: the normal response to perturbation and cyclical platelet disease. J Theor Biol. 2000;206:585–603.
90. Zhuge C, Lei J, Mackey MC. Neutrophil dynamics in response to chemotherapy and G-CSF. J Theor Biol. 2012;293:111–20.
91. Hannun Y. Apoptosis and the dilemma of cancer chemotherapy. Blood. 1997;89:1845–53.
92. Hammond WP, Price TH, Souza LM, Dale DC. Treatment of cyclic neutropenia with granulocyte colony stimulating factor. N Engl J Med. 1989;320:1306–11.
93. Koury MJ. Programmed cell death(apoptosis) in hematopoiesis. Exp Hematol. 1992;20:391–4.
94. Park JR. Cytokine regulation of apoptosis in hematopoietic precursor cells. Curr Opin Hematol. 1996;3:191–6.
95. Migliaccio AR, Migliaccio G, Dale DC, Hammond WP. Hematopoietic progenitors in cyclic neutropenia: effect of granulocyte colony stimulating factor in vivo. Blood. 1990;75:1951–9.
96. Williams G, Smith C. Molecular regulation of apoptosis: genetic controls on cell death. Cell. 1993;74:777–9.
97. Crawford J, Dale DC, Lyman GH. Chemotherapy-induced neutropenia: risks, consequences, and new directions for its management. Cancer. 2003;100:228–37.
98. Ozer H, Armitage JO, Bennett CL, et al. Update of recommendations for the use of hematopoietic colony-stimulating factors: evidence-based, clinical practice guidelines. J Clin Oncol. 2000;18:3558–85.

99. Clark OA, Lyman GH, Castro AA, Clark LG, Djulbegovic B. Colony-stimulating factors for chemotherapy-induced febrile neutropenia: a meta-analysis of randomized controlled trials. J Clin Oncol. 2005;23:4198–214.
100. Bennett CL, Weeks JA, et al. Use of hematopoietic colony-stimulating factors: comparison of the 1994 and 1997 American Society of Clinical Oncology surveys regarding ASCO clinical practice guidelines. Health Services Research Committee of the American Society of Clinical Oncology. J Clin Oncol. 1999;17:3676–81.

Chapter 15
Drug Resistance

Cristian Tomasetti

Abstract Drug resistance is a fundamental problem in the treatment of cancer since cancer that becomes resistant to the available drugs may leave the patient with no therapeutic alternatives. In this chapter, we consider the dynamics of drug resistance in blood cancer and the related issue of the dynamics of cancer stem cells. After describing the main types of chemotherapeutic agents available for cancer treatment, we review the different mechanisms of drug resistance development. Various mathematical models of drug resistance found in the literature are then reviewed. Given the well-known hierarchy of the hematopoietic system, it is critical to focus on those cells that have the ability to self-renew, since these will be the only cells able to induce long-term drug resistance. Thus, a recent mathematical model taking into account the complex dynamics of the leukemic stem-like cells is described. The chapter closes with a few applications of this model to chronic myeloid leukemia.

Keywords Drug resistance · Fluctuation analysis · Poisson process · Targeted therapy · Hematopoietic stem cells · Symmetric and asymmetric division · Leukemia · Hyperbolic PDE · Continuous-time branching processes · Probabilistic methods

Introduction

The development of drug resistance is a fundamental problem in the treatment of cancer. The beneficial effects induced by a drug at the beginning of cancer therapy are, more often than not, temporary and are typically lost due to the insurgence of drug resistance. In this chapter, we consider the dynamics of drug resistance in blood cancer and the related issue of the dynamics of cancer stem cells. We review the main biological mechanisms known to cause drug resistance as well as various mathematical models of drug resistance found in the literature. Subsequently, we focus on those leukemic cells that have the key ability to self-renew, that is, the

C. Tomasetti (✉)
Johns Hopkins School of Medicine, 550 North Broadway,
Suite 1103, Baltimore, MD 21205, USA
Tel.: 443-287-5683
e-mail: ctomasez@johnshopkins.edu

S. J. Corey et al. (eds.), *A Systems Biology Approach to Blood,*
Advances in Experimental Medicine and Biology 844, DOI 10.1007/978-1-4939-2095-2_15

leukemic stem cells. A recent mathematical model able to account for the complex dynamics of these cells is introduced and a few applications of this model to chronic myeloid leukemia are presented.

Chemotherapy

Aside for the possible surgical removal of a tumor, chemotherapy is the most common treatment against cancer (transplant, radiotherapy, and immunotherapy being the other major ones). While chemotherapeutic drugs are usually successful at reducing the initial tumor load, their effectiveness is often reduced or even completely lost after a few cycles of treatment. The main reason for their failure is the development of "drug resistance." To better address the issue of cell resistance to anticancer drugs, we will briefly review what chemotherapy is, how it has evolved in time, and how it works.

A chemotherapeutic drug is, broadly speaking, a chemical compound with the ability to reduce the tumor load. The first modern chemotherapy drug used against cancer was discovered in 1942, when the US government asked two Yale assistant professors, L. S. Goodman and A. Gilman, to study mustard gas, which had been developed and used as a chemical warfare agent in World War I. The two researchers found that nitrogen mustard, an alkylating agent derived from mustard gas, caused a dramatic regression of lymphoma in the patient under study. This discovery led in the late 1940s and early 1950s to the investigation and subsequent use in clinical practice of various compounds: A. Haddow used urethane in chronic myeloid leukemia (CML) patients, J. Burchenal used methotrexate to treat leukemia in children, and S. Farber utilized aminopterin in acute childhood [1, 2]. Since then, many more compounds have been found that are able of producing a significant regression of a tumor load. We can broadly divide the chemotherapeutic agents in the following categories [3]:

- *Alkylating agents* (and variations): By attaching alkyl groups to the guanine base of DNA double-helix strands, they chemically modify a cell's DNA, thus impairing its function. Indeed, by cross-linking guanine bases in DNA, they make the DNA strands unable to uncoil and separate. DNA replication is impaired and therefore the cell can no longer divide.
- *Antimetabolites*: Chemicals that are similar in structure to metabolites (molecules that are part of the normal cell metabolism) prevent these substances from being incorporated into the DNA during the S phase of the cell cycle (synthesis phase, when DNA replication occurs), consequently stopping normal cell development and division by damaging the DNA strand. In fact antimetabolites masquerade as purines or pyrimidines, molecules that are building blocks of DNA.
- *Plant alkaloids*: By inhibiting the assembly of microtubules they block cell division, due to the fundamental role played by microtubules in a cell.

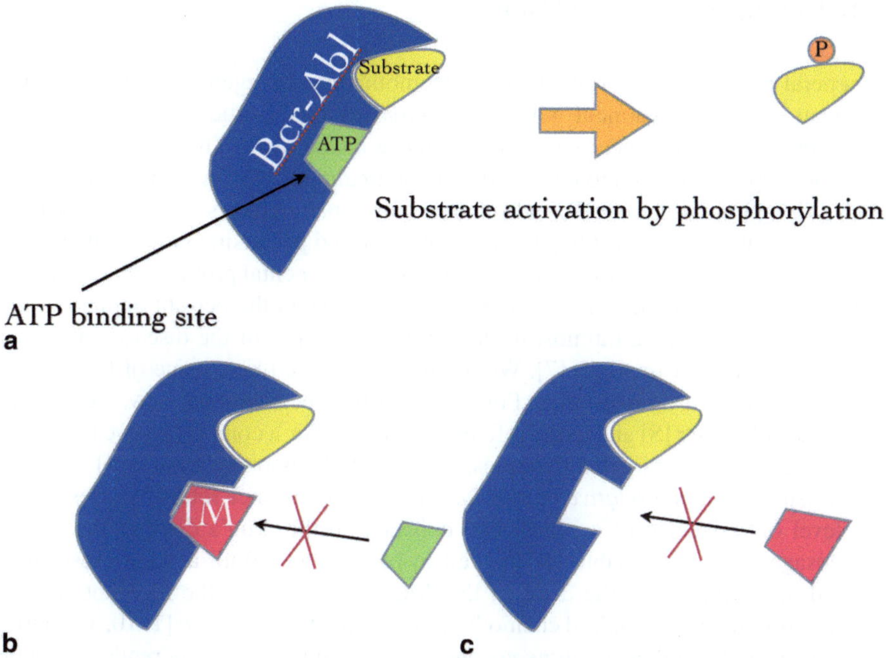

Fig. 15.1 A point mutation causing drug resistance. **a** The bcr-abl enzyme (*blue*) activates the substrate (*yellow*) via phosphorylation causing unregulated proliferation among the leukemic cells. **b** Imatinib (*red*) binds competitively to the ATP binding site inhibiting the enzyme's activity. **c** A point mutation induces a conformational change in the ATP binding site that does not allow imatinib to bind anymore, thus allowing ATP (*green*) to reach its binding site and therefore resulting in bcr-abl's activity. *ATP* adenosine triphosphate

- *Topoisomerase inhibitors*: By inhibiting topoisomerases, which are essential enzymes that maintain the topology of DNA, they cause damages to the transcription and replication of DNA (interfering with proper DNA supercoiling).
- All these types of drugs do not differentiate between cancer and normal cells, with the resulting well-known toxic effects. However, the inhibition of cell division or the interference against DNA replication is expected to harm cancer cells more than healthy cells, given that cancer cells should divide more often than normal.
- *Targeted therapies*: In the past decade, new and less toxic drugs have been designed to target only a particular molecular attribute that characterizes a given type of cancer cell and not its healthy counterpart. Imatinib mesylate, for example, works by inhibiting the bcr-abl enzyme activity (see Fig. 15.1), which characterizes CML cells [4, 5]. Because of the potential of these new therapies, we are witnessing the rise of a new discipline, molecular oncology, whose focus is to understand the genetic and biochemical mechanisms involved in cancer [6].

The Biology of Drug Resistance

A general phenomenon for all standard chemotherapeutic agents is that successive applications of the treatment will yield decreasing therapeutic benefits due to the development, in the cancer cells, of resistance to the drug. Thus, while standard chemotherapy has proven to be effective in the treatment of a few types of cancer—such as lymphomas, germ cell, and some pediatric malignancies—in the majority of cases the results are modest [6]. Even for the new and promising targeted therapies, the development of drug resistance constitutes a fundamental problem. For example, in CML patients, the occurrence of specific mutations in the bcr-abl domain of the leukemic cells will result almost invariably in the relapse of the disease due to the lost effectiveness of imatinib [7]. We briefly summarized here some of the various classifications of drug resistance. For a more comprehensive overview, we refer to the book by Teicher [8] and to the references therein. For a comprehensive treatment of drug resistance to targeted therapies, we refer to the book by Daniel [9].

Resistance may be *relative* or *absolute*. Relative resistance refers to cases where the level of resistance of a cancer cell depends on the drug's dosage: the higher the dosage the less probable for the cell to be resistant. With absolute resistance instead, no matter what the dose is, the drug will not affect the resistant cell. In general, resistance to standard chemotherapy appears to be relative [1, 10, 11]. Drug resistance can be both a *spontaneous* phenomenon (e.g., caused by random genetic mutations which occur independently and even before the drug is administered, as we will see) as well as an *induced* one, which means that drug resistance may originate as a consequence of taking a drug [12–14]. Drug resistance depends on many factors. We may divide them broadly into two categories [8, 15]:

- *Physiological resistance*, which depends on host factors, like, for example, the size, location, and growth rate of the cancer, the blood supply, the immune system status, the tumor microenvironment, the tumor pH, or the patient's intolerance to the effects of a drug. Location resistance is also known as diffusion resistance.
- *Biological resistance*, given by kinetic resistance or genetic and epigenetic alterations in the cancer cells.

In the following, we will focus only on the second kind. *Kinetic resistance* refers to the reduction in effectiveness of a drug caused by the cell division cycle. Such resistance is generally only temporary. As we have seen, many standard chemotherapeutic drugs (such as methotrexate, vincristine, and cytosine arabinoside, to name a few) are effective during only one specific phase of the cell cycle, e.g., during the S phase when the DNA is synthesized. Thus, in the case of a short exposure to the drug, the cancer cell will not be affected if it is found in a different phase. Even more importantly, the cell will be substantially invulnerable if it is out of the cell division cycle, i.e., in a "resting state" or G_0 state. This implies that the number of cells that are affected by the drug is lower for cell populations that have low proliferation rates.

Resistance to drugs may instead develop as a consequence of genetic events such as mutations, rather than developing due to kinetic reasons. This category includes both point mutations and chromosomal mutations, also known as gene amplifications.

Point mutations are genetic changes causing the replacement of a single base nucleotide or pair, with another nucleotide or pair in the DNA or RNA. These mutations may occur, for example, during DNA replication. If the mutation is nonsynonymous, it will change the cellular phenotype possibly making the cancer cell resistant to the drug. A major reason for the development of resistance to imatinib in CML patients is due to point mutations that alter the adenosine triphosphate (ATP) binding site of bcr-abl, which is the specific binding target of imatinib, thus inhibiting the activity of the drug [4, 5, 7]. This mechanism is depicted in Fig. 15.1.

Gene amplification is either the consequence of an abnormally large number of copies of a particular gene or the result of an overproduction of transcripts of particular gene. This means that a limited portion of the genome is reproduced to a much greater extent than the replication of DNA composing the remainder of the genome. Such a defect amplifies the phenotype that the gene confers on the cell, which, in turn, induces resistance by essentially providing the cells with more copies of a particular gene than the drug is able to cope with. The classical example is given by the amplification in the number of efflux transporters a cell is endowed with, thus increasing the ability of the cell to eliminate, expel the drug. While gene amplification (and consequently the resistance induced by it) may be a temporary phenomenon, point mutations appear to be permanent.

The cause for genetic mutations is not completely clear. Is it a random phenomenon, possibly occurring even before the therapy is started, or rather a drug-induced, directed one, perhaps both? Such a fundamental question has been the focus of the Nobel Prize winning work of Luria and Delbrück [16]. Using fluctuation analysis, Luria and Delbrück showed that drug resistance in in vitro bacterial cultures is primarily a random phenomenon rather than a drug-induced, directed one. Many further in vitro experiments with tumor cell lines confirmed this result. In vivo experiments have instead provided somewhat contradictory results, with some results supporting and others opposing the idea of drug resistance being a random phenomenon, see references [2, 17].

Mathematical Models of Drug Resistance

Drug resistance has been extensively studied in the mathematical literature. Originally, the modeling of resistance due to random point mutations was motivated by the experimental findings of Luria and Delbrückin 1943 on the development of resistance to antibiotics in bacteria due to mutations [16]. The mathematical model the authors formulated in order to answer that fundamental question was then used for estimating the rate at which mutations causing resistance occurred. In that model, Luria and Delbrück assumed that the process starts with one normal cell and no mutants. A random process, specifically a Poisson process with an intensity function, modeled the occurrence of mutations. Importantly, both normal and mutant cells were assumed to grow deterministically in an exponential fashion. The probability of no mutations was then calculated as well as the mean and the variance of the distribution of the number of mutants [16]. Given these estimates, it was possible to implement methods for the estimation of mutation rates from the data. This Nobel Prize winning work has been followed by a large literature on the study of the distribution of the number of mutants in a population which grows exponentially, known as the Luria and Delbrück distribution [18].

The first model of resistance to chemotherapy due to point mutations in cancer is the celebrated model by Goldie and Coldman and its extensions [2, 19–24].

In Goldie and Coldman [21], the growth of the drug-sensitive cancer cell population was approximated by using a deterministic exponential curve. At each division, it was assumed that there is a small positive probability that a cancer drug-sensitive cell may give rise to one drug-resistant cancer cell daughter because of a random point mutation. Such a mutant generated a clone growing according to a birth process. A Poisson distribution approximated the number of mutations occurring in the drug-sensitive population at any given time. It was also assumed that back mutations could not occur. Then, by using a filtered Poisson process, it was possible to calculate the number of mutants present in the cancer cell population. The probability of having no resistant cells present in a tumor was then calculated, where the nonexistence of resistant cells was assumed to be the condition for being cured. We note that in the model by Goldie and Coldman, the drug-sensitive population is modeled deterministically as in Luria and Delbrück [16], but the drug-resistant population is modeled stochastically rather than deterministically. The main results of Goldie and Coldman are that the probability of having no drug resistance present in a tumor is inversely related to the tumor size and that more frequent dosage repetitions are more successful in minimizing the risk of drug resistance development than less frequent doses administered for a longer period of time [21]. In Coldman et al., the authors extended the model to multi-drug resistance. It was assumed that multiple drug resistance occurred in single step where now a cancer cell may be sensitive to all drugs, resistant to only one of the drugs, to two given drugs, and so forth [23, 24]. The main conclusion of their study was that the best strategy was to use all available drugs simultaneously. Furthermore, it was shown that, if the simultaneous administration of all drugs is not possible, the sequential alternation of all drugs is optimal when these drugs are equally effective. Goldie and Coldman extended their mathematical model to consider also the development of drug resistance when the cancer stem cell hypothesis is considered [2, 22]. Unfortunately, they assumed that a stem cell could either renew symmetrically, producing two daughter stem cells, or differentiate symmetrically, producing two differentiated (not stem cells) daughters. In this way, the two division modes are simply equivalent to either a stem cell birth or death, leaving out from the stem cell dynamics the fundamental case of an asymmetric division, where a stem cell and a differentiated daughter cell are produced. Their model then reduces to the usual birth and death process of a growing population. Notwithstanding this limitation, their model was the first model on drug resistance that somehow considered the cancer stem cell dynamics.

A more recent study on random point mutations is by Komarova et al. [25–28]. For example, a model based on stochastic birth and death processes on a combinatorial mutation network was used to describe the development of resistance to multi-drug treatments [25, 27]. Thus, probabilistic methods and a hyperbolic partial differential equation were used to show how the pretreatment phase is more significant in the development of drug resistance than the treatment phase. This is a very natural, intuitive result given that the treatment will, in general, drastically reduce the cancer cell population and consequently also reduce the number of cell divisions from which random point mutations may arise. The main result obtained by the author is the following: In the case of a single-drug treatment, the probability to have resistant mutants generated before the beginning of the treatment and present, including their progeny, at some given time afterward, does not depend on the cancer turnover rate contrary to the multidrug scenario [25]. A consequence of such result would also be that the probability of single-drug treatment success would not depend on such a rate. In Tomasetti et al., however, it has been shown that this result does not hold for finite times (those of biological interest), especially if the difference between the death and the birth rate of the cancer cells is small [29]. Komarova et al. used the same methodology to analyze the development of resistance in CML [26, 27]. The authors suggested that a combination of three drugs with different specificities might overcome the problem of resistance [27]. At the same time, they observed that combining more than two current drugs may not provide any further therapeutic advantage, due to the problem of cross-resistance [26].

Another recent work on point mutations is by Iwasa et al. [29]. Continuous-time branching processes were used to calculate the probability of resistance at the time of detection of the cancer, as well as the expected number of mutants found at detection if resistance developed. These estimates were found both for the case where drug-sensitive and drug-resistant cells have the same birth and death rates as well as for the cases where the drug-resistant cells have a fitness advantage or disadvantage with respect to the wild-type cancer cells. Similarly to the results [21], the authors showed that the probability of resistance is an increasing function of the detection size and the mutation rate [29].

Finally, Durrett et al. used multi-type branching processes to study multi-drug resistance, obtaining estimates for the distribution of the first time when k mutations have accumulated in some cell, as well as for the growth of the various subpopulations of mutant cells [30].

While we will not consider the modeling of drug resistance due to gene amplification, kinetic resistance, drug-induced or physiological resistance, we would like to briefly comment on some of the works on these types of resistance and refer interested readers to the references therein. Modeling of resistance due to gene amplification can be found, e.g., in [31–33]. Drug resistance in these works is studied using stochastic branching models. Kinetic resistance has been mathematically studied in [34, 35]. The models in these papers are based on ordinary differential equations. An alternative approach on kinetic resistance, using age-structured models, can be found in [36–40]. Instead, for mathematical models and experimental findings on drug-induced resistance, we refer to [14, 41, 42].

A Model of Drug Resistance for Leukemias: Including the Dynamics of Leukemic Stem Cells

In the previously mentioned mathematical models of drug resistance, cancer cells are considered as a homogeneous population (aside from being or not drug resistant). In fact, cancer cells will typically differ in size, morphology, division rate, death rate, and resistance to a given drug [43, 44]. This diversity is caused by genetic differences, as well as by epigenetic plasticity [45].

Blood constitutes possibly the best example of a tissue with a large variety of cell types, and where homeostasis is maintained by a small subset of slowly replicating cells, known as the hematopoietic stem cells. These cells have the capacity of both self-renewal and differentiation into more mature—and much shorter lived—differentiated blood cells. From the point of view of drug resistance, this hematopoietic hierarchy implies that only the cancer cells that have the capacity for self-renewal can propagate long-term drug resistance. Therefore, these cancer cells should be taken into account in any model of drug resistance. In fact, only these cells should be taken into account. Note that this is true even if cancer stem cells were to be always drug resistant due to some intrinsic property related to their stemness.

Tomasetti et al. modeled drug resistance by focusing on the dynamics of a growing leukemic stem cell population where all their possible modes of division are included: symmetric self-renewal, asymmetric division, and symmetric differentiation [46]. We will review this model in some detail.

Consider the following question: What is the probability that by the time a patient is diagnosed with, e.g., CML some of the leukemic stem cells (LSCs) have been already hit by specific point mutations causing these cells and all their progeny to be resistant to a given targeted therapy? Since point mutations seem to represent the main cause of resistance to targeted therapies and since it appears that the main effect of these therapies on leukemic stem cells is cytostatic, it follows that an answer to the above question is the key in order to understand what is the probability, for a newly diagnosed CML patient, to incur in a later relapse of the disease due to the development of resistance to the treatment.

If we assume for simplicity that point mutations occur randomly during DNA replication, and letting u be the probability that when a leukemic stem cell divides one of its daughter cells will carry a point mutation, we can think about this problem as depicted in Fig. 15.2.

Denote by $S(t)$, the total number of wild-type LSCs present in a patient at time t. Denote by l and d, the rates at which LSCs divisions and deaths occur, respectively. Also, denote by a, b, and c the probabilities that a cell division will be asymmetric, symmetric differentiation, or symmetric self-renewal, respectively (thus $a+b+c=1$).

The average dynamics of the wild-type LSC population can then be described by the following equation:

$$\frac{\mathrm{d}S(t)}{\mathrm{d}t} = \left[l(1-u)(1-a-b) - \left(d + bl + \frac{ual}{2} \right) \right] S(t) \qquad (15.1)$$

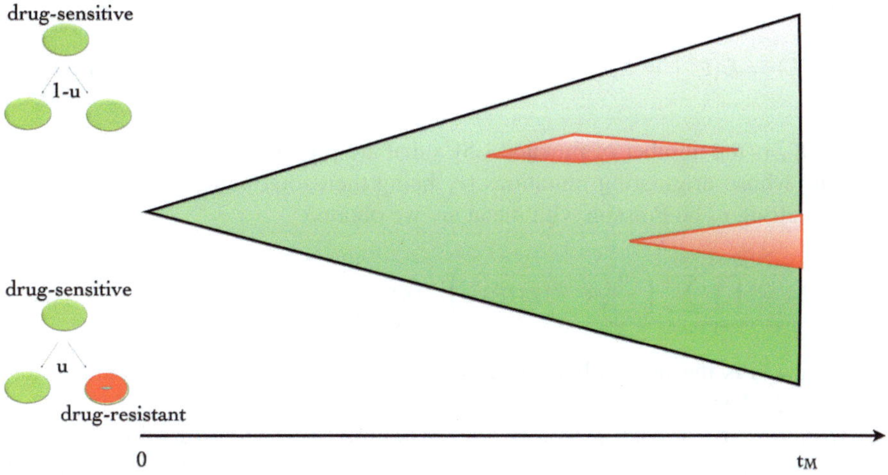

Fig. 15.2 Clonal expansions. The wild-type leukemic stem cells expansion (*in green*) from the first LSC, at time 0, up to the time of diagnosis t_M. Mutated LSCs subclones (*in red*) may appear and go subsequently extinct or grow and be present at the time of diagnosis. The model estimates the probability of having mutated LSCs (*red*) at time t_M. *LSC* leukemic stem cell

Indeed, the wild-type population grows only when a symmetric self-renewal without mutations occurs, an event of probability $(1-u)(1-a-b)$, while it decreases due to cell death, symmetric differentiation, or when the stem cell daughter of an asymmetric division is hit by a mutation. Since u is very small ($< 10^{-6}$ in CML), we can approximate the above equation by

$$\frac{\mathrm{d}S(t)}{\mathrm{d}t} = [l(1-a-2b)-d]\cdot S(t) \tag{15.2}$$

By solving Eq. (15.2), we find that the average time for which the wild-type LSC population will consist of x cells is

$$t_{x+1} - t_x = \frac{\ln(1+1/x)}{l(1-a-2b)-d}, \tag{15.3}$$

where t_x is the time at which the population reaches size x. Thus, the average number of mutations occurring while $S = x$, which we defined as m_x, is given by multiplying the number of cells present at that time by the mutation rate and by the average time for which $S = x$,

$$m_x = xul(1-a/2-b)\frac{\ln(1+1/x)}{l(1-a-2b)-d} \approx \frac{ul(1-a/2-b)}{l(1-a-2b)-d} \tag{15.4}$$

since $x \ln(1+1/x)$ 1. It is important to note that m_x is not a function of x.

Now, we will use branching processes [47] to model the mutant LSCs. Let M and T be the total number of wild-type and mutated LSCs present in a patient at the time of diagnosis, respectively, and let G_T be the probability generating function of T.

Then

$$G_T(\xi) = E[\xi^T] = E[\xi^{K_1+\ldots+K_{M-1}}] = E[E[\xi^{K_1}|r_1]]\cdots E[E[\xi^{K_{M-1}}|r_{M-1}]], \tag{15.5}$$

where K_x is the number of mutated LSCs that are present at the time of diagnosis, t_M, and whose originating mutations (r_x being their total number) occurred when $S = x$. Letting r_x be Poisson with mean m_x we obtain

$$G_T(\xi) = \prod_{x=1}^{M-1} \sum_{r_x=0}^{\infty} \left(\frac{m_x^{r_x}}{r_x!} e^{-m_x} g_x(\xi)^{r_x} \right) = \exp\left(-\sum_{x=1}^{M-1} m_x \left[1 - g_x(\xi)\right] \right), \tag{15.6}$$

where $g_x(\xi)$ is the probability generating function, at time t_M, of a mutant clone originated when $S = x$.

To find $g_x(\xi)$, we let $g_x(\xi,t)$ be the probability generating function, at time t, of a mutant clone originated when $S = x$. Here, time is measured from the time of occurrence of the originating mutation. Then, this generating function satisfies the Kolmogorov backward equation

$$\frac{\partial g}{\partial t} = l(1-a-b)g^2 + (d+bl) - (l(1-a-b)+d+bl)g. \tag{15.7}$$

Solving the above partial differential equation and noting that

$$M \approx x\mathrm{e}^{[l(1-a-2b)-d]t_{M-x}}, \tag{15.8}$$

where, $t_{M-\mathrm{x}}$ is the average time it takes for the wild-type LSCs to go from x to M, we find that

$$g_x(\xi) = g_x(\xi, t_{M-x}) \approx \frac{(\xi - 1)\left(\frac{d+bl}{l(1-a-b)}\right)\frac{M}{x} - \left(\xi - \frac{d+bl}{l(1-a-b)}\right)}{(\xi - 1)\frac{M}{x} - \left(\xi - \frac{d+bl}{l(1-a-b)}\right)}, \tag{15.9}$$

Plugging Eq. (15.9) in Eq. (15.6), we find that the probability, P_R, that at the time of detection, t_M, there are already present some mutated, drug resistance LSCs given by

$$P_R = 1 - G_T(0) \approx 1 - \exp\left(-uM\left(\frac{1-a/2-b}{1-a-b}\right)\frac{1}{C}\ln\frac{1}{1-C}\right), \tag{15.10}$$

where

$$C = \frac{d+bl}{l(1-a-b)}. \tag{15.11}$$

Moreover, it is now possible to calculate, from the probability generating function G_T, the various moments of the distribution of resistant LSCs. For example, the expected value at the time of diagnosis is given by

$$E[T] = G'_T(1) \approx uM\ \mathrm{In}(M)\left(\frac{1-a/2-b}{1-a-2b-d/l}\right). \tag{15.12}$$

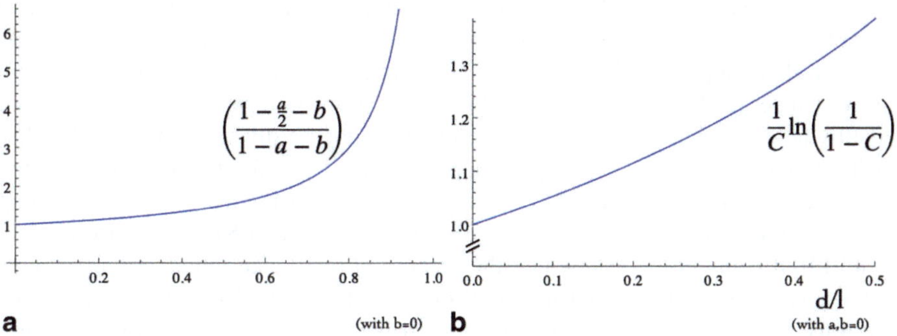

Fig. 15.3 Modes of division versus turnover rate. Plots of the given mathematical expressions, as functions of the parameter a (*left*), and of the turnover rate *d/l* (*right*)

In Fig. 15.3, where two key parts of the formula given in Eq. (15.10) are depicted, we can observe the important role played by the mode of division chosen by the LSCs in determining the amount of drug resistance generated before the start of the therapy. In fact, the first expression changes considerably, going from 1 to values above 6 (for realistic values of the parameter a), while the range for the second expression (accounting for the influence of the turnover rate *d/l*) is much smaller. Thus, it is fundamental to account for the mode of division chosen by the stem cells in any model of drug resistance development.

Applications of the Model

We will briefly consider two applications of the above model, both to CML.

A first straightforward application considers the effects that a targeted therapy, like imatinib, has on the LSCs. Assume that, when the therapy starts, the drug does not affect the LSC population dynamics at all, i.e., assume that the LSCs are insensitive to the drug. This has been the established point of view in the medical literature, and a mathematical model has been often used to support this hypothesis [48]. Interestingly, the mathematical model had that hypothesis included among its assumptions.

If this hypothesis were to be correct, then the LSC population will continue to grow also after the start of the therapy, beyond the number present at the time of detection given by M in Eq. (15.10). Therefore, the probability of developing LSCs resistant to the drug would continue growing, according to the formula found in Eq. (15.10). But this clearly contradicts the evidence observed in the data coming from long clinical trials [49]. Thus, it must be that imatinib also affects the LSC population [50].

A second nontrivial application considers the preferred mode of division of leukemic stem cells in CML. We have already seen that the mode of division chosen by the cancer stem cells will dramatically affect the dynamics of the tumor's growth as well as the generation of drug resistance due to random point mutations. Interestingly, there appears to be a link between symmetric self-renewal and an inherent risk of cancer [51], while it has been observed that the preferred mode of division of healthy hematopoietic stem cells is given by asymmetric division: Their probability to divide asymmetrically, given by the parameter a in Eq. (15.10), has been estimated to be close to one, and generally $a > 0.9$ [52, 53]. By using again the data coming from the International Randomized Study of Interferon and STI571 (IRIS) clinical trial [49], it is possible to show that the amount of resistance found among patients is too small for the parameter a in the formula of Eq. (15.10) to be close to one. We must rather have $a < 0.5$ (see [46] for further details). Thus, the model allows us to infer that the LSCs should have a much lower than normal tendency to divide asymmetrically hence indicating a substantial shift toward an increased symmetric self-renewal among leukemic stem cells.

Conclusions

We have reviewed a few elements of the basic biology behind the development of drug resistance in cancer and some of the mathematical models found in the literature. Given the key role played by cancer stem cells in carrying long-term drug resistance, we have analyzed in some detail a recent stochastic model that includes the dynamics of stem cells among its elements, and a few applications of this model to CML have been considered. Interestingly, the formulas obtained with this model have also been recently used in predicting drug resistance to targeted therapy in colorectal cancer [54]. In conclusion, mathematical modeling has proven to be a useful tool for analyzing the dynamics of drug resistance.

References

1. Perry MC. The chemotherapy source book. Philadelphia: Lippincott Williams & Wilkins; 2008.
2. Goldie JH, Coldman AJ. Drug resistance in cancer: mechanisms and models. Cambridge: Cambridge University Press; 1998.
3. Takimoto CH, Calvo E. Principles of oncologic pharmacotherapy. In: Pazdur R, Wagman LD, Camphausen KA, Hoskins WJ, editors. Cancer management: a multidisciplinary approach. UBM Medica; 2008.
4. Druker BJ, Tamura S, Buchdunger E, Ohno S, Segal GM, Fanning S, et al. Effects of a selective inhibitor of the Abl tyrosine kinase on the growth of Bcr-Abl positive cells. Nat Med. 1996;2:561–6.
5. Schindler T, Bornmann W, Pellicena P, Miller WT, Clarkson B, Kuriyan J. Structural mechanism for STI-571 inhibition of abelson tyrosine kinase. Science. 2000;289:1938–42.
6. Varmus H. The new era in cancer research. Science. 2006;312:1162–5.
7. McCormick F. New-age drug meets resistance. Nature. 2001;412:281–2.
8. Teicher BA. Cancer drug resistance. Totowa: Humana Press; 2006.
9. Daniel G. Targeted therapies: mechanisms of resistance. New York: Springer; 2011.

10. Frei E, Teicher BA, Holden SA, Cathcart KN, Wang YY. Preclinical studies and clinical correlation of the effect of alkylating dose. Cancer Res. 1988;48:6417–23.
11. Griswold DP, Trader MW, Frei E III, Peters WP, Wolpert MK, Laster WR. Response of drug-sensitive and -resistant L1210 leukemias to high-dose chemotherapy. Cancer Res. 1987;47:2323–7.
12. Schimke RT. Gene amplification, drug resistance, and cancer. Cancer Res. 1984;44:1735–42.
13. Schimke RT. Gene amplification in cultured cells. J Biol Chem. 1988;263:5989–92.
14. Souhami RL, Gregory WM, Birkhead BG. Mathematical models in high-dose chemotherapy. Antibiot Chemother. 1988;41:21–8.
15. Gottesman MM. Mechanisms of cancer drug resistance. Annu Rev Med. 2002;53:615–27.
16. Luria SE, Delbruck M. Mutations of bacteria from virus sensitivity to virus resistance. Genetics. 1943;28:491–511.
17. Skipper HE, Schabel FM, Lloyd H, editors. Dose-response and tumor cell repopulation rate in chemotherapeutic trials. New York: Marcel Dekker; 1979.
18. Zheng Q. Progress of a half century in the study of the Luria-Delbruck distribution. Math Biosci. 1999;162(1–2):1–32.
19. Coldman AJ, Goldie JH. Role of mathematical modeling in protocol formulation in cancer chemotherapy. Cancer Treat Rep. 1985;69:1041–8.
20. Coldman AJ, Goldie JH. A stochastic model for the origin and treatment of tumors containing drug-resistant cells. Bull Math Biol. 1986;48:279–92.
21. Goldie JH, Coldman AJ. A mathematic model for relating the drug sensitivity of tumors to their spontaneous mutation rate. Cancer Treat Rep. 1979;63:1727–33.
22. Goldie JH, Coldman AJ. Quantitative model for multiple levels of drug resistance in clinical tumors. Cancer Treat Rep. 1983;67:923–31.
23. Goldie JH, Coldman AJ. A model for resistance of tumor cells to cancer chemotherapeutic agents. Math Biosci. 1983;65:291–307.
24. Goldie JH, Coldman AJ, Gudauskas GA. Rationale for the use of alternating non-cross-resistant chemotherapy. Cancer Treat Rep. 1982;66:439–49.
25. Komarova N. Stochastic modeling of drug resistance in cancer. J Theor Biol. 2006;239:351–66.
26. Komarova N, Katouli AA, Wodarz D. Combination of two but not three current targeted drugs can improve therapy of chronic myeloid leukemia. PLoS ONE. 2009;4:e4423.
27. Komarova NL, Wodarz D. Drug resistance in cancer: principles of emergence and prevention. Proc Natl Acad Sci U S A. 2005;102:9714–9.
28. Komarova NL, Wodarz D. Effect of cellular quiescence on the success of targeted CML therapy. PLoS ONE. 2007;2:e990.
29. Iwasa Y, Nowak MA, Michor F. Evolution of resistance during clonal expansion. Genetics. 2006;172:2557–66.
30. Durrett R, Moseley S. Evolution of resistance and progression to disease during clonal expansion of cancer. Theor Popul Biol. 2010;77:42–8.
31. Harnevo LE, Agur Z. The dynamics of gene amplification described as a multitype compartmental model and as a branching process. Math Biosci. 1991;103:115–38.
32. Harnevo LE, Agur Z. Use of mathematical models for understanding the dynamics of gene amplification. Mutat Res. 1993;292:17–24.
33. Kimmel M, Axelrod DE. Mathematical models of gene amplification with applications to cellular drug resistance and tumorigenecity. Genetics. 1990;125:663–44.
34. Birkhead BG, Rakin EM, Gallivan S, Dones L, Rubens RD. A mathematical model of the development of drug resistance to cancer chemotherapy. Eur J Cancer Clin Oncol. 1987;23:1421–7.
35. Panetta JC, Adam J. A mathematical model of cycle-specific chemotherapy. Math Comput Model. 1995;22:67–82.
36. Cojocaru L, Agur Z. A theoretical analysis of interval drug dosing for cell-cycle-phase-specific drugs. Math Biosci. 1992;109:85–97.
37. Dibrov BF. Resonance effect in self-renewing tissues. J Theor Biol. 1998;192:15–33.

38. Gaffney EA. The application of mathematical modelling to aspects of adjuvant chemotherapy scheduling. J Math Biol. 2004;48:375–422.
39. Gaffney EA. The mathematical modelling of adjuvant chemotherapy scheduling: incorporating the effects of protocol rest phases and pharmacokinetics. Bull Math Biol. 2005;67:563–611.
40. Webb GF. Resonance phenomena in cell population chemotherapy models. Rocky Mt J Math. 1990;20:1195–216.
41. Gregory WM, Birkhead BG, Souhami RL. A mathematical model of drug resistance applied to treatment for small-cell lung cancer. J Clin Oncol. 1988;6:457–61.
42. Panetta JC. A mathematical model of drug resistance: heterogeneous tumors. Math Biosci. 1998;147(1):41–61.
43. Campbell LL, Polyak K. Breast tumor heterogeneity. Cell Cycle. 2007;6(19):2332–8.
44. Heppner GH. Tumor heterogeneity. Cancer Res. 1984;44:2259–65.
45. Park SY, Gonen M, Kim HJ, Michor F, Polyak K. Cellular and genetic diversity in the progression of in situ human breast carcinomas to an invasive phenotype. J Clin Invest. 2010;120(2):636–44.
46. Tomasetti C, Levy D. Role of symmetric and asymmetric division of stem cells in developing drug resistance. Proc Natl Acad Sci U S A. 2010;107(39):16766–71.
47. Kimmel M, Axelrod DE. Branching processes in biology. New York: Springer; 2002.
48. Michor F, Hughes TP, Iwasa Y, Branford S, Shah NP, Sawyers CL, et al. Dynamics of chronic myeloid leukaemia. Nature. 2005;435:1267–70.
49. Hochhaus A, O'Brien SG, Guilhot F, Druker BJ, Branford S, Foroni L, et al. Six-year follow-up of patients receiving imatinib for the first-line treatment of chronic myeloid leukemia. Leukemia. 2009;23:1054–61.
50. Tomasetti C. A new hypothesis: imatinib affects leukemic stem cells in the same way it affects all other leukemic cells. Blood Cancer J. 2011;e19. doi:10.1038/bcj.2011.17.
51. Morrison SJ, Kimble J. Asymmetric and symmetric stem-cell divisions in development and cancer. Nature. 2006;441:1068–74.
52. Giebel B, Zhang T, Beckmann J, Spanholtz J, Wernet P, Ho AD, et al. Primitive human hematopoietic cells give rise to differentially specified daughter cells upon their initial cell division. Blood. 2006;107:2146–52.
53. Wu M, Kwon HY, Rattis F, Blum J, Zhao C, Ashkenazi R, et al. Imaging hematopoietic precursor division in real time. Cell Stem Cell. 2007;1:541–54.
54. Diaz LA Jr, Williams RT, Wu J, Kinde I, Hecht JR. Berlin J, et al. The molecular evolution of acquired resistance to targeted EGFR blockade in colorectal cancers. Nature. 2012;486(7404):537–40.

Chapter 16
Etiology and Treatment of Hematological Neoplasms: Stochastic Mathematical Models

Tomas Radivoyevitch, Huamin Li and Rainer K. Sachs

Abstract Leukemias are driven by stemlike cancer cells (SLCC), whose initiation, growth, response to treatment, and posttreatment behavior are often "stochastic", i.e., differ substantially even among very similar patients for reasons not observable with present techniques. We review the probabilistic mathematical methods used to analyze stochastics and give two specific examples. The first example concerns a treatment protocol, e.g., for acute myeloid leukemia (AML), where intermittent cytotoxic drug dosing (e.g., once each weekday) is used with intent to cure. We argue mathematically that, if independent SLCC are growing stochastically during prolonged treatment, then, other things being equal, front-loading doses are more effective for tumor eradication than back loading. We also argue that the interacting SLCC dynamics during treatment is often best modeled by considering SLCC in microenvironmental niches, with SLCC–SLCC interactions occurring only among SLCC within the same niche, and we present a stochastic dynamics formalism, involving "Poissonization," applicable in such situations. Interactions at a distance due to partial control of total cell numbers are also considered. The second half of this chapter concerns chromosomal aberrations, lesions known to cause some leukemias. A specific example is the induction of a Philadelphia chromosome by ionizing radiation, subsequent development of chronic myeloid leukemia (CML), CML treatment, and treatment outcome. This time evolution involves a coordinated sequence of > 10 steps, each stochastic in its own way, at the subatomic, molecular, macromolecular, cellular, tissue, and population scales, with corresponding time scales ranging from picoseconds to decades. We discuss models of these steps and progress in integrating models across scales.

R. K. Sachs (✉) · H. Li
Department of Mathematics, Evans Hall, MC3840,
University of California at Berkeley, Berkeley, CA 94720, USA
Tel.: 510-658-5790
e-mail: sachs@math.berkeley.edu

T. Radivoyevitch
Epidemiology and Biostatistics, Case Western Reserve University,
Cleveland, OH, USA
e-mail: radivot@gmail.com; radivot@ccf.org

S. J. Corey et al. (eds.), *A Systems Biology Approach to Blood*,
Advances in Experimental Medicine and Biology 844, DOI 10.1007/978-1-4939-2095-2_16

Keywords Stochastic modeling · Monte Carlo simulations · Stemlike cancer cells (SLCC) · Acute myeloid leukemia (AML) · Chronic myeloid leukemia (CML) · Modeling protocols for fractionated dosing · Theorem on front-loading dose · Birth–death models · Extinction in stochastic systems · Clone niche model · Chromosomal aberrations · Ionizing radiation · Physical, chemical, cellular, clinical, and epidemiological time scales · Stochastic geometry · Markov chains · Poissonization · DNA double-strand breaks

Abbreviations

AML	Acute myeloid leukemia
AUC	Area under curve
CAS	Chromosome aberration simulator
CML	Chronic myeloid leukemia
DSB	DNA double-strand break(s)
Gy	Radiation dose unit = 1 J/kg
HSC	Normal hematopoietic stem cell(s)
LSS	Life span study of atomic bomb survivors
SEER	Cancer database (www.seer.cancer.gov)
SLCC	Stemlike cancer cell(s)
TCP	Tumor control probability
TKI	Tyrosine kinase inhibitor(s), e.g., imatinib

Introduction

Stochastics

We shall here often consider stemlike cancer cells (SLCC). Within a heterogeneous cancer, SLCC are cells putatively capable of reinitiating the cancer if placed in the right microenvironment [1]. Usually in this chapter, no other stemlike properties are implied. SLCC are sometimes also called "tumor clonogens" [2].

Many biological and medical situations involve "stochastic" effects. Stochasticity refers to situations that appear highly similar using currently optimal observational techniques, but that may nonetheless have very different outcomes. To model such situations, probabilistic methods are needed. "Deterministic" calculations refer to averages instead. Figure 16.1 gives a hypothetical example.

Mathematical and Computational Models of Stochastic Processes

In this chapter, we analyze the dynamics (i.e., time evolution) of neoplasms using the two main techniques for quantitative studies of stochastics: Monte Carlo simulation [3] and analytic stochastic process theory [4–6].

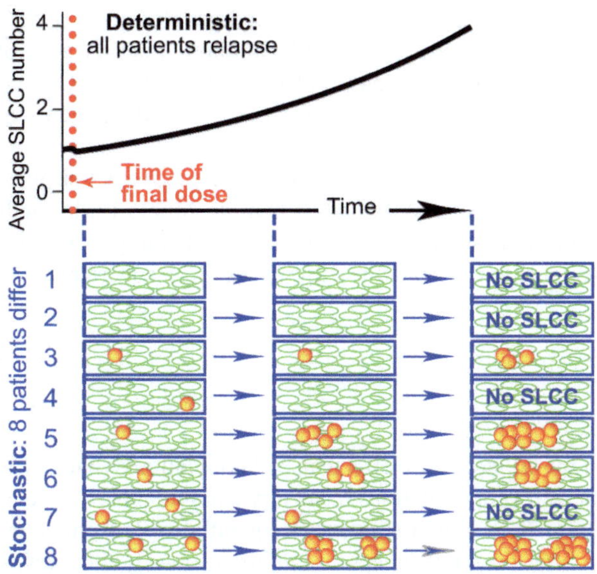

Fig. 16.1 Treatment discontinuation possibilities. To illustrate deterministic versus stochastic models, suppose that (1) Semlike cancer cells (SLCC) (*spheres*) have a growth advantage over normal stem cells in the absence of drug treatment, but a disadvantage under treatment and (2) treatment is discontinued when the average number of SLCC is just 1. What will happen? Assuming a deterministic model, SLCC will repopulate, e.g., with exponential growth as shown. However, a more realistic stochastic model can give different results, even if the average number of SLCC at each time equals the deterministic estimate for that time. Here snapshots, three times shown by the *dashed vertical lines*, given for eight patients. Of these eight, patients 1 and 2, responding stochastically to the treatment, were "lucky" enough to be fully SLCC-free when treatment was discontinued; afterwards, they remain disease-free. "Luck" refers to factors not systematically foreseeable with current techniques. Patient 4 was also lucky, and patient 7 was even luckier: both SLCC died out accidentally, one after the other, despite their growth advantage. On the other hand, patient 8 is substantially worse off than the deterministic model would predict, while patients 5 and 6 have also been unlucky. Whether any of the four SLCC-bearing patients will relapse clinically is also in part a matter of luck.

In Monte Carlo simulations, computer software uses random number generators, in effect rolling dice, to mimic the stochastic decisions a biological system makes. For example, when a SLCC undergoes mitosis in Fig. 16.1, there are three possibilities: a "birth" if both daughters are viable SLCC, no change in SLCC numbers if one daughter is an SLCC and the other daughter permanently leaves the SLCC subpopulation (e.g., undergoes apoptosis or differentiates), and a "death" if both daughters leave the SLCC subpopulation. Each of these three possibilities has a certain probability. Tracking SLCC dynamics for a specific simulated patient, Monte Carlo software will determine probabilistically at every successive mitosis which of these possibilities occurs and thereby track the integer number of SLCC in the patient. By simulating thousands of patients, one then gets an estimate of what percentage of patients is predicted to have 0, 1, 2, etc. SLCC at the end of the treatment.

Other subpopulations, e.g., drug resistant or emergent faster-growing SLCC subpopulations, are tracked similarly. Monte Carlo simulation is very generally applicable, is surprisingly simple to implement, and is clear intuitively. One common use of Monte Carlo simulations is in "individual-based" (including agent-based) modeling [7].

However, analytic methods, when available, often give better overall insight into the various possible dynamical outcomes than do Monte Carlo methods. We shall discuss this point in the section below on the front-loading theorem, when we have an example available to illustrate the differences more concretely.

Preview

As Fig. 16.1 suggests, stochastic effects in cell population dynamics are most important when some cell subpopulations have very small numbers—are teetering on the edge of extinction. This situation occurs frequently for leukemias: SLCC numbers are small soon after (cryptic) leukemia initiation; during later stages of prolonged leukemia treatments, e.g., just after treatment is discontinued; and after a bone marrow transplant. Moreover, new subpopulations that arise during progression are small at first, as are clones of drug-resistant mutants. Therefore, we shall here use neoplasms instead of normal hematopoiesis to illustrate stochastics.

One of our two main examples will concern modeling the eradication of leukemias which are proliferating during intermittent, fractionated drug dosing. The second concerns the induction of a *BCR–ABL* chromosome aberration (i.e., a Philadelphia chromosome) by ionizing radiation, subsequent development of chronic myeloid leukemia (CML), and CML response to tyrosine kinase inhibitor (TKI) treatment; this second example is instructive because it is highly multiscale, and because there are more than 10 interrelated steps involved, each with a somewhat different kind of stochastics.

Dose Timing

Fractionated Dosing

Doses of drugs (or ionizing radiation) are often "fractionated," i.e., given intermittently over a prolonged period (Fig. 16.2). One main reason is to control deleterious side effects. For fractionated treatment of SLCC that proliferate during the prolonged period, many clinical and mathematical questions arise, giving an instructive example of the difference between deterministic and stochastic analyses. We consider SLCC population dynamics during fractionated treatments that attempt to eradicate all SLCC and thereby achieve cure. We first discuss scenarios where SLCC do not interact with each other, only with their microenvironment, then scenarios where SLCC–SLCC interactions occur.

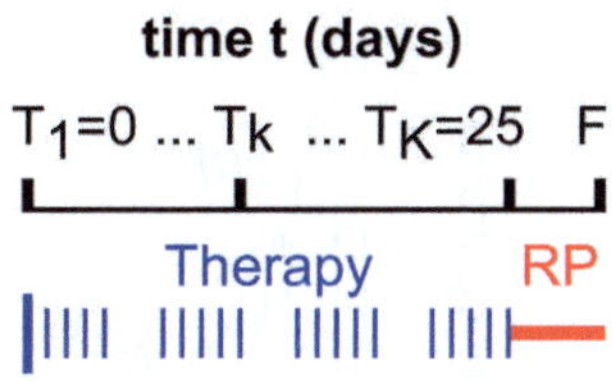

Fig. 16.2 A schematic example of fractionated dosing. Shown is a hypothetical 4-week cycle of one dose each weekday. The first dose fraction, given on Monday of week 1, is larger than the others. The total number of dose fractions and the time of the kth fraction are denoted by K and T_k, respectively, and RP denotes a recovery period during which the drug washes out of the patient (and short-term side effects ameliorate). The end of the RP is here taken as $F = 30$ days. Cycles may be repeated, but we here focus on a single cycle. Gaps and variable doses can occur within a cycle. Fractionated protocols are somewhat analogous to "metronomic" ones used, e.g., in antiangiogenic disease management [8]

Response to Fractionated Dosing: Deterministic Birth–Death Models for Non–Interacting SLCC

Consider SLCC in a large group of highly similar patients. Denote the (per SLCC) birthrate at time t by $b(t) \geq 0$, i.e., if $m(t)$ denotes the number of SLCC at time t averaged over patients, the average number of extra cells per patient born in a very short time interval dt is $b(t)m(t)dt$. The birthrate $b(t)$ might be time independent, but also might instead change in time, e.g., if treatment triggers a delayed biological response (see, e.g., Ref [9]). Here "birth" refers to the extra SLCC that arises when a symmetric mitosis without differentiation produces two viable SLCC daughters.

Similarly, denote the endogenous (treatment independent) death rate (per SLCC) by $d(t) \geq 0$ and the treatment death rate by $\alpha c(t)$. Here the drug blood plasma concentration $c(t)$, in μg/mL, is assumed to be that experienced by the SLCC, treatment death rate has been approximated as proportional to $c(t)$, and α is the constant needed to convert $c(t)$ to a death rate. For the kth dose D_k (in mg/kg), we take the incremental contribution to the concentration $c(t)$ to be $c_k(t) = s(t - T_k)D_k/V_d$. Here the shape $s(t)$, shown schematically in Fig. 16.3a, has been normalized to have an integral (AUC, area under curve) of 1 day and V_d is the volume of distribution. Table 16.1 summarizes the quantities we have introduced and defines some auxiliary quantities.

For fractionation, the total plasma concentration $c(t)$ has some pattern generally similar to the one shown in Fig. 16.3b. During treatment, the treatment death rate $\alpha c(t)$ will normally make the *net* death rate $\lambda(t) = d(t) + \alpha c(t) - b(t)$ positive, but if the treatment death rate is small or zero, $\lambda(t)$ can be negative—the number of SLCC usually increases in the absence of treatment. Note in Fig. 16.3b that the drug concentration can drop strongly on weekends, possibly allowing tumor repopulation. Thus, for rapidly growing neoplasms, weekend breaks are deprecated.

Assuming no SLCC–SLCC interactions, the standard deterministic differential equation for the average SLCC number $m(t)$ and its solution are the following, where the notation of Table 16.1 is used:

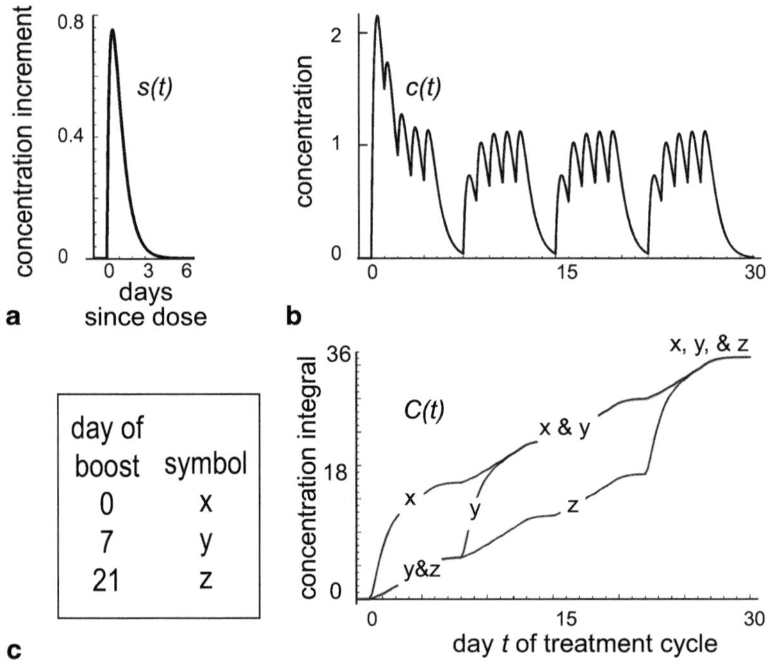

Fig. 16.3 The time course of drug concentration and its integral. **a** A schematic plot of the shape $s(t)$ of the drug concentration due to a dose given at $t = 0$; the total area under the curve (AUC) is taken to be 1 day. The numerical example here is intended only to illustrate the general behavior of the quantities in Eqs. 16.2 and 16.3. Table 16.1 gives the units for $s(t)$, $c(t)$, and $C(t)$. **b** A plot of $c(t)$ for the cycle shown in Fig. 16.2 and $s(t)$ as in **a**, with the same dose each weekday except for the first Monday where the dose is higher. Throughout this section, we assume that $c(t)$ has dropped back down to zero by the time the recovery period ends. **c** The concentration integral C(t) for three slightly different cycles: the same dose each weekday for 4 weeks except on one Monday (day 0, 7, or 21), when the dose is 10-fold (an unusually high boost, chosen here for visual clarity in this schematic example). The curve for an earlier boost is always at least as high as for a later boost and is higher at some times. This feature is general; its implications for stochastic proliferation lead to a front-loading theorem

$$\text{(A)}\ \mathrm{d}m/\mathrm{d}t = -\lambda m; \text{the solution is}$$

$$\text{(B)}\ m(t) = m(0)\exp\left[-\int_0^t \lambda(y)\mathrm{d}y\right]; \text{here}$$

$$\text{(C)}\ \int_0^F \lambda(t)\mathrm{d}t = \int_0^F [d(t) - b(t)]\mathrm{d}t + \alpha C(F)$$

$$= \int_0^F [d(t) - b(t)]\mathrm{d}t + \alpha \sum\nolimits_{k=1}^{K} D_k. \qquad (16.1)$$

Here Eq. 16.1 states that on average, the increase in SLCC number in a short time dt is the negative of the net death rate $\lambda(t)$ multiplied by the average cell number

Table 16.1 Symbols used

	Symbol	Definition or example	Name	Units
1	K	Fig. 16.2	Total number of fractions	None
2	k	$k = 1, 2, \ldots, K$	Fraction number	None
3	D_k	Text below Fig. 16.2	Dose for kth fraction	mg/kg
4	T_k	Fig. 16.2	Time of kth fraction	Days
5	F	Fig. 16.2	Time at end of recovery	Days
6	$s(t)$	Fig. 16.3	Concentration increment curve	None
7	$c_k(t)$	$c_k(t) = D_k\ s(t - T_k)$	kth concentration increment	μg/mL
8	$c(t)$	$\sum_{k=1}^{k} c_k(t)$; Fig. 16.3	Concentration at time t	μg/mL
9	$C(t)$	$C(t) = \int_0^t c(y)dy$	Concentration integral; AUC $= C(F)$	μg-day/mL
10	$b(t) \geq 0$	Text below Fig. 16.2	Birthrate per cell	Per day
11	$d(t) \geq 0$	Text below Fig. 16.2	Endogenous death rate per cell	Per day
12	$\alpha > 0$	Text below Fig. 16.2	SLCC drug sensitivity	mL/(μg-day)
13	$\alpha c(t) \geq 0$	8 and 12 above	Treatment death rate per cell	Per day
14	$\lambda(t)$	$\lambda(t) = d(t) + \alpha c(t) - b(t)$	Net death rate per cell	Per day
15	$m(t)$	Text below Fig. 16.2	Average number of SLCC at time t	None
16	$q(F)$	Eq. (16.3)	Survival probability	None
17	V_d	Text near Fig. 16.3	Volume of drug distribution	L/kg

$m(t)$. The solution Eq. 16.1 of this differential equation is familiar from calculus (e.g., Ref. [10]; Box 6.2). If the net death rate λ happens to be constant in time, i.e., $\lambda(t) = a > 0$, then Eq. 16.1 is just the equation of exponential decay $m(t) = m(0)e^{-at}$; in general, the exponent involves an integral of $\lambda(t)$ over time. Eq. 16.1 follows from the definitions in Table 16.1 of λ, $C(t)$, and $c_k(t)$ together with properties of $s(t)$ and $c(t)$ described in the caption to Fig. 16.3. The summation in Eq. 16.1 is merely the total dose, independent of the way this dose is distributed among the dose fractions. This fact implies, since α is fixed, the standard, important (and at first surprising) result that the deterministic estimate of the average number $m(F)$ of SLCC at the final time F is independent of the order in which different doses are administered, provided that (a) the drug pharmacokinetics are linear (e.g., doubling the dose doubles $c(t)$) and (b) "other things are equal," namely that dosing does not alter SLCC dose sensitivity α or the time dependence of the endogenous rates $b(t)$ and $d(t)$. In contrast, stochastic estimates are sensitive to dose delivery timing, as will be discussed below.

When the SLCC do not interact with each other, we can get full information by considering a single clone that has exactly one cell at time $t = 0$. We now temporarily confine attention to this case. Then, for all cases of interest $m(F)$ in Eq. 16.1, the average number of SLCC at the end of recovery, assuming the initial number was 1, obeys $m(F) << 1$ since we are considering treatments designed to eradicate. When

$m(F) << 1$, this average number can usually be considered the deterministic estimate of the probability that a clone started by a single cell at time $t = 0$ has been eradicated by time $t = F$. Two ways to see this fact are the following: If, as is usually an excellent approximation, no such clones have more than 1 SLCC at time $t = F$, then $m(F)$ is exactly the eradication probability; if the cell number per clone is Poisson distributed, the survival probability is $1 - \exp[-m(F)] \cong m(F)$.

Response to Fractionated Dosing: Stochastic Birth–Death Models

In this section, we continue to assume no SLCC interactions; we focus attention entirely on a single clone that at time $t = 0$ has only one cell, assume $m(F) << 1$, and interpret $m(F)$ as the deterministic estimate of the survival probability for such a clone. Following seminal work by Tucker and coworkers (reviewed in Ref. [11]), stochastic birth–death models (reviewed in [5, 6, 12]) have generally been used for estimating disease persistence in the situation described deterministically by Eq. 16.1. The stochastic models calculate the time-dependent probabilities that a clone has no SLCC, exactly 1 SLCC, exactly 2 SLCC, etc. They can therefore be used to calculate the survival probability $q(F)$ that the clone has at least 1 SLCC at the end of the recovery period. For birth and death rates linearly proportional to m, corresponding to independent SLCC, and to exponential growth or decay, the result (Ref. [5] example 4.8) is as follows:

$$q(F) = \frac{m(F)}{1 + \int_0^F b(t) \exp\left[-\int_t^F \lambda(y) \mathrm{d}y\right] \mathrm{d}t}. \tag{16.2}$$

We analyze and interpret this useful equation via six remarks and a front-loading theorem. The six remarks are:

1. The numerator is the deterministic estimate of the clone's survival probability, as discussed above.
2. If the (per cell) birthrate $b(t)$ is zero, no stochastic correction is needed; $q(F) = m(F)$.
3. Otherwise, the extra term in the denominator of Eq. 16.2 is positive, i.e., stochastic fluctuations always decrease $q(F)$ below $m(F)$, and in this sense, deterministic estimates tend to be overly pessimistic. An intuitive interpretation is that a stochastic fluctuation can "accidentally" eradicate a small clone, and thereafter, no stochastic fluctuation or birthrate, however large, can resurrect the clone. On the other hand, a stochastic fluctuation that increases the SLCC number merely changes a surviving clone into a larger surviving clone which, with luck, can still be eradicated.
4. Suppose, for a given protocol, we compare two patients who have the same net death rate $\lambda(t)$ and suppose the second patient has $b(t)$ greater than the first (and thus necessarily a greater endogenous death rate $d(t)$ to keep $\lambda(t)$ fixed). Then the second patient has a better prognosis—the estimated SLCC survival probability,

Eq. 16.2, is smaller since the integral in the denominator of Eq. 16.2 contains $b(t)$, and everything else in Eq. 16.2 depends only on the net death rate. The intuitive interpretation is that increasing the birth and endogenous death rates by the same amount corresponds to a more fluctuating situation and thus a higher chance of reaching a zero-SLCC state accidentally.

5. The exponential in the denominator of Eq. 16.2 is an "influence function" that, intuitively speaking, carries the effects of having an extra cell born at time t forward to the later time F.
6. Many other results can be obtained from Eq. 16.2, e.g., estimates of doses needed to cure half of a patient population. Stochastic process theory (e.g., [5]) also gives many further equations similar to Eq. 16.2, e.g., for the survival probability $q(t)$ at times earlier than F, for the probability that a clone has exactly one SLCC at any time, not just the probability $q(t)$ that it has at least one, etc. These further results are not needed in this chapter but are often useful in detailed calculations.

The Front-Loading Theorem

This section describes a theorem based on the stochastic formalism above. Often in fractionated treatments, some doses are "boosted," i.e., larger than others (e.g., Fig. 16.2). Typically "front loading" or "back loading" is used, boosting early or late doses, respectively. We now show that if "other things"—SLCC sensitivity α and per cell birth and endogenous death rates $b(t)$ and $d(t)$—are equal, stochastic fluctuations favor front loading as far as SLCC eradication is concerned in the present scenario. This result substantially generalizes a previous result on front loading [13].

Theorem Consider a cycle where two doses obey $D_k < D_j$, with fraction k earlier than fraction j. Suppose the birthrate obeys $b(t) > 0$. Then switching D_k and D_j (so that the larger dose is earlier instead of later) always decreases the SLCC clone survival probability $q(F)$, provided "other things are equal," as defined above.

Remarks Thus, stochastic effects on eradication of SLCC favor front loading. It should be emphasized that the theorem does not imply front loading is always the preferred protocol. One main reason is that front loading may increase temporary side effects so severely that patients may need or choose to discontinue or postpone treatments, which could be very counterproductive. In fact, a routine extension of the theorem states that for stochastically proliferating cancers, the best way to eradicate SLCC with a given total dose is to concentrate all the dose in one single treatment given as soon as possible—intuitively speaking, the sooner we drive the average SLCC number to low levels, the longer the period where chance events might eradicate the SLCC population entirely. But usually, the main reason for fractionation in the first place is that such a 1-dose protocol causes intolerable side effects.

Additional caveats concern the assumption that individual dose sizes do not influence the sensitivity α or the time course of the endogenous per cell birth and death rates $b(t)$ and $d(t)$. There are many known exceptions, some of which tend to discourage front loading while others favor it. For example, if front loading results in earlier reoxygenation of hypoxic niches for SLCC, it may have the deleterious effect of increasing $b - d$ but also the favorable effect of increasing α. Furthermore, recall that Eq. 16.2 does not apply to cases where SLCC interact with each other.

Protocols that strike a balance between tumor control and side effects, and account for changes in "other things" rather than just considering tumor control with other things being equal, are beyond the scope of this chapter, whose emphasis is on interpreting stochasticity. However, it is useful to have the general results above available: In many cases, stochastic fluctuations favor eradication and front boosting; in particular, the idea of boosting the last few doses to clean up remaining SLCC rather than hitting early and hard may often be a fallacy.

Idea of the Theorem's Proof Using the notations in Table 16.1, basic properties of integrals, and basic properties of exponentials, Eq. 16.2 can be rewritten as follows:

$$\text{A)}\; q(F) = \frac{m(F)}{1 + m(F)\int_0^F [b(t)/m(t)]\mathrm{d}t}, \quad \text{where } m(F) = \exp\left[-\int_0^F \lambda(t)\mathrm{d}t\right] \text{ and}$$

$$\text{B)}\; \frac{1}{m(t)} = \exp\left[\int_0^t \lambda(y)\mathrm{d}y\right] = \exp\left[\int_0^t [d(y) - b(y)]\mathrm{d}y\right]\exp[\alpha C(t)]. \qquad (16.3)$$

Here $m(F)$ in the numerator of Eq. 16.3 is the deterministic result, and, as discussed in connection with Eq. 16.2, is independent of the individual dose sizes, provided the total dose is fixed. In the denominator, $m(F)$ also appears in this rewritten form of Eq. 16.2. Thus, the only quantity in the Eq. 16.3 for $q(F)$ that differs between the two alternative protocols is $1/m(t)$ in the integral. Eq. 16.3 shows that actually only the term $\exp[\alpha C(t)]$ differs. As illustrated in Fig. 16.3c, and readily proved by calculus, this term in the denominator of Eq. 16.3 is larger for the protocol where a boost is given earlier so that, with $b(t) > 0$ in Eq. 16.3, the SLCC clone survival probability $q(F)$ is smaller for the more nearly front-loaded protocol. If the drug is completely cleared between doses, the theorem and its proof take on a particularly simple form that might be relevant, for example, in the treatment of acute myeloid leukemia (AML) with ara-C, a drug that is cleared rapidly [14].

The theorem furnishes an instructive example of the advantages and disadvantages of analytic versus Monte Carlo stochastic calculations. On the one hand, Monte Carlo techniques can almost never give sufficient overall insights to prove a general theorem, and sometimes do not even hint at such general results. On the other hand, as soon as modeling assumptions are generalized to be more realistic, there are often no explicit solutions such as those given in Eqs. 16.1 and 16.2, while Monte Carlo methods continue to be applicable with only minor modifications.

We next consider an example of this difference by dropping our assumption above that SLCC do not interact with each other. In reality, SLCC do interact with each other [15]: by approaching a set point when the number in the body is too large or too small, by paracrine signaling, by competing for local resources such as oxygen or supportive microenvironmental niches, etc.

SLCC–SLCC Interactions via Total SLCC Control

Especially for chronic as opposed to acute leukemias, feedback control of the total SLCC number in the body is likely to occur. Feedback models have been considered by various mathematical modeling groups (e.g., in the case of CML [16–20]). Here a simplified version will be used to highlight differences between a deterministic model analyzed with a differential equation versus its stochastic counterpart analyzed by Monte Carlo methods. Consider the following deterministic birth–death model, somewhat similar to Eq. 16.1, for the total number of SLCC $m(t)$ in the body:

$$\mathrm{d}m/\mathrm{d}t = -\lambda m; \quad \text{where } \lambda = d + \alpha c(t) - b(m). \tag{16.4}$$

The key difference is that the birthrate b now depends on the total number $m(t)$, due to feedback control mediated by intercellular interactions, which complicates the solutions, especially for the stochastic counterpart of Eq. 16.4. In this illustrative example, the endogenous death rate $d > 0$ is taken as constant (i.e., feedback control on d is neglected for simplicity). The remaining term in λ is the treatment death rate as before. To get an explicit example, one should require that the net endogenous birthrate (i.e., the so-called Malthusian rate, $b - d$, which, as is typical, is the only combination that matters to this deterministic equation), have the well-known logistic form analyzed, e.g., in [21]. Then $b(m) - d$ has the form $b - d = (1/n)\{1\text{-}[m(t)/\mathrm{K}]\}d$. Here K is the set point number ("carrying capacity") in the absence of treatment, the solutions of interest obey $m(t) \leq K$, and $(1/n) > 0$ is a dimensionless parameter that determines, intuitively speaking, how fast $m(t)$ snaps back when pulled below its set point by a transient drug treatment.

For the corresponding stochastic model, one must specify the birth and endogenous death rates separately and can then use Monte Carlo methods. In terms of the constant endogenous death rate d and the snapback parameter $1/n$, the birthrate is given as $b(m) = d[(n+1)/n][1 - m/[\,(n+1K)\,]$. Thus, $b(m)$ is a decreasing linear function of m with a slope $-dK/n$ that becomes flatter as $1/n$ decreases toward 0. As $n \to \infty$, b becomes independent of m, i.e., $db/dm = 0$. In any case, when $m = K$, $b = d$.

The stochastic version of the model was implemented using the R package adaptivetau [22]. Adaptivetau gives a Monte Carlo implementation for a discrete-time Markov chain "embedded" [4] in the continuous-time Markov chain corresponding to Eq. 16.4, as follows. In effect, Eq. 16.4 is broken into a sequence of state-changing time steps randomly drawn from an appropriate exponential distribution, and changes in the integer SLCC number $m(t)$ at the corresponding times are tracked. At each step, a random number generator is used to decide if the state change at that step was a birth or a death, as governed by the birth and death rates. "Markov" refers to the fact that the system is assumed to have no memory. For example, if there are three cells at a given time, the birth and death rates are the same whether the system has always contained three cells or started at some other number and arrived at three only via some complicated sequence of births and deaths.

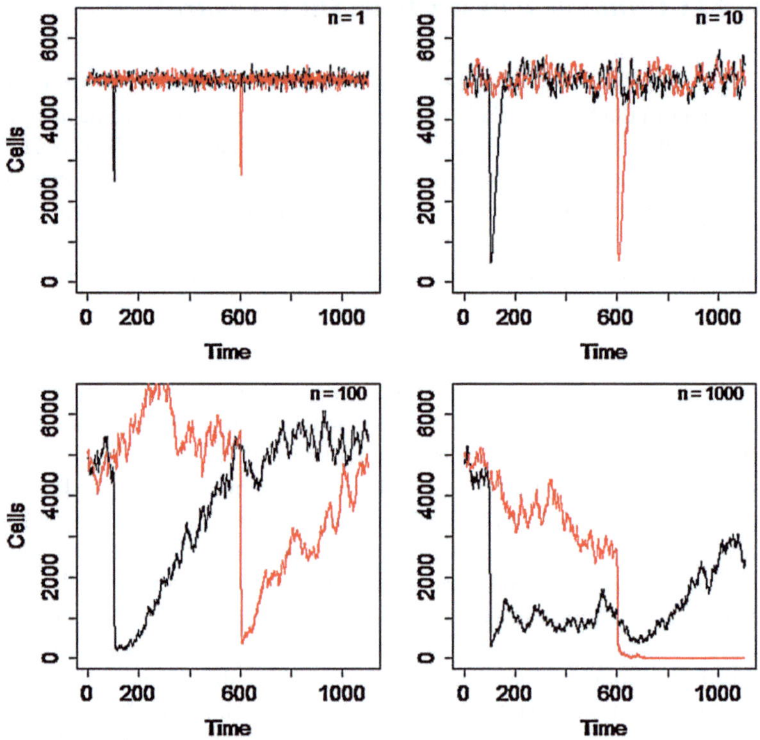

Fig. 16.4 Patient trajectories for different levels of SLCC control looseness. A 50 % increase in death rate was applied for 5 days either early at $t = 100$ (*black*) or late at $t = 600$ (*red*). Increasing levels of control system looseness (higher n values) imply increasing time constants with which the system returns to set point after perturbation by the drug, and also imply increases in the variance about the set point. Out of 100 such early-dosed (*black*) and 100 late-dosed (*red*) simulations at each n value, $n = 1$ and 10 did not yield a single cure, $n = 100$ yielded 1 cure, and $n = 1000$ yielded 52 early-dose cures and 44 late-dose cures (e.g., *red trajectory* in the bottom right panel)

In these illustrative calculations, patients were simulated with $d = 1$, $K = 5000$, $m(0) = 5000$ (i.e., the system always started at its set point), $c(t) = 0$ except during a 5-day dosing interval over which $\alpha c(t) = 0.5$, and with dosing starting either early at day 100 or late at day 600. Simulations were run through day 1105. Such simulations may have relevance to chronic leukemias or myeloid dysplastic syndrome wherein malignant cells may be under looser control than normal cells, but still be under more control than in acute leukemias (that may have $1/n = 0$). To illustrate the range of possibilities of posttreatment kinetic recoveries to set point, representative examples of simulations with n values of 1, 10, 100, and 1000 are shown in Fig. 16.4.

Figure 16.4 shows that looser control causes increase in both the variance about the set point (e.g., before perturbation by the drug) and in the amount of time needed for cell numbers to return to set point once the exposure to the drug has stopped. Based

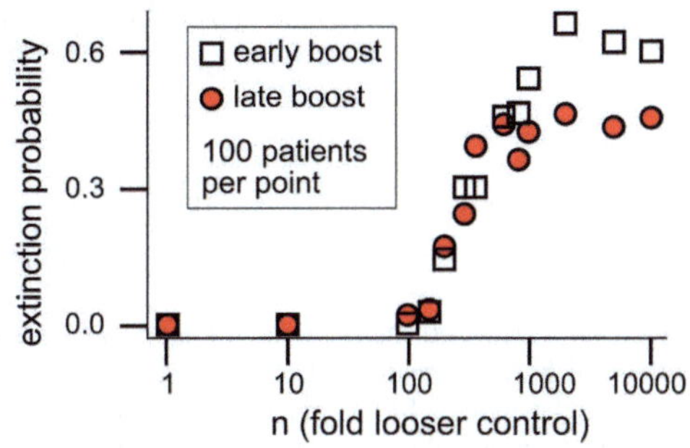

Fig. 16.5 Loose control ($n \geq 1000$) is required for an early- versus late-boost advantage

on Fig. 16.4, the probability of malignant population extinction might be predicted to be very low when $n < \sim 100$, and indeed, examining 100 early- and late-dosed patient trajectories at each n over a broad range (Fig. 16.5) confirms this fact and further indicates that the advantage of giving doses earlier rather than later requires considerable SLCC control looseness. The intuitive reason is that the endogenous per cell birthrate is larger for smaller cell numbers; front boosting means longer times with smaller cell numbers and thus a tendency, opposite to that of stochasticity, to favor back boosting. The implication is that front boosting may be more applicable to acute than to chronic leukemias.

Local SLCC–SLCC Interactions: The Niche Model and Poissonization

Cumbersome features of Monte Carlo calculations can be very substantially reduced if the main SLCC–SLCC interactions are local. Specifically, as in [23], we can emphasize *subpopulations* of SLCC rather than individual SLCC. The SLCC subpopulations are visualized as comparatively distant from each other, with each subpopulation localized in its own microenvironmental "niche" and intercellular interactions among SLCC being confined to cells in the same niche (Fig. 16.6). Here a niche can be an actual biological structure, such as a crypt in the case of colon cancer, or a bone marrow niche in the case of leukemias [24], or it may merely reflect low cell mobility with dominant intercellular interactions localized.

The niche model is flexible to the extent that all kinds of SLCC–SLCC interactions are allowed, though only for SLCC within the same niche. In some ways, the model is simpler than even a model of independent SLCC, because in some estimates (e.g., [25]), niches can be treated as the basic units, without needing to analyze how many SLCC each surviving SLCC niche contains after recovery.

In designing the niche model, the attitude was that simplifying assumptions are inevitable but implicit assumptions are to be avoided. The main assumptions are the following.

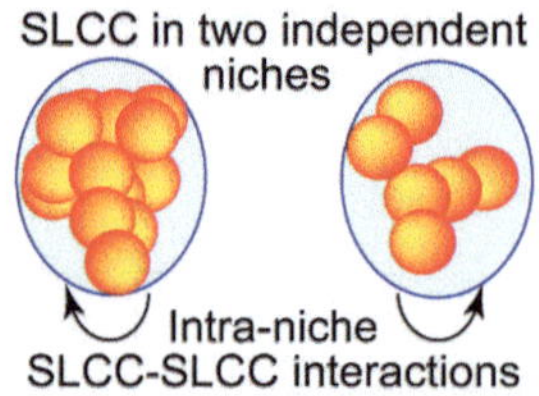

Fig. 16.6 Localized interactions among SLCC. Using Monte Carlo methods to model situations where all SLCC can interact, even if far apart, can lead to formidable computational problems. In the niche model, SLCC are grouped into different, noninteracting subpopulations in different microenvironmental niches, facilitating estimates of how local SLCC–SLCC interactions influence growth.

1. SLCC-occupied niches are "identically distributed"—probabilistically identical. For example, two niches need not have the same number of SLCC at a given time, but if they do, the per unit time *probabilities* at that time of their numbers increasing or decreasing by a given amount are the same.
2. At time t, a patient has $\mathbf{N}(t)$ SLCC-occupied niches which are independent, where $\mathbf{N}(t)$ is a stochastic process, i.e., is characterized by the time-dependent probabilities that a patient has no niches occupied by any SLCC, exactly one such niche, exactly two such niches, etc.
3. It is not realistic to assume the initial niche number $\mathbf{N}(0)$ is a given constant; it is somewhat more realistic to assume $\mathbf{N}(0)$ is a Poisson-distributed stochastic variable with average N. This assumption is also more convenient mathematically and leads to simple, general results for the probability distribution of $\mathbf{N}(t)$ for $t > 0$ (in contrast to the probability distribution for the number of SLCC per niche which can be complicated for $t > 0$). The assumption that the initial probability distribution is Poisson is called "Poissonization" in probability theory. It is used frequently (reviewed in [26]). Its advantages as regards both realism and mathematical convenience are analogous to those achieved by using Gaussian rather than fixed step sizes in random walk polymer calculations [27], or by using canonical rather than petit canonical ensembles in statistical physics. Technically speaking, the assumption implies that there will be some patients who, initially, are free of SLCC but nevertheless symptomatic; for $N > 5$, as will be the case for almost any diagnosable cancer, the fraction of such patients ($< 1\,\%$) is too small to influence the overall conclusions. In any case, such a situation is not impossible: it could occur, e.g., if less stemmy (i.e., progenitor) tumor cells die off slowly, continuing meanwhile to create a burden of differentiated cancer cells, as suggested for CML in [28].
4. On the time scales of interest (e.g., a few months; see Fig. 16.2), niches populated by SLCC do not give "birth," i.e., the number of niches containing SLCC cannot increase. This assumption is crucial to the formalism. As mentioned above, the assumption still allows any behavior for the SLCC birthrates within one niche. For a discussion of empirical evidence that on longer time scales niches can "infect"

neighboring niches (to use a solid tumor example, stemlike colon cancer cells within one intestinal crypt can invade neighboring crypts), see [25].

We now show that with these assumptions, the tumor dynamics during treatment and recovery is described by (a) a relatively simple (time inhomogeneous Poisson) stochastic process, $\mathbf{N}(t)$, the number of independent, identically distributed SLCC niches at time t; and (b) a (usually more complicated) time-dependent probability distribution of SSLC within a niche undergoing a birth–death process with interactions.

For one niche, which in general may have many cells before treatment starts, let Q denote the probability that the niche still contains at least one SLCC at the end of recovery, i.e., till $t = F$, and let tumor control probability (TCP) denote the probability that a patient is "cured," i.e., is entirely free of SLCC at time F.

Theorem 2 $\mathbf{N}(F)$ is Poisson distributed with mean NQ; therefore, $\text{TCP} = \exp[-NQ]$.

Proof $\mathbf{N}(0)$ is Poisson distributed with mean N. Extinction of all SLCC in a niche gives a random thinning of $\mathbf{N}$ with survival probability Q. The basic Poisson color theorem [4], that a random thinning of a Poisson process is a Poisson process, implies $\mathbf{N}(F)$ is Poisson distributed with mean NQ; the Poisson probability that $\mathbf{N}(F) = 0$ is thus $\text{TCP} = \exp[-NQ]$.

Despite the simplicity of the theorem and its proof, it is a quite useful result because it shows that all attention can be focused on a single representative niche and (apart from some generalizations of our TCP criterion) focused specifically on the probability Q that a single niche remains populated by at least one SLCC.

For intraniche SLCC interdependence, the main formal advantage of the new formalism comes into play: Typically, the average number $\mathbf{N}(0)$ is large; consequently, the average number of SLCC per niche is much smaller than the total number of SLCC per patient. With SLCC–SLCC interactions, which frequently require Monte Carlo approaches, confined to SLCC within a niche, and with all niches probabilistically identical, the computer-intensive part of the calculations becomes much easier to implement.

Summary

"Stochastic" usually refers to differences, e.g., among similar patients, whose underlying causes are not observable with the best methods presently available, so that quantifying the differences involves probabilities. Stochastics are especially important when dealing with small numbers, e.g., a cell subpopulation which is likely to have only one or two cells. "Stochastic processes" track the time evolution of a system, such as a leukemia, for which stochastics are important. Techniques for analyzing stochastic processes include explicit analytic results and Monte Carlo computer simulations, with the former usually preferable in the comparatively few cases where they are available and the latter unusually flexible.

Fractionated drug treatments designed to eradicate stemlike leukemic cells have important stochastic aspects. Stochastics always favor front loading, to the (limited) extent that confounding factors such as SLCC–SLCC interactions, normal tissue toxicities, or treatment-induced changes in drug response can be ignored. If interacting SLCC can be divided into subpopulations, each occupying a spatially distinct niche, tracking their time development and possible eradication computationally becomes simpler. SLCC global interactions via control of total SLCC numbers in the body are also amenable to calculations. Such calculations suggest greater stochastic effects on the cancer stem cell extinction probability as stochasticity increases and control loosens, e.g., as in transitions from chronic-phase CML to blast crisis.

Chromosomal Aberrations and Hematological Neoplasms: An Example

Chromosomal aberrations are strongly *associated* with cancers [30] and can *cause* some hematological malignancies. This section considers Philadelphia (i.e., *BCR–ABL*) chromosome formation by ionizing radiation in a primitive hematopoietic cell, resulting in CML—the best understood example of cancer initiation, development, and treatment. Characterization of CML involves an unusually rich and instructive example that illustrates how different stochastic mathematical and computational methods can be used to study stochastic systems in general. It may eventually help us understand other, more common but less well understood, cancers.

Some Examples of Chromosome Aberrations

We here consider structural chromosomal aberrations that occur among unduplicated (G0/G1) chromosomes due to misrejoining of DNA double-strand break (DSB) free ends (Fig. 16.7), e.g., by the nonhomologous end joining repair process [31]. Detailed studies of such aberrations are based on various modern versions of fluorescent in situ hybridization [32]. Chromosome aberrations drive both lymphoid and myeloid malignancies. Aberrations that drive lymphoid malignancies are in part due to DSB of recombinations that diversify antigen specific T cell receptors and B cell antibodies in G0/G1 cells [33]. Examples are ones that cause MYC overexpression due its translocation to an immunoglobulin enhancer in Burkitt's lymphoma [34] or similarly BCL2-overexpression in follicular lymphomas [35].

Myeloid malignancies (e.g., CML or AML) are also caused by chromosomal aberrations, but in these cases, it is expected that both DSB are formed by processes that are not DNA sequence specific, e.g., DSB created by reactive oxygen species, certain cytotoxic anticancer drugs, or ionizing radiation. The rest of this section focuses on ionizing radiation induction of myeloid leukemias, especially CML.

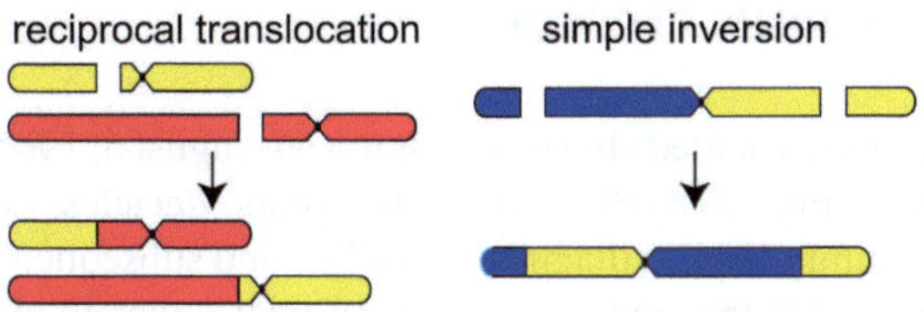

Fig. 16.7 Simple aberrations. Chromosomes are shown with DSB (*gaps*) and centromeres (*constrictions*). Each DSB has two free ends. The free ends can misrejoin to make the rearranged chromosomes shown. These are among the simplest chromosome aberrations since only two DSBs are involved. There are other ways to misrejoin here, but for translocations and inversions, cell proliferation is often not impaired, basically because each rearranged chromosome has one and only one centromere. Some cancers are caused by such "transmissible" misrejoinings.

Ionizing Radiation

Radiobiology is highly suited for dissecting mechanistic CML studies because ionizing radiation, which is known to induce CML [36] that appears indistinguishable from background CML [37], is an unusually well-understood carcinogen, especially at small time and length scales. One reason is that by using highly controllable dose changes, one can verify the presence of radiation action and analyze its details—dose–response relations are central in radiobiology. The following radiation properties have been characterized and mathematically modeled with relatively high precision: single-particle radiation track structure (reviewed in [38]); micro- and nanodosimetry (reviewed in [39, 40]); inter- and intratissue dose distributions, such as human dose–volume histograms (e.g., [41]); radiochemistry action at submillisecond times (reviewed in [42]); subsequent DNA damage–repair–misrepair mechanisms (e.g., [43]); DNA damage-processing outcomes (e.g., chromosomal aberrations [44], especially important here); gene mutations [45]; transcriptome changes [46]; and many additional relevant end points [47]. Overall, radiation perturbations in organisms are frequently more informative than chemical perturbations due to (a) better-known dose localization, both spatially and temporally; (b) thorough knowledge of particle-track physics; and (c) alterations not only in the dose but also in the ionizing particle type and/or energy [48], with resulting *changes* in response giving extra information. For example, the specific initiation time for a radiation-induced neoplasm is often precisely known.

A gold standard for analyses of ionizing radiation-induced cancers is the ongoing life span study (LSS, [49, 50]) of the Japanese survivors of the atomic bombs. The population affected was large and more nearly representative of normal human demographics than in other studies of radiation carcinogenesis. Irradiation was whole-body. Much of this dataset is publicly available. It shows significant radiogenic excess relative risk at many cancer sites. This excess relative risk is larger for leukemias than for most solid cancers.

A Deluge of Stochastic Processes

When an atomic bomb exploded above Hiroshima on August 6, 1945, it set off a chain of stochastic events (Fig. 16.8) that culminated years, decades, and probably even more than a half-century later in diagnoses of CML and subsequent CML dynamics. We shall here summarize the comparatively very-well-understood physicochemical stochastics involved (Fig. 16.8a–16.8c); emphasize the chromosome aberration formation stochastics and their biological implications (Fig. 16.8d–16.8f); and comment briefly on the later stochastics, e.g., during drug treatments (Fig. 16.8g–16.8j).

CML also has stochastic aspects not shown in Fig. 16.8, including the following: age dependence of sporadic (i.e., background) CML, modeled, e.g., by the stochastic two-stage clonal expansion model [51]; stochastic mass action kinetics during progression of CML to blast crisis [52, 53]; stochastic competition between SLCC and normal hematopoietic stem cell (HSC) [24]; the probability distribution for the latency time from CML initiation (Fig. 16.8f) to diagnosis (Fig. 16.8h), a stochastic variable [54]; stochastic interactions of SLCC with the immune system (reviewed, e.g., in [55]), etc.

Femtoseconds to Microseconds: Physical and Chemical Stochastics

A first step toward carcinogenesis after the atomic bomb exploded was when one of its photons entered a cell nucleus and in $\sim 10^{-14}$ s underwent a Compton collision [56] with an electron (Fig. 16.8a). The photon energy was stochastic, but for the sake of simplifying the present example, we suppose its energy was exactly 500 keV, corresponding to a small wavelength ($\sim 3 \times 10^{-12}$ m) and high frequency ($\sim$1 cycle in 10^{-20} s). Even knowing these numbers, the energy of the scattered electron, which is biologically the more relevant quantity, is still a stochastic variable, as we still do not know how close the photon came to the electron; indeed, quantum uncertainties exist related to the fact that photon and electron are wavelike and thus cannot in principle be located exactly. The probability distribution of this stochastic electron energy is known quite accurately: It is given to excellent approximation by one of the oldest results of relativistic quantum electrodynamics, the Klein–Nishina formula [56]. On average, the scattered electron energy is $\sim$175 keV. To continue this analysis, let us suppose that the ejected electron starts with a kinetic energy of exactly 175 keV.

High-energy ejected electrons initiate "tracks" in cells (Fig. 16.8b). Tracks are intricate stochastic spatial patterns of energy depositions. More specifically, they are the locations and energy amounts of thousands of ionizations and molecular excitations created along or near the path of the ejected/primary electron. Ionizations occur when the primary electron encounters other electrons in the biological material and gives them enough energy to leave their molecule. Meanwhile, excitations occur when the energy is insufficient for the struck electron to leave its molecule, yet enough to raise it to a higher energy molecular orbital, and as a result increase its chemical

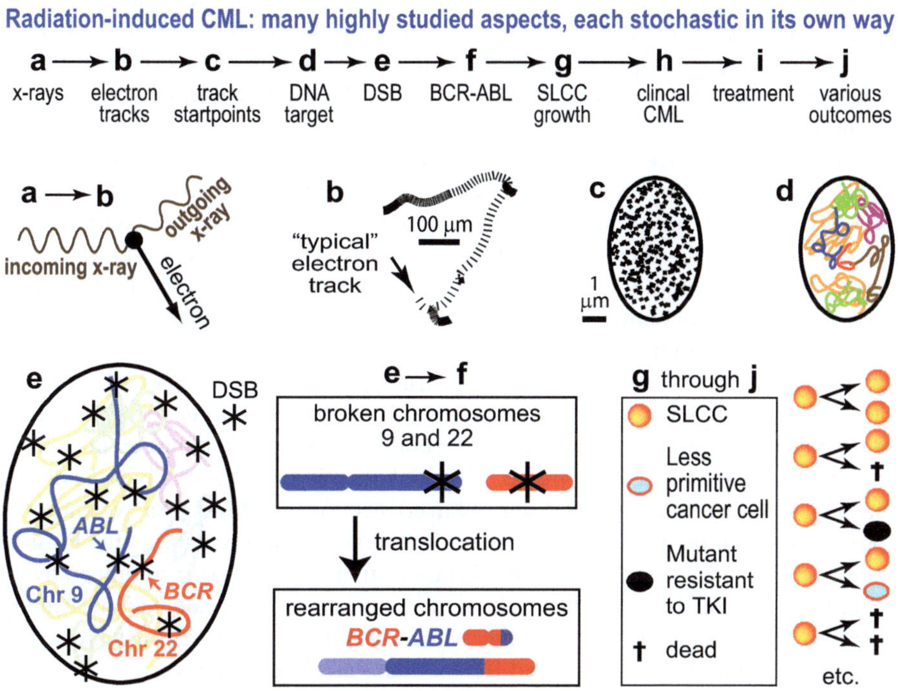

Fig. 16.8 Ionizing radiation induction of CML involves many steps, each stochastic in its own way. The figure schematically shows the induction and consequences of a BCR–ABL translocation. *Top row*: almost every item (**a–j**) and every arrow between items is stochastic in its own way, requiring specially tailored mathematical analysis. The other panels are used in the text to discuss details. Surprisingly, many different kinds of interrelated stochastic effects occur and have been modeled mathematically with various techniques. The process is multiscale in space (subnanometers to > meters) and in time (subpicoseconds to decades). Different ways of calculating the probabilities for the transition **c→d→e→f** will be emphasized in the text. The simplest calculations use the following *randomness assumptions:* all DNA has the same radiation sensitivity; DNA is located at random within a cell nucleus; DSB are consequently located at random, both in space and in the genome; DSB free ends (Fig. 16.7) misrejoin at random aside from a bias for partners formed nearby; transmissible misrejoinings involving more than two DSB are negligible at the doses of interest; and misrejoining is far less frequent than is "restitution," i.e., restoration of the DNA with at most a few local base pair changes that do not impact the fitness of the cell.

reactivity. Figure 16.8b shows the ionizations schematically, exaggerating the track thickness for visibility's sake. The central path is tortuous because when the primary electron imparts energy into its surroundings, it recoils. Some energies can be so large that "secondary" electrons start their own tracks (dark spurs in Fig. 16.8b). Eventually, as the electron slows down, it imparts more energy (not less) to its surroundings per unit contour length of its track (Fig. 16.8b toward top).

For well over 60 years, a gradually accelerating research effort has been devoted to developing Monte Carlo software to analyze such stochastic tracks for various particles and energies. Several powerful, new, primarily physics, packages have been developed in the last few years (e.g., [57]; [*geant4.cern.ch/*]) and are now being mined for biological implications (e.g., [38]). 175 keV electron tracks are considered sparsely ionizing: except near the track end, they deposit much less energy per unit contour length than do, e.g., 1 MeV alpha particles.

With electron tracks accurately characterized, one can then take into account, with a much simpler stochastic calculation, that there were many different tracks, not just one. Fig. 16.8c schematically shows the starting points of the electron tracks in one cell nucleus after a substantial dose. Each such track will typically soon leave the nucleus (note the difference in scale between Fig. 16.8b, 16.8c); other electron tracks (not shown) starting outside the nucleus will enter it. The starting points, to excellent approximation, form a three-dimensional stationary Poisson point process, i.e., the stochastic geometric pattern appropriate for completely randomly located and completely independent points ([58], Chap. 2). Moreover, since we do not know the orientation of the incoming photons with respect to the cell nucleus, it is appropriate to assume isotropy, i.e., all initial directions being equally probable for each of the electrons. Since, as discussed in connection with Fig. 16.8a, we know the probabilities for the various initial energies that the electrons can have, the Poisson point process gives a complete stochastic description of how to go, in a comparatively simple way, from (difficult) one-track calculations to multitrack calculations. Among the most important conclusions is that, while ionizations due to one track are highly correlated, ionizations due to two different tracks are independent.

Once all these energy depositions have occurred, stochastic radiochemistry is needed to estimate their subsequent effects at submicrosecond times. For example, one needs the probability that an OH radical produced by a track diffuses a given distance, perhaps toward a DNA strand that it could attack. These radiochemical aspects have also been comparatively well explored [42].

Microseconds to Kiloseconds: Stochastic Aspects of BCR–ABL Formation

Quantitative estimates have been developed for the number of CML "SLCC predecessors"—normal cells at risk of a CML-initiating *BCR–ABL* chromosome translocation. The results are relevant not only to CML dynamics but also to understanding stem cells, since comparing the estimated predecessor number with estimates of the HSC number indicates that either HSC numbers are larger than previously thought, or that SLCC predecessors are less primitive than HSC, meaning that CML initiation involves some dedifferentiation (reviewed in [54]). The SLCC predecessor number estimates can be carried out separately for ionizing radiation-induced and for age-driven CML. Here only the ionizing radiation case will be described.

The calculations, stochastic throughout, involve the following: superimposing radiation tracks geometrically on their main target, DNA (Fig. 16.8d); estimating the locations of the DSB produced (Fig. 16.8e, 16.8f); and estimating the number and type of chromosome aberrations that result from DSB misrejoinings (Figs. 16.7 and 16.8f).

Relative to biological system uncertainties, radiation track structure, and other physicochemical aspects outlined in the above section (Femtoseconds to Microseconds) can be regarded as known; a further simplification is that we are here interested in situations where, relative to biological time scales, the radiation exposure is essentially instantaneous. As yet less well understood than the physicochemical aspects are the targets (stochastic geometric patterns of the DNA strands in an interphase cell nucleus, here analyzed for their configurations during the G0/G1 phase of the cell cycle), and the stochastic dynamics of misrejoining DSB. Considerable progress, however, has been made recently (reviewed in [38]) and the problems are now being attacked in many different ways. This section reports on different estimated dose–response relations for chromosome translocations (e.g., formation of the Philadelphia chromosome in Fig. 16.8f), depending on different stochastic models of two aspects of forming a translocation: how the DNA is distributed in an interphase cell nucleus and how DSB misrejoin.

A pioneering paper [59] used a theorem in stochastic geometry [58] to derive the dose dependence of chromosome translocation production under various assumptions. The simplest calculation uses the randomness assumptions listed in the caption to Fig. 16.8; it is described next.

Multiple radiation tracks, averaged as outlined in the above section (Femtoseconds to Microseconds), determine a macroscopic radiation dose D (energy deposited per unit mass, measured in Gy, "gray," where 1 Gy = 1 J/kg; radiation dose is an intensive quantity). A related quantity is $D(r)$, the expected amount of energy per unit mass deposited at distance r from an arbitrary, energy-weighted, energy deposition point. $D(r)$ thus refers to energy deposition point pairs; it is a conditional dose, the condition being that the origin is an energy deposition point. Kellerer and Rossi [59] showed that $D(r)=D+[t(r)/\rho 4\pi r^2]$, where ρ is the mass density and $t(r)$ is a one-track point pair function that depends only on the radiation type, not on the dose D; the extra term in square brackets arises because different energy depositions from a single track are correlated.

Now, making the randomness assumptions (Fig. 16.8), let $s(r)\mathrm{d}r$ be the expected volume of DNA within a spherical shell (volume $4\pi r^2 \mathrm{d}r$) centered at an arbitrary point in the DNA, and define $S(r) = s(r)/V$, where V is the total volume of DNA in the cell nucleus. $S(r)$ is another point pair function. Furthermore, let $g(r)$ be the probability density for two DSBs to misrejoin if they are created at a Euclidean distance r from each other; g is known to be biased toward small distances; one possibility is $g = a \exp[-ar]$ for $a > 0$. In any case, for a given $g(r)$, the expected yield $E(T|D)$ of chromosome translocations as a function of average dose D is then [59]:

$$E(T\,|\,D) = \frac{1}{4} YD \int_0^{\infty} YD(r)S(r)g(r)\mathrm{d}r. \tag{16.5}$$

Here Y is the DSB yield per Gy per human G0/G1 cell. A characteristic feature of this formulation is that $g(r)$ maps the initial state directly into the final state without explicitly tracking the intervening evolution in time.

The idea behind the proof [58] of Eq. 16.5 is that radiation and DNA are statistically independent, so probabilities of having point pairs a distance r apart, and with each point being both in a track and in DNA, can be described using products $D(r)S(r)$ of corresponding quantities for point pairs in tracks and DNA, separately. A more detailed intuitive interpretation of Eq. 16.5 is reviewed in [54].

Substituting the result discussed above, $D(r) = D + [t(r)/\rho 4\pi r^2]$, into Eq. 16.5 yields

$$E(T|D) = \frac{1}{4}Y^2 D\left(\int_0^\infty \frac{t(r)}{\rho 4\pi r^2} S(r)g(r)\mathrm{d}r + D\int_0^\infty S(r)g(r)\mathrm{d}r\right), \qquad (16.6)$$

which gives, in approximate agreement with observations on chromosome aberrations, a linear–quadratic dose–response for translocations $E(T|D) = \alpha_T D + \beta_T D^2$ with

$$\alpha_T = \frac{1}{4}Y^2 \int_0^\infty \frac{t(r)}{\rho 4\pi r^2} S(r)g(r)\mathrm{d}r \text{ and } \beta_T = \frac{1}{4}Y^2 \int_0^\infty S(r)g(r)\mathrm{d}r. \qquad (16.7)$$

Given two genes “a” and “b,” an adaptation of Eqs. 16.6 and 16.7 can be developed, using in place of $S(r)$ a gene-specific version $S_{ab}(r)$ based on the lengths, in base pairs, of the relevant portions of the two genes involved. The adaptation can then be used to predict the gene pair translocation dose–response, e.g., the *BCR–ABL* dose–response relevant to radiation-induced CML risk estimation [60].

In Eq. 16.7, the track structure descriptor $t(r)$ is known to very high precision relative to $S(r)$ and $g(r)$. Meanwhile, the DSB yield Y has been estimated by several independent assays to be ~25–40 per Gy for human G0/G1 cells subjected to sparsely ionizing radiation (reviewed, e.g., in [47, 61, 62]); $g(r)$ can be estimated by fitting Eq. 16.7 to chromosome aberration data across different radiation types (e.g., [60]), or by using the Markov chain calculation described below, which in effect replaces $g(r)$ by a stochastic process calculation.

$S(r)$ and $S_{ab}(r)$ can be estimated by combining theory with nonaberration data. They can be calculated directly from the shape and size of the cell nucleus when the randomness assumptions (Fig. 16.8 caption) hold. Otherwise, corrections are needed. For example, $S_{ab}(r)$ may be abnormally large for small r if the corresponding two loci have an abnormal bias toward being geometrically close together. It is clear from Eq. 16.7 that with $g(r)$ in any case biased toward small distances, the two loci are then predicted to participate in more translocations than average, especially since $t(r)$ is also biased toward small distances because it refers to point pairs in a single track. Studying interphase large-scale DNA geometry is now an unusually active area of research, sparked in part by the cancer genome project. The results, described next, can help in estimating S and its loci-specific counterparts.

In studying chromosome aberrations, it has long been considered that large-scale DNA geometry during cell cycle interphase involves some stochastic motifs, mixed in with systematic ones (e.g., [63–65], reviewed in [66]). For example, certain chromosomes systematically tend to be located toward the nuclear membrane more often than other chromosomes, but there is much intercellular variation, caused, e.g., by constrained Brownian motion of segments of a chromosome. These results are inferred from various kinds of observations, including the following: fluorescent staining (e.g., [67, 68], reviewed in [69]); details of chromosome aberration spectra (e.g., [70], reviewed in [44]); and, more recently, high-throughput assays such as "Hi-C" [66, 71].

Of special interest here are estimates of whether, prior to irradiation, problematical gene pairs such as *BCR* and *ABL* (or, for example, *PML* and *RARA,* whose translocation causes acute promyelocytic leukemia) show a bias toward unusually close spatial proximity. Here again, early data come from fluorescence in situ hybridization imaging ([72–74], reviewed in [75]). Very comprehensive information on gene pair association probabilities is now becoming available with next-generation sequencing technology. It raises the intriguing possibility that the "repair factories" inferred from studies of radiogenic chromosome aberrations [76] and more recently from radiation-induced DNA damage foci [43] could be related to the transcription organizing centers suggested by work on nuclei not damaged by radiation [77].

Methods of applying such DNA geometry observations to radiogenic chromosome aberration frequency estimation are available (e.g., [78]). In particular, the results were used to improve estimates of the *BCR–ABL* loci-specific two-point function S_{ab} discussed in section above and thereby improve estimates of Philadelphia chromosome yields/risks after exposures to ionizing radiation [60]. The improved translocation yield estimate was important because it reduced the estimated number of CML SLCC predecessor cell numbers enough to achieve marginal consistency with the largest current estimates of HSC numbers (reviewed in [54]). Hopefully, Hi-C data will continue to be made publically available (http://hic.umassmed.edu/welcome/welcome.php), as it is likely that such data will enable major improvements in chromosome aberration formation models.

However, even if $S(r)$ in Eq. 16.6 can eventually be understood as well as is $t(r)$, the remaining quantity, the misrejoining probability density $g(r)$, remains at present more ad hoc and phenomenological. One attack on this drawback is to think of misrejoining as a sequence of steps [79, 80], not as a single step that transforms an entire initial situation such as those shown in Fig. 16.9 (left column) into an entire final situation (right column). Chromosome aberration simulator (CAS) software for studying DSB induction and misrejoining implements this idea [44, 81], circumvents many of the simplifying randomness assumptions, and can identify cells that have only transmissible rearrangements (compare Figs. 16.7 and 16.9).

CAS implementation is via Monte Carlo methods. A number of repair factories are assigned to a cell, with the average number taken as an adjustable constant to be determined for the particular cell type by examining the aberration data themselves; often about 10 repair factories per cell are estimated, but sometimes the number is substantially larger. The greater the number of repair factories, the greater the dose range where stochastic, rather than deterministic, models are needed. Chromosomes are assigned at random to each repair factory, using Poisson distributions. A Poisson distribution with average proportional to dose, the proportionality constant being the model's second (and last) adjustable constant, is then used to assign DSB to each chromosome in accordance with its length (in Mbp) and its radiation sensitivity. Thereafter, it is assumed that all pairwise misrejoinings among DSB for a particular repair factory are equally likely (i.e., each repair factory acts as a well-mixed reaction vessel) and DSB free ends for different repair factories never misrejoin. An embedded Markov chain, similar to adaptivetau discussed above except that it does not estimate the time taken for each step, gives detailed output for the DSB restitution/misrepair process, which can be parsed, e.g., to estimate the (tiny) fraction of cells with transmissible damage that includes a Philadelphia chromosome (Fig. 16.9).

Kiloseconds to Gigaseconds: Stochastics of CML Latency and Treatment Response

After establishment of a Philadelphia chromosome clone, the malignant cell population evolves dynamically over a period of many years, the latter part of which typically includes medical, e.g. TKI, treatment (Fig. 16.8g–16.8i). Recently, many groups have studied stochastic aspects of this cell population dynamics; surveys include [54, 82, 83]; examples include [29, 53, 84–86]. Mathematical techniques used include Monte Carlo simulations, stochastic birth–death processes that are often generalized to track several related subpopulations, and the Moran model [5], a stochastic process that assumes the total number of stem cells, HSC + SLCC, is fixed at a single deterministic number during subpopulation evolution.

For example, one of the clinically important malignant cell population dynamical topics analyzed stochastically is the development of drug resistance, before or during TKI treatment. Two seminal papers [87, 88] modeled imatinib-resistant mutants, multidrug-resistant mutants, and the use of TKI cocktails to combat such resistance. Also seminal in these papers was the emphasis on point mutations within the ABL domain of leukemic cells, and on mutation rate per cell division as the key parameter.

More recent papers, e.g., [82, 89–91], review, extend, and in some cases modify, these analyses. There is considerable agreement in these later papers, e.g., on the following points. Only mutations in CML SLCC are important, not mutations in more differentiated cells. Back mutations and resistance due to two or more successive mutations in one SLCC subclone are so rare they can be neglected, but mutations that confer resistance to more than one drug cannot. A substantial fraction of the resistant mutants are already present, perhaps undetectably, at diagnosis. And a cocktail of two different TKI should be considered for first-line treatment.

Finally (Fig. 16.8j), on still larger time and length scales, there are statistical studies of human CML population dynamics. For example, the LSS [50] is still giving new information on CML almost 70 years after the A-bomb.

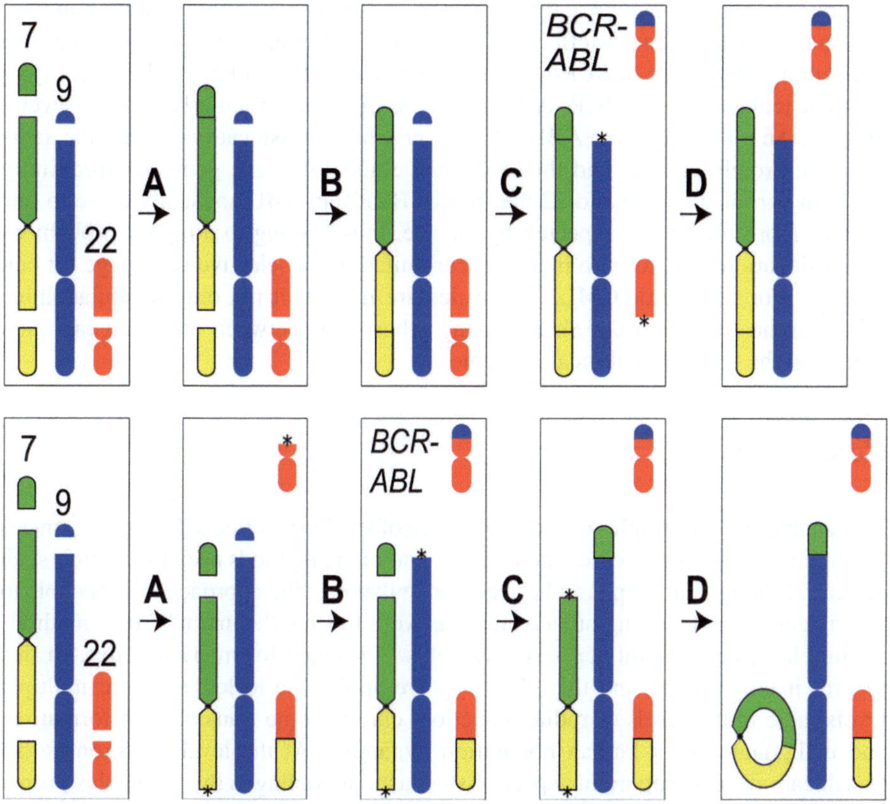

Fig. 16.9 Transmissible aberration patterns. The figure shows chromosomes 7, 9, and 22 in an interphase cell nucleus starting (*left*) with four DNA DSBs (*gaps*) whose free ends can restitute or misrejoin stepwise. Top row: restitutions A and B first restore Chr. 7 to its undamaged state; then a translocation (C and D) makes a Philadelphia chromosome (labeled *BCR–ABL*). The entire final rearrangement (*top right*) is most likely transmissible to daughter cells and therefore liable to initiate CML. The two small asterisks in the fourth rectangle show DSB free ends destined to misrejoin or remain single, because their partners have already been unfaithful. In the bottom row, the starting point is the same, and a Philadelphia chromosome also forms. However, due to other misrejoinings, the final state (*bottom right*) is not transmissible. The ring (*left*) and its replicate can link during the S phase of the cell cycle, causing severe problems at mitosis; the acentric fragment (*right bottom*) and its replicate can get lost or be incorrectly distributed between daughters even if mitosis succeeds. These problems tend to reduce the clonal fitness and thus make the Philadelphia chromosome less dangerous. In general, the possible number of initial states is very large indeed; the number of possible final states for a given initial state with many breaks is also large. CAS systematically tracks all the possibilities and assigns them probabilities.

Summary

Chromosome aberrations can be found in many cancers, but they are particularly important in hematologic malignancies. In the case of lymphoid malignancies, DSB are created endogenously in antigen-specific immunity processes, and this can have unwanted translocation side effects [92]. Myeloid leukemias are also associated with aberrations. In the case of *BCR–ABL* causing CML and 20–30 % of adult acute lymphocytic leukemias [93], BCR–ABL prevalence may be particularly high for several reasons: An unusually large ABL intron over which translocations create the same chimeric protein product and thus the same clinical disease; geometric proximity between chromosomes 9 and 22 that places BCR and ABL unusually close to one another more often than expected by chance, thus leading to a higher probability of translocation; and, as with many cancer mutations, a selective advantage for cell survival. For radiogenic CML, stochastics are important not only in Philadelphia chromosome formation but also at many other steps between an initial radiation event and the long-term clinical outcome.

Conclusions

This chapter gave examples of the importance of stochastic effects during the genesis and treatment of hematological neoplasms, and of the methods used to quantify such effects. Although our emphasis has been on leukemias, the approach is applicable to understanding and treating other cancers as well. Indeed, the mathematical methods applied have been useful across many fields of science for many decades. In this regard, it is not surprising then that they are applicable to leukemias at multiple levels, here at the levels of radiation action, chromosome translocation formation, and malignant cell or human population dynamics. At the level of chromosome translocations, one important aspect of stochastic modeling is that it enables proper accounting of DSB restitution, particularly at low doses where proper DSB free ends find one another largely because there are no other local options available that might cause a misrejoining. At the higher level of cell numbers, stochastic modeling enables representation of malignant clone extinction probabilities. The importance of stochastic models in the genesis and treatment of leukemias will only increase in the future as mathematical models of cancer are applied more and more to the design of clinical trials.

References

1. Carlesso N, Cardoso AA. Stem cell regulatory niches and their role in normal and malignant hematopoiesis. Curr Opin Hematol. 2010;17(4):281–6.
2. Zaider M, Hanin L. Tumor control probability in radiation treatment. Med Phys. 2011;38(2):574–83.
3. Manly BFJ. Randomization, Bootstrap, and Monte Carlo methods in biology. 3rd Edn. Boca Raton: Chapman & Hall/CRC 2006.
4. Grimmett G, Stirzaker D. Probability and random processes. 3rd Edn. Oxford: Oxford University Press; 2001.
5. Tan W. Stochastic models with applications to genetics, cancers, AIDS, and other biomedical systems. London: World Scientific; 2002.
6. Kimmel M, Axelrod DE. Branching processes in biology. New York: Springer; 2002.

7. Rejniak KA, Anderson AR. Hybrid models of tumor growth. Wiley Interdiscip Rev Syst Biol Med. 2011;3(1):115–25.
8. Hahnfeldt P, Folkman J, Hlatky L. Minimizing long-term tumor burden: the logic for metronomic chemotherapeutic dosing and its antiangiogenic basis. J Theor Biol. 2003;220(4):545–54.
9. Fowler JF. 21 years of biologically effective dose. Br J Radiol. 2010;83(991):554–68.
10. Otto S, Day T. A biologist's guide to mathematical modeling in ecology and evolution. Princeton: Princeton University Press; 2007.
11. Tucker SL, Taylor JM. Improved models of tumour cure. Int J Radiat Biol. 1996;70(5):539–53.
12. Hanin LG. A stochastic model of tumor response to fractionated radiation: limit theorems and rate of convergence. Math Biosci. 2004;191(1):1–17.
13. Sachs RK, Heidenreich WF, Brenner DJ. Dose timing in tumor radiotherapy: considerations of cell number stochasticity. Math Biosci. 1996;138(2):131–46.
14. Capizzi RL, Yang JL, Cheng E, et al. Alteration of the pharmacokinetics of high-dose ara-C by its metabolite, high ara-U in patients with acute leukemia. J Clin Oncol. 1983;1(12):763–71.
15. Box C, Rogers SJ, Mendiola M, Eccles SA. Tumour-microenvironmental interactions: paths to progression and targets for treatment. Semin Cancer Biol. 2010;20(3):128–38.
16. Ainseba B, Benosman C. Optimal control for resistance and suboptimal response in CML. Math Biosci. 2010;227(2):81–93.
17. Wodarz D. Stem cell regulation and the development of blast crisis in chronic myeloid leukemia: implications for the outcome of imatinib treatment and discontinuation. Med Hypotheses. 2008;70(1):128–36.
18. Nanda S, Moore H, Lenhart S. Optimal control of treatment in a mathematical model of chronic myelogenous leukemia. Math Biosci. 2007;210(1):143–56.
19. Kim PS, Lee PP, Levy D. Modeling regulation mechanisms in the immune system. J Theor Biol. 2007;246(1):33–69.
20. Foo J, Drummond MW, Clarkson B, Holyoake T, Michor F. Eradication of chronic myeloid leukemia stem cells: a novel mathematical model predicts no therapeutic benefit of adding G-CSF to imatinib. PLoS Comput Biol. 2009;5(9):e1000503.
21. Edelstein-Keshet L. Mathematical models in biology. Philadelphia: SIAM; 2005.
22. Radivoyevitch T. stochFeedbackEarlyLate.R. 2012. http://epbi-radivot.cwru.edu/EPBI473/files/wk13tumorTherapy/. Accessed 18 March 2012.
23. Shuryak I, Hahnfeldt P, Hlatky L, Sachs RK, Brenner DJ. A new view of radiation-induced cancer: integrating short- and long-term processes. Part I: approach. Erratum in: Radiat Environ Biophys 2011 Nov;50(4):607–8. Radiat Environ Biophys. 2009;48(3):263–74.
24. Catlin SN, Guttorp P, Abkowitz JL. The kinetics of clonal dominance in myeloproliferative disorders. Blood. 2005;106(8):2688–92.
25. Shuryak I, Sachs RK, Brenner DJ. A new view of radiation-induced cancer. Radiat Prot Dosimetry. 2010;143(2–4):358–64.
26. Jacquet P, Szpankowski W. Analytical depoissonization and its applications. Theor Comput Sci. 1998;201(1):1–62.
27. Doi M, Edwards S. The theory of polymer dynamics. NY: Clarendon; 1986.
28. Lenaerts T, Pacheco JM, Traulsen A, Dingli D. Tyrosine kinase inhibitor therapy can cure chronic myeloid leukemia without hitting leukemic stem cells. Haematologica. 2010;95(6):900–7.
29. Fakir H, Hlatky L, Li H, Sachs R. Repopulation of interacting tumor cells during fractionated radiotherapy: stochastic modeling of the tumor control probability. Med Phys. 2013;40(12):121716.
30. Mitelman F, Johansson B, Mertens F. The impact of translocations and gene fusions on cancer causation. Nat Rev Cancer. 2007;7(4):233–45.
31. Jeggo P. The role of the DNA damage response mechanisms after low-dose radiation exposure and a consideration of potentially sensitive individuals. Radiat Res. 2010;174(6):825–32.
32. Hada M, Wu H, Cucinotta FA. mBAND analysis for high- and low-LET radiation-induced chromosome aberrations: a review. Mutat Res. 2011;711(1/2):187–92.

33. Desiderio S. Temporal and spatial regulatory functions of the V(D)J recombinase. Semin Immunol. 2010;22(6):362–9.
34. Bernheim A, Berger R, Lenoir G. Cytogenetic studies on African Burkitt's lymphoma cell lines: t(8;14), t(2;8) and t(8;22) translocations. Cancer Genet Cytogenet. 1981;3(4):307–15.
35. Bakhshi A, Wright JJ, Graninger W, et al. Mechanism of the t(14;18) chromosomal translocation: structural analysis of both derivative 14 and 18 reciprocal partners. Proc Natl Acad Sci U S A. 1987;84(8):2396–400.
36. Gilbert ES. Ionising radiation and cancer risks: what have we learned from epidemiology? Int J Radiat Biol. 2009;85(6):467–82.
37. Waller CF, Fetscher S, Lange W. Treatment-related chronic myelogenous leukemia. Ann Hematol. 1999;78(8):341–54.
38. Friedland W, Dingfelder M, Kundrat P, Jacob P. Track structures, DNA targets and radiation effects in the biophysical Monte Carlo simulation code PARTRAC. Mutat Res. 2011;711(1/2):28–40.
39. Grosswendt B, Pszona S, Bantsar A. New descriptors of radiation quality based on nanodosimetry, a first approach. Radiat Prot Dosimetry. 2007;126(1–4):432–44.
40. Hei TK, Ballas LK, Brenner DJ, Geard CR. Advances in radiobiological studies using a microbeam. J Radiat Res (Tokyo). 2009;50(Suppl A):A7–A12.
41. Hodgson DC, Koh ES, Tran TH, et al. Individualized estimates of second cancer risks after contemporary radiation therapy for Hodgkin lymphoma. Cancer. 2007;110(11):576–86.
42. Wardman P. The importance of radiation chemistry to radiation and free radical biology (The 2008 Silvanus Thompson Memorial Lecture). Br J Radiol. 2009;82(974):89–104.
43. Costes SV, Chiolo I, Pluth JM, Barcellos-Hoff MH, Jakob B. Spatiotemporal characterization of ionizing radiation induced DNA damage foci and their relation to chromatin organization. Mutat Res. 2010;704(1–3):78–87.
44. Sachs RK, Levy D, Hahnfeldt P, Hlatky L. Quantitative analysis of radiation-induced chromosome aberrations. Cytogenet Genome Res. 2004;104:142–8.
45. Nakachi K, Hayashi T, Hamatani K, Eguchi H, Kusunoki Y. Sixty years of follow-up of Hiroshima and Nagasaki survivors: current progress in molecular epidemiology studies. Mutat Res. 2008;659(1/2):109–17.
46. Chaudhry MA. Biomarkers for human radiation exposure. J Biomed Sci. 2008;15(5):557–63.
47. NRC. Health risks from exposure to low levels of ionizing radiation: BEIR VII phase 2. Washington, DC: National Academy of Sciences (www.nap.edu); 2005.
48. Durante M, Cucinotta FA. Heavy ion carcinogenesis and human space exploration. Nat Rev Cancer. 2008;8(6):465–72.
49. Ozasa K, Shimizu Y, Sakata R, et al. Risk of cancer and non-cancer diseases in the atomic bomb survivors. Radiat Prot Dosimetry. 2011; 146(1–3):272–5.
50. Douple EB, Mabuchi K, Cullings HM, et al. Long-term radiation-related health effects in a unique human population: lessons learned from the atomic bomb survivors of Hiroshima and Nagasaki. Disaster Med Public Health Prep. 2011;5(Suppl 1):122–33.
51. Moolgavkar SH, Luebeck EG. Multistage carcinogenesis and the incidence of human cancer. Genes Chromosomes Cancer. 2003;38(4):302–6.
52. Michor F. Chronic myeloid leukemia blast crisis arises from progenitors. Stem Cells. 2007;25(5):1114–8.
53. Sachs RK, Johnsson K, Hahnfeldt P, Luo J, Chen A, Hlatky L. A multicellular basis for the origination of blast crisis in chronic myeloid leukemia. Cancer Res. 2011;71(8):2838–47.
54. Radivoyevitch T, Hlatky L, Landaw J, Sachs RK. Quantitative modeling of chronic myeloid leukemia: insights from radiobiology. Blood. 2012;119(19):4363–71.
55. Paquin D, Kim PS, Lee PP, Levy D. Strategic treatment interruptions during imatinib treatment of chronic myelogenous leukemia. Bull Math Biol. 2011;73(5):1082–100.
56. Turner JE. Atoms, radiation, and radiation protection. 3rd Edn. New York: Wiley-VCH; 2007.
57. Plante I, Ponomarev A, Cucinotta FA. 3d visualisation of the stochastic patterns of the radial dose in nano-volumes by a Monte Carlo simulation of Hze Ion track structure. Radiat Prot Dosimetry. 2011;143(2–4):156–61.

58. Stoyan D, Kendall WS, Mecke J. Stochastic geometry and its applications. 2nd Edn. Chichester: Wiley; 2008.
59. Kellerer A, Rossi H. A generalized formulation of dual radiation action. Radiat Res. 1978;75:471–88.
60. Radivoyevitch T, Kozubek S, Sachs R. Biologically based risk estimation for radiation-induced CML—Inferences from BCR and ABL geometric distributions. Radiat Environ Biophys. 2001;40:1–9.
61. Holley WR, Mian IS, Park SJ, Rydberg B, Chatterjee A. A model for interphase chromosomes and evaluation of radiation-induced aberrations. Radiat Res. 2002;158(5):568–80.
62. Neumaier T, Swenson J, Pham C, et al. Evidence for formation of DNA repair centers and dose-response nonlinearity in human cells. Proc Natl Acad Sci U S A. 2012;109(2):443–8.
63. Hahnfeldt P, Hearst JE, Brenner DJ, Sachs RK, Hlatky LR. Polymer models for interphase chromosomes. Proc Natl Acad Sci U S A. 1993;90(16):7854–8.
64. Sachs RK, van den Engh G, Trask B, Yokota H, Hearst JE. A random-walk/giant-loop model for interphase chromosomes. PNAS. 1995;92(7):2710–4.
65. Cremer T, Kurz A, Zirbel R, et al. Role of chromosome territories in the functional compartmentalization of the cell nucleus. Cold Spring Harb Symp Quant Biol. 1993;58:777–92.
66. Fudenberg G, Mirny LA. Higher-order chromatin structure: bridging physics and biology. Curr Opin Genet Dev. 2012;22(2):115–24.
67. Yokota H, van den Engh G, Hearst JE, Sachs RK, Trask BJ. Evidence for the organization of chromatin in megabase pair-sized loops arranged along a random walk path in the human G0/G1 interphase nucleus. J Cell Biol. 1995;130(6):1239–49.
68. Cremer T, Cremer M, Dietzel S, Muller S, Solovei I, Fakan S. Chromosome territories–a functional nuclear landscape. Curr Opin Cell Biol. 2006;18(3):307–16.
69. Rouquette J, Cremer C, Cremer T, Fakan S. Functional nuclear architecture studied by microscopy: present and future. Int Rev Cell Mol Biol. 2010;282:1–90.
70. Hlatky L, Sachs RK, Hahnfeldt P. The ratio of dicentrics to centric rings produced in human lymphocytes by acute low-LET radiation. Radiation Res. 1992;129(3):304–8.
71. Lieberman-Aiden E, van Berkum NL, Williams L, et al. Comprehensive mapping of long-range interactions reveals folding principles of the human genome. Science. 2009;326(5950):289–93.
72. Kozubek S, Lukasova E, Jirsova P, et al. 3D Structure of the human genome: order in randomness. Chromosoma. 2002;111(5):321–31.
73. Kozubek S, Lukasova E, Mareckova A, et al. The topological organization of chromosomes 9 and 22 in cell nuclei has a determinative role in the induction of t(9,22) translocations and in the pathogenesis of t(9,22) leukemias. Chromosoma (Berlin). 1999;108(7):426–35.
74. Neves H, Ramos C, Silva MGD, Parreira A, Parreira L. The nuclear topography of ABL, BCR, PML, and RARalpha genes: Evidence for gene proximity in specific phases of the cell cycle and stages of hematopoietic differentiation. Blood. 1999;93(4):1197–207.
75. Falk M, Lukasova E, Kozubek S. Higher-order chromatin structure in DSB induction, repair and misrepair. Mutat Res. 2010;704(1–3):88–100.
76. Savage JRK. Enhanced perspective: proximity matters. Science. 2000;290:62–3.
77. Rajapakse I, Groudine M. On emerging nuclear order. J Cell Biol. 2011;192(5):711–21.
78. Friedland W, Paretzke HG, Ballarini F, Ottolenghi A, Kreth G, Cremer C. First steps towards systems radiation biology studies concerned with DNA and chromosome structure within living cells. Radiat Environ Biophys. 2008;47(1):49–61.
79. Radivoyevitch T, Hoel DG, Chen AM, Sachs RK. Misrejoining of double-strand breaks after X irradiation: relating moderate to very high doses by a Markov model. Radiation Res. 1998;149(1):59–67.
80. Radivoyevitch T. Time course solutions of the Sax-Markov binary eurejoining/misrejoining model of DNA double-strand breaks. Radiat Environ Biophys. 2000;39(4):265–73.
81. Chen AM, Lucas JN, Simpson PJ, et al. Computer simulation of data on chromosome aberrations produced by X rays or alpha particles and detected by fluorescence in situ hybridization. Radiation Res. 1997;148(5 Suppl):S93–101.

82. Roeder I, d'Inverno M. New experimental and theoretical investigations of hematopoietic stem cells and chronic myeloid leukemia. Blood Cells Mol Dis. 2009;43(1):88–97.
83. Whichard ZL, Sarkar CA, Kimmel M, Corey SJ. Hematopoiesis and its disorders: a systems biology approach. Blood. 2010;115(12):2339–47.
84. Michor F, Iwasa Y, Nowak MA. The age incidence of chronic myeloid leukemia can be explained by a one-mutation model. Proc Natl Acad Sci U S A. 2006;103(40):14931–4.
85. Kim PS, Lee PP, Levy D. Modeling imatinib-treated chronic myelogenous leukemia: reducing the complexity of agent-based models. Bull Math Biol. 2008;70(3):728–44.
86. Roeder I, Herberg M, Horn M. An "age"-structured model of hematopoietic stem cell organization with application to chronic myeloid leukemia. Bull Math Biol. 2009;71(3):602–26.
87. Michor F, Hughes TP, Iwasa Y, et al. Dynamics of chronic myeloid leukaemia. Nature. 2005;435(7046):1267–70.
88. Komarova NL, Wodarz D. Drug resistance in cancer: principles of emergence and prevention. Proc Natl Acad Sci U S A. 2005;102(27):9714–9.
89. Leder K, Foo J, Skaggs B, Gorre M, Sawyers CL, Michor F. Fitness conferred by BCR-ABL kinase domain mutations determines the risk of pre-existing resistance in chronic myeloid leukemia. PLoS ONE. 2011;6(11):e27682.
90. Tomasetti C, Levy D. Role of symmetric and asymmetric division of stem cells in developing drug resistance. Proc Natl Acad Sci U S A. 2010;107(39):16766–71.
91. Katouli AA, Komarova NL. Optimizing combination therapies with existing and future CML drugs. PLoS ONE. 2010;5(8):e12300.
92. Robbiani DF, Bothmer A, Callen E, et al. AID is required for the chromosomal breaks in c-myc that lead to c-myc/IgH translocations. Cell. 2008;135(6):1028–38.
93. Moorman AV, Harrison CJ, Buck GA, et al. Karyotype is an independent prognostic factor in adult acute lymphoblastic leukemia (ALL): analysis of cytogenetic data from patients treated on the Medical Research Council (MRC) UKALLXII/Eastern Cooperative Oncology Group (ECOG) 2993 trial. Blood. 2007;109(8):3189–97.

Chapter 17
Assessing Hematopoietic (Stem-) Cell Behavior During Regenerative Pressure

Thomas Stiehl, Anthony D. Ho and Anna Marciniak-Czochra

Abstract Hematopoiesis is a complex and strongly regulated process. In case of regenerative pressure, efficient recovery of blood cell counts is crucial for survival of an individual. We propose a quantitative mathematical model of white blood cell formation based on the following cell parameters: (1) proliferation rate, (2) self-renewal, and (3) cell death. Simulating this model we assess the change of these parameters under regenerative pressure. The proposed model allows to quantitatively describe the impact of these cell parameters on engraftment time after stem cell transplantation. Results indicate that enhanced self-renewal during the posttransplant period is crucial for efficient regeneration of blood cell counts while constant or reduced self-renewal leads to delayed recovery or graft failure. Increased cell death in the posttransplant period has a similar impact. In contrast, reduced proliferation or pre-homing cell death causes only mild delays in blood cell recovery which can be compensated sufficiently by increasing the dose of transplanted cells.

Keywords Stem cell, Hematopoiesis, Bone marrow transplantation, Stem cell transplantation, Quantitative mathematical modeling, Dynamical systems, Engraftment failure, Blood regeneration, Self-renewal, Proliferation

Introduction

Hematopoiesis is a key example for tissue maintenance and regeneration throughout the lifespan of an organism. The efficiency of the hematopoietic system is demonstrated by a daily output of more than 10^{11} cells in human adults [20, 48]. The

A. M.-Czochra (✉) · T. Stiehl
Interdisciplinary Center for Scientific Computing (IWR),
University of Heidelberg, Heidelberg, Germany
e-mail: Anna.Marciniak@iwr.uni-heidelberg.de

A. D. Ho
Department of Hematology and Oncology,
University Hospital of Heidelberg, Heidelberg, Germany

A. M.-Czochra
BIOQUANT Center, University of Heidelberg, Heidelberg, Germany

S. J. Corey et al. (eds.), *A Systems Biology Approach to Blood*,
Advances in Experimental Medicine and Biology 844, DOI 10.1007/978-1-4939-2095-2_17

hematopoietic system not only replaces the steady state daily loss of mature blood cells but in addition it is also able to respond to dramatic changes such as blood loss, immune defense, or bone marrow transplantation.

A large number of hematopoietic disorders [17] demonstrate the complex interplay between its different components, which in case of impairment lead to detrimental effects. Due to the vital importance of blood formation it is not surprising that hematopoiesis is governed by complex regulatory mechanisms, such as negative feedback loops. Over the last decades a large number of different hematopoietic cytokines have been characterized [28] and their effects on blood formation have been investigated. Important examples are erythropoietin (EPO) and granulocyte colony-stimulating factor (G-CSF, pegfilgastrim) with their therapeutic applications. Nevertheless, the interaction of different regulatory molecules and the signalling cues invoked by them seem to be highly complicated [32] and are not fully understood.

In addition to the molecular mechanisms it seems crucial to obtain a better understanding of the processes taking place at the level of cells such as frequency of divisions, probability of cell death, and probability to differentiate. Due to experimental limitations and changes in cell behavior under culture conditions [5] it is difficult to directly observe the effect of cytokine stimulation on the cellular level. Mathematical modeling is an effective tool to close this gap since it allows to compare the impact of possible regulatory mechanisms and modes of cell behavior on observable system behavior. Formulation of different possible scenarios in terms of mathematical models and comparison of model simulations to experimental or clinical data provide insights into underlying cellular behavior.

Mathematical models have a long history of being used to better understand biological and medical problems. Depending on the question of interest, different mathematical approaches can be chosen. Large cell populations can be reasonably modelled by differential equations describing the change of cell concentrations over time. Depending on the structure of the considered population and on spatial effects, either ordinary or partial differential equations are obtained. A necessary assumption for the application of differential equations is that a given event happens to each individual of a given subpopulation with the same probability (subpopulations behave as a "well mixed tank"). If cell populations can be divided into a finite number of discrete subpopulations, as it is done in the classical understanding of the hematopoietic system, and if spatial interactions can be neglected a system of ordinary differential equations is obtained [26, 27, 42]. In case of a continuous structure, consisting of many infinitely small subpopulations, partial differential equations of the transport type can be used [10]. If interaction of cells in space is important, partial differential equations describing change of cell concentrations in time and space can be formulated [15]. To incorporate stochastic effects, branching processes [18] or individual-based models can be applied [22, 33, 34]. Individual-based models are valid for small populations but become computationally intensive in case of large numbers of individuals.

The current work is devoted to a more rigorous understanding of the changes of hematopoietic cell behavior under regenerative pressure. To characterize the impact of regenerative pressure on system behavior we chose hematopoietic stem cell transplantation as a scenario distant from steady-state conditions. Hematopoietic stem cell transplantation is a clinical intervention allowing to treat otherwise untreatable diseases, such as leukemias [13], hematologic [24], or metabolic disorders [47, 50]. Nevertheless, hematopoietic stem cell transplantation can be linked to life threatening complications. During the period from infusion of the transplant until recovery of leukocyte counts, patients are prone to infections. Other delicate but rare complications are delayed engraftment [46] and graft failure [17, 21], where engraftment is delayed or absent. Especially, in case of allogeneic transplantation where donor and recipient of the transplant are different persons, complications may be severe. One example is the so called graft-versus-host disease, where host's tissues are attacked by immune cells originating from the graft.

In the current work we use an ordinary differential equation model to describe the formation of white blood cells (leukopoiesis) and their dynamics after stem cell transplantation. Model choice is based on the classical understanding of the hematopoietic system, with all lineages originating from the hematopoietic stem cells (HSC) that give rise to different discrete populations of progenitor cells producing mature blood cells [14, 16, 17]. The model we apply has been calibrated to granulopoiesis and is able to describe neutrophil engraftment after stem cell transplantation quantitatively [41]. In the model, cell behavior is characterized by the following parameters: (1) proliferation rate, which describes how often a cell of a given type divides per unit of time; (2) fraction of self-renewal, describing the probability that a daughter cell belongs to the same sub-population as the mother cell; and (3) death rate, describing which fraction of a cell population dies per unit of time.

We will investigate the following questions:

- Which of these cell parameters may decrease or increase under regenerative pressure?
- What is the quantitative impact of the change of these cell parameters on the time needed for regeneration of normal blood cell counts?
- Which cellular behavior might lead to impairment of blood regeneration and delayed engraftment after transplantation?
- How do different changes of cell behavior influence engraftment kinetics on the level of patient populations?

In the section "Model Description" we shortly describe the model and present the model parameters we use for simulations. In the section "Impact of Cytokine-Dependent Self-Renewal and Proliferation on Neutrophil Engraftment" we investigate the impact of proliferation and self-renewal on reconstitution after bone marrow transplantation. In the section "Reasons for Delayed Engraftment" we investigate how different impairments of transplant cell function, for example, in case of allogeneic transplantation, might influence the time needed for engraftment. We then conclude with a discussion of results and final remarks.

Model Description

We use a refined version of the model proposed by Marciniak-Czochra et al. [25, 26, 42]. The refined and calibrated version of the model can be applied to quantitatively describe neutrophil engraftment after stem cell transplantation [41]. In the current work we restrict ourselves to a short description of the model.

Assumptions

Lineage Structure

Based on the classical knowledge of hematopoiesis [14, 16, 17] we assume that each hematopoietic lineage can be subdivided into a finite sequence of subpopulations traversed sequentially starting from the hematopoietic stem cell (HSC) and ending at the mature cell population (e.g., granulocytes). We refer to these subpopulations as "compartments" or "maturation stages." Some of the compartments may belong to more than one cell lineage. For simplicity the model is restricted to one cell lineage. Since, infections constitute a major complication after hematopoietic stem cell transplantation (HSCT), we are interested in regeneration of the white blood cell population the most of which is constituted by neutrophils. Therefore, the model is based on the architecture of the neutrophil lineage. We distinguish eight maturation stages, denoted as c_i, $1 \leq i \leq 8$:

- Primitive hematopoietic stem cells (HSC, stage $i = 1$),
- Long-term-culture initiating cells (LTC-IC, stage $i = 2$),
- Colony-forming-unit-granulocyte-macrophage (CFU-GM, stage $i = 3$),
- Colony-forming-unit-granulocyte (CFU-G, stage $i = 4$),
- Myeloblasts (stage $i = 5$),
- Promyelocytes (stage $i = 6$),
- Myelocytes (stage $i = 7$),
- Mature neutrophil granulocytes in circulation (stage $i = 8$).

Cell Properties

Cell behavior of each maturation stage is described by proliferation rate, fraction of self-renewal, and death-rate of cells belonging to the respective stage.

- Proliferation rate of stage i at time t is denoted as $p_i(t)$. It describes how often a cell of the given type divides per unit of time. If T_i is the average time between two divisions of cells of stage i, then the proliferation rate is $\ln(2)/T_i$. In accordance with literature [17], we assume that mature neutrophils are postmitotic, i.e., $p_8 \equiv 0$.

- Fraction of self-renewal of stage i at time t is denoted as $a_i(t)$. It describes the probability that a daughter cell originating from division remains in the same maturation stage as the mother cell. The fraction of self-renewal determines how often a cell divides on average before it moves to a more mature state. It can be proven mathematically [26, 42] that the fraction of self-renewal of HSC has to be larger than 0.5 and that of all other cell types.
- Death rate of cells at stage i at time t is denoted as $d_i(t)$. It describes which fraction of a cell population dies or disappears per unit of time. We assume, in accordance with biological data [4], that the rate of disappearance of neutrophils from bloodstream, d_8, is greater than zero.

Cytokine Effects

Based on literature we assume that increased cytokine concentrations enhance proliferation of immature cells [35, 44]. We restrict this effect to $i \leq 4$, due to findings in different progenitor populations [6, 44]. It has been reported that proliferation may increase fourfold upon cytokine stimulation [44]. Furthermore, half-life of neutrophils in blood stream may increase at most fourfold. We assume that the fraction of self-renewal of all stages increases in augmented cytokine concentrations. This choice is motivated by a qualitative study of mathematical models [26]. We will return to this assumption in the next section.

Cytokine Dynamics

For simplicity we restrict ourselves to one cytokine, such as G-CSF. We assume that the cytokine is produced at a constant rate α and degraded by receptor mediated endocytosis. Since the density of G-CSF receptors is maximal on mature cells [38], we neglect the effect of immature stages on cytokine clearance. We furthermore assume a constant in time cytokine degradation at rate γ by a cell-independent mechanism, for example, by liver or kidney. These assumptions result in the following equation for cytokine concentration $S(t)$:

$$\frac{d}{dt}S(t) = \alpha - \beta c_8(t)S(t) - \gamma S(t).$$

Based on studies of cytokine kinetics during infections or injuries [2], we assume that the time scale of cytokine dynamics is much shorter than that of cell cycle. We therefore apply a quasi-steady-state approximation ($\frac{d}{dt}S(t) = 0$, for all t). Rescaling leads to the relative cytokine effect

$$s(t) = \frac{1}{1 + kc_8(t)} \in (0,\ 1],$$

where k depends on the rates of cell dependent and cell independent degradation [26].

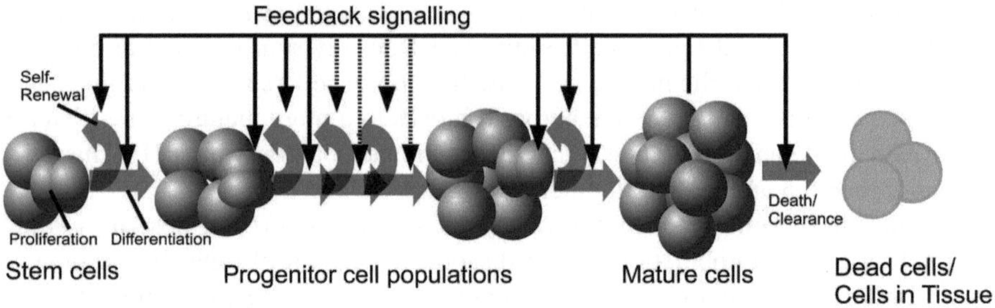

Fig. 17.1 Model visualization: Self-renewal and differentiation of all progenitor populations is regulated by cytokine feedback. Cytokine signalling also regulates clearance of mature cells and proliferation rates of stem cells and early progenitors. Mature cells are postmitotic. Cytokine levels depend on mature cell count

For simplicity we assume that $p_i(t)$ and $a_i(t)$ depend linearly on $s(t)$. Since $p_i(t)$ increases at most fourfold upon cytokine stimulation and $d_8(t)$ decreases at most to 1/4 of its steady state value, we assume, that $d_8(t)$ depends linearly on $1/s(t)$.

Model Formulation

In each compartment, except for the mature cell compartment, the flux to mitosis at time t is given by $p_i(t)c_i(t)$. Since we neglect cell cycle duration, outflux from mitosis at time t is given by $2p_i(t)c_i(t)$, from which $2a_i(t)p_i(t)c_i(t)$ stay in stage i and $2(1 - a_i(t))p_i(t)c_i(t)$ move to stage $i + 1$. Taking into account that the last compartment is postmitotic and that HSC are the most primitive cells, we obtain the following model system [26, 41, 42].

$$
\begin{aligned}
\frac{dc_1(t)}{dt} &= (2a_{1,max}\tilde{s}(t)-1)p_{1,max}s(t)c_1(t)\\
\frac{dc_2(t)}{dt} &= (2a_{2,max}\tilde{s}(t)-1)p_{2,max}s(t)c_2(t) + 2(1-a_{1,max}\tilde{s}(t))p_{1,max}s(t)c_1(t)\\
\frac{dc_3(t)}{dt} &= (2a_{3,max}\tilde{s}(t)-1)p_{3,max}s(t)c_3(t) + 2(1-a_{2,max}\tilde{s}(t))p_{2,max}s(t)c_2(t)\\
\frac{dc_4(t)}{dt} &= (2a_{4,max}\tilde{s}(t)-1)p_{4,max}s(t)c_4(t) + 2(1-a_{3,max}\tilde{s}(t))p_{3,max}s(t)c_3(t)\\
\frac{dc_5(t)}{dt} &= (2a_{5,max}\tilde{s}(t)-1)p_{5,max}c_5(t) + 2(1-a_{4,max}\tilde{s}(t))p_{4,max}s(t)c_4(t)\\
\frac{dc_6(t)}{dt} &= (2a_{6,max}\tilde{s}(t)-1)p_{6,max}c_6(t) + 2(1-a_{5,max}\tilde{s}(t))p_{5,max}c_5(t)\\
\frac{dc_7(t)}{dt} &= (2a_{7,max}\tilde{s}(t)-1)p_{7,max}c_7(t) + 2(1-a_{6,max}\tilde{s}(t))p_{6,max}c_6(t)\\
\frac{dc_8(t)}{dt} &= 2(1-a_{7,max}\tilde{s}(t))p_{7,max}c_7(t) - \frac{\tilde{d}_8}{s(t)}c_8(t)\\
s(t) &= \frac{1}{(1+kc_8(t))}\\
\tilde{s}(t) &= \frac{1}{(1+\tilde{k}c_8(t))}
\end{aligned}
$$

The model is summarized in Fig. 17.1.

Table 17.1 Steady-state proliferation rates and half-life times found in literature

Population	Generation time	Reference
prim HSC	23–67 weeks, mean: 47 weeks	[37]
Myeloblast	16–18 h	[8, 11]
Promyelocyte	18–24 h	[8, 11]
Myelocyte	52 h	[8, 11]
Population	*Half-life-time*	*Reference*
Granulocytes	6.7 (4–10) h	[4]

Table 17.2 Composition of the transplant of 3.6×10^6 cells per kg of body weight used for simulations

Cell type	Number of transplanted cells per kg of body weight
prim HSC	$\approx 3 \times 10^3 - 4 \times 10^3$
LTC-IC	$\approx 36 \times 10^3$
CFU-GM	$\approx 155 \times 10^3$
CFU-G	$\approx 54 \times 10^4$
Myeloblast	0
Promyelocyte	0
Myelocyte	0
Mature neutrophil	0

Model Simulations

Numerical solutions of the system are obtained using an explicit Runge–Kutta-Scheme provided by the Matlab solver ode23t (MathWorks, Natick, MA). Model parameters are chosen in accordance with measurements and steady-state values from literature as described by Stiehl et al. [41]. The parameters used are summarized in Table 17.3. Neutrophil engraftment is defined as achievement of 5×10^8 neutrophils per liter of blood. To relate cells per kg of body weight to cells per liter of blood we assume a body weight of 70 kg and a blood volume of 5 l. Initial conditions are set based on the composition of $CD34^+$ peripheral blood cells after G-CSF mobilization. The choice of the initial condition is discussed in a recent article [41] and presented in Table 17.2. It is based on the data from literature summarized in Table 17.3. For simulations of large patient populations we used random parameter sets uniformly distributed within the ranges from reference [41] which are presented in Table 17.1. To exclude pathological cases we only include parameter sets leading to neutrophil engraftment within at most 3 months for a transplant dose of 10^6 $CD34^+$ cells per kg of body weight and leading to steady state neutrophil counts of at least 10^9 per liter of blood. If not indicated differently, we simulate patient groups of 1000 patients.

Table 17.3 Data on peripheral stem cells grafts

Cell Type	Mobilization	Frequency in CD34$^+$-enriched subpopulation
CFU-GM	G-CSF	43: ∅: 43.2/10^3 cells, SD: 21/10^3 cells
	G-CSF + chemotherapy	43: ∅: 78.4/10^3 cells, SD: 37.8/10^3 cells
LTC-IC	G-CSF	3: $\approx 10/10^3$
CFU-G	G-CSF	3: $\approx 150/10^3$
CFU-GM	G-CSF	3: $\approx 10/10^3$

∅ mean, *SD* standard deviation

Table 17.4 Parameter ranges used for simulation of large patient groups after autologous transplantation without posttransplant cytokine administration. To account for interindividual heterogeneity ranges were taken larger than in literature

Parameter	Value	Parameter	Corresponding division frequency	Parameter	Value
$a_{1,max}$	> 0.5 and < 1.0	$p^*_{1,max}$	$\frac{1}{15\text{ weeks}} - \frac{1}{100\text{ weeks}}$	d_8	$0.35\frac{1}{\text{day}} - 1.0\frac{1}{\text{day}}$
$a_{2,max}$	> 0 and $< a_{1,max}$	$p^*_{2,max}$	$\frac{1}{10\text{ days}} - \frac{1}{15\text{ weeks}}$	k_1	$(3 \text{ to } 9) \times 10^{-9}$
$a_{3,max}$	> 0 and $< a_{1,max}$	$p^*_{3,max}$	$\frac{1}{10\text{ days}} - \frac{1}{15\text{ weeks}}$	k_2	$(6.4 \text{ to } 19.2) \times 10^{-10}$
$a_{4,max}$	> 0 and $< a_{1,max}$	$p^*_{4,max}$	$\frac{1}{4\text{ days}} - \frac{1}{1\text{ day}}$		
$a_{5,max}$	> 0 and $< a_{1,max}$	$p_{5,max}$	$\frac{1}{60\text{ h}} - \frac{1}{15\text{ h}}$		
$a_{6,max}$	> 0 and $< a_{1,max}$	$p_{6,max}$	$\frac{1}{60\text{ h}} - \frac{1}{15\text{ h}}$		
$a_{7,max}$	> 0 and $< a_{1,max}$	$p_{7,max}$	$\frac{1}{60\text{ h}} - \frac{1}{15\text{ h}}$		

* Proliferation rates $p_{1,max}$ to $p_{4,max}$ are maximal proliferation rates. The steady state values are approximately one fourth of the given maximal values

Impact of Cytokine-Dependent Self-Renewal and Proliferation on Neutrophil Engraftment

To investigate the impact of cytokine-dependent self-renewal and proliferation we proceed as follows. We simulate the calibrated model described in the previous Section for 1000 virtual patients, each of them characterized by an individual, random set of model parameters. This scenario will serve as a reference scenario and the virtual patients as reference group. The reference group has the following properties:

- Proliferation rates of stages 1–4 increase at most fourfold if cytokine concentrations increase.
- Clearance of stage 8 decrease at most fourfold if cytokine concentrations increase.
- Self-renewal of all stages increases if cytokine concentrations increase.

We compare engraftment kinetics of the reference group to the following hypothetical scenarios:

1. Proliferation rates are independent of cytokine concentration and constant in time for all stages. In the reference scenario proliferation rates of stages 1–4 depend on cytokine concentration. We fix these proliferation rates to the steady-state values of the reference scenario, $p_{i,max}/(2a_{1,max})$, $i = 1, \ldots, 4$ and set $k = 0$ to achieve independence of cytokine levels. This choice guarantees that steady-state proliferation and self-renewal are the same as in the steady state of the reference scenario. For the simulations we use the parameter set of the reference group subjected to the transformation $p_{i,max} \mapsto p_{i,max}/(2a_{1,max})$, $i = 1, \ldots, 4$.
2. Self-renewal of committed progenitors is independent of cytokine concentration and constant in time. We fix the fractions of self-renewal of stages 3–7 to their corresponding steady state values in the reference scenario. Self-renewal of stages 1 and 2, corresponding to long- and short-term repopulating stem cells, is assumed to be cytokine dependent. Regulated self-renewal of the stem cell population is important to allow expansion of the stem cell compartment after transplantation and maintenance of its size in steady state in absence of apoptosis. This assumption is supported by evidence for repopulation of the hematopoietic system by one single transplanted cell [29, 30]. For all non-stem cell populations we set the fraction of self-renewal to $a_{i,max}/(2a_{1,max})$ and remove multiplication by $\tilde{s}(t)$. This guarantees that in the steady state self-renewal is identical to the reference scenario. For the simulations we use the parameter set of the reference group subjected to the transformation $a_{i,max} \mapsto a_{i,max}/(2a_{1,max})$, for all non-stem cell populations.

Simulation results are depicted in Fig. 17.2. The impact of cytokine-dependent proliferation is relatively small. Average engraftment is extended to 25 days, compared to the 13 days in the reference scenario. In comparison the impact of cytokine-independent self-renewal of non-stem cells on reconstitution time is dramatic. Average engraftment time increases from 13 days in the reference scenario to 2798 days, which is unrealistically long.

Regulation of Self-Renewal

Simulations indicate that regulation of self-renewal behavior has a significant impact on engraftment kinetics. Enhancement of self-renewal leads to reduction of the period of aplasia in comparison to fixed self-renewal probability. Another strategy to supply bloodstream with mature cells might be enhanced differentiation of immature cells resulting in fast transition from the immature to the mature state. On the other hand this scenario could lead to a depletion of immature cells, especially stem cells, if the differentiating cells cannot reestablish high enough mature blood cell counts. To obtain insight if enhancement of differentiation during shortage of mature cells has beneficial effects due to immediate transition from the immature to the mature state or detrimental effects due to exhaustion of the stem cell pool, we propose the following modification of the model.

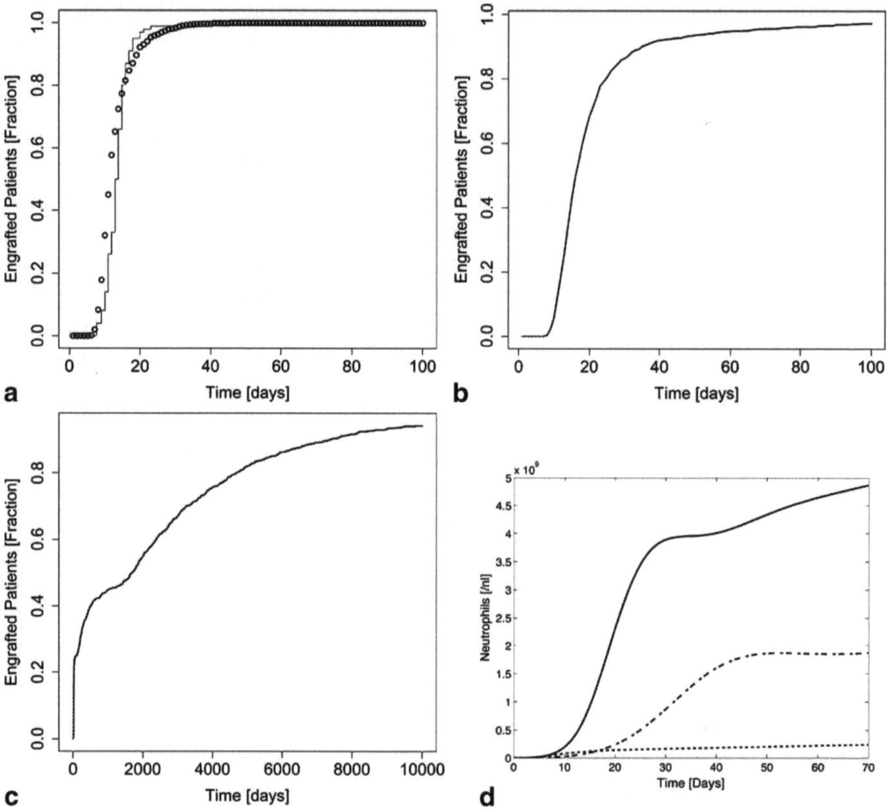

Fig. 17.2 Impact of cytokine dependent proliferation and self-renewal on engraftment kinetics. **a** Fraction of engrafted patients versus time for a population of 1000 virtual patients (reference group) and transplantation of 9×10^6 $CD34^+$ cells per kg of body weight (circles). Proliferation and self-renewal depend on the cytokine concentration as described in the section "Model Description". Model output is compared to patient data (*lines*), taken from Lowenthal et al. [23]. **b** Fraction of engrafted patients versus time for the same population of 1000 virtual patients and 9×10^6 $CD34^+$ cells per kg of body weight. Proliferation rates are independent of cytokine concentrations. In steady state proliferation rates are the same as in **a**. **c** Engraftment for the same virtual population, with cytokine dependent proliferation rates but fixed self-renewal of maturation stages 3–7. In steady state proliferation rate and self-renewal are identical in each of the presented scenarios. **d** Example for neutrophil recovery in one virtual patient. *Solid line* corresponds to **a**, *dashed-dotted line* to **b** and *dotted line* to **c**

$$\frac{dc_1(t)}{dt} = (2a_{1,max}\{1-\hat{s}(t)\}-1)p_{1,max}s(t)c_1(t)$$

$$\frac{dc_2(t)}{dt} = (2a_{2,max}\{1-\hat{s}(t)\}-1)p_{2,max}s(t)c_2(t) + 2(1-a_{1,max}\{1-\hat{s}(t)\})p_{1,max}s(t)c_1(t)$$

$$\frac{dc_3(t)}{dt} = (2a_{3,max}\{1-\hat{s}(t)\}-1)p_{3,max}s(t)c_3(t) + 2(1-a_{2,max}\{1-\hat{s}(t)\})p_{2,max}s(t)c_2(t)$$

$$\frac{dc_4(t)}{dt} = (2a_{4,max}\{1-\hat{s}(t)\}-1)p_{4,max}s(t)c_4(t) + 2(1-a_{3,max}\{1-\hat{s}(t)\})p_{3,max}s(t)c_3(t)$$

$$\frac{dc_5(t)}{dt} = (2a_{5,max}\{1-\hat{s}(t)\}-1)p_{5,max}c_5(t) + 2(1-a_{4,max}\{1-\hat{s}(t)\})p_{4,max}s(t)c_4(t)$$

$$\frac{dc_6(t)}{dt} = (2a_{6,max}\{1-\hat{s}(t)\}-1)p_{6,max}c_6(t) + 2(1-a_{5,max}\{1-\hat{s}(t)\})p_{5,max}c_5(t)$$

$$\frac{dc_7(t)}{dt} = (2a_{7,max}\{1-\hat{s}(t)\}-1)p_{7,max}c_7(t) + 2(1-a_{6,max}\{1-\hat{s}(t)\})p_{6,max}c_6(t)$$

$$\frac{dc_8(t)}{dt} = 2(1-a_{7,max}\{1-\hat{s}(t)\})p_{7,max}c_7(t) - \frac{\tilde{d}_8}{s(t)}c_8(t)$$

$$s(t) = \frac{1}{(1+kc_8(t))}$$

$$\hat{s}(t) = \frac{1}{(1+\hat{k}c_8(t))}, \ \hat{k} = \frac{\tilde{k}}{(2a_{1,max}-1)^2}$$

This system has the same steady-state population sizes, proliferation, and self-renewal as the system from the previous section. Figure 17.3 shows time evolution of neutrophils for one patient. Neutrophil counts rise immediately after transplantation but after a relatively short pulse cell counts decline toward zero. A similar behavior is observed for all patients from the reference scenario. To investigate if this result depended on parameter choice of the reference group, we repeated simulations for another group of 1000 patients with the following broader parameter ranges for division rates: Division rate of stem cells between once per 15 and and once per 100 weeks, division rate of stages 2–4 between once per 15 weeks and once per day, division rates of stages 5–7 between once per 3 days and once per 15 h. We only considered parameters leading to steady-state cell counts of more than 10^9 per liter and a transplant dose of 9×10^6 cells per kg of body weight. Again none of the patients showed reconstitution. For the reference model patients engrafted after a median time of 13.8 days. This demonstrates that reduction of self-renewal during aplasia is not compatible with the calibrated model from the previous section and with clinical data. We therefore assume enhanced self-renewal during aplasia.

Reasons for Delayed Engraftment

In the previous section we investigated the impact of regulation of cell function by cytokines on engraftment kinetics. In this section we focus on possible reasons for delays during engraftment. Since patients are prone to infections during aplasia, delayed engraftment may result in severe complications. Delayed engraftment is a well-known complication after stem cell transplantations [46]. Various data suggest that average recovery after allogeneic transplantation takes longer than autologous transplantation of comparable cell doses. The latter can be explained due to immunologic and nonimmunologic processes affecting transplanted cells [39]. Damages of microenvironment due to chemotherapy might be another reason for delayed engraftment. Although there exist hypotheses which factors might lead to delays in reconstitution it is unknown how cell behavior might change in an impaired

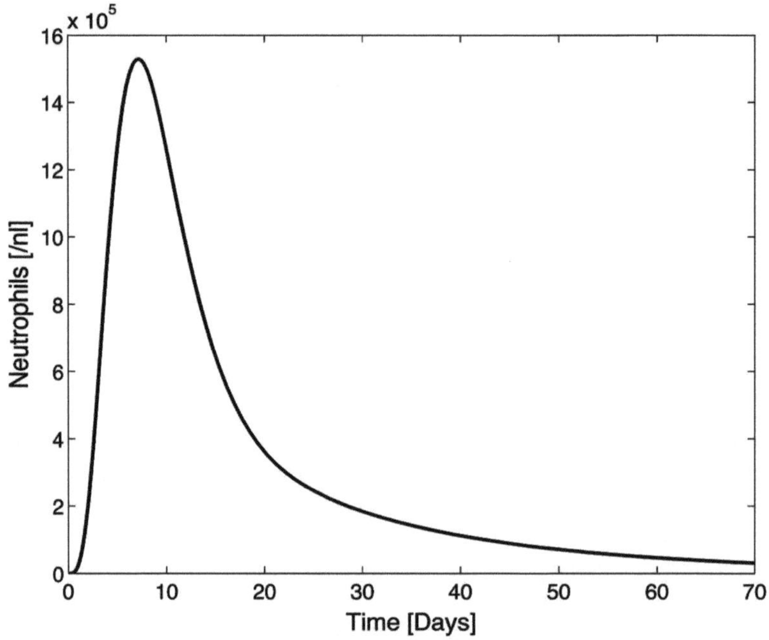

Fig. 17.3 Example for failing engraftment if self-renewal decreases in case of aplasia. If self-renewal decreases in case of aplasia, cells show a rapid transition form the immature stages to the mature stages. This results in a pulse of mature cells shortly after transplantation. If this pulse is not sufficient to compensate for the lack of mature cells persisting stimulation of cell differentiation leads to exhaustion of the immature cell pool and decline of mature cell counts

environment. In the following, we investigate the influence of the following processes on delays of engraftment:

- Cell death before homing
- Cell death in the posttransplantation period
- Reduced proliferation of transplanted cells
- Reduced self-renewal of transplanted cells

One approach to overcome complications of immunologic mechanisms is enlargement of the transplant [19, 31]. For this reason we investigate the impact of the mentioned mechanisms on engraftment kinetics for different transplant doses.

Cell Death Before Homing

Cell death before homing is simulated as a reduction of transplant dose. This is justified by the fact that the duration for the homing process is short in comparison to the time needed for engraftment [9, 49]. We assume that all stages are affected

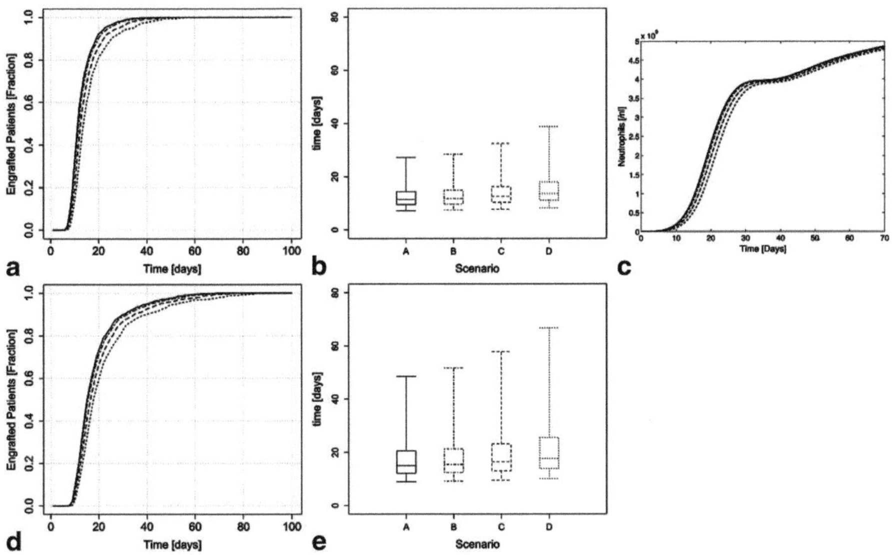

Fig. 17.4 Impact of pre-homing cell death on engraftment kinetics: *Solid line* shows outcome for the reference group, that is, without cell death before homing (*Scenario A*), *dashed-dotted line* shows outcome if 10 % of transplanted cells die before homing (*Scenario B*), *dashed line* if 30 % (*Scenario C*) and *dotted* if 50 % of transplanted cells die before homing (*Scenario D*). *Upper row*: transplant dose 9×10^6 CD 34^+ cells per kg of body weight; **a** Kaplan–Meier plot for neutrophil engraftment; **b** box–whisker plot, whiskers denote 97.5th and 2.5th percentile respectively; **c** example for time dynamics of neutrophil counts for one fixed set of cell properties and the different fractions of transplant dying before homing. *Lower row*: The same for a transplant dose of 3×10^6 CD 34^+ cells per kg of body weight; **d** Kaplan–Meier plot; **e** box–whisker plot

equally. Figure 17.4 depicts the influence of pre-homing cell death on patient recovery for the 1000 patients of the reference scenario. Simulations show moderate effect of pre-homing cell death decreasing with increasing transplant dose. For a transplant of 3×10^6 cells per kg of body weight, engraftment time increases by 3 % if pre-homing cell death increases from 0 to 10 %. This is in accordance with the observation that for most patients reconstitution time weakly depends on transplant dose above certain thresholds of transplanted cells [45].

Cell Death in the Posttransplant Period

To study the effect of enhanced cell death in the posttransplantation period we include death rates constant in time for all cell stages but the HSCs. Since we are only interested in cases where the stem cell population does not die out, we set the death rate of the stem cell population to a constant fraction of the maximal influx to the stem cell compartment which is given by $(2a_{1,max} - 1)p_1$. We assume that the death rate is the same for all non-stem cells in the bone marrow. We assume

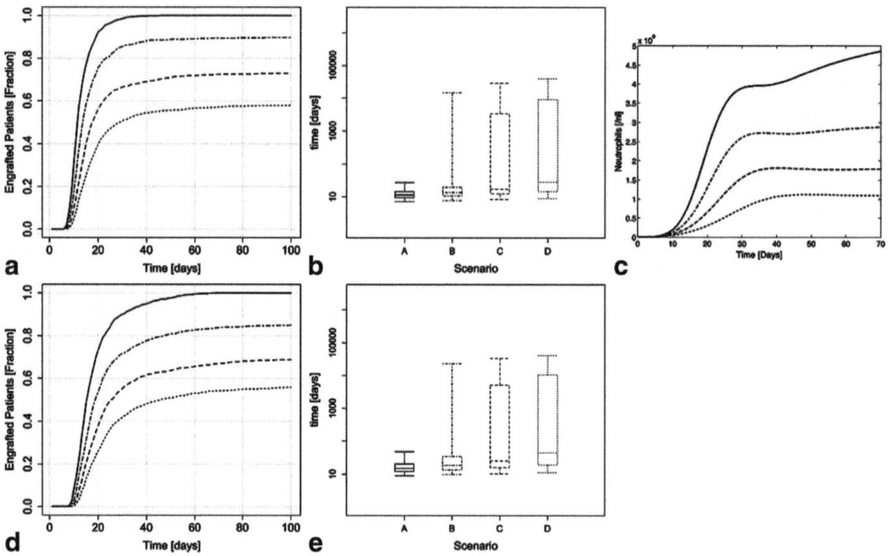

Fig. 17.5 Impact of enhanced cell death during early posttransplantation period: *Solid line* shows outcome for the reference group, i.e. without death of immature cells in the posttransplantation period (*Scenario A*), *dashed-dotted line* shows outcome if $d_2 = \cdots = d_7 = 0.05$, $d_1 = 0.05 \cdot (2a_{1,max} - 1)p_1$ (*Scenario B*), *dashed line* if $d_2 = \cdots = d_7 = 0.1$, $d_1 = 0.1 \cdot (2a_{1,max} - 1)p_1$ (*Scenario C*) and *dotted* if $d_2 = \cdots = d_7 = 0.15$, $d_1 = 0.15 \cdot (2a_{1,max} - 1)p_1$ (*Scenario D*). *Upper row*: transplant dose 9×10^6 CD 34^+ cells per kg of body weight; **a** Kaplan-Meier plot for neutrophil engraftment; **b** box–whisker plot, whiskers denote 97.5th and 2.5th percentile resp.; **c** example for time dynamics of neutrophil counts for one fixed set of cell properties and the different death rates in the posttransplantation period. *Lower row*: The same for a transplant dose of 3×10^6 CD 34^+ cells per kg of body weight; **d** Kaplan–Meier plot; **e** box–whisker plot

clearance of neutrophils in blood stream (half-life less than 1 day) is not affected. Long time cell death changes steady-state cell counts. Enhanced cell death increases interindividual variation. If we increase cell death for non-stem cells from 0 to 15 % of the death rate of mature neutrophils, median recovery time increases by about 300 % and interpercentile ranges increase even stronger. The latter indicates strong interindividual variations. Simulation results are depicted in Fig. 17.5. Result indicate that cell death in the posttransplant period has a strong impact on reconstitution time, which is hardly influenced by transplanted cell doses.

Reduced Proliferation

The transplanted cells are influenced by the host's microenvironment. One possible influence might be impairment of proliferation. Since all mitotic cells are situated in the bone marrow, we assume that the relative loss of proliferative activity is the same for all stages. To investigate its influence on engraftment, we simulate engraftment

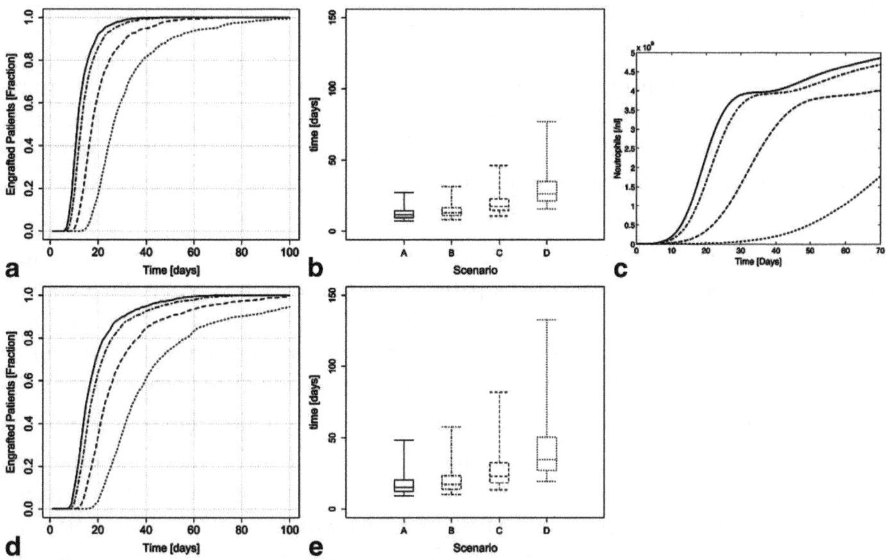

Fig. 17.6 Impact of impaired proliferation on engraftment time: *Solid line* shows outcome for the reference group (*Scenario A*), *dashed-dotted line* shows outcome if proliferation rates of all stages are reduced by 10 % w.r.t. the reference scenario (*Scenario B*), *dashed line* shows outcome if proliferation rates of all stages are reduced by 30 % w.r.t. the reference scenario (*Scenario C*) and *dotted line* shows outcome if proliferation rates of all stages are reduced by 50 % w.r.t. the reference scenario (*Scenario D*). *Upper row*: transplant dose 9×10^6 CD 34^+ cells per kg of body weight; **a** Kaplan–Meier plot for neutrophil engraftment; **b** box–whisker plot, whiskers denote 97.5th and 2.5th percentile resp.; **c** example for time dynamics of neutrophile counts for one fixed set of cell properties and the different proliferation rates in the posttransplantation period. *Lower row*: The same for a transplant dose of 3×10^6 CD 34^+ cells per kg of body weight; **d** Kaplan–Meier plot; **e** box–whisker plot

of the reference population and compare it to populations with reduced proliferation rates. We proceed as follows: Let denote by $p^j_{i,max}$ the maximal proliferation rate of cells of stage i in patient j of the reference population ($i = 1, \dots, 7$). We simulate engraftment for populations where the maximal proliferation rate of cells of stage i in patient j is $\alpha p^j_{i,max}$, with $\alpha < 1$. All other cell parameters are identical to those of the respective patient of the reference population. The results for $\alpha = 0.9$, $\alpha = 0.7$, and $\alpha = 0.5$ are depicted in Fig. 17.6. Reduction of proliferation rates by 10 % leads to increase of engraftment time by 15 %. In summary, figures show moderate increase of engraftment time in case of impaired proliferation. This effect can easily be compensated by enlarged doses of transplanted cells.

Reduced Self-Renewal

Self-renewal of the transplanted cells might also be influenced by the host's microenvironment. We study its impact on engraftment kinetics by reducing self-renewal of the reference population. As above we are only interested in cases where stem cells are able to survive, therefore we impose that $a_{1,max} > 0.5$. We proceed as follows: Let $a^j_{i,max}$, $(j = 2, \ldots, 7)$ denote the maximal possible fraction of self-renewal of cells in non-stem stage i of patient j of the reference group. Let $a^j_{1,max}$ denote maximal possible self-renewal of stem cells in patient j of the reference group. We compare engraftment of the reference group to engraftment of populations, where maximal possible fraction of self-renewal of cells in non-stem stage i of patient j is given by $\alpha a^j_{i,max}$, $(j = 2, \ldots, 7)$ and maximal possible fraction of self-renewal of stem cells in patient j by $0.5 + \alpha(a^j_{1,max} - 0.5)$, and $\alpha < 1$. The latter choice is motivated by the fact that $a^j_{1,max}$ has to be larger than 0.5 to obtain engraftment. All other cell parameters are chosen identical to the corresponding patients in the reference group. In Fig. 17.7 we show engraftment of populations with $\alpha = 0.98$, $\alpha = 0.95$, and $\alpha = 0.9$. A reduction of self-renewal by 2 % leads to an increase of mean engraftment time by 350 % for a transplant of 3×10^6 cells per kg of body weight. The results in Fig. 17.7 imply that self-renewal has a strong impact on engraftment, which cannot be compensated sufficiently by transplant enlargement.

Discussion

Dynamics of blood regeneration after stem cell transplantation depends on cellular behavior during the posttransplantation period. In the current work we use mathematical modelling and computer simulations to quantify the impact of cell proliferation, self-renewal, and cell death on engraftment time. Since hematopoiesis is a strongly regulated process [14, 16, 17] different cell properties might change in the posttransplantation period due to cytokine feedback. Furthermore, preceding chemotherapies [12], diseases [1, 36], or immunologic mechanisms [40] might impair functionality of the transplanted cells and therefore cause delayed engraftment. To optimize clinical outcome after hematopoietic stem cell transplantation it is helpful to know which cell properties lead to fast engraftment and which functional impairments may be responsible for observed delays.

In the current work we use a calibrated model of granulopoiesis to investigate the impact of different cell properties on post transplant neutrophil engraftment. The proposed model allows to quantify the effect of distinct changes of cell behavior on post transplant blood cell dynamics. We quantify the effect of cell behavior on regeneration dynamics simulating engraftment of a large group of individual patients with stochastic interindividual differences in cell parameters. We use this group as a reference group. We then vary distinct parameters for each individual in the reference group. Comparing outcome of the group with modified cell parameters to outcome of the reference group, we are able to quantify the impact of cell behavior on reconstitution dynamics. Such studies are not possible in experimental settings.

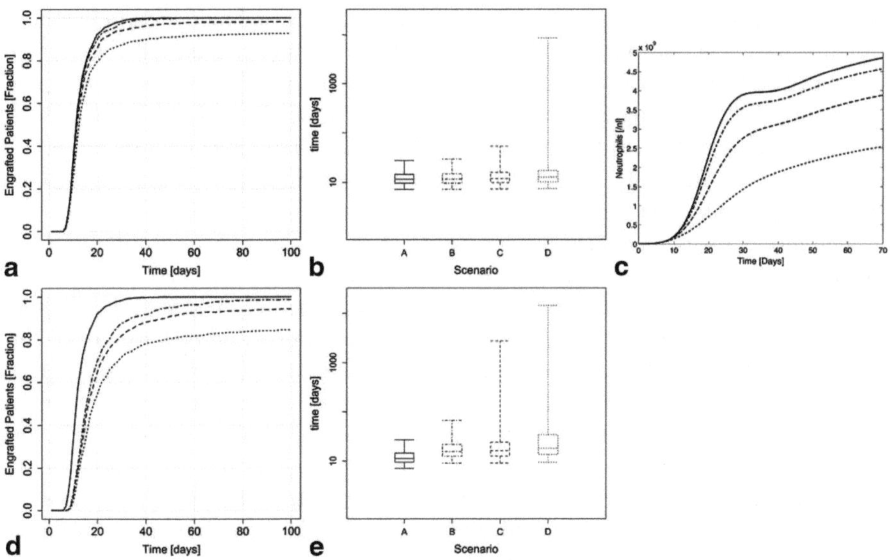

Fig. 17.7 Impact of reduced self-renewal on engraftment: *Solid line* shows outcome for the reference group (*Scenario A*), *dashed-dotted line* shows outcome if self-renewal of all non-stem stages is reduced by 2 % w.r.t reference population and the distance from stem cell self-renewal to 0.5 is reduced by 2 % w.r.t. to the reference population (*Scenario B*). *Dashed line* shows outcome if the respective quantities are reduced by 5 % (*Scenario C*) and *dotted line* if they are reduced by 10 % (Scenario D). *Upper row*: transplant dose 9×10^6 CD 34^+ cells per kg of body weight; **a** Kaplan–Meier plot for neutrophil engraftment; **b** box–whisker-plot, whiskers denote 97.5th and 2.5th percentile resp.; **c** example for time dynamics of neutrophil counts for one fixed set of cell properties and the different self-renewal fractions in the posttransplantation period. *Lower row*: The same for a transplant dose of 3×10^6 CD 34^+ cells per kg of body weight; **d** Kaplan–Meier plot; **e** box–whisker plot

We investigated whether self-renewal and proliferation of stem and progenitor cells may change in the posttransplant period. It has been described in literature that progenitor cells can proliferate four times faster under cytokine stimulation. Our simulations indicate that this behavior leads to a reduction of mean engraftment time by a factor of 2. In comparison to this the impact of enhanced self-renewal during the posttransplant period reduces average engraftment time by more than a factor of 100, a similar result has been described in [26]. We also demonstrated that enhanced differentiation during the posttransplant period is not compatible with sustained engraftment due to exhaustion of the stem cell compartment.

To better understand possible reasons for delayed engraftment we investigated the impact of the following scenarios on engraftment time:

- Cell death before homing
- Cell death in the posttransplantation period
- Reduced proliferation of transplanted cells
- Reduced self-renewal of transplanted cells

Simulation results indicate that cell death before homing and reduction of proliferation rate have a minor impact on blood cell recovery while reduced self-renewal and enhanced cell death in the posttransplant period lead to major delays in engraftment. Interindividual heterogeneity between patients increases in each of the considered scenarios. This might explain higher rates of graft failure or patient death in case of impaired cell function. The observed increased interindividual variation is in line with the clinically observed heterogeneity.

Our model focuses on phenomenological cell properties such as proliferation and death. Immunologic mechanisms are not directly included. Instead it is assumed that they influence the considered functional cell properties. The same is assumed for interaction of hematopoietic cells and the surrounding microenvironment. Our model does not directly include exogenous complications such as infections. Since, in reality patients are not able to survive arbitrarily long with low leukocyte counts we may interpret very long engraftment times (e.g., > 100 days) in model simulations as graft failures.

During recent years the concept of megadose transplantation has been elaborated and it has been shown that megadose transplants are able to circumvent problems during engraftment in different settings [19, 31]. For this reason we investigated how far large cell doses allow to compensate for changes in stem- and progenitor cell function. Whenever such a compensation is possible megadose transplants might be an appropriate means to reduce complications. Our simulations indicate that cell death before homing and reduced proliferation rates can be compensated by enlarged transplants while ongoing cell death and reduced self-renewal cannot be efficiently compensated. This observation might lead to the speculation that in scenarios where megadose transplants are beneficial either pre-homing cell death or reduced proliferation can play an important role while in cases where mega-dose transplantations fail to improve patient outcome ongoing cell death, for example, due to immune reactions or reduced self-renewal may dominate the posttransplant period. The fact that different impairments of cell function have different impacts on reconstitution dynamics in computer simulations might suggest that clinical outcome depends on pretreatment and diagnosed disease. This may explain the benefit from reduced intensity conditioning regimens [7]. Optimal treatment requires eradication of malignant cells with minimal damage to the micro-environment. Treatment regimens in clinical practice are a compromise between these two requirements. A better understanding of the impact of impaired cell function might therefore be helpful to optimize treatment regimens.

In summary, our results indicate that enhanced self-renewal in the posttransplant period is crucial for efficient engraftment. Delayed engraftment can be caused by various mechanisms. The effect of impaired self-renewal and ongoing cell death is most detrimental for clinical outcome and cannot be improved by megadose transplants.

References

1. Abe M. Targeting the interplay between myeloma cells and the bone marrow microenvironment in myeloma. Int J Hematol. 2011;94(4):334–43.
2. Bogner V, Keil L, Kanz KG, Kirchhoff C, Leidel BA, Mutschler W, Biberthaler P. Very early posttraumatic serum alterations are significantly associated to initial massive RBC substitution, injury severity, multiple organ failure and adverse clinical outcome in multiple injured patients. Eur J Med Res. 2009;14(7):284–91.
3. Boiret N, Kanold J, Bons JM, Rapatel C, Halle P, Tournilhac C, Guilhouard L, Guerin JJ, Travade P, Demeocq F, Bonhomme J, Berger MG. Granulocyte colony-stimulating factor-mobilized peripheral blood CD34+ cells from children contain the same levels of long-term culture-initiating cells producing the same numbers of colony-forming cells as those from adults, but display greater in vitro monocyte/macrophage potential. Br J Haematol. 2001;112(3):806–13.
4. Cartwright GE, Athens JW, Wintrobe MM. The kinetics of granulopoiesis in normal man. Blood. 1964;24(6):780–803.
5. Choi YS, Noh SE, Lim SM, Kim DI. Optimization of ex vivo hematopoietic stem cell expansion in intermittent dynamic cultures. Biotechnol Lett. 2010;32(12):1969–75.
6. Craig W, Kay R, Cutler RL, Lansdorp PM. Expression of thy-1 on human hematopoietic progenitor cells. J Exp Med. 1993;177(5):1331–42.
7. Cremer B, Sandmaier BM, Bethge W, Lange T, Goede V, Holtick U, Hallek M, Hubel K. Reduced-intensity conditioning in allogeneic stem cell transplantation for hematological malignancies: a historical perspective. Onkologie. 2011;34(12):710–5.
8. Cronkite EP. Kinetics of granulopoiesis. Clin Haematol. 1979;8(2):351–70.
9. D'Hondt L,Wuu J, Andre M, Lerberghe C, Guillaume T, Feyens AM, Humblet Y, Dromelet A, Chatelain B, Longueville J, Stewart F, D'Hondt V, Symann M. Clearance kinetics of CD34+ cells from peripheral blood: an independent predictor of hematologic recovery after high-dose chemotherapy and hematopoietic stem cell transplantation. Bone Marrow Transplant. 1999;24(5):483–9.
10. Doumic M, Marciniak-Czochra A, Perthame B, Zubelli J. Structured population model of stem cell differentiation. SIAM J Appl Math. 2012;71:1918–40.
11. Fliedner TM. Kinetik und regulationsmechanismen des granulozytenumsatzes. Schweiz Med Wochenschr. 1974;104(4):98–107.
12. Galotto M, Berisso G, Del L, Podesta M, Ottaggio L, Dallorso S, Dufour C, Ferrara GB, Abbondandolo A, Dini G, Bacigalupo A, Cancedda R, Quarto, R. Stromal damage as consequence of high-dose chemo/radiotherapy in bone marrow transplant recipients. Exp Hematol. 1999;27(9):1460–66.
13. Hamilton BK, Copelan EA. Concise review: the role of hematopoietic stem cell transplantation in the treatment of acute myeloid leukemia. Stem Cells. 2012;30(8):1581–86.
14. Handin RI, Lux SE, Stossel TP. Blood: principles and practice of hematology, 2nd ed. Phiadelphia: Lippincott; 2003.
15. Hillen T, Painter, KJ. A user's guide to PDE models for chemotaxis. J Math Biol. 2009;58(1–2):183–217.
16. Jandl JH. Blood. Boston: Little, Brown and Company; 1996.
17. Kaushansky K, Lichtman AM, Beutler LE, Kipps TJ, Prchal J, Seligsohn U. Williams Hematology. 8th ed. New York: Mcgraw-Hill Professional; 2010.
18. Kimmel M, Axelrod, DE. Branching processes in biology. New York: Springer; 2002.
19. Koh LP, Chao, NJ. Nonmyeloablative allogeneic hematopoietic stem cell transplant using mismatched/haploidentical donors: a review. Blood Cells Mol Dis. 2008;40(1):20–4.
20. Lansdorp, PM. Stem cell biology for the transfusionist. Vox Sang. 1998;74(Suppl 2):91–4.
21. Leung AY, Kwong YL. Haematopoietic stem cell transplantation: current concepts and novel therapeutic strategies. Br Med Bull. 2009;93:85–103.
22. Loeffler M, Roeder I. Tissue stem cells: de, plasticity, heterogeneity, selforganization and models—a conceptual approach. Cells Tissues Organs. 2002;171:8–26.

23. Lowenthal RM, Faberes C, Marit G, Boiron JM, Cony-Makhoul P, Pigneux A, Agape P, Vezon G, Bouzgarou R, Dazey B, Fizet D, Bernard P, Lacombe F, Reiffers J. Factors in haemopoietic recovery following chemotherapy-mobilised autologous peripheral blood progenitor cell transplantation for haematological malignancies: a retrospective analysis of a 10-year single institution experience. Bone Marrow Transplant. 1998;22(8):763–70.
24. MacMillan ML, Walters MC, Gluckman E. Transplant outcomes in bone marrow failure syndromes and hemoglobinopathies. Semin Hematol. 2010;47(1):37–45.
25. Marciniak-Czochra A, Stiehl T. Mathematical models of hematopoietic reconstitution after stem cell transplantation. In: Bock HG, Carraro T, Jager W, Korkel S, Rannacher R, Schloder JP, editors. Model based parameter estimation: theory and applications. Heidelberg: Springer; 2013.
26. Marciniak-Czochra A, Stiehl T, Jager W, Ho AD, Wagner W. Modeling of asymmetric cell division in hematopoietic stem cells—regulation of self-renewal is essential for efficient repopulation. Stem Cells Dev. 2009;18(3):377–85.
27. Marciniak-Czochra A, Stiehl T, Wagner W. Modeling of replicative senescence in hematopoietic development. Aging (Albany NY). 2009;1(8):723–32.
28. Metcalf D. Hematopoietic cytokines. Blood. 2008;111:485–91.
29. Notta F, Doulatov S, Laurenti E, Poeppl A, Jurisica I, Dick JE. Isolation of single human hematopoietic stem cells capable of long-term multilineage engraftment. Science. 2011;333(6039):218–21.
30. Osawa M, Hanada K, Hamada H, Nakauchi H. Long-term lymphohematopoietic reconstitution by a single CD34-low/negative hematopoietic stem cell. Science. 1996;273(5272):242–5.
31. Reisner Y. Hematopoietic stem cell transplantation across major genetic barriers. Immunol Res. 2007;38(1–3):174–90.
32. Roeder I, de Haan G, Engel C, Nijhof W, Dontje B, Loeffler M. Interactions of erythropoietin, granulocyte colony-stimulating factor, stem cell factor, and interleukin-11 on murine hematopoiesis during simultaneous administration. Blood. 1998;91(9):3222–9.
33. Roeder I, Horn K, Sieburg HB, Cho R, Muller-Sieburg C, Loeffler M. Characterization and quanti of clonal heterogeneity among hematopoietic stem cells: a model-based approach. Blood. 2008;112(13):4874–83.
34. Roeder I, Kamminga LM, Braesel K, Dontje B, de Haan G, Loeffler M. Competitive clonal hematopoiesis in mouse chimeras explained by a stochastic model of stem cell organization. Blood. 2005;105(2):609–16.
35. Roelofs H, de Pauw ES, Zwinderman AH, Opdam SM, Willemze R, Tanke HJ, Fibbe WE. Homeostasis of telomere length rather than telomere shortening after allogeneic peripheral blood stem cell transplantation. Blood. 2003;101(1):358–62.
36. Calado RT. Immunologic aspects of hypoplastic myelodysplastic syndrome. Semin Oncol. 2011;38(5):667–72.
37. Shepherd BE, Guttorp P, Lansdorp PM, Abkowitz JL. Estimating human hematopoietic stem cell kinetics using granulocyte telomere lengths. Exp Hematol. 2004;32(11):1040–50.
38. Shinjo K, Takeshita A, Ohnishi K, Ohno R. Granulocyte colony-stimulating factor receptor at various stages of normal and leukemic hematopoietic cells. Leuk Lymphoma. 1997;25(1–2):37–46.
39. Shizuru JA, Bhattacharya D, Cavazzana-Calvo M. The biology of allogeneic hematopoietic cell resistance. Biol Blood Marrow Transplant. 2010;16(1 Suppl):S2–7.
40. Shono Y, Ueha S, Wang Y, Abe J, Kurachi M, Matsuno Y, Sugiyama T, Nagasawa T, Imamura M, Matsushima K. Bone marrow graft-versus-host disease: early destruction of hematopoietic niche after mhc-mismatched hematopoietic stem cell transplantation. Blood. 2010;115(26):5401–11.
41. Stiehl T, Ho AD, Marciniak-Czochra A. The impact of CD34 + cell dose on engraftment after SCTs: personalized estimates based on mathematical modeling. Bone Marrow Transplant. 2014;49(1): 30–7.
42. Stiehl T, Marciniak-Czochra A. Characterization of stem cells using mathematical models of multistage cell lineages. Math Comput Model. 2011;53(7–8):1505–17.

43. Suzuki T, Muroi K, Amemiya Y, Miura Y. Analysis of peripheral blood CD34+ cells mobilized with granulocyte colony-stimulating factor (G-CSF) using a long-term culture system. Bone Marrow Transplant. 1998;21(8):751–7.
44. Thornley I, Sutherland DR, Nayar R, Sung L, Freedman MH, Messner HA. Replicative stress after allogeneic bone marrow transplantation: changes in cycling of CD34+CD90+ and CD34+CD90- hematopoietic progenitors. Blood. 2001;97(6):1876–8.
45. To LB, Levesque JP, Herbert KE. How I treat patients who mobilize hematopoietic stem cells poorly. Blood. 2011;118(17):4530–40.
46. Trebeden-Negre H, Rosenzwajg M, Tanguy ML, Lefrere F, Azar N, Heshmati F, Belhocine R, Vernant JP, Klatzmann D, Norol F. Delayed recovery after autologous peripheral hematopoietic cell transplantation: potential effect of a high number of total nucleated cells in the graft. Transfusion. 2010;50:2649–59.
47. Valayannopoulos V, Wijburg FA. Therapy for the mucopolysaccharidoses. Rheumatology (Oxford) 2011;50(Suppl 5):v49–59.
48. Vaziri H, Dragowska W, Allsopp RC, Thomas TE, Harley CB, Lansdorp PM. Evidence for a mitotic clock in human hematopoietic stem cells: loss of telomeric DNA with age. Proc Natl Acad Sci U S A. 1994;91(21):9857–60.
49. Whetton AD, Graham GJ. Homing and mobilization in the stem cell niche. Trends Cell Biol. 1999;9(6):233–8.
50. Wynn R. Stem cell transplantation in inherited metabolic disorders. Hematology (Am Soc Hematol Educ Program). 2011; 1(1)285–91.

Chapter 18
Engineered Cell-Based Therapies: A Vanguard of Design-Driven Medicine

Rachel M. Dudek, Yishan Chuang and Joshua N. Leonard

Abstract Engineered cell-based therapies are uniquely capable of performing sophisticated therapeutic functions *in vivo*, and this strategy is yielding promising clinical benefits for treating cancer. In this review, we discuss key opportunities and challenges for engineering customized cellular functions using cell-based therapy for cancer as a representative case study. We examine the historical development of chimeric antigen receptor (CAR) therapies as an illustration of the engineering design cycle. We also consider the potential roles that the complementary disciplines of systems biology and synthetic biology may play in realizing safe and effective treatments for a broad range of patients and diseases. In particular, we discuss how systems biology may facilitate both fundamental research and clinical translation, and we describe how the emerging field of synthetic biology is providing novel modalities for building customized cellular functions to overcome existing clinical barriers. Together, these approaches provide a powerful set of conceptual and experimental tools for transforming information into understanding, and for translating understanding into novel therapeutics to establish a new framework for design-driven medicine.

Keywords Cancer · Systems biology · Synthetic biology · Immunotherapy · CAR therapy · Genetic engineering · Computational biology

J. N. Leonard (✉) · R. M. Dudek · Y. Chuang
Northwestern University, 2145 Sheridan Road,
Technological Institute, Rm. E136, Evanston, IL 60208-3120, USA
Tel.: 847-491-7455
e-mail: j-leonard@northwestern.edu

R. M. Dudek
Tel.: 847-467-2725
e-mail: rmdudek@northwestern.edu

Y. Chuang
Tel.: 847-467-2725
e-mail: yishanchuang2013@u.northwestern.edu

S. J. Corey et al. (eds.), *A Systems Biology Approach to Blood*,
Advances in Experimental Medicine and Biology 844, DOI 10.1007/978-1-4939-2095-2_18

Abbreviations

APC	Antigen presenting cell(s)
CAR	Chimeric antigen receptor(s)
CLL	Chronic lymphocytic leukemia
CTL	Cytotoxic T lymphocyte (CTL)
CTLA4	Cytotoxic T lymphocyte antigen 4
DPLSR	Discriminant partial least squares regression
EBV	Epstein-Barr virus
EGFR	Epidermal growth factor receptor
GVHD	Graft versus host disease
HLA	Human leukocyte antigen
IDO	Indoleamine 2,3-dioxygenase
IL	Interleukin
iPSC	Induced pluripotent stem cell(s)
MDSC	Myeloid derived suppressor cell(s)
MHC	Major histocompatibility complex
NK	Natural killer
ODE	Ordinary differential equation(s)
PCA	Principal component analysis
PD1	Programmed cell death protein 1
PDE	Partial differential equation(s)
scFv	Single chain variable fragment
STAT3	Signal transducer and activator of transcription 3
TAA	Tumor associated antigen(s)
Th1	Helper T cell, type 1
Th2	Helper T cell, type 2
TIL	Tumor infiltrating lymphocyte(s)
TCR	T cell receptor(s)
TNP	2,4,6-trinotrophenol
Treg	Regulatory T cell
V_H	Variable heavy chain
V_L	Variable light chain

Introduction

Cell-based therapies are uniquely able to sense and respond to their environment, synthesize multiple bioactive molecules, and confer multifactorial effector functions in vivo. In this way, cells can be considered "devices" that carry out sophisticated functions that cannot be achieved with small molecule drugs or biomolecular therapeutics. However, the development of novel cell-based therapies is inherently complex, due to challenges in understanding and controlling both the cell-based therapeutic itself and the manner in which it interacts with host biology. Consequently, the process of developing effective cell-based therapies may be summarized by two interrelated challenges:

1. Identify a therapeutic strategy: Which desired effector functions are predicted to confer the desired therapeutic effects *in vivo*?
2. Implement a therapeutic strategy: How should a cell-based therapy be manipulated or engineered to manifest the desired therapeutic effector functions?

Here, we explore the challenge of developing effective cell-based therapies using immunotherapy for cancer as an illustrative case study. We consider how transformative advances in our abilities to acquire extensive biological data and to construct novel biomolecular "parts" may enable and accelerate the development of cell-based therapies. In particular, we discuss the complementary approaches of systems biology and synthetic biology, which combine to create a powerful design cycle for engineering safe and effective cell-based therapies.

Origins of Cell-Based Immunotherapy for Cancer

Targeted pharmacological agents can extend the lives of cancer patients by months or even years. Particularly effective examples include monoclonal antibodies (e.g., trastuzumab for breast cancer and ipilimumab for metastatic melanoma) [1, 2] and selective kinase inhibitors (e.g., imatinib for chronic myelogenous leukemia) [3, 4]. However, these targeted approaches sometimes fail to eradicate rare resistant cancer cells (perhaps including cancer "stem cells") [5, 6], and over time, other tumor cells may evolve resistance by mutating or downregulating the protein targeted by the therapeutic [7, 8]. Either mechanism may lead to tumor escape and relapse, typically rendering the cancer refractory to the previously effective therapeutic. Thus, additional approaches are required to meet the needs of patients for whom molecular therapeutics are not yet available or are no longer beneficial.

A promising alternative strategy is to harness a patient's own immune system to eradicate or control tumor growth. Such an immunotherapy approach generally takes one of two forms: (1) Induction of an antitumor immune response through therapeutic cancer vaccines (reviewed in [9, 10]) or (2) adoptive transfer of tumor-reactive T cells that directly mediate tumor killing and control. Immunotherapy may provide an additional option for patients whose disease is refractory to pharmacological agents, or it may complement other therapeutic approaches [11] and protect against relapse by inducing immunological memory of tumor-associated antigens (TAA). In this review, we focus on the development of adoptive T cell-based strategies as a representative case study of approaches for engineering cell-based therapies.

Many immunotherapy strategies are originally motivated by considering presumed natural mechanisms for controlling tumor growth. The *immunosurveillance* theory, which was first formulated in the mid-twentieth century, posits that during homeostasis, the adaptive arm of the immune system controls nascent tumors by recognizing mutant protein antigens expressed by tumor cells and targets these cells for killing [12, 13]. In updated versions of this model, cytotoxic T lymphocytes (CTL) expressing antigen receptors that recognize specific TAA presented on tumor cells in the context of major histocompatibility complexes (i.e., MHC-I) are stimulated to proliferate and drive targeted cell killing via the CTL arsenal of mechanisms [14]. Such CTL may also differentiate into a memory phenotype that also prevents tumor recurrence. This conceptual model has been refined and expanded over half a century to now propose that disease results from a gradual escape from immunological control during cancer progression, via processes collectively termed *immunoediting* [15]. Since natural killer (NK) T cells kill tumors that simply downregulate MHC to evade CTL recognition via the "missing self" mechanism [16], immune evasion generally involves both reduced tumor immunogenicity and active suppression of CTL function in the tumor microenvironment. This local immune suppression may involve a number of suppressors of CTL function, including programmed cell death protein 1 (PD1), cytotoxic T lymphocyte antigen 4 (CTLA4), and indoleamine 2,3 dioxygenase (IDO), which are expressed by both tumor cells and immune cells including regulatory T cells (Treg) and myeloid-derived suppressor cells (MDSC) in the tumor microenvironment [17–19]. This overall conceptual model is supported by the common observation of anergic or otherwise dysfunctional tumor-specific T cells in the vicinity of established tumors.

Based on this understanding of tumor–immune interactions, early immunotherapy approaches were motivated by the hypothesis that naturally occurring tumor-specific CTL may be harnessed therapeutically. In this approach, autologous tumor infiltrating lymphocytes (TIL) are isolated from a surgically accessible tumor, expanded, and activated *ex vivo,* and re-infused into the patient [20]. In clinical trials, autologous TIL therapy has shown promise for treating melanoma, but efficacy has largely been limited to this type of cancer. The reasons for this restriction are not yet clear, but proposed explanations include high rates of mutation in melanoma and unique modes of immune dysfunction in different cancers [14]. More generally, experience with autologous TIL highlighted the importance of generating sufficient quantities of T cells having both tumor antigen specificity and the capacities to persist, proliferate, and induce cytotoxic functions at the tumor site upon re-infusion.

The advent of technologies for genetically modifying human cells opened the door to potentially programming desired functionalities into a cell-based therapy. An important first step was the demonstration by Rosenberg and colleagues that T cells engineered with a retroviral vector expressing a reporter gene could be administered to human cancer patients without inducing significant toxicity [21]. One approach for applying genetic engineering to circumvent the challenge of isolating and expanding TIL is to identify a T cell receptor (TCR) that is specific for a given TAA and then clone and express this TCR as a transgene in autologous T cells [22]. Such a model TCR is generally isolated from TIL of a patient with a good response to TIL therapy. The hypothesis motivating this approach is that when the engineered T cell encounters a tumor cell expressing the TAA, the transgenic TCR will induce downstream signaling through native pathways, resulting in proliferation and induction of cytotoxicity. Since this approach relies upon native TCR, it is limited in that the transgenic TCR only recognizes TAA presented in the context of a compatible MHC-I. Thus, a given cloned TCR is only effective in HLA-matched patients (i.e., those expressing compatible MHC-I), and overall efficacy is diminished by low MHC-I expression on tumor cells. In addition, since the transgenic TCR may be expressed alongside a native TCR within a single-engineered cell, mispairing between TCR chains creates receptors with hybrid specificity, potentially limiting recognition of the tumor or raising the risk of off-target activation and induction of harmful autoimmunity [23]. Mispairing between transgenic TAA-specific TCR chains and native TCR chains has been implicated in the development of lethal graft-versus-host disease (GVHD) in mice [24]. However, prior to these animal studies, similar transgenic TCR had been expressed in human subjects, with none developing GVHD and some experiencing tumor regression [25, 26]. Thus, the actual risk of toxicity due to hybrid TCR remains unclear, but caution is nonetheless warranted.

Although transgenic TCR-based approaches demonstrated the feasibility of genetically modifying cells to create customized therapeutics, the challenges and limitations associated with this particular strategy also motivated the development of a new technology platform that is amenable to modular incorporation of specific functionalities. For these reasons, we posit that this next wave of therapies that utilize chimeric antigen receptors (CAR) represents a fundamental shift in strategy from recapitulating natural functionalities to designing novel therapeutics that may be described as cell-based "devices." Thus, the historical development of CAR therapies (Fig. 18.1) illustrates the engineering design cycle and may guide the future development of other cell-based therapies.

Development of CAR Therapy: The Engineering Design Cycle in Action

The central innovation of the CAR is genetically fusing the ligand-binding variable domain of an antibody recognizing a TAA to the intracellular domains of a TCR capable of inducing T cell activation [27]. This ligand-binding domain typically consists of a single chain variable fragment (scFv), in which variable heavy chain (V_H) and variable light chain (V_L) domains are joined by a linker to form a single chain. Since antibody recognition of the TAA does not require presentation of the TAA in the context of MHC, an engineered T cell expressing the appropriate CAR may be activated by contacting the TAA in an MHC-independent fashion, thus overcoming a key limitation of the transgenic TCR approach.

The first CAR was successfully expressed in T cells in 1989 by Eshhar and colleagues [28] (Fig. 18.1). In this design, the intracellular architecture of the native TCR α (alpha) and β (beta) chains were maintained, but the ligand-binding domain was derived from an antibody specific for the model antigen, 2,4,6-trinitrophenol (TNP). When presented with target TNP-A.20 cells (a TNP-modified B cell lymphoma from BALB/c mice), T cells transfected with the chimeric TCR secreted IL-2 and killed these target cells in an antigen-specific manner. However, since this design is subject to mispairing with native TCR chains just as in the transgenic TCR

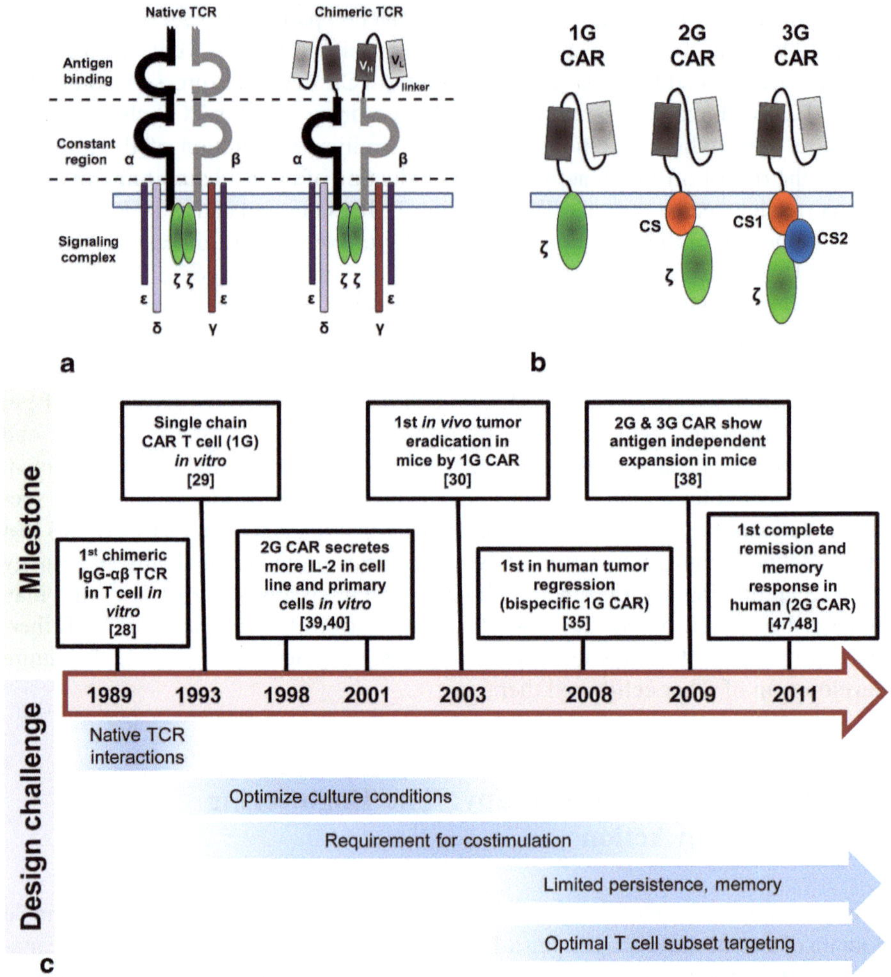

Fig. 18.1 Development of CAR therapy through the engineering design cycle. This figure provides a historical perspective on the development of cell-based therapy for cancer as an example of engineering design in practice. **a** The *native TCR* comprises *alpha* and *beta* chains that form a signaling complex with *delta*, *gamma*, *epsilon*, and *zeta* chains to transmit a signal after the TCR binds its target antigen presented on the MHC of another cell. Specificity for a novel antigen may be conferred by creating a *chimeric TCR*, in which the TCR variable region is replaced by an antibody-derived single chain variable fragment (*scFv*), comprising variable heavy (V_H) and variable light (V_L) chains joined by a linker. **b** First generation (*1G*) CAR comprised a *scFv* domain fused to a transmembrane domain and a signal transduction domain (typically the TCR zeta chain). Second and third generation (*2G*, *3G*) CAR incorporated an additional one or two tandem costimulatory domains (CS or CS1/CS2, respectively). **c** Historical timeline of CAR development illustrating iterative refinement through identification of design challenges and then engineering innovations to address these challenges. The milestones selected are representative of this process but are not exhaustive. *CAR* chimeric antigen receptor, *TCR* T cell receptor, V_H variable heavy domain, V_L variable light domain, *1G* first generation, *2G* second generation, *3G* third generation, *MHC* major histocompatibility locus

approach, the same group revised their design in 1993 by inventing a construct then termed a T-body, comprising an immunoglobulin ligand-binding domain fused to the CD3γ (gamma) or CD3ζ (zeta) T cell activation domain [29]. This T-body scheme is now considered the first generation (1G) CAR. Efforts to translate this 1G CAR to an *in vivo* model identified several key challenges, including the achievement of adequate T cell proliferation and survival *ex vivo* and activity *in vivo*. A combination of improved culture conditions, incorporating IL-15 to induce expression of antiapoptotic molecules, and costimulation provided by CD80 expressed on CD19$^+$ Raji target tumor cells *in vivo* was credited with enabling the success of the first in vivo tumor regression in mice using 1G CAR T cells [30]. However, since costimulatory molecules are not expressed by many tumors, a key result of this study was identifying the need for costimulation *in vivo* as an important design challenge for CAR engineering.

Endogenous T cells receive costimulatory signals through either (a) engagement of the T cell's CD28 receptor by B7 ligands (i.e., CD80/B7-1 or CD86/B7-2) on a professional antigen-presenting cell (APC; i.e., costimulation in *cis*) or (b) by binding interleukin 2 (IL-2), which is typically secreted by an activated helper CD4$^+$ T cell (i.e., Th1 or Th2 cells) responding to the binding of antigen to its own TCR (i.e., costimulation in *trans*). However, since CAR recognize TAA in an MHC-independent context, *cis*-costimulation through interaction with APC presenting TAA does not occur, and *trans*-costimulation is required. Although IL-2 can easily be provided during *ex vivo* culture, subsequent lack of IL-2 costimulation *in vivo* may lead to poor proliferation and persistence. Indeed, early pilot studies with 1G CAR T cells demonstrated persistence of only a few days [31]. Although IL-2 can be administered systemically and has been shown to extend CAR T cell persistence [32], sustained IL-2 infusion is technically challenging, costly, and can cause severe toxicities associated with sepsis in some patients [33]. Each of these challenges motivated another iterative refinement of CAR design to directly incorporate provision of costimulation.

CAR designs incorporating costimulation may be divided into two general strategies. In the first strategy, T cells expressing a native TCR specific for a persistent antigen (such as an antigen associated with a chronic viral infection) are engineered to also express the CAR. Thus, the resulting "bispecific" CAR T cell receives costimulation when its endogenous TCR binds the persistent viral antigen presented on APC in the vicinity of the tumor, where it is activated when its CAR binds TAA on the tumor [34]. The first example of tumor regression in a CAR clinical trial was achieved using a bispecific 1G CAR T cell expressing a native TCR specific for the Epstein-Barr virus (EBV) and a CAR specific for the GD2 antigen expressed by neuroblastoma cells [35]. Patients enrolled in this trial all had persistent EBV infection, as evidenced by the presence of EBV-specific IgG in their blood. A follow-up to this study 4 years later indicated that the bispecific CAR T cells persisted in some patients, and that disease remission correlated strongly with persistence of the engineered T cells [36]. In the second strategy, the ability to initiate costimulatory signaling is incorporated into the CAR itself by appending additional intracellular domains to the CAR, such as domains from CD28, CD134, and/or CD137 [37, 38]. CAR incorporating two intracellular domains are termed second generation (2G) and those incorporating three intracellular domains are termed third generation (3G). In both Jurkat immortalized T cells [39] and primary T cells [40], 2G CAR incorporating dual CD3ς (zeta) and CD28 signaling domains in tandem led to improved IL-2 secretion and cytotoxicity *in vitro*.

Continual development of CAR platforms over the last decade produced abundant evidence that this approach yielded superior cytokine secretion and proliferation [41, 42], and even resistance to suppression by Treg [43]. By 2010, a complete lymphoma remission was reported in a human patient treated with 2G CAR therapy although by 7 months post-therapy the patient relapsed following the failure of the 2G CAR T cells to persist and establish memory [44]. Strategies to increase persistence included both adjunctive therapies, such as IL-2 administration [45], and redesign of the CAR to incorporate the endodomain of CD137 (also termed 4-1BB) in place of or in addition to the endodomain of CD28. By 2010, two independent preclinical studies demonstrated that T cells expressing either a CD137-CD3ς (zeta) 2G CAR or a CD28-CD137-CD3ς (zeta) 3G CAR directed against the CD19 antigen achieved antigen-independent T cell expansion and conferred a survival benefit in mouse xenograft models [38, 46]. Taking this CD137-CD3ς (zeta) 2G CAR into a phase I clinical trial, June and colleagues reported in 2011 that out of three chronic lymphocytic leukemia (CLL) patients treated, one exhibited a partial response and two exhibited complete responses and the establishment of memory CAR T cells [47, 48]. Following this potential breakthrough, a number of clinical trials have subsequently commenced to investigate other promising CAR designs including 3G CAR [49], to extend this promising therapy to larger and younger patient cohorts [50, 51], and to pave the way for phase II clinical trials.

Realizing the promise of CAR therapy for diverse patient groups now requires systematic investigations into (1) identifying which cancers are amenable to immunotherapy and corresponding TAA [52, 53], (2) identifying which T cell subsets to engineer with CAR and what transduction methods and culture conditions to employ [54], and (3) identifying effective dosing regimens and adjunctive therapies during administration of CAR T cells to patients [55]. Adjunctive therapies appear particularly promising, as was highlighted by a 2013 clinical trial from Memorial Sloan-Kettering in which a 2G CAR against CD19 was administered to patients with refractory B cell acute lymphoblastic leukemia (B-ALL) following chemotherapy preconditioning to the point of minimal residual disease and subsequent allogeneic hematopoietic stem cell transplantation (allo-HSCT). In this trial, 88 % of patients (14 out of 16) experienced a complete remission [51, 56]. A detailed discussion of these design considerations is beyond the scope of this review, but a summary of current approaches, considerations, and relevant reviews is presented in Table 18.1.

While achieving long-term memory and complete remission with 2G CAR therapy is a major milestone, a persistent challenge in CAR therapy is optimizing the balance between efficacy and safety. In one notable case, a patient with metastatic colon cancer died after being treated with a 3G CAR against the HER2 antigen [57]. It was hypothesized that the CAR recognized low levels of HER2 expressed on lung epithelial cells, leading to excessive pulmonary infiltration and cytokine secretion in the lungs that resulted in respiratory failure. This case highlights the fact that achieving an optimal balance between safety and efficacy will depend strongly on the

Table 18.1 Key design parameters for CAR therapy

Design parameter	Options	Ref.
Disease	B cell malignancies, neuroblastoma, ovarian cancer, renal cell carcinoma, melanoma, etc.	[52, 53]
Antigen	α-folate receptor, CD19, CEA, CAIX, GD_2, ERBB2/ HER2/neu	[52, 53, 126]
T cell subset	αβ, γδ, iNKT, $CD8^+$, $CD4^+$, antigen (e.g., EBV) specific	[54, 127]
Transduction method	γ retrovirus, lentivirus, transposon, electroporation	[54, 127]
Culture conditions	APC cell line, artificial APC, OKT3, IL-2, IL-15, other cytokines	[54]
CAR model	1G, 2G, 3G, bispecific	[52, 55, 127]
Dosing	Regimen (single vs. split), number of cells	[58]
Conditioning/ adjunctive thereapy	lymphodepletion, myeloablation; IL-2, dendritic cells co- or post-administration	[55]

CAR chimeric antigen receptor, *APC* antigen presenting cell, *EBV* Epstein–Barr virus

type of disease targeted and the TAA selected. For example, in B cell malignancies or melanoma, ablation of healthy cells sharing the TAA (i.e., B cells and melanocytes) is not lethal, and so therapeutic benefit is optimized with highly active CAR responses. In contrast, the case of colon cancer expressing HER2 discussed above suggests that the therapeutic window is more restricted, due to the risks of CAR-mediated killing of healthy cells. In fact, excessive CAR-mediated killing of the targeted tumor cells can also restrict the therapeutic window. In the anti-CD19 CAR trial performed by June and colleagues, one of three patients receiving a CAR T cell infusion experienced life-threatening toxicities due to excessive tumor lysis syndrome and cytokine storm, and this patient was treated with a systemic corticosteroid to attenuate the tumor lytic activity of the CAR T cells [47]. It is possible that this steroid administration explains why this patient did not experience the same degree of tumor remission as did the other patients enrolled in the trial. Thus, early toxicity due to excessive on-target effects may be a general challenge for CAR safety, and potential solutions based upon dosing strategies have been proposed [58]. More generally, the identification of these challenges and exploration of potential solutions also demonstrates the progression of CAR therapy through another iteration of the design cycle.

The history of CAR development is a compelling illustration of the power of the engineering design cycle. Grand challenges in this field now include efficiently and expediently capitalizing on these early successes to extend these approaches to treat other types of cancer, to design therapies that benefit large and potentially diverse patient groups, and to simultaneously maximize efficacy and safety. Meeting these goals will require employing emerging tools for both understanding these biological phenomena at a systems level and new technologies for translating this understanding into viable therapeutic strategies. In the following sections, we discuss how the complementary approaches of systems biology and synthetic biology may create a new framework for successfully addressing these challenges (Fig. 18.2).

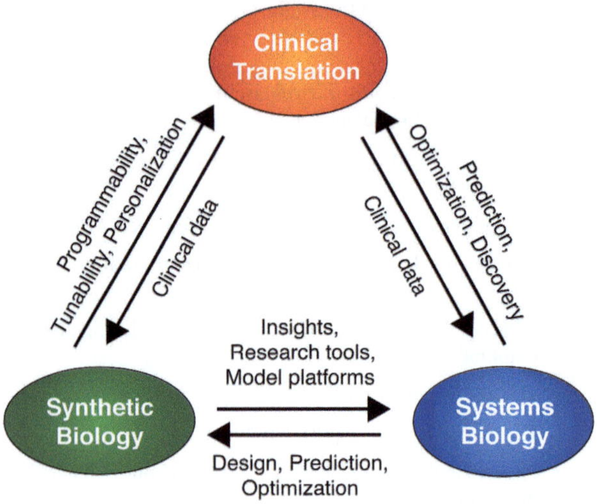

Fig. 18.2 Emerging framework for design-driven medicine. The synergistic relationships between *systems biology*, *synthetic biology*, and *clinical translation* together provide a framework for translating information into understanding, and for translating understanding into improved therapies through the engineering design cycle

Harnessing Systems Biology to Understand and Overcome Barriers to Cell-Based Immunotherapy

Realizing the potential of cell-based therapies for cancer will require improved understanding of both the functions of engineered cells and the manners in which these therapeutics interact with host physiology. In particular, clinical experience to date highlights two general challenges that pose significant barriers to achieving broad clinical efficacy. First, tumors frequently evolve the capacity to modulate local immune responses, and this local immune dysfunction is broadly considered a central impediment to both natural and therapeutically induced immune control of cancer [9, 59–61]. Although suppression of CTL responses is a general feature of many cancers, many aspects of immune dysfunction vary widely between different types of cancer. Second, patient-to-patient variability in both tumor biology and host immune function poses a challenge for (a) gaining understanding from immunotherapy clinical investigations and for (b) identifying immunotherapy strategies that are likely to be effective for a particular patient. Each of these challenges comprises a systems-level problem, and the expanding field of systems biology provides a toolbox of conceptual and research tools that may be applied to addressing such questions.

Systems biology is a scientific approach that seeks to explain how ensembles of biological entities interact to generate aggregate behaviors that govern the function of living systems (reviewed in [62–64]). This approach may be applied at multiple scales, ranging from investigating the operation of genetic and biochemical networks in individual cells to explaining the origins of variability between individual humans within a population. Systems biology provides a framework for pairing the power of numerous new technologies for collecting vast quantities of biological information with the computational and analytical tools required to make use of data of this type and scale.

The systems biology toolbox provides a range of computational methods that may be applied to distinct challenges in understanding tumor–immune interactions and enabling cell-based therapy for cancer [64–67]. At one end of the spectrum, statistical methods such as clustering and principal component analysis (PCA) may be employed to characterize datasets without requiring prior knowledge of the underlying biology [68]. Discriminant partial least squares regression (DPLSR) is related to PCA and enables one to identify combinations of measured variables that best predict a given outcome, such as the intracellular signaling events that predict tumor necrosis factor (TNF)-induced apoptosis of tumor cells [69, 70]. Bayesian inference can be used to infer causal relationships, such as the topology of intracellular signaling networks, as was demonstrated in naïve $CD4^{+}$ T cells [71, 72]. Each of these methods may be paired with experimental investigations into tumor–immune interactions and could be a useful tool for harnessing large or complicated datasets to discover patterns or generate novel hypotheses regarding potential causal relationships.

At the other end of the computational methods spectrum, mechanistic modeling is a powerful approach for investigating specific questions or phenomena. Mechanistic models describe specific events and biochemical transformations such as gene expression, protein phosphorylation, and complex formation, and the dynamics of these processes are described by constructing mathematical expressions (typically ordinary or partial differential equations, ODE or PDE, respectively) based upon known or postulated mechanisms [66]. Because such models are based upon prior knowledge of the system, this approach is often especially useful for (a) exploring the behavior of a putative conceptual model and formulating novel hypotheses, (b) determining whether a single conceptual model can be consistent with a series of observations, or (c) determining which of a set of hypothetical mechanisms is most consistent with experimental observations.

Mechanistic models are especially useful for investigating and understanding complex systems, such as immune function, which often defy explanation by a simple narrative. For example, signaling in response to interleukins 6 and 10 (IL-6 and IL-10) is somewhat paradoxical, in that the receptor for each cytokine signals through the same transcription factor (signal transducer and activator of transcription 3, STAT3), yet IL-6 and IL-10-induced responses are pro-inflammatory or anti-inflammatory, respectively. To potentially explain this paradox, Fowler and colleagues generated an ODE-based model of this system that explains a range of experimental observations by hypothesizing the existence of a simple but previously uncharacterized mechanism for crosstalk between these pathways [73]. Tumor biology has also been widely modeled using methods ranging from ODE to agent-based simulations (reviewed in [74]), and a number of approaches have also incorporated tumor–immune interactions [75–78]. Mechanistic models of *in vivo* cancer scenarios are particularly useful as tools for investigating conceptual questions, due to the difficulty of observing and modulating tumor dynamics in living animals or patients. As history has demonstrated, the emergence of powerful new experimental methods will drive the development and need for computational models that enable researchers to utilize these data.

Several new experimental systems biology tools might be especially useful for elucidating complex tumor–immune interactions by enabling robust characterizations at the level of individual cells. For example, using "mass cytometry," one can simultaneously quantify the expression of up to 100 markers or components in single cells, far exceeding the capacity of multiparametric flow cytometry [79]. In a demonstration of this method, patient-derived bone marrow cells were analyzed using 34 parameters including both cell surface markers and functional markers such as intracellular cytokine content and transcription factor phosphorylation states. This high-dimension characterization revealed subtle cell type-specific differences in signaling during hematopoietic development and in response to drugs. Other methods enable the characterization of time-dependent behavior at this single-cell level of resolution. For example, microengraved subnanoliter-scale wells have been used to profile the dynamics of cytokine secretion by individual T cells, which identified the existence of a number of distinct functional programs within a single population of human primary T cells [80].

An exciting and largely open grand challenge in experimental systems biology is the characterization of human immune variability. Although immune function is understood to vary widely between individuals, the nature, consequences, and underlying causes of this variability remain poorly understood. This challenge is particularly applicable to cell-based therapy for cancer, since the primary mode of action for this approach (cell-mediated tumor killing) depends strongly upon host immune function, which may both promote therapeutic action and duration of benefits or pose a barrier to efficacy. Several studies have characterized variability at the genetic level, including profiles of variation in the human MHC locus [81] or sequencing of adaptive immune receptor repertoires [82]. However, to date we lack similarly comprehensive *functional* characterizations of immune variability, the interpretation of which is particularly challenging because of noise arising from patient history, both recent and long term. Thus, a new approach that could address this need is the functional profiling of mature cells differentiated from patient-derived induced pluripotent stem cells (iPSC), which can be generated from cells easily collected via a cheek swab. This method has already proven useful for elucidating mechanisms of disease-associated differences in neuronal function [83], and this approach could help to decouple genetically encoded aspects of immune variability from those that depend on patient history. Ultimately, such insights would contribute fundamental understanding of tumor–immune function and would be useful for guiding clinical translation of cell-based therapeutics.

Systems biology could serve the mission of clinically translating cell-based therapy by enabling physicians to best match patients with a therapy that is likely to provide clinical benefits. Such an approach has been proposed to improve the design of clinical trials with pharmacological anticancer therapy [84–88], and computational tools provide a framework for analyzing preclinical patient samples to target clinical trials to patient groups most likely to respond to a given drug [89, 90]. Computational tools could also prove useful for monitoring and predicting patient responses during clinical trials already underway. This concept was illustrated in the context of prophylactic vaccination for yellow fever, in which high throughput screening was combined with clustering analysis to identify early response biomarkers that predicted long-term protective effects of vaccination [91]. Finally, computational systems biology approaches could be used to design more effective therapeutic regimens [84, 92–94]. For example, a mathematical model describing the potential of lung cancer tumor cells to acquire function-altering mutations in epidermal growth factor receptor (EGFR) was used to design a novel therapeutic strategy predicted to prolong clinical benefit of tyrosine kinase inhibitor (TKI) drugs targeting this pathway [92]. Such insights could be difficult or impossible to glean from patient data or in vitro experiments in the absence of mechanistic, quantitative models.

Increasingly powerful models describing tumor–immune interactions in cell-based therapy are also proving useful for understanding therapeutic efficacy and limitations and for designing improved therapies [75, 94–100]. Mathematical models have proven particularly useful for interpreting data from preclinical and clinical cell-based therapy experiments, and for resolving counterintuitive or seemingly contradictory observations [75, 94, 96, 97]. For instance, a computational investigation was used to develop a novel hypothesis explaining the surprising experimental observation that in murine models of melanoma, Th2 cells (which generally promote antibody-mediated immunity) are better able to promote tumor rejection than are Th1 cells (which generally promote CTL-mediated immunity) [97]. Modeling may also help to explain mixed clinical results, as was demonstrated for adoptive CTL transfer treatment of glioblastoma [94]. Importantly, by potentially explaining the origins of treatment failure, these investigations also generated suggestions as to how such failures may be overcome.

Ultimately, computational models could enable physicians to customize cell-based therapies to individual patients. For example, a model that was "personalized" with patients' prostate-specific antigen levels was able to predict individual responses to vaccination for treatment of prostate cancer [101]. Personalized models could ultimately be used to simulate patient responses to different therapeutic regimens. The increasing power of computational tools to predict and evaluate patient responses to therapeutics has inspired a number of provocative suggestions as to how clinical trials may be restructured [102–105]. Given the unique challenges associated with cell-based therapy for cancer, computational systems biology is certain to play an increasingly important role in both fundamental research and in maximizing both safety and efficacy in the clinic.

Synthetic Biology: A New Frontier in Design-Driven Medicine

The emerging technology of synthetic biology is a potentially transformative approach to engineering biological systems and could enable the rapid translation of fundamental understanding into novel and effective therapies. In combination with tools and information from clinical science and systems biology, synthetic biology could therefore enable a new approach to design-driven medicine (Fig. 18.2). The central objective of the field of synthetic biology may be summarized as developing the knowledge and capabilities necessary to engineer novel biological systems in a robust fashion [106–109]. Ultimately, this approach should enable one to start with an abstract functional goal, and then provide tools and methods for building a biological system that carries out this function. In the example of cell-based therapies for cancer, synthetic biology could bridge the gap between formulating a hypothesis describing a desirable cellular function—"I predict that a cell that did X would have Y effect"—and being able to test or implement that hypothesis by engineering a cell-based "device" accordingly.

Although much of the foundational work in synthetic biology was performed in microbial systems, an increasing number of applications are targeted directly at clinical applications, including those that engineer mammalian cells to serve as cell-based therapies (reviewed in [107, 108]). Several early proof-of-principle demonstrations illustrate potential new therapeutic modalities that may be constructed using synthetic biology. First, synthetic biology may enable physicians to externally control or modulate the function of a therapeutic post-administration (Fig. 18.3a). In one such example, Smolke and colleagues engineered RNA devices that are activated by the drug theophylline [110]. These RNA devices were expressed in engineered T cells, such that upon oral administration of theophylline, the adoptively transferred T cells expressed IL-2 and proliferated. If implemented with CAR therapy, such a strategy could reduce the need for systemic administration of IL-2 and potentially decrease toxicity associated with high cytokine levels in the blood. Hypothetically, this control strategy could also be inverted to tunably downregulate T cell proliferation in the event of toxicities such as cytokine storm or tumor lysis syndrome. In another strategy to temporarily and reversibly put the brakes on excessive CAR T cell activation, Lim et al. reported a "pause switch" comprising components from bacterial virulence proteins and yeast MAP kinase pathways, which dampened T cell proliferation upon activation of the switch by the antibiotic, tetracycline [111]. In another example, Fussenegger and colleagues developed a system in which externally applied light could control gene expression *in vivo* [112]. In this system, immortalized human embryonic kidney (HEK) cells were engineered to produce a peptide regulating glucose homeostasis in response to blue light. These engineered cells were then microencapsulated in a hydrogel matrix and implanted subcutaneously, and exposing these animals to blue light pulses successfully conferred regulation of serum glucose levels. Recently, externally tunable synthetic biology technology has reached human

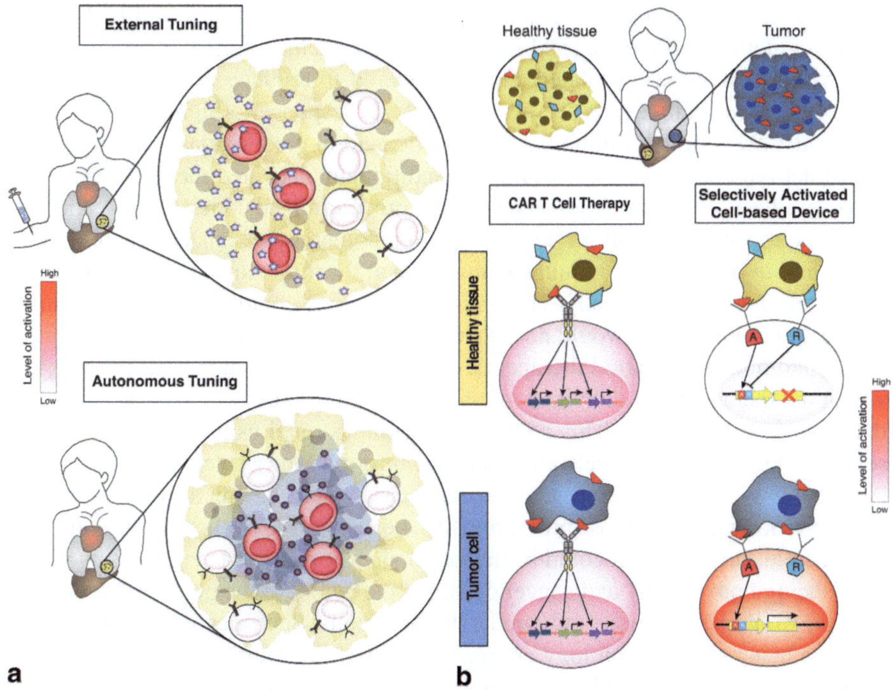

Fig. 18.3 Opportunities for harnessing synthetic biology to enhance cell-based therapy for cancer. Synthetic biology could provide novel modalities for enhancing cell-based therapies to improve both safety and efficacy. **a** These modalities include enabling physicians to tune therapeutic activity post-administration using an externally administered cue (*top*) or "programming" cell-based therapies to autonomously regulate their own activity in response to environmental cues (*bottom*). **b** Example of how a synthetic biology gene circuit can enhance specificity of a cell-based therapy for a tumor versus healthy tissue. In this example, the engineered cell expresses two receptors that each detect one of two external ligands, either a TAA shared by the tumor and a healthy tissue or an antigen found only on the healthy tissue, and each receptor releases a transcriptional activator (*A*) or a transcriptional repressor (*R*), upon detecting its respective ligand. In this way, the cell performs a multiparametric logical evaluation of the two ligands, such that the engineered cell becomes "activated" (e.g., triggering it to express a potent immune stimulant or toxin) only when contacting a tumor cell and not when contacting a healthy cell sharing the TAA **b**. *A* transcriptional activator, *R* transcriptional repressor, TAA tumor associated antigen

clinical trials in the form of externally tunable expression of IL-12 from an adenoviral vector that is injected intratumorally [113]. This technology utilizes Intrexon's RheoSwitch system, which may be activated by an orally administered activator ligand [114, 115]. Enabling physicians to control the activity of cell-based therapies could facilitate both personalized treatment strategies and optimization of safety and efficacy in response to real-time monitoring of patient responses.

A second attractive modality that could be realized through synthetic biology is the construction of autonomously regulating therapeutics (Fig. 18.3a). In general, the therapeutic would survey its surroundings and perform some function in response to this information. One application of this capability would be to create a self-regulating therapeutic, and this concept was demonstrated using an implant comprised of microencapsulated cells engineered to regulate uric acid homeostasis [116]. These cells expressed urate oxidase in response to high serum levels of uric acid, successfully eliminating excess uric acid in urate oxidase-deficient mice. A distinct type of application would be conditionally activated therapeutics, which perform a desired function only in a specific, predetermined context. This capability would be especially useful for cell-based immunotherapy of cancer, where it might be desired to deliver local immune stimulation at the tumor site at a level that could not be tolerated systematically. Thus, synthetic biology could enable one to *create* a therapeutic window that would not otherwise exist using current approaches.

Autonomously regulated therapeutics could potentially function as prodrugs, in which the therapeutic is delivered to many cells, but it only becomes "activated" in the target cell of interest. This concept was demonstrated using a gene circuit that sensed cellular miRNA levels and used this information to selectively activate the circuit only when it was expressed in specific cancer cells having miRNA expression "fingerprints" matching those programmed to be recognized by the gene circuit [117]. In principle, such a circuit could be delivered via a nontargeted gene therapy vector, transducing both healthy and cancer cells, but the output of the gene circuit (such as a toxin) would be expressed only in the diseased cells. Such a capability is uniquely possible using synthetic biology approaches to perform multiparametric evaluation of cellular features, and both analog and digital (e.g., Boolean) evaluation have been performed in mammalian cells using "parts" composed of RNA elements [118–120], gene transcription networks [121], and other protein-based elements [122].

An alternative approach to autonomous regulation that is particularly attractive for cell-based cancer therapy would be to program an engineered cell to evaluate its environment and then become "activated" only under prespecified conditions. Such a cell-based therapy could be programmed to travel throughout the body and deliver a potent immune stimulant only when the engineered cell enters the tumor microenvironment. For example, engineered logical evaluation could be used to prevent activation in healthy tissue by programming the therapeutic to survey for both a TAA and a second antigen that is expressed only on healthy tissue that might also express the TAA at low levels (Fig. 18.3b). One version of such a strategy has been implemented in the form of the inhibitory CAR (iCAR), which combines an extracelluar antigen-binding domain fused to the intracellular domain of a native T cell inhibitory receptor, PD-1 or CTLA-4 [123]. Expression of iCAR effectively dampened T cell activation via either a native TCR primed against a model antigen or a transduced CAR when both the activating and inhibitory receptor were engaged, without impeding activation when only the activating antigen was present. Similarly, specificity could be achieved by engineering the cell-based therapy to become activated only when it encounters two TAA, neither of which is uniquely expressed by tumor cells. One approach to implementing this strategy has been to transduce a T cell with both a suboptimal CAR specific for one antigen and also with a costimulatory receptor (CCR) specific for a second antigen, thereby making full T cell activation conditional upon binding to both antigens [124]. In addition to providing specificity, combinatorial antigen recognition strategies could also be employed to circumvent tumor escape by antigen downregulation [50].

While CAR engineering is a promising approach, realizing the promise of programmable cell-based therapies will require the development of new synthetic biology "parts." For example, robustly interfacing engineered cells with host physiologywill require cell-surface sensors that detect exclusively extracellular species and relay these detection events into engineered intracellular gene circuits. To address this need, Daringer et al. recently reported a modular approach to synthetic receptor engineering, which couples the sensing of a target extracellular analyte to the induction of a signaling pathway that is orthogonal to native signaling pathways in the cell [125]. Such orthogonality facilitates the construction of complex cellular programs, such as the multiparameteric evaluation of environmental cues to enhance or restrict CAR T cell activation in particular milieus. It is already clear that synthetic biology dovetails with advances in systems biology and clinical science to provide a novel approach to therapeutic development and an ensemble of capabilities that are particularly well-suited to overcoming current barriers to safe and effective immunotherapy of cancer.

Closing Thoughts

Engineered cell-based therapy for cancer is a strategy that is poised to realize true clinical benefits. Grand challenges for this field include extending early successes to broader patient groups and to multiple types of disease, while maximizing both safety and efficacy. The synergistic technologies of systems and synthetic biology promise to accelerate the achievement of these goals by enabling researchers to transform information into understanding, and to translate understanding into novel therapeutic strategies. Harnessing these approaches may prove pivotal for advancing this field from one driven by heuristics to one driven by knowledge and even patient-specific information. Ultimately, immunotherapy of cancer may prove to be the first of many engineered cell-based therapies, ushering in a new wave of design-driven medicine.

Acknowledgments RMD and JNL received support from the Defense Advanced Research Projects Agency, Award number W911NF-11-2-0066.

References

1. Romond EH, Perez EA, Bryant J, et al. Trastuzumab plus adjuvant chemotherapy for operable HER2-positive breast cancer. N Engl J Med. 2005;353(16):1673–84.
2. Hodi FS, O'Day SJ, McDermott DF, et al. Improved survival with ipilimumab in patients with metastatic melanoma. N Engl J Med. 2010;363(8):711–23.
3. Druker BJ, Talpaz M, Resta DJ, et al. Efficacy and safety of a specific inhibitor of the BCR-ABL tyrosine kinase in chronic myeloid leukemia. N Engl J Med. 2001;344(14):1031–7.

4. Druker BJ, Sawyers CL, Kantarjian H, et al. Activity of a specific inhibitor of the BCR-ABL tyrosine kinase in the blast crisis of chronic myeloid leukemia and acute lymphoblastic leukemia with the Philadelphia chromosome. N Engl J Med. 2001;344(14):1038–42.
5. Lapidot T, Sirard C, Vormoor J, et al. A cell initiating human acute myeloid leukaemia after transplantation into SCID mice. Nature. 1994;367(6464):645–8.
6. Bonnet D, Dick JE. Human acute myeloid leukemia is organized as a hierarchy that originates from a primitive hematopoietic cell. Nat Med. 1997;3(7):730–7.
7. Gorre ME, Mohammed M, Ellwood K, et al. Clinical resistance to STI-571 cancer therapy caused by BCR-ABL gene mutation or amplification. Science. 2001;293(5531):876–80.
8. Scaltriti M, Rojo F, Ocaña A, et al. Expression of p95HER2, a truncated form of the HER2 receptor, and response to anti-HER2 therapies in breast cancer. J Natl Cancer Inst. 2007;99(8):628–38.
9. Goldman B, DeFrancesco L. The cancer vaccine roller coaster. Nat Biotechnol. 2009;27(2):129–39.
10. Palucka K, Ueno H, Banchereau J. Recent developments in cancer vaccines. J Immunol. 2011;186(3):1325–31.
11. Blank CU, Hooijkaas AI, Haanen JB, Schumacher TN. Combination of targeted therapy and immunotherapy in melanoma. Cancer Immunol Immunother. 2011;60(10):1359–71.
12. Burnet M. Cancer: a biological approach. III. Viruses associated with neoplastic conditions. IV. Practical applications. Br Med J. 1957;1(5023):841–7.
13. Thomas L Cellular and humoral aspects of the hypersensitive states. Paper presented at: Cellular and humoral aspects of the hypersensitive states: a symposium held at the New York Academy of Medicine. New York; 1959.
14. Restifo NP, Dudley ME, Rosenberg SA. Adoptive immunotherapy for cancer: harnessing the T cell response. Nat Rev Immunol. 2012;12(4):269–81.
15. Schreiber RD, Old LJ, Smyth MJ. Cancer immunoediting: integrating immunity's roles in cancer suppression and promotion. Science. 2011;331(6024):1565–70.
16. Höglund P, Glas R, Ohlén C, Ljunggren HG, Kärre K. Alteration of the natural killer repertoire in H-2 transgenic mice: specificity of rapid lymphoma cell clearance determined by the H-2 phenotype of the target. J Exp Med. 1991;174(2):327–34.
17. Ahmadzadeh M, Johnson LA, Heemskerk B, et al. Tumor antigen-specific CD8 T cells infiltrating the tumor express high levels of PD-1 and are functionally impaired. Blood. 2009;114(8):1537–44.
18. Friberg M, Jennings R, Alsarraj M, et al. Indoleamine 2,3-dioxygenase contributes to tumor cell evasion of T cell-mediated rejection. Int J Cancer. 2002;101(2):151–5.
19. Gabrilovich DI, Ostrand-Rosenberg S, Bronte V. Coordinated regulation of myeloid cells by tumours. Nat Rev Immunol. 2012;12(4):253–68.
20. Rosenberg SA, Packard BS, Aebersold PM, et al. Use of tumor-infiltrating lymphocytes and interleukin-2 in the immunotherapy of patients with metastatic melanoma. A preliminary report. N Engl J Med. 1988;319(25):1676–80.
21. Rosenberg SA, Aebersold P, Cornetta K, et al. Gene transfer into humans–immunotherapy of patients with advanced melanoma, using tumor-infiltrating lymphocytes modified by retroviral gene transduction. N Engl J Med. 1990;323(9):570–8.
22. Clay TM, Custer MC, Sachs J, Hwu P, Rosenberg SA, Nishimura MI. Efficient transfer of a tumor antigen-reactive TCR to human peripheral blood lymphocytes confers anti-tumor reactivity. J Immunol. 1999;163(1):507–13.
23. van Loenen MM,deB, Amir AL, et al. Mixed T cell receptor dimers harbor potentially harmful neoreactivity. Proc Natl Acad Sci U S A. 2010;107(24):10972–7.
24. Bendle GM, Linnemann C, Hooijkaas AI, et al. Lethal graft-versus-host disease in mouse models of T cell receptor gene therapy. Nat Med. 2010;16(5):565–70, 561 p following 570.
25. Rosenberg SA. Of mice, not men. no evidence for graft-versus-host disease in humans receiving T-cell receptor-transduced autologous T cells. Mol Ther. 2010;18(10):1744–5.
26. Johnson LA, Morgan RA, Dudley ME, et al. Gene therapy with human and mouse T-cell receptors mediates cancer regression and targets normal tissues expressing cognate antigen. Blood. 2009;114(3):535–46.

27. Irving BA, Weiss A. The cytoplasmic domain of the T cell receptor zeta chain is sufficient to couple to receptor-associated signal transduction pathways. Cell. 1991;64(5):891–901.
28. Gross G, Waks T, Eshhar Z. Expression of immunoglobulin-T-cell receptor chimeric molecules as functional receptors with antibody-type specificity. Proc Natl Acad Sci U S A. 1989;86(24):10024–8.
29. Eshhar Z, Waks T, Gross G, Schindler DG. Specific activation and targeting of cytotoxic lymphocytes through chimeric single chains consisting of antibody-binding domains and the gamma or zeta subunits of the immunoglobulin and T-cell receptors. Proc Natl Acad Sci U S A. 1993;90(2):720–4.
30. Brentjens RJ, Latouche JB, Santos E, et al. Eradication of systemic B-cell tumors by genetically targeted human T lymphocytes co-stimulated by CD80 and interleukin-15. Nat Med. 2003;9(3):279–86.
31. Park JR, Digiusto DL, Slovak M, et al. Adoptive transfer of chimeric antigen receptor re-directed cytolytic T lymphocyte clones in patients with neuroblastoma. Mol Ther. 2007;15(4):825–33.
32. Till BG, Jensen MC, Wang J, et al. Adoptive immunotherapy for indolent non-Hodgkin lymphoma and mantle cell lymphoma using genetically modified autologous CD20-specific T cells. Blood. 2008;112(6):2261–71.
33. Atkins MB, Lotze MT, Dutcher JP, et al. High-dose recombinant interleukin 2 therapy for patients with metastatic melanoma: analysis of 270 patients treated between 1985 and 1993. J Clin Oncol. 1999;17(7):2105–16.
34. Rossig C, Bollard CM, Nuchtern JG, Rooney CM, Brenner MK. Epstein-Barr virus-specific human T lymphocytes expressing antitumor chimeric T-cell receptors: potential for improved immunotherapy. Blood. 2002;99(6):2009–16.
35. Pule MA, Savoldo B, Myers GD, et al. Virus-specific T cells engineered to coexpress tumor-specific receptors: persistence and antitumor activity in individuals with neuroblastoma. Nat Med. 2008;14(11):1264–70.
36. Louis CU, Savoldo B, Dotti G, et al. Antitumor activity and long-term fate of chimeric antigen receptor-positive T cells in patients with neuroblastoma. Blood. 2011;118(23):6050–6.
37. Finney HM, Akbar AN, Lawson AD. Activation of resting human primary T cells with chimeric receptors: costimulation from CD28, inducible costimulator, CD134, and CD137 in series with signals from the TCR zeta chain. J Immunol. 2004;172(1):104–13.
38. Milone MC, Fish JD, Carpenito C, et al. Chimeric receptors containing CD137 signal transduction domains mediate enhanced survival of T cells and increased antileukemic efficacy in vivo. Mol Ther. 2009;17(8):1453–64.
39. Finney HM, Lawson AD, Bebbington CR, Weir AN. Chimeric receptors providing both primary and costimulatory signaling in T cells from a single gene product. J Immunol. 1998;161(6):2791–97.
40. Hombach A, Wieczarkowiecz A, Marquardt T, et al. Tumor-specific T cell activation by recombinant immunoreceptors: CD3 zeta signaling and CD28 costimulation are simultaneously required for efficient IL-2 secretion and can be integrated into one combined CD28/CD3 zeta signaling receptor molecule. J Immunol. 2001;167(11):6123–31.
41. Maher J, Brentjens RJ, Gunset G, Rivière I, Sadelain M. Human T-lymphocyte cytotoxicity and proliferation directed by a single chimeric TCRzeta/CD28 receptor. Nat Biotechnol. 2002;20(1):70–5.
42. Pulè MA, Straathof KC, Dotti G, Heslop HE, Rooney CM, Brenner MK. A chimeric T cell antigen receptor that augments cytokine release and supports clonal expansion of primary human T cells. Mol Ther. 2005;12(5):933–41.
43. Loskog A, Giandomenico V, Rossig C, Pule M, Dotti G, Brenner MK. Addition of the CD28 signaling domain to chimeric T-cell receptors enhances chimeric T-cell resistance to T regulatory cells. Leukemia. 2006;20(10):1819–28.
44. Kochenderfer JN, Wilson WH, Janik JE, et al. Eradication of B-lineage cells and regression of lymphoma in a patient treated with autologous T cells genetically engineered to recognize CD19. Blood. 2010;116(20):4099–102.

45. Kochenderfer JN, Dudley ME, Feldman SA, et al. B-cell depletion and remissions of malignancy along with cytokine-associated toxicity in a clinical trial of anti-CD19 chimeric-antigen-receptor-transduced T cells. Blood. 2012;119(12):2709–20.
46. Zhong XS, Matsushita M, Plotkin J, Riviere I, Sadelain M. Chimeric antigen receptors combining 4-1BB and CD28 signaling domains augment PI3kinase/AKT/Bcl-XL activation and CD8+ T cell-mediated tumor eradication. Mol Ther. 2010;18(2):413–20.
47. Kalos M, Levine BL, Porter DL, et al. T cells with chimeric antigen receptors have potent antitumor effects and can establish memory in patients with advanced leukemia. Sci Transl Med. 2011;3(95):95ra73.
48. Porter DL, Levine BL, Kalos M, Bagg A, June CH. Chimeric antigen receptor-modified T cells in chronic lymphoid leukemia. N Engl J Med. 2011;365(8):725–33.
49. Till BG, Jensen MC, Wang J, et al. CD20-specific adoptive immunotherapy for lymphoma using a chimeric antigen receptor with both CD28 and 4-1BB domains: pilot clinical trial results. Blood. 2012;119(17):3940–50.
50. Grupp SA, Kalos M, Barrett D, et al. Chimeric antigen receptor-modified T cells for acute lymphoid leukemia. N Engl J Med. 2013;368(16):1509–18.
51. Davila ML, Riviere I, Wang X, et al. Efficacy and toxicity management of 19-28z CAR T cell therapy in B cell acute lymphoblastic leukemia. Sci Transl Med. 2014;6(224):224ra5.
52. Sadelain M, Brentjens R, Rivière I. The promise and potential pitfalls of chimeric antigen receptors. Curr Opin Immunol. 2009;21(2):215–23.
53. Jena B, Dotti G, Cooper LJ. Redirecting T-cell specificity by introducing a tumor-specific chimeric antigen receptor. Blood. 2010;116(7):1035–44.
54. June CH, Blazar BR, Riley JL. Engineering lymphocyte subsets: tools, trials and tribulations. Nat Rev Immunol. 2009;9(10):704–16.
55. Kohn DB, Dotti G, Brentjens R, et al. CARs on track in the clinic. Mol Ther. 2011;19(3):432–8.
56. Brentjens RJ, Davila ML, Riviere I, et al. CD19-targeted T cells rapidly induce molecular remissions in adults with chemotherapy-refractory acute lymphoblastic leukemia. Sci Transl Med. 2013;5(177):177ra38.
57. Morgan RA, Yang JC, Kitano M, Dudley ME, Laurencot CM, Rosenberg SA. Case report of a serious adverse event following the administration of T cells transduced with a chimeric antigen receptor recognizing ERBB2. Mol Ther. 2010;18(4):843–51.
58. Ertl HC, Zaia J, Rosenberg SA, et al. Considerations for the clinical application of chimeric antigen receptor T cells: observations from a recombinant DNA Advisory Committee Symposium held June 15, 2010. Cancer Res. 2011;71(9):3175–81.
59. Gajewski TF. Failure at the effector phase: immune barriers at the level of the melanoma tumor microenvironment. Clin Cancer Res. 2007;13(18 Pt 1):5256–61.
60. Whiteside TL. The tumor microenvironment and its role in promoting tumor growth. Oncogene. 2008;27(45):5904–12.
61. Sica A, Bronte V. Altered macrophage differentiation and immune dysfunction in tumor development. J Clin Invest. 2007;117(5):1155–66.
62. Kitano H. Computational systems biology. Nature. 2002;420(6912):206–10.
63. Kitano H. Systems biology: a brief overview. Science. 2002;295(5560):1662–4.
64. Arkin AP, Schaffer DV. Network news: innovations in 21st century systems biology. Cell. 2011;144(6):844–9.
65. van Riel NA. Dynamic modelling and analysis of biochemical networks: mechanism-based models and model-based experiments. Brief Bioinform. 2006;7(4):364–74.
66. Aldridge BB, Burke JM, Lauffenburger DA, Sorger PK. Physicochemical modelling of cell signalling pathways. Nat Cell Biol. 2006;8(11):1195–203.
67. Price ND, Shmulevich I. Biochemical and statistical network models for systems biology. Curr Opin Biotechnol. 2007;18(4):365–70.
68. Janes KA, Yaffe MB. Data-driven modelling of signal-transduction networks. Nat Rev Mol Cell Biol. 2006;7(11):820–8.
69. Janes KA, Kelly JR, Gaudet S, Albeck JG, Sorger PK, Lauffenburger DA. Cue-signal-response analysis of TNF-induced apoptosis by partial least squares regression of dynamic multivariate data. J Comput Biol. 2004;11(4):544–61.

70. Janes KA, Albeck JG, Gaudet S, Sorger PK, Lauffenburger DA, Yaffe MB. A systems model of signaling identifies a molecular basis set for cytokine-induced apoptosis. Science. 2005;310(5754):1646–53.
71. Sachs K, Gifford D, Jaakkola T, Sorger P, Lauffenburger DA. Bayesian network approach to cell signaling pathway modeling. Sci STKE. 2002;2002(148):pe38.
72. Sachs K, Perez O, Pe'er D, Lauffenburger DA, Nolan GP. Causal protein-signaling networks derived from multiparameter single-cell data. Science. 2005;308(5721):523–9.
73. Fowler KD, Kuchroo VK, Chakraborty AK. A model for how signal duration can determine distinct outcomes of gene transcription programs. PLoS ONE. 2012;7(3):e33018.
74. Edelman LB, Eddy JA, Price ND. In silico models of cancer. Wiley Interdiscip Rev Syst Biol Med. 2010;2(4):438–59.
75. de Pillis LG, Radunskaya AE, Wiseman CL. A validated mathematical model of cell-mediated immune response to tumor growth. Cancer Res. 2005;65(17):7950–8.
76. Mallet DG, De Pillis LG. A cellular automata model of tumor-immune system interactions. J Theor Biol. 2006;239(3):334–50.
77. Eikenberry S, Thalhauser C, Kuang Y. Tumor-immune interaction, surgical treatment, and cancer recurrence in a mathematical model of melanoma. PLoS Comput Biol. 2009;5(4):e1000362.
78. Woelke AL, Murgueitio MS, Preissner R. Theoretical modeling techniques and their impact on tumor immunology. Clin Dev Immunol. 2010;2010:271794.
79. Bendall SC, Simonds EF, Qiu P, et al. Single-cell mass cytometry of differential immune and drug responses across a human hematopoietic continuum. Science. 2011;332(6030):687–96.
80. Han Q, Bagheri N, Bradshaw EM, Hafler DA, Lauffenburger DA, Love JC. Polyfunctional responses by human T cells result from sequential release of cytokines. Proc Natl Acad Sci U S A. 2012;109(5):1607–12.
81. de Bakker PI, McVean G, Sabeti PC, et al. A high-resolution HLA and SNP haplotype map for disease association studies in the extended human MHC. Nat Genet. 2006;38(10):1166–72.
82. Benichou J, Ben-Hamo R, Louzoun Y, Efroni S. Rep-Seq: uncovering the immunological repertoire through next-generation sequencing. Immunology. 2012;135(3):183–91.
83. Pasca SP, Portmann T, Voineagu I, et al. Using iPSC-derived neurons to uncover cellular phenotypes associated with Timothy syndrome. Nat Med. 2011;17(12):1657–62.
84. Swierniak A, Kimmel M, Smieja J. Mathematical modeling as a tool for planning anticancer therapy. Eur J Pharmacol. 2009;625(1–3):108–21.
85. Roeder I, Horn M, Glauche I, Hochhaus A, Mueller MC, Loeffler M. Dynamic modeling of imatinib-treated chronic myeloid leukemia: functional insights and clinical implications. Nat Med. 2006;12(10):1181–4.
86. Adjei AA, Christian M, Ivy P. Novel designs and end points for phase II clinical trials. Clin Cancer Res.;15(6):1866–72.
87. Dancey JE, Dobbin KK, Groshen S, et al. Guidelines for the development and incorporation of biomarker studies in early clinical trials of novel agents. Clin Cancer Res. 2010;16(6): 1745–55.
88. Freedman AN, Sansbury LB, Figg WD, et al. Cancer pharmacogenomics and pharmacoepidemiology: setting a research agenda to accelerate translation. J Natl Cancer Inst. 2010;102(22):1698–705.
89. Cohen AL, Soldi R, Zhang H, et al. A pharmacogenomic method for individualized prediction of drug sensitivity. Mol Syst Biol. 2011;7:513.
90. Gardner SN. Modeling multi-drug chemotherapy: tailoring treatment to individuals. J Theor Biol. 2002;214(2):181–207.
91. Querec TD, Akondy RS, Lee EK, et al. Systems biology approach predicts immunogenicity of the yellow fever vaccine in humans. Nat Immunol. 2009;10(1):116–25.
92. Chmielecki J, Foo J, Oxnard GR, et al. Optimization of dosing for EGFR-mutant non-small cell lung cancer with evolutionary cancer modeling. Sci Transl Med. 2011;3(90):90ra59.
93. Haeno H, Gonen M, Davis MB, Herman JM, Iacobuzio-Donahue CA, Michor F. Computational modeling of pancreatic cancer reveals kinetics of metastasis suggesting optimum treatment strategies. Cell. 2012;148(1-2):362–75.

94. Kronik N, Kogan Y, Vainstein V, Agur Z. Improving alloreactive CTL immunotherapy for malignant gliomas using a simulation model of their interactive dynamics. Cancer Immunol Immunother. 2008;57(3):425–39.
95. Nani F, Freedman HI. A mathematical model of cancer treatment by immunotherapy. Math Biosci. 2000;163(2):159–99.
96. Leon K, Garcia K, Carneiro J, Lage A. How regulatory CD25+CD4+ T cells impinge on tumor immunobiology: the differential response of tumors to therapies. J Immunol. 2007;179(9):5659–68.
97. Eftimie R, Bramson JL, Earn DJ. Modeling anti-tumor Th1 and Th2 immunity in the rejection of melanoma. J Theor Biol. 2010;265(3):467–80.
98. Eftimie R, Bramson JL, Earn DJ. Interactions between the immune system and cancer: a brief review of non-spatial mathematical models. Bull Math Biol. 2011;73(1):2–32.
99. Joshi B, Wang X, Banerjee S, Tian H, Matzavinos A, Chaplain MA. On immunotherapies and cancer vaccination protocols: a mathematical modelling approach. J Theor Biol. 2009;259(4):820–7.
100. Kiran KL, Lakshminarayanan S. Global sensitivity analysis and model-based reactive scheduling of targeted cancer immunotherapy. Biosystems. 2010;101(2):117–26.
101. Kronik N, Kogan Y, Elishmereni M, Halevi-Tobias K, Vuk-Pavlovic S, Agur Z. Predicting outcomes of prostate cancer immunotherapy by personalized mathematical models. PLoS ONE. 2010;5(12):e15482.
102. Grove A. Rethinking clinical trials. Science. 2011;333(6050):1679.
103. Carome M, Wolfe S. Rethinking clinical trials: phase 1 studies insufficient. Science. 2011;334(6061):1346.
104. Marcus NA. Rethinking clinical trials: change is coming. Science. 2011;334(6061):1346.
105. Borhani DW, Butts JA. Rethinking clinical trials: biology's mysteries. Science. 2011;334(6061):1346–7.
106. Fritz BR, Timmerman LE, Daringer NM, Leonard JN, Jewett MC. Biology by design: from top to bottom and back. J Biomed Biotechnol. 2010;2010:232016.
107. Ruder WC, Lu T, Collins JJ. Synthetic biology moving into the clinic. Science. 2011;333(6047):1248–52.
108. Chen YY, Smolke CD. From DNA to targeted therapeutics: bringing synthetic biology to the clinic. Sci Transl Med. 2011;3(106):106ps42.
109. Purnick PE, Weiss R. The second wave of synthetic biology: from modules to systems. Nat Rev Mol Cell Biol. 2009;10(6):410–22.
110. Chen YY, Jensen MC, Smolke CD. Genetic control of mammalian T-cell proliferation with synthetic RNA regulatory systems. Proc Natl Acad Sci U S A. 2010;107(19):8531–6.
111. Wei P, Wong WW, Park JS, et al. Bacterial virulence proteins as tools to rewire kinase pathways in yeast and immune cells. Nature. 2012;488(7411):384–8.
112. Ye H, Daoud-El Baba M, Peng RW, Fussenegger M. A synthetic optogenetic transcription device enhances blood-glucose homeostasis in mice. Science. 2011;332(6037):1565–8.
113. Schwartzentruber DJ, Kirkwood JM, Guarino MJ, et al. Immunotherapy of advanced melanoma by intratumoral injections of autologous, purified dendritic cells transduced with gene construct of interleukin-12, with dose-dependent expression under the control of an oral activator ligand. J Clin Oncol. 2011;29(suppl):abstr 2540.
114. Murugesan S, Chan T, Reed C, et al. Combined direct intratumoral adenoviral delivery and production of Rheoswitch®-regulated IL-12 and IFNα enhances antitumor activity in lung and breast cancer models. In: Proceedings of the 103rd Annual Meeting of the American Association for Cancer Research; 2012 March 31–April 4; Chicago, IL. Philadelphia (PA): AACR; Cancer Res 2012;72(8 Suppl):Abstract nr 1546.
115. Herberman RB, Bi MX, Moreno M, et al. Local and systemic anti-tumor immunity is induced by Rheoswitch regulated IL-12 production after intra-tumoral injection of adenovirus vector as well as vector-transduced dendritic cells. Mol Ther. 2011;19(7):1380.
116. Kemmer C, Gitzinger M, Daoud-El Baba M, Djonov V, Stelling J, Fussenegger M. Self-sufficient control of urate homeostasis in mice by a synthetic circuit. Nat Biotechnol. 2010;28(4):355–60.

117. Xie Z, Wroblewska L, Prochazka L, Weiss R, Benenson Y. Multi-input RNAi-based logic circuit for identification of specific cancer cells. Science. 2011;333(6047):1307–11.
118. Rinaudo K, Bleris L, Maddamsetti R, Subramanian S, Weiss R, Benenson Y. A universal RNAi-based logic evaluator that operates in mammalian cells. Nat Biotechnol. 2007;25(7):795–801.
119. Culler SJ, Hoff KG, Smolke CD. Reprogramming cellular behavior with RNA controllers responsive to endogenous proteins. Science. 2010;330(6008):1251–5.
120. Bayer TS, Smolke CD. Programmable ligand-controlled riboregulators of eukaryotic gene expression. Nat Biotechnol. 2005;23(3):337–43.
121. Kramer BP, Fischer C, Fussenegger M. BioLogic gates enable logical transcription control in mammalian cells. Biotechnol Bioeng. 2004;87(4):478–84.
122. Lohmueller JJ, Armel TZ, Silver PA. A tunable zinc finger-based framework for boolean logic computation in mammalian cells. Nucleic Acids Res. 2012;40(11):5180–7.
123. Fedorov VD, Themeli M, Sadelain M. PD-1- and CTLA-4-based inhibitory chimeric antigen receptors (iCARs) divert off-target immunotherapy responses. Sci Transl Med. 2013;5(215):215ra172.
124. Kloss CC, Condomines M, Cartellieri M, Bachmann M, Sadelain M. Combinatorial antigen recognition with balanced signaling promotes selective tumor eradication by engineered T cells. Nat Biotechnol. 2013;31(1):71–5.
125. Daringer NM, Dudek RM, Schwarz KA, Leonard JN. Modular extracellular sensor architecture for engineering mammalian cell-based devices. ACS Synth Biol. 2014. [Epub ahead of press].
126. Park TS, Rosenberg SA, Morgan RA. Treating cancer with genetically engineered T cells. Trends Biotechnol. 2011;29(11):550–7.
127. Dotti G, Savoldo B, Brenner M. Fifteen years of gene therapy based on chimeric antigen receptors: "are we nearly there yet?". Hum Gene Ther. 2009;20(11):1229–39.

Part IV
Epilogue

Chapter 19
A Systems Approach to Blood Disorders

Pankaj Qasba

Abstract A systems approach to blood diseases can help make essential contributions to our ability to diagnose, treat, and perhaps even prevent common diseases in humans. Using blood as a window, one can study health and disease through this unique tool box with reactive biological fluids that mirrors the prevailing hemodynamics of the vessel walls and the various blood cell types. Many blood diseases, rare and common, can and have been exploited using systems biology approaches with successful results and therefore ideal models for systems medicine. More importantly, hematopoiesis offers one of the best studied systems with insight into stem cell biology, cellular interaction, development; linage programming and reprogramming that are influenced every day by the most mature and understood regulatory networks.

Keywords Hematopoiesis · Regulatory networks · Blood system · Erythropoiesis · Fanconi anemia (FA) · Diamond–Blackfan anemia (DBA) · Bone marrow failure syndromes (BMFs) · In silco · System biology · System medicine · Omics · Multiscale · Microfluidics · Hemoglobinopathies · Fetal hemoglobin · Induced pluripotent stem (iPS) cells

The hematopoietic system serves as a paradigm for understanding tissue stem cells, their biology, involvement in aging, and disease. Investigations in the field of hematopoiesis have yielded critical insights into the areas of stem cell biology; the role of cellular interactions in development and in tissue homeostasis; lineage programming and reprogramming by transcription factors; and stage- and age-specific

The views expressed by PQ in this chapter are personal and do not necessarily represent those of the US government.

P. Qasba (✉)
Blood Diseases Branch, Division of Blood Diseases and Resources,
National Heart, Lung, and Blood Institute, NIH, 6701 Rockledge Dr,
MSC 7950, Bethesda, MD, 20892-7950, USA
Tel.: 301-435-0050
e-mail: qasbap@nhlbi.nih.gov

S. J. Corey et al. (eds.), *A Systems Biology Approach to Blood,*
Advances in Experimental Medicine and Biology 844, DOI 10.1007/978-1-4939-2095-2_19

differences in cellular phenotypes [1, 2]. These critical insights provide us the tools to further elucidate and understand the transcriptional network that controls lineage choices and reprogramming of cellular lineages [3, 4]. The elucidation of regulatory networks is the most mature of the various systems biology components for hematopoiesis and serves as the first, and an essential step in building hypotheses and understanding biology at the system level [5]. However, our understanding of these molecular interactions and complex networks is still minimal. We must examine how genes, proteins, and other cellular components interact during normal and disease states to capture how cell activity is coordinated and controlled. The blood system truly provides a unique "tool box": a heterogeneous cellular compartment, and a vascular component comprising the vessel wall with a reactive biological fluid. So the time is right for us to take a broader systems biology approach to understanding erythropoiesis and blood diseases.

Systems biology methodologies have been successfully exploited in several blood diseases. Fanconi anemia (FA) is an autosomal recessive disorder characterized by congenital abnormalities, bone marrow failure, chromosome fragility, and cancer susceptibility. A remarkably high clinical variability exists among FA patients [6] possibly contributed by many regulatory mechanisms that affect the expression level of FA proteins. A logical systems approach by Tategu et al. [7] identified cross talk between E2F transcription factors and the FA pathway. Their in-silico mining of a transcriptome database and promoter analyses revealed that most of the FA gene members were regulated by E2F transcription factors, known to be pivotal regulators of cell cycle progression. Similarly, using systems approaches, one can probe the cross talk between the propensity of head and neck cancers, bone marrow failure, and DNA repair failure in FA patients and possibly make predictions of early onset of such cancers [8]. One can model FA cells or induced pluripotent stem (iPS) cells to delineate the role of human papillomavirus in the pathogenesis of these malignancies. Diamond–Blackfan anemia (DBA) is a bone marrow disorder that is characterized by impaired production of red blood cells. Approximately, half of all cases can be attributed to ribosomal protein gene mutations [9]. More recently, a new mutation in gene encoding the hematopoietic transcription factor GATA1 was linked to the disease [10–12]. Again, a systems approach will help to decipher the expanded network of genes involved in the remaining 50 % of the DBA patients without a genetic lesion and help understand the interplay between the ribosomal genes, hematopoietic transcription factors, and impaired erythroid development in this disease.

Other bone marrow failure syndromes (BMFs) such as Shwachman–Diamond syndrome (SDS), dyskeratosis congenita, myelodysplastic syndrome (MDS) and aplastic anemias present opportunities as models for a systems approach that will result in a better understanding of the phenotypic variability that is a striking feature of these syndromes [13–17]. The studies of many of these rare disorders are facilitated by the existence of established registries that link biospecimens with extensive phenotypic and genotypic data [18]. Registries enable the applications of computational biology and modeling to narrow down the parameters for potential drug screens and drug candidates for specific patient populations. The use of *in-silico* clinical trials may be appropriate for rare diseases where small numbers have prevented undertaking traditional clinical trials [19]. The use of these technologies in selected rare diseases may generate sufficient data to support the development of systems biology methodologies for "virtual" clinical trials that could reduce the time for the development of therapeutic interventions and the number of subjects required for their testing. Many of rare diseases have strong patient advocacy organizations and have developed sophisticated and inclusive registries that could be employed in support of such a systems-based approach to translational research.

Vascular biology and hemostasis provide another example amenable to systems approaches, Diamond et al. utilized high-throughput liquid handling and microfluidics to develop an integrated view of blood function under flow in response to multiple stimuli encountered during thrombosis and hemostasis [20]. This system allows multiscale prediction of patient-specific platelet function under flow and pharmacologic modulation, providing a possible risk assessment for arterial thrombotic events based on a patient's unique platelet phenotype. It is possible to provide a realistic description of intracellular signaling, platelet membrane receptor function, and the assembly of extracellular processes on platelet membranes or in plasma by exploiting a systems model of platelet and plasma function [21, 22]. Similarly, when stored platelets and red blood cells develop storage lesions during storage in plastic blood containers, the use of systems biology and integrated "omics" represents a tangible solution in the study of blood cell storage lesions by providing comprehensive biochemical descriptions of organisms through quantitative measurements and data integration in mathematical models. The biological knowledge collected from a target organism can be translated into a mathematical format and used to compute physiological properties. Hence the approach can help optimize the development of methods media that may lead to extended storage periods for blood components [23]. More importantly, the synergy between experimental and computational sciences provides a critical insight into better understanding of the molecular mechanism of red cell storage lesions.

Systems approaches have been employed in the study of the hemoglobinopathies (sickle cell disease (SCD) and thalassemia). Using Bayesian networks, Sebastiani et al. analyzed 108 SNPs in 39 candidate genes in 1398 individuals with SCD and generated 31 SNPs in 12 genes interacting with fetal hemoglobin to modulate risk of stroke, a severe complication affecting individuals with SCD. This group produced an algorithm that utilized SNPs from candidate modifier genes to define an individual subject's susceptibility to stroke. Using this predictive model, a high degree of accuracy was obtained to define the risk of stroke in 114 subjects [24].

More recently, Karniadkis laboratory has been developing a multiscale model for quantifying biophysical characteristics of SCD [25]. The model(s) would predict cell shape from hemoglobin polymerization, blood rheology based on cellular shape and mechanics, and the risk of vascular occlusion due to sickle cell adhesion to other blood cells and the vascular endothelium. Such work represents an example of unusually comprehensive integration from molecular- to tissue-level modeling of a major blood disease, again demonstrating that blood diseases provide the "tool sets" to apply systems biology approaches. One of the best studied examples in hemoglobinopathies is the transcription factor gene BCL11A, first identified by genome-wide association studies (GWAS) as a locus associated with the fetal hemoglobin level and disease severity [26], but its biological function was unclear. Orkin, Sankaran, and other laboratories have now established BCL11A as the key transcription factor regulating the fetal-to-adult hemoglobin switch in red blood cells, and genetic variation identified through GWAS is located in an enhancer element regulated by GATA1, TAL1, and other closely related transcription factors [27, 28]. This provides an opportunity to use systems biology approaches to predict and validate the developmental stage-specific gene regulatory networks in human primary erythroid precursors by integrating genomic and epigenomic data types. Furthermore, one can interrogate these networks' responses to external perturbations with implications for the robust developmental control in an erythroid gene regulatory model [18, 29, 30].

Rothenberg et al. has taken advantage of T cell development that is characterized by an experimentally tractable system for dynamical modeling of a lineage choice process [31, 32]. This is yet another example of systems approach to study transcriptional control of cell-fate choice and subsequent lineage specification. Finally, the iPS cells derived from blood will be complemented by iPS cells derived from cells of the immune system that have undergone T or B cell receptor recombination, opening the possibility for treating antigen-specific autoimmunity and allergies. Also, iPS cells from individual patients will be useful in exploring mechanisms of disease initiation, progression, and development in a test tube [33].

In a recent interview, Lee Hood, a biotech pioneer and cofounder of institute for systems biology said, "I believe that a systems approach to medicine will increasingly lead to a better understanding of disease mechanisms, *will make blood a window into studying health and disease…*" [33]. Blood system truly provides a unique tool box with a reactive biological fluid whose function is dictated by prevailing hemodynamics, vessel wall characteristics, heterogeneous cell types—red and white, platelets and their precursors, T and B cells, and p—human blood presents itself as an ideal multicellular "organ" for systems biology analysis.

References

1. Kim SI, Bresnick EH. Transcriptional control of erythropoiesis: emerging mechanisms and principles. Oncogene. 2007;26:6777–94.
2. Rothenberg EV. Negotiation of the T lineage fate decision by transcription-factor interplay and microenvironmental signals. Immunity. 2007;26:690–702.
3. Rossi DJ, Bryder D, Zahn JM, Ahlenius H, Sonu R, Wagers AJ, Weissman IL. Cell intrinsic alterations underlie hematopoietic stem cell aging. Proc Natl Acad Sci U S A. 2005;102:9194–9.
4. Rossant J. Stem cells: the magic brew. Nature. 2007;448:260–2.
5. Davidson EH, McClay DR, Hood L. Regulatory gene networks and the properties of the developmental process. Proc Natl Acad Sci U S A. 2003;100:1475
6. Kee Y, D'Andrea AD. Molecular pathogenesis and clinical management of Fanconi anemia. J Clin Invest. 2012;122(11):3799–806.
7. Tategu M, Arauchi T, Tanaka R, et al. Systems biology-based identification of crosstalk between E2F transcription factors and the Fanconi anemia pathway. Gene Regul Syst Bio. 2007;1:1–8.

8. Garaycoechea JI, Crossan GP, Langevin F, Daly M, Arends MJ, Patel KJ. Genotoxic consequences of endogenous aldehydes on mouse hematopoietic stem cell function. Nature. 2012;489:571–5.
9. Draptchinskaia N, Gustavsson P, Andersson B, et al. The gene encoding ribosomal protein S19 is mutated in Diamond-Blackfan anemia. Nat Genet. 1999;21:169–75.
10. Narla A, Vlachos A, Nathan DG. Diamond Blackfan anemia treatment: past, present, and future. Semin Hematol. 2011;48(2):117–23.
11. Ball S. Diamond Blackfan anemia. Hematology Am Soc Hematol Educ Program. 2011; 2011:487.
12. Sankaran VG, Ghazvinian R, Do R, Thiru P, et al. Exome sequencing identifies GATA1 mutations resulting in Diamond-Blackfan anemia. J Clin Invest. 2012;122:2439–43.
13. Du HY, Mason PJ, Bessler M, Wilson DB. TINF2 mutations in children with severe aplastic anemia. Pediatr Blood Cancer. 2009;52:687.
14. Yamaguchi H, Baerlocher GM, Lansdorp PM, Chanock SJ, Nunez O, Sloand E, et al. Mutations of the human telomerase RNA gene (TERC) in aplastic anemia and myelodysplastic syndrome. Blood. 2003;102:916–8.
15. Alter BP, Young NS. The bone marrow failure syndromes. In: Nathan DG, Orkin HS, editors. Hematology of infancy and childhood, vol. 1. Philadelphia: Saunders; 1998. pp. 237–335.
16. Kirwan M, Dokal I. Dyskeratosis congenita, stem cells and telomeres. Biochim Biophys Acta. 2009;1792:371–9.
17. Agarwal S, Loh YH, McLoughlin EM, Huang J, Park IH, Miller JD, et al. Telomere elongation in induced pluripotent stem cells from dyskeratosis congenita patients. Nature. 2010;464:292–6.
18. Hood L, Qiang T. Systems approaches to biology and disease enable translational systems medicine. Genomics Proteomics Bioinform. 2012;10(4):10181–5.
19. Leroy H, James RH, Michael EP, Biaoyang L. Systems biology and new technologies enable predictive and preventative medicine. Science. 2004;306:640–3.
20. Diamond SL. Systems biology to predict blood function. J Thromb Haemost. 2009;7:177–80.
21. Flamm MH, Colace TV, Chatterjee MS, Jing H, et al. Multiscale prediction of patient-specific platelet function under flow. Blood. 2012;120:190–8.
22. Matthew HF, Thomas VC, et al. Multiscale prediction of patient-specific platelet function under flow. Blood. 2012;120:190–8.
23. Giuseppe P, Bernhard ØP, Olafur ES, et al. Systems biology of stored blood cells: can it help to extend the expiration date? J Proteomics. 2012;76:163–7.
24. Sebastiani P, Ramoni MF, Nolan V, Baldwin CT, Steinberg MH. Genetic dissection and prognostic modeling of overt stroke in sickle cell anemia. Nat Genet. 2005;37:435–40.
25. Lei H, Karniadakis GE. Quantifying the rheological and hemodynamic characteristics of sickle cell anemia. Biophys J. 2012;102(2):185–94.
26. Sankaran VG, Menne TF, Xu J, Akie TE, et al. Human fetal hemoglobin expression is regulated by the developmental stage-specific repressor BCL11A. Science. 2008;322(5909):1839–42.
27. Stuart HO, Leonard IZ. Hematopoiesis: an evolving paradigm for stem cell biology. Cell. 2008;132:631–44.
28. Vijay GS, Stuart HO. Genome-wide association studies of hematologic phenotypes: a window into human hematopoiesis. Curr Opin Genet Dev. 2013;23:1–6.
29. Joseph L, Albert-Laszlo B. Systems biology and the future of medicine, Wiley Interdiscip. Rev Syst Biol Med. 2011;3(6):619–27.
30. Carlson B. Biotechnol Healthc. Spring. 2010;7(1):12–7.
31. Rothenberg EV, Zhang J, Li L. Multilayered specification of the T-cell lineage fate. Immunol Rev. 2010;238:150–68.
32. Georgescua C, Longabaughb WJR, Scripture-Adams DD, David-Fung ES, et al. A gene regulatory network armature for T lymphocyte specification. Proc Natl Acad Sci U S A. 2008;105:20100–105.
33. Hood L. Deciphering complexity: a personal view of systems biology and the coming of "Big" science. Genet Eng Biotechnol News. 2011;31:131.

Index

S. J. Corey et al. (eds.), *A Systems Biology Approach to Blood,*
Advances in Experimental Medicine and Biology 844, DOI 10.1007/978-1-4939-2095-2

MIX
Papier aus verantwortungsvollen Quellen
Paper from responsible sources
FSC® C105338

If you have any concerns about our products,
you can contact us on
ProductSafety@springernature.com

In case Publisher is established outside the EU,
the EU authorized representative is:
Springer Nature Customer Service Center GmbH
Europaplatz 3, 69115 Heidelberg, Germany

Printed by Libri Plureos GmbH
in Hamburg, Germany